Red Snapper Ecology and Fisheries in the U.S. Gulf of Mexico

Funding for the publication of this book was provided by

Sea Grant LOUISIANA

NATIONAL OCEANIC AND ATMOSPHERIC ADMINISTRATION NOAA U.S. DEPARTMENT OF COMMERCE

University of West Florida

USA UNIVERSITY OF SOUTH ALABAMA

Louisiana State University Coastal Fisheries Institute

Red Snapper
Ecology and Fisheries in the U.S. Gulf of Mexico

Edited by

William F. Patterson, III
Department of Biology
University of West Florida
11000 University Parkway
Pensacola, Florida 32514 USA

James H. Cowan, Jr.
Department of Oceanography and Coastal Sciences
School of Coast and Environment
Louisiana State University
Baton Rouge, Louisiana 70803 USA

Gary R. Fitzhugh
National Marine Fisheries Service
Panama City Laboratory
3500 Delwood Beach Road
Panama City, Florida 32408 USA

David L. Nieland
Louisiana Sea Grant College Program
Louisiana State University
Baton Rouge, Louisiana 70803 USA

American Fisheries Society Symposium 60

Proceedings of the Symposium

Fisheries Ecology of Gulf of Mexico Red Snapper

Held in San Antonio, Texas
February 10–12, 2006

American Fisheries Society
Bethesda, Maryland
2007

The American Fisheries Society Symposium series is a registered serial.
A suggested citation format for this book follows.

Entire Book

Patterson, W. F., III, J. H. Cowan, Jr., G. R. Fitzhugh, and D. L. Nieland, editors. 2007. Red snapper ecology and fisheries in the U.S. Gulf of Mexico. American Fisheries Society, Symposium 60, Bethesda, Maryland.

Chapter within the Book

Lyczkowski-Shultz, J., and D. S. Hanisko. 2007. A time series of observations of red snapper larvae from SEAMAP surveys 1982–2003: seasonal occurrence, distribution, and abundance size. Pages 3–23 *in* W. F. Patterson, III, J. H. Cowan, Jr., G. R. Fitzhugh, and D. L. Nieland, editors. Red snapper ecology and fisheries in the U.S. Gulf of Mexico. American Fisheries Society, Symposium 60, Bethesda, Maryland.

Printed in the United States of America on acid-free paper.

Library of Congress Control Number 2007942375
ISBN 978-1-888569-97-1
ISSN 0892-2284

American Fisheries Society Web site address: *www.fisheries.org*

American Fisheries Society
5410 Grosvenor Lane, Suite 100
Bethesda, Maryland 20814
USA

Table of Contents

Preface

Red snapper is among the most ecologically and economically important reef fishes in the northern Gulf of Mexico (GOM). Fisheries management for the species also happens to be among the most controversial in the U.S. The GOM red snapper stock has been estimated to be overfished and undergoing overfishing since at least the late 1980s, even in the face of persistent warnings from the scientific community that the risk of stock collapse is real. Management is complicated, however, because the greatest source of mortality for red snapper is estimated to come from shrimp trawl bycatch, not the direct fisheries. Despite all the efforts to solve the bycatch problem and otherwise recover red snapper, including direct Congressional intercession in management and earmarked funding for various research and development programs, the stock remains significantly overfished.

Few other species or assemblages have had as many financial resources motivated to improve knowledge of basic population biology, engineer solutions to management issues such as shrimp trawl bycatch, develop state-of-the-art assessment techniques, or institute novel management approaches as has GOM red snapper. This book results from a symposium that was aimed at gathering scientists of various affiliations and expertise to present and discuss the latest research on red snapper ecology and fisheries. The three-day symposium was part of the 2006 Mid-Year Meeting of the Southern Division of the American Fisheries Society held in San Antonio, Texas. Beyond the scientific presentations and related discussions of sometimes controversial aspects of red snapper biology and management, a central goal of the symposium was to produce an edited volume that might serve as a peer-reviewed compendium of the current state of knowledge for red snapper in U.S. waters of the GOM. This volume contains 22 chapters that do just that. We are hopeful that fellow scientists and fishery managers will find value in the papers contained herein, whether they are working on red snapper issues or not.

Chapters are organized in three sections: Life History and Ecology, Population Dynamics and Structure, and Fisheries Management and Conservation. Much research effort has been focused in recent years on understanding the basic biology of red snapper, and subjects of papers in the Life History and Ecology section range from larval distributions to site fidelity of adults to petroleum platforms in the western GOM. The Population Dynamics and Structure section contains a variety of manuscripts whose subjects range from adult red snapper movement inferred from tagging studies to regional differences in population demographics to a review of the genetic structure of GOM red snapper. The last section of the book, Fisheries Management and Conservation, begins with a review of the history of GOM red snapper fisheries management. Papers that follow include inputs for stock assessment, such as developing a fishery-independent index of population biomass from larval surveys and a reconstruction of historic landings back to the late 19th century, as well as two papers that examine the potential population effects of minimum size limits in the fishery. The last two chapters of the book are an assessment of red snapper stock status and a review of recent management actions and future challenges to rebuilding Gulf red snapper.

We have many people and organizations to thank for making the symposium and this book a reality. We thank members of the 2006 Southern Division of the American Fisheries Society Mid-Year Meeting Organizing Committee for being receptive to the idea of hosting the symposium. In particular, general chair Dave Terre and program chairs Tim Bonner and Dick Luebke bent over backwards to make the symposium a reality, even to the point of providing a meeting room for a third day when symposium presentations (n = 37) would not fit in two days. Several other members of the local organizing committee and student volunteers made sure the symposium ran smoothly and without incident, and San Antonio was a great host city. We thank scores of reviewers for providing timely critical reviews of papers submitted. We thank NOAA Fisheries, Louisiana Sea Grant, Louisiana State University's Coastal Fisheries Institute, the University of South Alabama's Department of Marine Sciences, and the University of West Florida's Office of Graduate Studies for generously contributing funds to cover publication costs. And, we thank Aaron Lerner and the staff of AFS Books for patiently shepherding this project to completion.

Kim and Ian Workman kindly donated the cover art, which is a stunning example of the Gyotaku art form they are well-known for producing (www.kimiansart.com). The direct fish print was made by applying black ink to a 410 mm total red snapper, overlaying handmade banana bark paper, and pressing the paper onto the fish to obtain a mirror image. The image was subsequently enhanced with water colors.

Will Patterson
Jim Cowan
Gary Fitzhugh
Dave Nieland

Symbols and Abbreviations

The following symbols and abbreviations may be found in this book without definition. Also undefined are standard mathematical and statistical symbols given in most dictionaries.

A	ampere
AC	alternating current
Bq	becquerel
C	coulomb
°C	degrees Celsius
cal	calorie
cd	candela
cm	centimeter
Co.	Company
Corp.	Corporation
cov	covariance
DC	direct current; District of Columbia
D	dextro (as a prefix)
d	day
d	dextrorotatory
df	degrees of freedom
dL	deciliter
E	east
E	expected value
e	base of natural logarithm (2.71828…)
e.g.	(exempli gratia) for example
eq	equivalent
et al.	(et alii) and others
etc.	et cetera
eV	electron volt
F	filial generation; Farad
°F	degrees Fahrenheit
fc	footcandle (0.0929 lx)
ft	foot (30.5 cm)
ft^3/s	cubic feet per second (0.0283 m3/s)
g	gram
G	giga (109, as a prefix)
gal	gallon (3.79 L)
Gy	gray
h	hour
ha	hectare (2.47 acres)
hp	horsepower (746 W)
Hz	hertz
in	inch (2.54 cm)
Inc.	Incorporated
i.e.	(id est) that is
IU	international unit
J	joule
K	Kelvin (degrees above absolute zero)
k	kilo (103, as a prefix)
kg	kilogram
km	kilometer
l	levorotatory
L	levo (as a prefix)
L	liter (0.264 gal, 1.06 qt)
lb	pound (0.454 kg, 454g)
lm	lumen
log	logarithm
Ltd.	Limited
M	mega (10^6, as a prefix); molar (as a suffix or by itself)
m	meter (as a suffix or by itself); milli (10^{-3}, as a prefix)
mi	mile (1.61 km)
min	minute
mol	mole
N	normal (for chemistry); north (for geography); newton
N	sample size
NS	not significant
n	ploidy; nanno (10^{-9}, as a prefix)

o	ortho (as a chemical prefix)
oz	ounce (28.4 g)
P	probability
p	para (as a chemical prefix)
p	pico (10^{-12}, as a prefix)
Pa	pascal
pH	negative log of hydrogen ion activity
ppm	parts per million
qt	quart (0.946 L)
R	multiple correlation or regression coefficient
r	simple correlation or regression coefficient
rad	radian
S	siemens (for electrical conductance); south (for geography)
SD	standard deviation
SE	standard error
s	second
T	tesla
tris	tris(hydroxymethyl)-aminomethane (a buffer)
UK	United Kingdom
U.S.	United States (adjective)
USA	United States of America (noun)
V	volt
V, Var	variance (population)
var	variance (sample)
W	watt (for power); west (for geography)
Wb	weber
yd	yard (0.914 m, 91.4 cm)
α	probability of type I error (false rejection of null hypothesis)
β	probability of type II error (false acceptance of null hypothesis)
Ω	ohm
μ	micro (10^{-6}, as a prefix)
′	minute (angular)
″	second (angular)
°	degree (temperature as a prefix, angular as a suffix)
%	per cent (per hundred)
‰	(per thousand)

Life History and Ecology Overview

GARY R. FITZHUGH

The papers in this symposium session were varied in regard to problems, approaches, and applications of red snapper research but are linked by the theme of ontogeny of red snapper from the larval to the young adult stage. They come broadly under the heading of life history and ecology. In aggregate, these papers recognize that a better understanding of behavior and habitat use during early life history is needed to inform some key policy and management issues. A further theme to emerge from these studies is that developments in gear and technology have been, and remain to be, instrumental in advancing this understanding.

Gulf of Mexico fisheries management and policy information needs are extending well beyond basic data feeding single species stock assessments. For example, new questions are being asked such as what is the potential for larval entrainment at liquefied natural gas facilities (LNG), and can LNG activities become a significant source of additive mortality? Lyczkowski-Shultz and Hanisko present a larval survey time-series (SEAMAP) helping to address such questions but further point to current limitations in the survey. There continues to be a scientific discussion on the ecological and fisheries role of artificial reefs. Do artificial reefs that mimic natural low relief hard bottom habitat and much larger petroleum platforms, which fill the water column, contribute to fisheries production and by how much? Increasingly, the retirement age of petroleum platforms is being reached in the Gulf and decisions will need to be made regarding whether to remove or 'reef' these structures. This attraction-production debate is being informed by several papers in this session including Geary et al., McCawley and Cowan, Westmeyer et al., and Wells and Cowan. One of the most important and controversial commercial gears in the Gulf—shrimp trawls—presents a large challenge for red snapper management and contributions by Geary et al., Parsons and Foster, and Wells and Cowan will help to inform policy here. Efforts are required to further minimize bycatch mortality. What are the stage-based habitat transitions that inform our knowledge of gear effects? In particular, the studies by Geary et al., Parsons and Foster, and Wells and Cowan ask what are the habitats and life history stages to be avoided and behavioral traits to be capitalized on in designing fishing gear and bycatch reduction devices? Rummer's paper in this section focuses our attention on the realization that as fishing pressure increases in some sectors, management methods relying on catch and release may be underestimating the mortality impacts of discards; if so, more direction towards spatial management (rather than catch limits) could be warranted.

It is appropriate that our attention is turning more towards spatial aspects in managing red snapper and associated fisheries. These studies show that our understanding of spatial complexity is evolving on multiple scales—regionally, by noting higher red snapper larval concentration (and presumably spawning potential) in the western versus the eastern Gulf (Lyczkowski-Shultz and Hanisko)—down to small local scales by recognizing the amount and distinct types of important habitat (Geary et al., Wells and Cowan). At early settlement and for young juvenile stages, "open" bottom types (mud, sand, shell) may be suitable as nursery areas with structured habitat (shell ridges, reefs, ledges) and artificial habitat becoming more important for older juveniles and young adults (1–2 years). Some findings suggest inter-year class interactions may be complex, affecting residence and survival (McCawley and Cowan, Wells and Cowan). For instance, the presence/abundance of older stages on reefs may preclude younger stage colonization. But the finding by McCawley and Cowan that older juvenile and young adult red snapper show little or no dependencies on reef associated food items further reveals to us the nuanced complexity of habitat function and the challenge of issues such as the attraction-production debate.

Our progress in addressing these issues and the scientific value of our information depends on advancing technology and methods development. In two studies censusing red snapper and reported in this section, natural habitat was first characterized using either digital side-scan or multibeam sonar methods (Geary et al., Wells and Cowan). It is important to note that much greater information content and precision will be gained when the seafloor is mapped first and surveys can then be stratified by habitat. To date, habitat characterization efforts are still too few and far between. Survey and censusing tools continue to develop and details of deployment and attributes can matter. Swimming performance and behavior clearly change with ontogeny which in turn affects catch rates (Parsons and Foster). There continues to be a role for active survey methods using nets and towed gear for youngest stages (e.g., Lyczkowski-Shultz and Hanisko, Geary et al.) while passive visual methods (scuba, cameras) are needed to census older red snapper (Wells and Cowan). Acoustic tagging offers great potential for continuous monitoring of movements but challenges in the offshore environment are substantial (Westmeyer et al.). Advances in addressing issues such as the attraction-production debate will likely depend on results from multiple approaches and survey tools.

Our future success may also lie in adopting a framework to merge our increasing understanding of changing habitat use during ontogeny and deal with the apparent complexity. For instance, attributes of habitat use, fidelity, prey demand, abundance dynamics, and species associations by life history stage and within appropriate spatial scales are the underpinnings of what is needed to examine red snapper population- and community-dynamics within the context of foraging arena theory; an integrated view of the process and likelihood of eating and being eaten (Walters and Martell 2004). Adopting such a framework may be one of the most promising ways to move forward to ecosystem-scale implications and management. No doubt, our increased understandings will also inform our stock assessment approaches. As Porch (this volume) assures us—future assessments will likely be using more information on movement and recruitment patterns in red snapper as it becomes available.

Reference

Walters, C. J. and S. J. D. Martell. 2004. Fisheries ecology and management. Princeton University Press. Princeton and Oxford.

American Fisheries Society Symposium 60:3–23, 2007

A Time Series of Observations on Red Snapper Larvae from SEAMAP Surveys, 1982–2003: Seasonal Occurrence, Distribution, Abundance, and Size

JOANNE LYCZKOWSKI-SHULTZ[1] AND DAVID S. HANISKO
NOAA Southeast Fisheries Science Center
National Marine Fisheries Service
Mississippi Laboratories
3209 Frederic Street, Pascagoula, Mississippi 39567 USA

Abstract.—Eleven taxa of snappers were found among the 20,301 snapper larvae examined from over 14,000 bongo and neuston samples collected during SEAMAP surveys, 1982 to 2003. During that time series, a total of 639 red snapper *Lutjanus campechanus* larvae were identified in bongo samples and 1053 larvae in neuston samples. Red snapper larvae first appeared in May and were present as late as November. Months of highest occurrence and abundance were July and September when larvae were taken in 12.7% and 11.0% of bongo samples, and in 7.6% and 8.4% of neuston samples. Mean abundance in those months was 1.18 and 0.82 larvae under 10 m^2 of sea surface for bongo samples and 0.36 larvae per 10 min for neuston samples. By November, percent occurrence was less than or equal to 0.2% and mean abundance was less than 0.01 larvae in samples from either gear. Larvae identifiable as red snapper ranged in body length from 2.4 to 19.2 mm (mean = 5.12, median = 4.60) in bongo net samples; and 2.7–24.0 mm (mean = 4.37, median = 4.00) in neuston net samples. Over 95% of larvae in bongo samples were less than or equal to 8.3 mm and in neuston samples were less than or equal to 5.6 mm. Larvae were captured throughout the survey area but were consistently observed in greatest abundance at stations on the mid-continental shelf west of the Mississippi River, especially off western Louisiana and central Texas. This time series of observations is the data set from which annual estimates of larval red snapper abundance were derived for use as a fishery independent index of adult stock size.

Introduction

The Southeast Area Monitoring and Assessment Program (SEAMAP) has supported collection and analysis of ichthyoplankton samples from resource surveys in the Gulf of Mexico (GOM) since 1982 with the goal of producing a long-term database on the early life stages of fishes. Specimens from these collections have been used to describe the larval development of a wide variety of fishes in the western central North Atlantic (Ditty and Shaw 1992, 1993; Ditty et al. 1994; Drass et al. 2000; Richards 2005). Spawning seasonality of Gulf fishes have been inferred from SEAMAP data (Ditty et al. 1988), as well as, the potential 'sources

[1] Corresponding author: joanne.lyczkowski-shultz@noaa.gov.

and sinks' of recruits to reefs in the northern GOM (Hanisko and Lyczkowski-Shultz 2003). Estimates of abundance of fish eggs and larvae from SEAMAP surveys are being used to assess the impact to Gulf fisheries of entrainment mortality at proposed liquefied natural gas terminals using open rack vaporizers (*www.uscg.mil/hq/g-m/mso/mso5.htm*). Annual indices of larval abundance from SEAMAP surveys are being used in stock assessments of Atlantic bluefin tuna, *Thunnus thynnus* (Scott et al. 1993) and Gulf king mackerel, *Scomberomorus cavalla* (Gledhill and Lyczkowski-Shultz 2000).

Larval indices based on the SEAMAP time series of ichthyoplankton data were examined at recent Southeast Data, Assessment and Review Workshops (SEDAR7 Stock Assessment Report 2004; SEDAR9 Stock Assessment Report 2005; available at *www.sefsc.noaa.gov/sedar/*) for use in stock assessments of red snapper *Lutjanus campechanus*, vermilion snapper *Rhomboplites aurorubens* and gray triggerfish *Balistes capriscus*. A summary of seasonal occurrence, abundance, distribution and size of red snapper larvae collected during twenty-two years of SEAMAP surveys in the GOM is presented in this paper. These are the data from which annual estimates of larval red snapper abundance were derived for use as a fishery independent index of adult stock size (Hanisko et al. 2007, this volume).

Methods and Materials

Surveys and Collections:

SEAMAP resource surveys have been conducted by the National Marine Fisheries Service since 1982 in cooperation with the states of Florida, Alabama, Mississippi, and Louisiana. Plankton sampling is conducted during these surveys at predetermined SEAMAP stations arranged in a fixed, systematic grid across the U.S. Exclusive Economic Zone of the GOM (Figure 1A). Most but not all SEAMAP stations (designated by a unique SEAMAP or 'b' number) are located at ~56 km or ½ degree intervals along this grid. Some SEAMAP stations are located at under 56 km intervals especially along the continental shelf edge, while others have been moved to avoid obstructions, navigational hazards or shallow water. Most SEAMAP plankton samples are taken during both dedicated plankton and shrimp/bottomfish (trawl) surveys but over the years additional samples have been taken using SEAMAP gear and collection methods at locations other than designated SEAMAP stations or outside established SEAMAP surveys, e.g. during Louisiana seasonal trawl surveys, SEAMAP Squid/Butterfish survey, and other serendipitous or special projects.

Although other plankton sampling gear types and mesh sizes have been used over the SEAMAP time series, the gear and methodology considered as standard for SEAMAP surveys are those described in Kramer et al. (1972), Smith and Richardson (1977) and Posgay and Marak (1980). A 61 cm (outside diameter) bongo frame fitted with 0.333 (0.335)[†] mm mesh netting is fished in a double-oblique tow path from a maximum depth of 200 m or 2–5 m off the bottom at depths less than 200 m. It is assumed that the net does not fish or fishes very little during descent to maximum tow depth because towing cable payout is rapid (~40–50 m/min) while cable retrieval rate is slow (~20 m/min). A mechanical flowmeter is mounted off-center in the mouth of each bongo net to record the volume of water filtered. Volume filtered ranges from ~20–600 m^3 but is typically 30–40 m^3 at the shallowest stations and 300–400 m^3 at the deepest stations. A single or double 2x1 m pipe frame neuston net fitted with 0.947 (0.950)[†] mm mesh netting is towed at the surface with the frame half-sub-

[†]Mesh size change in database does not represent an actual change in gear but only a change in the accuracy at which plankton mesh aperture size can be measured by the manufacturer.

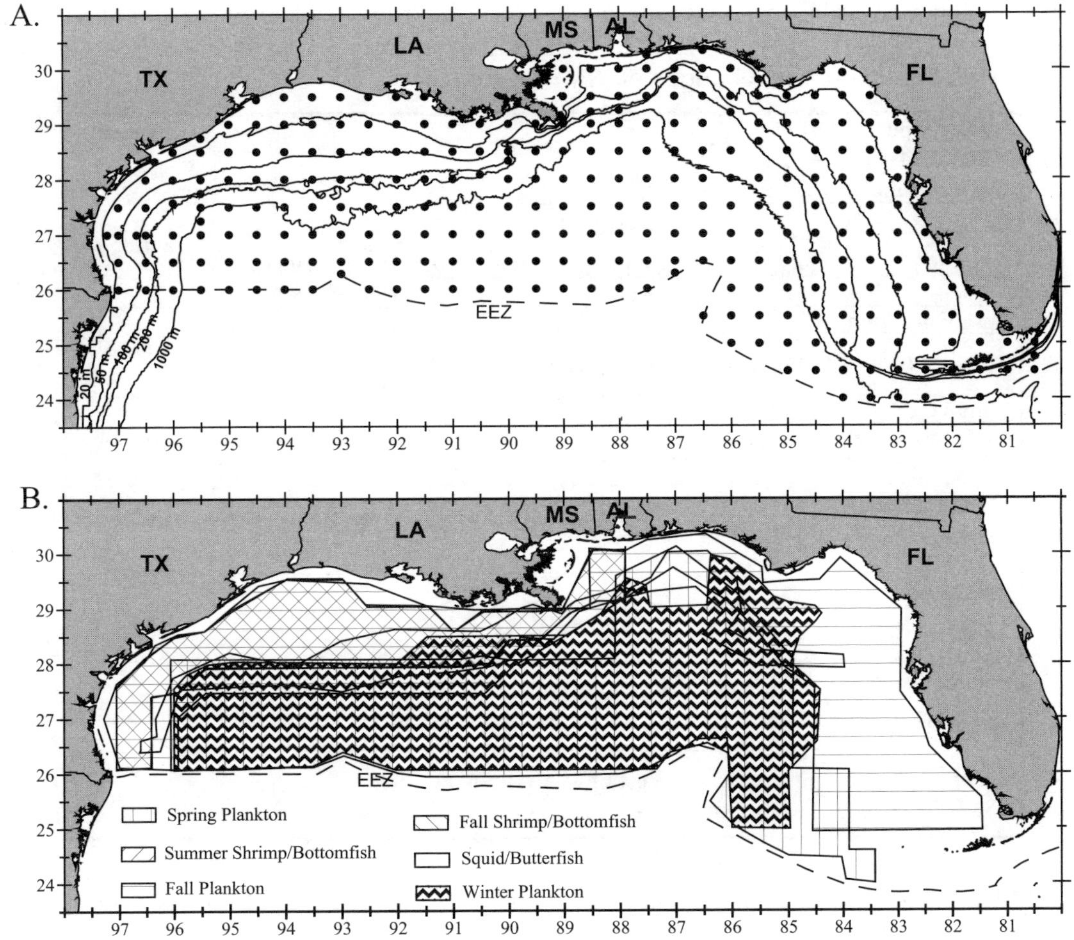

Figure 1. A. Location of SEAMAP stations arranged in a fixed, systematic grid across the Gulf of Mexico. Most but not all stations are located at ~56 km or ½ degree intervals along the grid. B. Generalized areas of the Gulf of Mexico where plankton sampling was conducted during SEAMAP resource surveys.

merged for 10 min. Samples are taken upon arrival on station regardless of time of day. At each station, either a bongo and/or neuston tow are made depending on the specific survey. Samples are routinely preserved in 5–10% formalin and later transferred after 48 h to 95% ethanol for long term storage. During some surveys selected samples are preserved initially in 95% ethanol and within 24–36 h transferred to fresh ethanol.

Catches of larvae from bongo nets are standardized to account for volume filtered and depth of the sampled water column and expressed as number of larvae under 10 m^2 of sea surface. This is accomplished by dividing the number of larvae of each taxon caught in a sample by the volume of water filtered during the tow; and then multiplying the resultant by the maximum depth of the tow in meters and the factor 10. Catches of larvae from neuston nets are standardized to account for net tow duration and expressed as number of larvae per 10 min tow. During SEAMAP surveys, environmental parameters including temperature, salinity, dissolved oxygen and optical transmission are measured in situ with a SEA-

Table 1. Temporal and spatial coverage of SEAMAP plankton sampling effort during resource surveys in the U.S. Gulf of Mexico, 1982 to 2003. Months in bold account for > 70% of samples taken during the survey. *BN* = bongo net; *NN* = neuston net.

Survey type Number of BN/NN samples	Months	Primary survey area	Time period	% Total BN/NN samples
Winter plankton 332 / 289	**Dec** Jan, Feb	Coastal LA; Shelf edge to U.S. EEZ	1982–1997; 1983 & 1984, 1993 & 1996	4.7 / 3.6
Spring plankton 2196 / 3209	Mar, **Apr**, **May,** Jun	Coastal LA; Shelf edge to U.S. EEZ	1982–1995; 1982 to present	31.1 / 40.3
Summer trawl 1052 / 977	**Jun**, **Jul**	5 to 50 fm, south TX to Mobile Bay	1982 to present	14.9 / 12.3
Squid/butterfish 88 / 85	May, **Aug**	Shelf edge northern Gulf	Aug 1985, May 1986	1.2 / 1.1
Fall plankton 2273 / 2413	Aug, **Sep**, Oct	Coastal & shelf waters, south TX to south FL	1986 to present	32.2 / 30.3
Fall trawl 867 / 762	**Oct**, **Nov**, Dec	5 to 50 fm, south Texas to Mobile Bay	1982 to present	12.3 / 9.6
Other 253 / 221	**Mar**, Apr–Nov	West of Mississippi River off LA	Mostly pre-1986	3.6 / 2.8

BIRD SBE 25 or SBE 911, conductivity temperature depth (CTD) profiler at each station. Although complete CTD profiles of the water column are recorded at sea, only observations from three depth levels; surface, mid and max (near bottom at depths <200 m) were entered in the SEAMAP database. Environmental data from stations where red snapper larvae were captured during SEAMAP surveys in 2000 through 2003 are presented. Dissolved oxygen data were summarized only for surveys in 2001 to 2003 because the oxygen sensor used during the 2000 field season was defective.

Assignment of samples to either day, night or twilight periods for diel comparisons of larval abundance and size followed Seidelmann (1992) and were based on sample date, time and location (coordinates). Nautical twilight, not civil twilight (*www.weather.gov/glossary*) was used as the reference datum for determining which samples had been taken during the transitional period between night and day.

This review of the early life history of red snapper in the GOM is based on over 7,000 bongo and 7,900 neuston collections from SEAMAP plankton sampling during the period 1982 to 2003. The original intent of SEAMAP was to sample both the open (shelf edge to U.S. Exclusive Economic Zone or EEZ) and continental shelf (10–200 m) portions of the Gulf in their entirety at least once during each season. This goal was never achieved and thus the preponderance of SEAMAP plankton samples have been collected during only four annual resource surveys; three in shelf and coastal waters and one in deep Gulf waters (Table 1; Figure 1B).

Two dedicated plankton surveys account for 60–70% of all SEAMAP plankton samples (bongo and neuston combined). The

Spring Plankton survey has been conducted in U.S. EEZ open Gulf waters primarily in April and May since 1982, although in several years during the time series sampling was conducted on the west Florida shelf. The second dedicated plankton survey is the Fall Plankton survey. It has been conducted in coastal and continental shelf waters from south Texas to south Florida as a Gulfwide survey during late August to mid October since 1986. Starting in 1999 the area covered by the Fall Plankton survey was expanded to include stations beyond the continental shelf in the western Gulf. Sampling conducted during a plankton survey of coastal and continental shelf waters of the U.S. GOM during the month of August in 1984 is not considered to be part of the annual SEAMAP Fall Plankton survey. Samples taken during the SEAMAP Summer (mid June through July) and Fall (mid October through November) Shrimp/Bottomfish (trawl) surveys each make up 10–15% of all SEAMAP plankton samples. The area covered by these two surveys includes coastal and continental shelf waters west of 88°W longitude; although during the period 1982–1988, sampling was also irregularly conducted off northwest and western Florida.

Sampling in winter months accounts for less than 5% of all SEAMAP plankton samples. There have been only four winter plankton surveys (1983, 1984, 1993 and 1996) all in open Gulf waters. The exception being the Louisiana seasonal trawl survey of state waters that was conducted typically in December, from 1982 to 1997. Approximately 1% of all SEAMAP plankton samples were taken during a special SEAMAP Squid/Butterfish trawl survey of shelf edge waters in the northern Gulf in August 1985 and May 1986. Additional SEAMAP plankton samples (~3%) were taken prior to 1986 and mostly in the northern Gulf west of the Mississippi River.

Complete descriptions of survey methodologies, data collection and sampling effort by year and survey type can be found in the SEAMAP Environmental and Biological Atlases of the Gulf of Mexico, 1982 to 2001 published by Gulf States Marine Fisheries Commission, Ocean Springs, MS (available at *www.gsmfc.org*).

Sample Processing and Identification of Snapper Larvae:

Plankton samples were sorted for fish eggs and larvae and initial identifications and measurements of larvae were made at the Sea Fisheries Institute, Plankton Sorting and Identification Center (ZSIOP), in Gdynia and Szczecin, Poland. Plankton samples, collected by the Louisiana Department of Wildlife and Fisheries during the period 1989–2002, were processed by Louisiana state biologists. Vials of eggs and identified larvae, plankton displacement volumes, total egg counts; and counts and measurements of identified larvae were sent to the SEAMAP Archive at the Fish and Wildlife Research Institute, Florida Fish and Wildlife Conservation Commission, St. Petersburg, Florida. These data were entered into the SEAMAP database at the SEAMAP Archive where specimens are curated and loaned to researchers upon request. Data files containing specimen identifications and lengths were sent to the NMFS Mississippi Laboratories where these data were combined with field collection data, edited and maintained in the SEAMAP database. Data are available on request from the SEAMAP Data Manager at the Mississippi Laboratories.

Body length (BL) as defined by Leis and Rennis (1983) was measured to the nearest 0.1 mm with an ocular micrometer fitted to a dissecting microscope. Body length is equivalent to notochord length or standard length depending on the stage of development (Richards 2005). The standard measurement in preflexion and flexion stage larvae is notochord length, i.e. the straight line distance from the tip of the snout to the posterior tip of the notochord. The standard measurement in postflexion larvae is standard length, i.e. the distance from the tip of the snout along the body midline to a vertical line through the posterior edge of the hypural plate.

In order to assure consistent identifications over the SEAMAP time series, all snapper larvae were examined and identified by ichthyoplankton specialists at the Southeast Fisheries Science Center, Mississippi Laboratories using an identification protocol based on descriptions in Drass et al. (2000) and Lindeman et al. (2005). The level of identification achievable under this protocol depended on the extent of first dorsal fin development, as well as the following morphological traits: presence or absence of melanistic pigment on the throat (sternohyoideus muscle), and on the anterior surface of the visceral mass or gut; and whether preopercular spines or dorsal spines were smooth or serrated (Table 2). Specimens were identified as red snapper only when a minimum of five dorsal spines were present, those spines were smooth, not serrated and melanistic pigmentation on the body and fins matched the description and illustrations of reared and wild caught red snapper larvae in Rabalais et al. (1980), Collins et al. (1980), and Drass et al. (2000).

Red snapper are among six of the twelve snapper species of the subfamily Lutjaninae found in the GOM whose larvae have been described. Despite these descriptions snapper larvae can be distinguished from each other only after dorsal and pelvic spines have begun to develop using a combination of morphological characters (Lindeman et al. 2005). Red snapper larvae prior to dorsal and pelvic spine formation are generally under 3.5 mm BL and cannot be confidently identified in field collections because of the lack of estab-

Table 2. Key morphological characters used to identify snapper larvae in SEAMAP collections (Lindeman et al. 2005). Melanistic pigmentation on the body and fins of specimens identified as *L. campechanus* matched the description and illustrations of reared red snapper larvae in Drass et al. (2000). shs = sternohyoideus muscle; avm = anterior visceral mass.

Dorsal fin development	Pigment on: throat (shs) and/or gut (avm)	Preopercular spines	Dorsal spines	Identification
little or none	none	smooth		Lutjanidae
little or none	shs	smooth or serrated		*R. aurorubens*
little or none	avm	smooth		*Lutjanus* spp.
2 – 4 spines	none	smooth	smooth	Lutjanidae
2 – 4 spines	shs	serrated	smooth	*R. aurorubens*
2 – 4 spines	avm	smooth	smooth	*Lutjanus* spp.
5 spines	avm	smooth	smooth	*L. campechanus*
≥ 5 spines	shs + avm	serrated	serrated	*R. aurorubens*
≥ 5 spines	avm + shs	smooth	smooth	*L. campechanus*
≥ 5 spines	avm + shs	smooth	serrated	*Lutjanus* spp.

Table 3. Snapper larvae identified to the lowest taxon in SEAMAP ichthyoplankton samples from the Gulf of Mexico, 1982–2003.

Taxon	Number of specimens	% of total
Etelinae	7	<1
Etelis oculatus	10	<1
Lutjanidae	8,516	42
Lutjanus spp.	2,678	13
L. analis	6	<1
L. apodus	1	<1
L. campechanus	1692	8
L. griseus	250	1
L. synagris	173	1
Pristipomoides aquilonaris	2060	10
Rhompoblites aurorubens	4908	24
	$\sum$ = 20,301	

lished characteristics that permit early stage larvae of the lutjanines to be distinguished from each other. The few specimens identifiable as red snapper in SEAMAP collections that were under 3.5 mm BL resulted from variability in size at developmental stage and/or shrinkage during capture and preservation.

The question arises as to the potential for misidentification of red snapper larvae in SEAMAP collections since the larvae of all snappers found in the region have not been described. It is unlikely that this caused extensive misidentification of red snapper larvae considering how much the larvae of species whose larval development has been described differ from each other and red snapper in pigmentation and body shape (Drass et al. 2000). Most of the snappers whose larvae remain undescribed inhabit coral reefs and reef associated ledges as adults, and clear shallow waters or mangrove areas as juveniles (Anderson 2003); biotopes of limited extent in the northern GOM (Parker et al. 1983). No adults or juveniles of the six snapper species whose larvae are undescribed were taken during annual summer and fall SEAMAP shrimp/bottomfish (trawl) surveys from 1982 to 2005 (G. Pellegrin, NOAA/SEFSC Mississippi Laboratories, personal communication). Fewer than five individuals per year of these species were ever observed during ten years of NMFS reef fish video surveys of reef and hard bottom habitat from Brownsville, Texas to the Florida Keys (K. Rademacher, NMFS/SEFSC Mississippi Laboratories, personal communication).

Results

Eleven taxa of snappers were found among the 20,301 snapper larvae examined from over 14,000 bongo and neuston samples collected during SEAMAP surveys from 1982 to 2003 (Table 3). Forty-two per cent of snapper larvae could not be identified beyond the family level and 13% could only be identified to the genus level, *Lutjanus* spp. due to lack of development of distinguishing characters. The annual percentage of *Lutjanus* spp. varied from 6 to 26% over the 22 year time series. The proportion of larvae that could be identified as red snapper varied with size-class (Figure 2). A subset of larval identifications from samples taken in the north-central GOM in June and July best serves to illustrate the relationship between size and achievable identification lev-

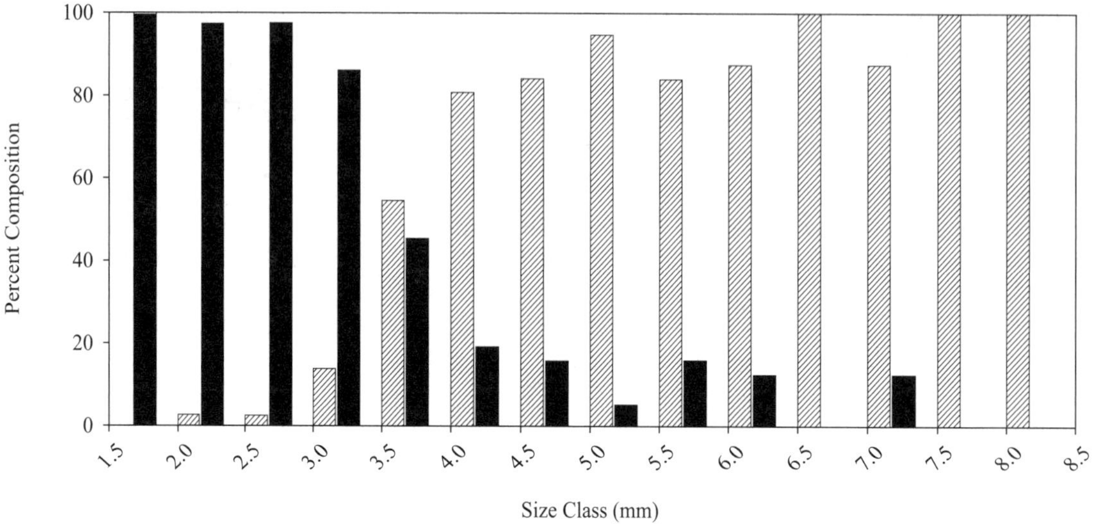

Figure 2. Proportion by size-class of larvae identifiable as *Lutjanus* spp., *n* = 397 (solid bars) and *Lutjanus campechanus, n* = 453 (shaded bars) in June and July SEAMAP plankton samples from the north-central Gulf of Mexico.

el because *L. campechanus* is the dominant species of *Lutjanus* in that region. Slightly over 50% of snapper larvae in the 3.5–3.9 mm BL size-class could be identified as red snapper while all larvae over 4 mm BL were developed enough to be confidently identified as red snapper. Larvae over 4 mm BL identified as *Lutjanus* spp. are the larvae of other species of snappers.

The larvae of red snapper were collected in plankton samples from 95 of 279 state and federal cruises from 1982 through 2003. During that time series a total of 639 red snapper larvae were identified in bongo samples and 1053 larvae in neuston samples from May to November. Months of highest occurrence and abundance with all surveys combined were July and September when larvae were taken in 12.7% and 11.0% of bongo samples; and 7.6% and 8.4% of neuston samples (Table 4). Mean abundance in those months was 1.18 and 0.82 larvae per 10 m^2 for bongo samples, and 0.36 larvae per 10 min for neuston samples. The surveys that accounted for most red snapper larvae were the Summer Shrimp/Bottomfish and Fall Plankton surveys; conducted, principally, in June & July, and September, respectively (Table 5). When captures of red snapper larvae are compared using only samples from the SEAMAP survey area west of 87.75°W longitude, i.e. the survey area common to both these surveys, per cent occurrence and mean abundance were highest in the month of September (Figure 3). The apparent decline in occurrence and abundance of larvae in August may be an artifact of sampling. Only 668 plankton samples were taken during that month over the SEAMAP time series as compared to over 1000 and 3000 samples in July and September, respectively (Table 5). Most of the August samples came from one, Gulfwide survey in 1984. By November red snapper larvae occurred in less than or equal to 0.2% of SEAMAP plankton samples and their mean abundance was less than 0.01 larvae.

Red snapper larvae were captured over a wide range of water depths, 9–2000 m, with mean station depth = 91 m and median depth = 45.5 m during SEAMAP Summer Shrimp/Bottomfish and Fall Plankton surveys. Since depth at capture of red snapper larvae taken in bongo net samples is not known, environmental data from mid and max water column depths only define the range in conditions experienced by red snapper larvae in subsurface waters (Table 6). Surface conditions, however, do reflect the habitat of red snapper larvae captured by SEAMAP neuston nets in the upper 0.5 m of the water column. Average surface environmental conditions indicate that red snapper larvae consistently inhabit warm (mean temperature = 29°C), saline (mean salinity = 34 psu), clear (median optical transmission = 87%) waters. Larvae were taken at stations where minimum oxygen values at mid and near bottom depths were indicative of hypoxia.

Table 4. Summary of larval red snapper catches by month during SEAMAP surveys in the Gulf of Mexico, 1982–2003. no. occ. = number of samples with red snapper larvae; FO = frequency of occurrence; CV = coefficient of variation of mean abundance. Abundance in bongo samples = larvae under 10 m^2 sea surface. Abundance in neuston samples = larvae per 10 min.

Bongo samples

Month	No. samples	No. occ.	No. larvae	% FO	Mean abundance ± SE	Max station abundance	Summed abundance	CV
Jan	72	0	-	-	-	-	-	-
Feb	35	0	-	-	-	-	-	-
Mar	185	0	-	-	-	-	-	-
Apr	607	0	-	-	-	-	-	-
May	1410	3	5	0.2	0.02 ± 0.01	10.32	23.21	60
Jun	684	60	91	8.8	0.55 ± 0.10	53.33	374.61	19
Jul	590	75	147	12.7	1.18 ± 0.20	82.61	695.71	17
Aug	337	23	25	6.8	0.44 ± 0.10	13.69	146.94	22
Sep	1749	193	352	11.0	0.82 ± 0.08	43.02	1431.80	10
Oct	684	11	18	1.6	0.14 ± 0.06	28.69	97.07	39
Nov	472	1	1	0.2	0.01 ± 0.01	2.89	2.89	100
Dec	236	0	-	-	-	-	-	-

Neuston samples

Month	No. samples	No. occ.	No. larvae	% FO	Mean abundance ± SE	Max station abundance	Summed abundance	CV
Jan	76	0	-	-	-	-	-	-
Feb	33	0	-	-	-	-	-	-
Mar	50	0	-	-	-	-	-	-
Apr	890	0	-	-	-	-	-	-
May	2136	9	12	0.4	<0.01 ± <0.01	2.00	13.00	35
Jun	768	31	90	4.0	0.12 ± 0.04	30.00	91.82	36
Jul	536	41	185	7.7	0.36 ± 0.13	55.46	194.75	35
Aug	331	22	75	6.7	0.23 ± 0.08	23.80	74.76	37
Sep	1887	159	676	8.4	0.36 ± 0.05	53.00	671.26	15
Oct	658	11	14	1.7	0.02 ± 0.01	1.97	13.42	31
Nov	403	1	1	0.3	<0.01 ± <0.01	1.00	1.00	100
Dec	188	0	-	-	-	-	-	-

Table 5. Summary of larval red snapper catches in bongo and neuston net samples by month and SEAMAP survey type based on all surveys in the Gulf of Mexico, 1982–2003. No. occ. = number of samples with red snapper larvae; FO = frequency of occurrence. Abundance in bongo samples = larvae under 10 m^2 sea surface. Abundance in neuston samples = larvae per 10 min. CV = coefficient of variation of mean abundance. SP = Spring Plankton; SG = Summer Shrimp/Bottomfish; FG = Fall Shrimp/Bottomfish; FP = Fall Plankton; AS and AF = Alabama Summer and Fall Plankton; SQ = Squid/Butterfish. **denotes sampling outside established SEAMAP surveys.

Bongo samples

Month	Survey type	No. samples	No. occ.	No. larvae	% FO	Mean abundance ± SE	Max. station abundance	Summed abundance	CV
Jan	**	72	0	-	-	-	-	-	-
Feb	**	35	0	-	-	-	-	-	-
Mar	**	137	0	-	-	-	-	-	-
Jun	**	50	1	1	2.0	0.07 ± 0.07	3.57	3.57	100
Jul	**	22	1	1	4.6	0.16 ± 0.16	3.48	3.48	100
Aug	**	13	0	-	-	-	-	-	-
Nov	**	31	1	1	3.2	0.09 ± 0.09	2.89	2.89	100
Dec	**	225	0	-	-	-	-	-	-
Mar	SP	48	0	-	-	-	-	-	-
Apr	SP	607	0	-	-	-	-	-	-
May	SP	1394	3	5	0.2	0.02 ± 0.01	10.32	23.21	60
Jun	SP	147	0	-	-	-	-	-	-
Jun	AS	2	0	-	-	-	-	-	-
Jun	SG	485	59	90	12.2	0.77 ± 0.14	53.33	371.04	19
Jul	SG	565	74	146	13.1	1.23 ± 0.21	82.61	692.23	17
May	SQ	16	0	-	-	-	-	-	-
Jul	SQ	3	0	-	-	-	-	-	-
Aug	SQ	69	7	8	10.1	0.88 ± 0.33	13.69	60.53	38
Sept	AF	6	0	-	-	-	-	-	-
Aug	FP	255	16	17	6.3	0.34 ± 0.09	11.43	86.41	27
Sept	FP	1743	193	352	11.1	0.82 ± 0.08	43.02	1431.80	10
Oct	FP	269	6	7	2.2	0.11 ± 0.05	9.09	28.76	46
Oct	FG	415	5	11	1.2	0.16 ± 0.09	28.69	68.31	52
Nov	FG	441	0	-	-	-	-	-	-
Dec	FG	11	0	-	-	-	-	-	-

Mean length and overall size range did not change appreciably over the seven months when red snapper larvae were present in SEAMAP plankton samples (Table 7). There was no indication in monthly size frequency distributions of increasing modal size as the spawning season progressed (Figure 4). Larvae captured in bongo nets ranged from 2.4 to 19.2 mm BL with a mean of 5.1 mm (median = 4.6); size range in neuston samples was 2.7–24.0 mm BL with a mean of 4.4 mm (median = 4.0). Although size range of red snapper larvae captured in bongo and neuston samples was similar, cumulative frequencies indicate that overall, smaller larvae were found in neuston net samples. Ninety-five per cent of larvae captured in neuston samples were less than or

Table 5. (Continued)

Neuston samples

Month	Survey type	No. samples	No. occ.	No. larvae	% FO	Mean abundance ± SE	Max. station abundance	Summed abundance	CV
Jan	**	76	0	-	-	-	-	-	-
Feb	**	33	0	-	-	-	-	-	-
Mar	**	13	0	-	-	-	-	-	-
Apr	**	5	0	-	-	-	-	-	-
May	**	84	0	-	-	-	-	-	-
Jun	**	72	0	-	-	-	-	-	-
Jul	**	3	0	-	-	-	-	-	-
Aug	**	13	0	-	-	-	-	-	-
Nov	**	31	0	-	-	-	-	-	-
Dec	**	180	0	-	-	-	-	-	-
Mar	SP	37	0	-	-	-	-	-	-
Apr	SP	885	0	-	-	-	-	-	-
May	SP	2038	9	12	0.4	0.01 ± <0.01	2.00	13.00	35
Jun	SP	249	3	33	1.2	0.13 ± 0.12	30.00	32.91	91
Jun	AS	4	0	-	-	-	-	-	-
Jun	SG	443	28	57	6.3	0.13 ± 0.03	10.02	58.90	25
Jul	SG	530	41	185	7.7	0.37 ± 0.13	55.46	194.75	34
May	SQ	14	0	-	-	-	-	-	-
Jul	SQ	3	0	-	-	-	-	-	-
Aug	SQ	68	10	59	14.7	0.86 ± 0.39	23.80	58.76	45
Sep	AF	134	0	-	-	-	-	-	-
Oct	AF	18	0	-	-	-	-	-	-
Aug	FP	250	12	16	4.8	0.06 ± 0.02	3.00	16.00	31
Sep	FP	1753	159	676	9.1	0.38 ± 0.06	53.00	671.26	14
Oct	FP	258	3	3	1.2	0.01 ± 0.01	1.00	2.91	57
Oct	FG	382	8	11	2.1	0.03 ± 0.01	1.97	10.50	36
Nov	FG	372	1	1	0.3	<0.01 ± <0.01	1.00	1.00	100
Dec	FG	8	0	-	-	-	-	-	-

equal to 5.6 mm BL while that cumulative percentage of larvae in bongo net samples were less than or equal to 8.3 mm BL. One possible explanation for this observation is differential avoidance of bongo and neuston nets by snapper larvae. The difference between nighttime and daytime catch rates is a measure of sampling gear avoidance with avoidance being directly related to size of larvae (Morse 1989). Not only were red snapper larvae caught more frequently and in greater numbers at night regardless of gear (Table 8), but the night to day ratio of mean abundance (all larvae combined) for neuston samples (13.6) was seven times greater than the ratio (1.9) for bongo samples.

Capture locations and mean station abundances of red snapper larvae taken during four ongoing SEAMAP surveys and one special project survey yielded a generalized depiction of seasonal distribution of red snapper larvae in the Gulf of Mexico (Figures 5–9). Larvae were captured in the months of May and June during the SEAMAP Spring Plankton survey that is conducted in Gulf waters primarily beyond the continental shelf (Figure 5). The greatest number of captures and

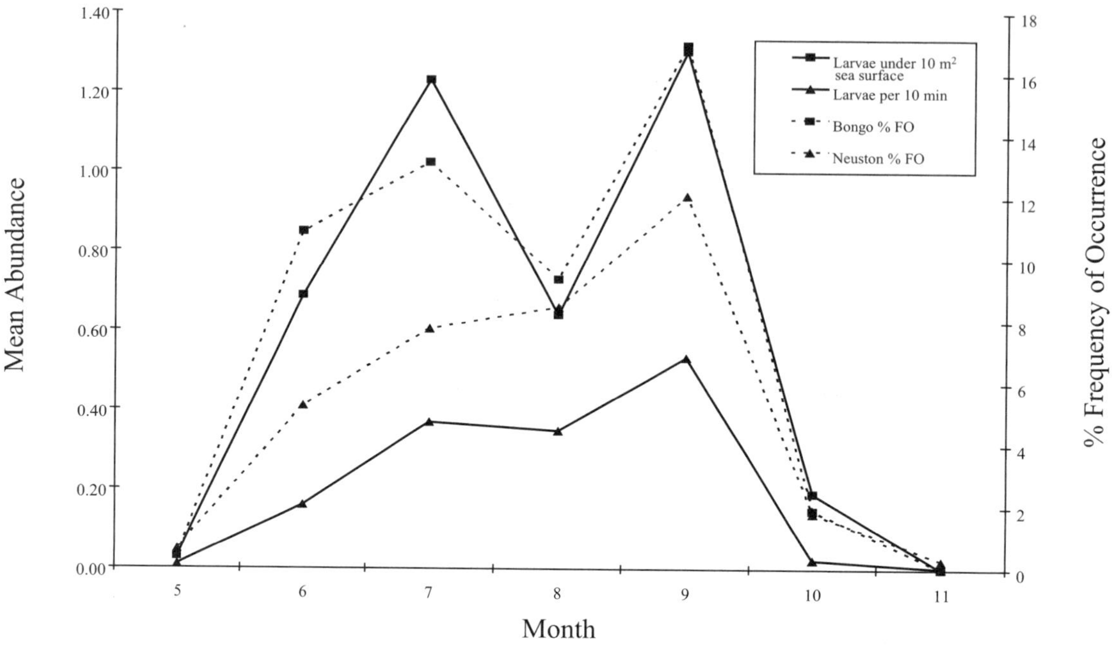

Figure 3. Monthly mean abundance and percent frequency of occurrence (FO) of red snapper larvae in SEAMAP bongo (□) and neuston (Δ) samples west of 87.75°W longitude in the Gulf of Mexico.

highest mean station abundances during this survey were observed in the westernmost region of the survey area, although larvae also were taken at stations in the north-central and extreme southeastern corner of the survey area. Predictably, occurrence and abundance of larvae was much greater during the summer Shrimp/Bottomfish survey of continental shelf waters conducted mainly from Mobile Bay to south Texas in mid June through July (Figure 6). Most occurrences and highest mean station abundances were observed off central and western Louisiana where water depths ranged from 50 to 100 m. Red snapper larvae were also consistently taken off south Texas, Mississippi and Alabama but mean abundances east of the Mississippi River were lower than west of the River. Mean station abundances were higher during the Gulfwide SEAMAP Fall (late August to mid October) Plankton survey then during the summer survey (Figure 7). Larvae were concentrated in the same general area as the summer survey but mean station abundances were markedly higher from 50 to beyond 100 m depth off central and south Texas during the fall than the summer survey. The Fall Plankton survey is the only survey that has, since 1986, consistently covered the eastern Gulf on the continental shelf off northwest and west Florida. Red snapper larvae occurred there much less frequently and in lower numbers than in the western Gulf. The Squid/Butterfish trawl survey was conducted as a SEAMAP special project for only two years but the coverage of that survey, along the continental shelf edge and during the month of August, added observations in an area and time that is underrepresented in the established surveys (Figure 8). No red snapper larvae were captured in May 1986 in collections made predominantly around the mouth of the Mississippi River. There were few captures of red snapper larvae in August 1985 despite extensive sampling along the 100–200 m contours over the northern Gulf from south Texas to

Table 6. Summary of environmental data gathered at stations where *Lutjanus campechanus* were caught during SEAMAP Summer Shrimp/Bottomfish (SG) and Fall Plankton (FP) surveys, 2000 to 2003. Measurements were taken at three levels in the water column; surface (surf); middle of water column (mid); and maximum sampling depth (max). Water temperature (temp) was measured in °C; salinity (sal) in psu; dissolved oxygen (oxy) in mg/l; and optical transmission (trans) in %. *Oxygen data from 2001 to 2003 surveys only.

	N	Mean	Std. Dev.	Median	Min	Max
SG surf temp	42	29.36	0.97	29.41	27.10	31.03
FP surf temp	92	29.20	0.71	29.35	25.40	30.25
SG mid temp	41	27.21	2.55	27.66	18.47	30.43
FP mid temp	92	26.50	3.62	28.47	17.79	30.09
SG max temp	41	23.94	3.15	23.97	15.36	28.92
FP max temp	92	22.28	4.64	22.63	13.30	29.73
SG surf sal	41	32.53	2.78	33.43	24.18	35.74
FP surf sal	92	34.67	2.16	35.49	27.15	36.57
SG mid sal	41	34.75	2.21	35.31	25.46	36.90
FP mid sal	91	35.65	1.67	36.36	27.66	36.65
SG max sal	41	35.70	1.20	36.06	30.02	36.90
FP max sal	91	35.99	1.11	36.39	30.06	36.57
SG surf oxy*	27	5.66	0.45	5.70	4.70	6.70
FP surf oxy*	65	6.07	0.19	6.10	5.50	6.60
SG mid oxy*	27	5.62	0.91	5.70	2.30	6.90
FP mid oxy*	65	5.82	0.86	6.00	2.50	7.30
SG max oxy*	27	3.92	1.52	4.00	1.10	6.70
FP max oxy*	65	4.56	0.99	4.40	1.90	6.70
SG surf trans	37	81.76	14.09	86.45	24.67	90.27
FP surf trans	70	86.37	2.66	87.12	74.72	88.69
SG mid trans	37	83.04	14.29	87.31	24.57	91.24
FP mid trans	70	85.45	5.29	87.04	59.27	88.50
SG max trans	37	55.64	19.84	61.34	5.99	82.04
FP max trans	70	73.97	11.50	75.81	37.26	88.80

Table 7. Size of red snapper larvae by month captured in SEAMAP bongo and neuston net samples. BL = body length; *N* = number of larvae measured.

	May	Jun	Jul	Aug	Sep	Oct	Nov
BL (mm)			Bongo				
Mean	4.7	5.1	5.2	5.3	5.1	5.0	4.8
Median	4.5	4.6	4.5	4.8	4.7	4.2	-
Range	3.8–6.5	3.3–10.0	2.4–19.2	3.6–10.2	3.1–18.2	3.5–8.1	-
95% Quantile	6.5	8.0	10.7	8.7	8.2	8.1	-
N	5	89	147	25	352	18	1
BL (mm)			Neuston				
Mean	4.7	5.0	5.1	4.3	4.1	4.3	4.8
Median	4.5	4.3	4.2	4.1	4.0	4.2	-
Range	3.1–6.5	3.3–21.0	3.1–24.0	3.4–12.6	2.7–10.5	3.4–5.0	-
95% Quantile	6.5	6.3	10.9	5.6	5.2	5.0	-
N	14	90	158	73	622	14	1

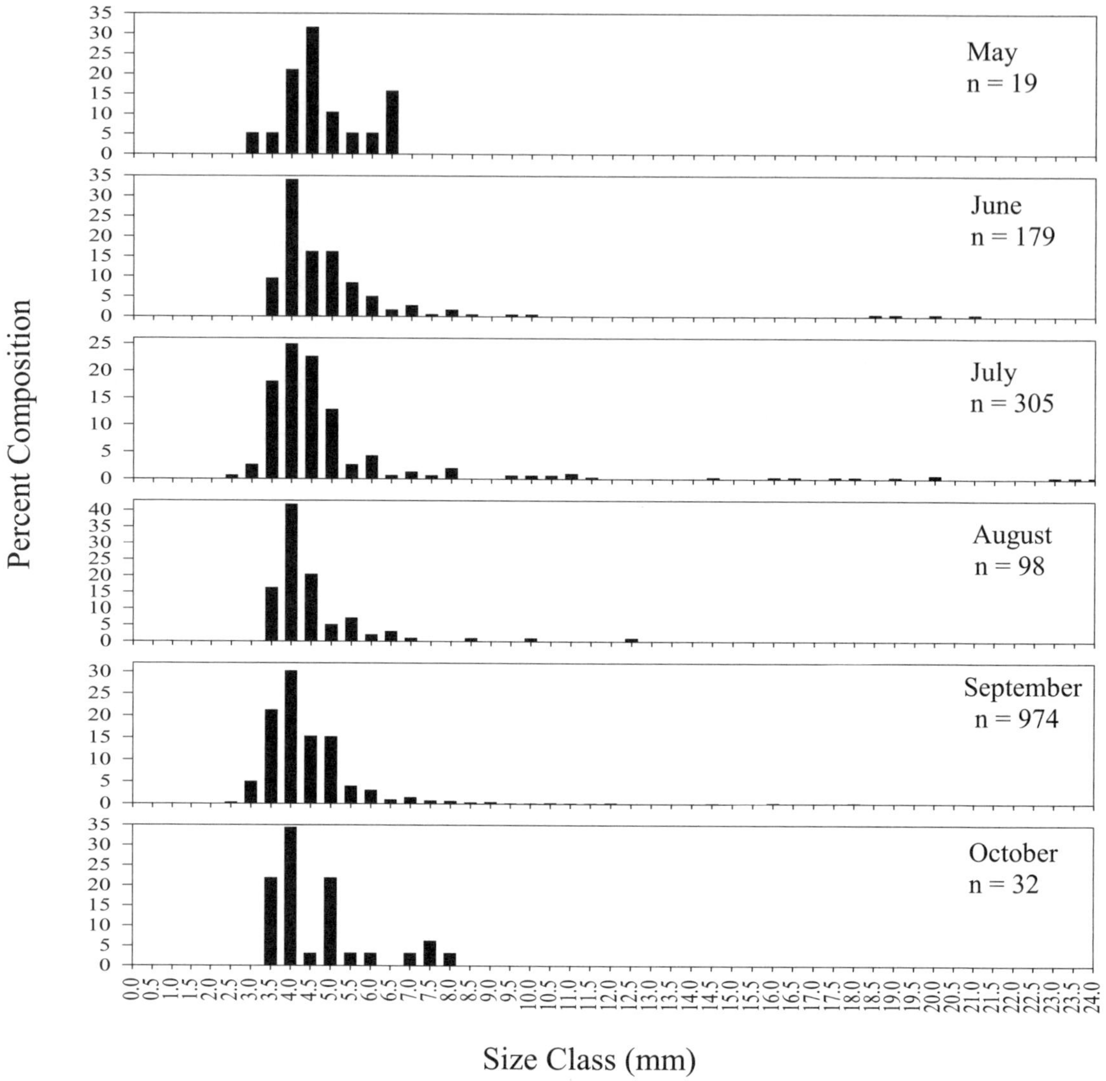

Figure 4. Monthly size frequency distributions of red snapper larvae in SEAMAP plankton samples. *N* = number of larvae measured. Individual y-axes are not to same scale.

midway down the west Florida shelf. Red snapper larvae were rarely taken during the Fall Shrimp/Bottomfish survey of continental shelf waters conducted mainly from Mobile Bay to south Texas in mid October through November (Figure 9). All captures were made in October except for two specimens taken in November; all captures were made in the western Gulf, and except for one, west of 94° west longitude.

Discussion

Temporal and spatial coverage of SEAMAP surveys for the most part encompass times and areas of known red snapper spawning (Collins et al. 1996, 2001; Woods 2003). The presence of red snapper larvae in SEAMAP plankton collections indicates continuous spawning within the spawning season, April–May and September–October, as determined from examination of adult go-

Table 8. Mean abundance and percent occurrence of red snapper larvae captured in bongo and neuston samples from day, night and twilight periods during SEAMAP Summer Shrimp/Bottomfish and Fall Plankton surveys, 1982–2003. Bongo abundance = larvae under 10 m² sea surface; neuston abundance = larvae per 10 min.

	Bongo			Neuston		
	Day	Night	Twilight	Day	Night	Twilight
Mean abundance	0.60	1.13	0.53	0.05	0.68	0.12
Percent occurrence	8.8	13.8	7.3	2.7	15.4	3.3
No. samples	1785	1218	314	1732	1201	301

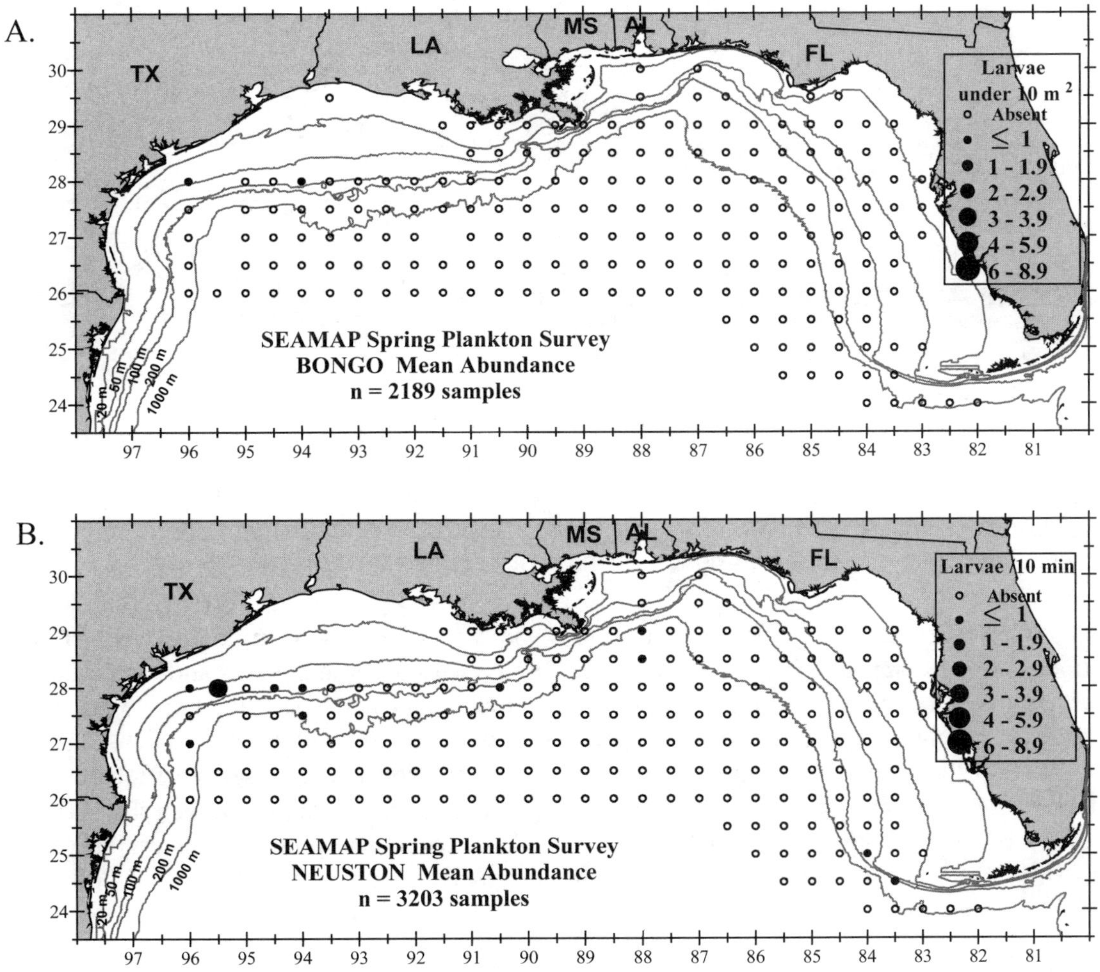

Figure 5. Mean abundance of red snapper larvae in bongo (A) and neuston (B) net samples during SEAMAP Spring Plankton surveys, 1982–2003. Abundance in bongo samples = larvae under 10 m² sea surface. Abundance in neuston samples = larvae per 10 min.

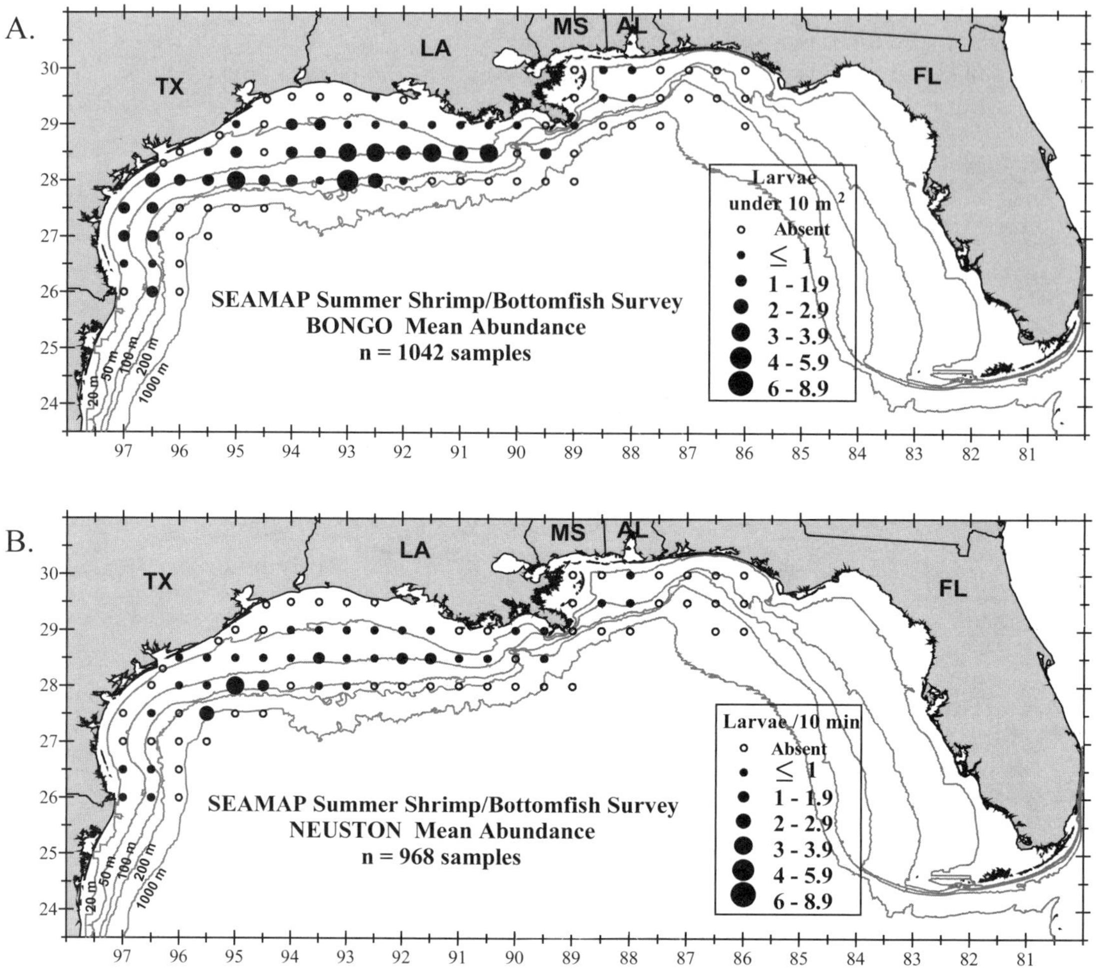

Figure 6. Mean abundance of red snapper larvae in bongo (A) and neuston (B) net samples during SEAMAP Summer Shrimp/Bottomfish surveys, 1982–2003. Abundance in bongo samples = larvae under 10 m^2 sea surface. Abundance in neuston samples = larvae per 10 min.

nads (Collins et al. 2001). The sharp drop in number of red snapper larvae caught in October may be indicative of abrupt termination of spawning in this species as was suggested by Woods (2003) based on the low incidence during the spawning season of red snapper ovaries exhibiting more than 50% atresia. It is likely that larvae were not found in SEAMAP samples from April because most sampling during the springtime survey is conducted in deep Gulf waters far from red snapper spawning sites. Station depths and bottom temperatures where red snapper larvae were captured during SEAMAP surveys agree with observed depths and temperatures at locations where spawning female red snapper were found (Collins et al. 2001). Furthermore, the consistent presence of red snapper larvae in samples taken between the 100 and 200 m contour in both the western and eastern Gulf supports the contention of Collins et al. (2001) that red snapper may spawn over a wide depth range, i.e. from mid-shelf to the continental slope. Settlement of juvenile red snapper has been reported on the outer continental shelf in the northwestern Gulf

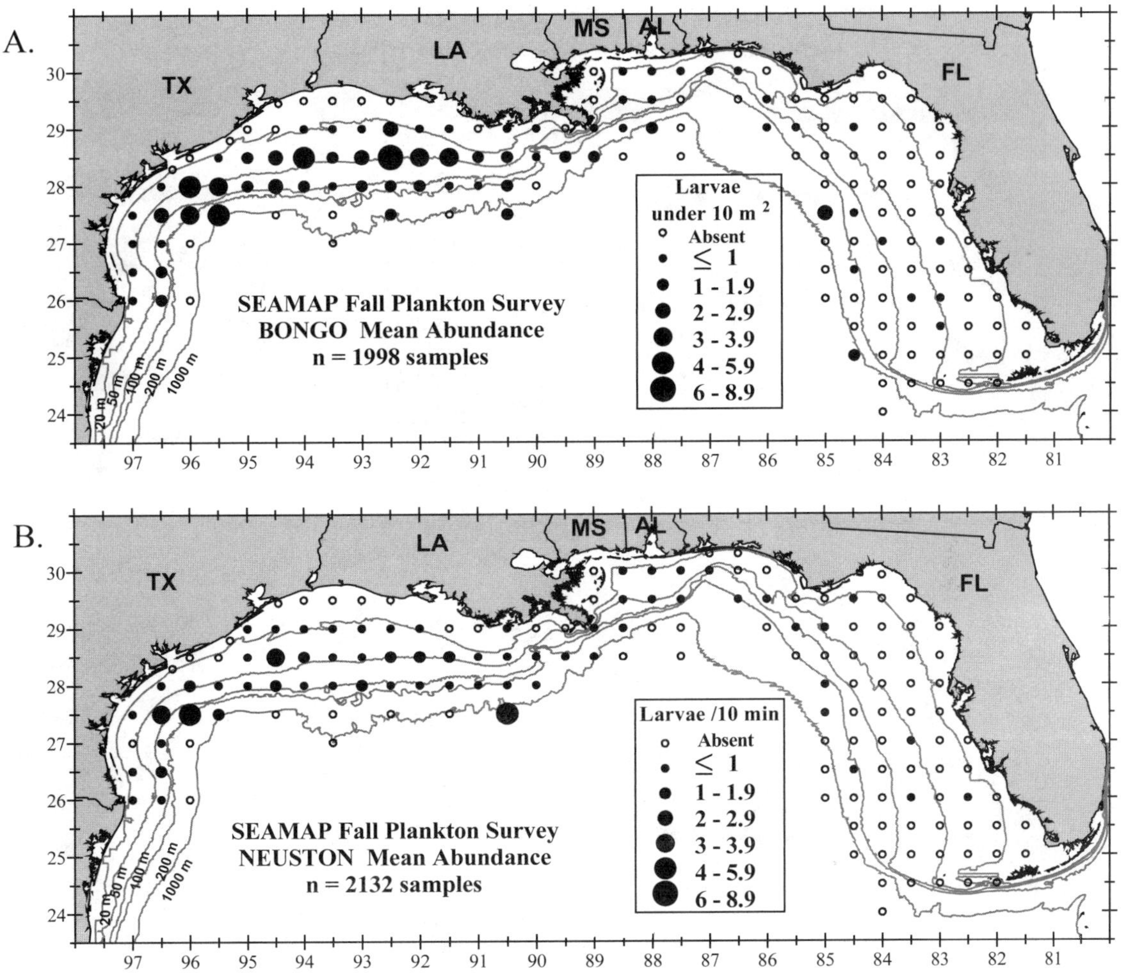

Figure 7. Mean abundance of red snapper larvae in bongo (A) and neuston (B) net samples during SEAMAP Fall Plankton surveys, 1986–2003. Abundance in bongo samples = larvae under 10 m² sea surface. Abundance in neuston samples = larvae per 10 min.

from June through October (Holt and Arnold 1982).

Data from SEAMAP samples present little information on vertical distribution of red snapper larvae, other than they are slightly more common below than at the surface. Recently, a critical need has arisen for information on vertical distribution of fish larvae, especially in the northern Gulf of Mexico, so that submerged warming water intakes at Liquefied Natural Gas (LNG) facilities might be placed at depths where entrainment of fish larvae would be minimized. There is some evidence that snapper larvae are vertically stratified in shelf waters less than 55 m in depth. Lutjanid larvae (not identified beyond the family level) captured at three discrete depths in coastal waters off east Louisiana, Mississippi and Alabama during September 1984, 1986 & 1987 were, in general, captured more frequently and in greater numbers at depths of 5 and 11 m than at 1 m (Lyczkowski-Shultz, J., NMFS Mississippi Laboratories, unpublished data).

Avoidance of SEAMAP bongo and, especially, neuston nets was clearly evident

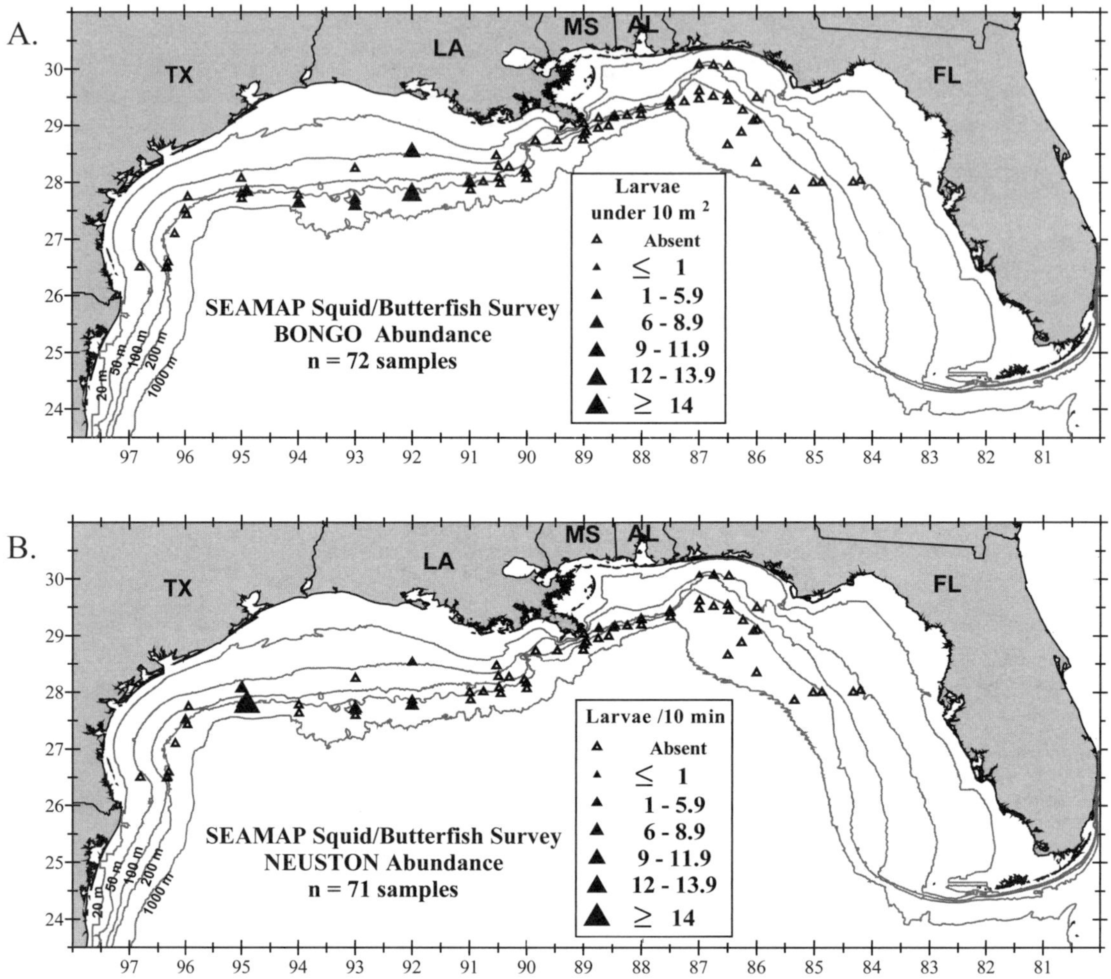

Figure 8. Station abundance of red snapper larvae in bongo net (A) and neuston (B) net samples during the SEAMAP Squid/Butterfish survey in August 1985. Abundance in bongo samples = larvae under 10 m^2 sea surface. Abundance in neuston samples = larvae per 10 min.

among red snapper larvae over 5 mm BL. Relatively few specimens greater than 6 mm BL were ever captured in plankton samples even though settlement from the planktonic to the benthic habitat first occurs at 16–19 mm and 26–30 d (Szedlmayer and Conti 1999; Rooker et al. 2004). Another reason why few postlarval and juvenile red snapper were taken in plankton samples may have been because the area within a few meters of the bottom is not effectively sampled using SEAMAP methods. If presettlement snappers become concentrated in the lower reaches of the water column they would not be susceptible to capture by bongo nets. Regardless of the cause (avoidance or undersampled habitat), the near absence of larvae from over half the expected planktonic size range and duration in SEAMAP samples preclude the use of these data to generate meaningful estimates of larval mortality rates. Unfortunately, actual not relative estimates of natural mortality are needed to effectively assess the impact of larval mortality caused by entrainment in LNG facilities in terms of loss to red snapper fisheries in the Gulf of Mexico. The inabil-

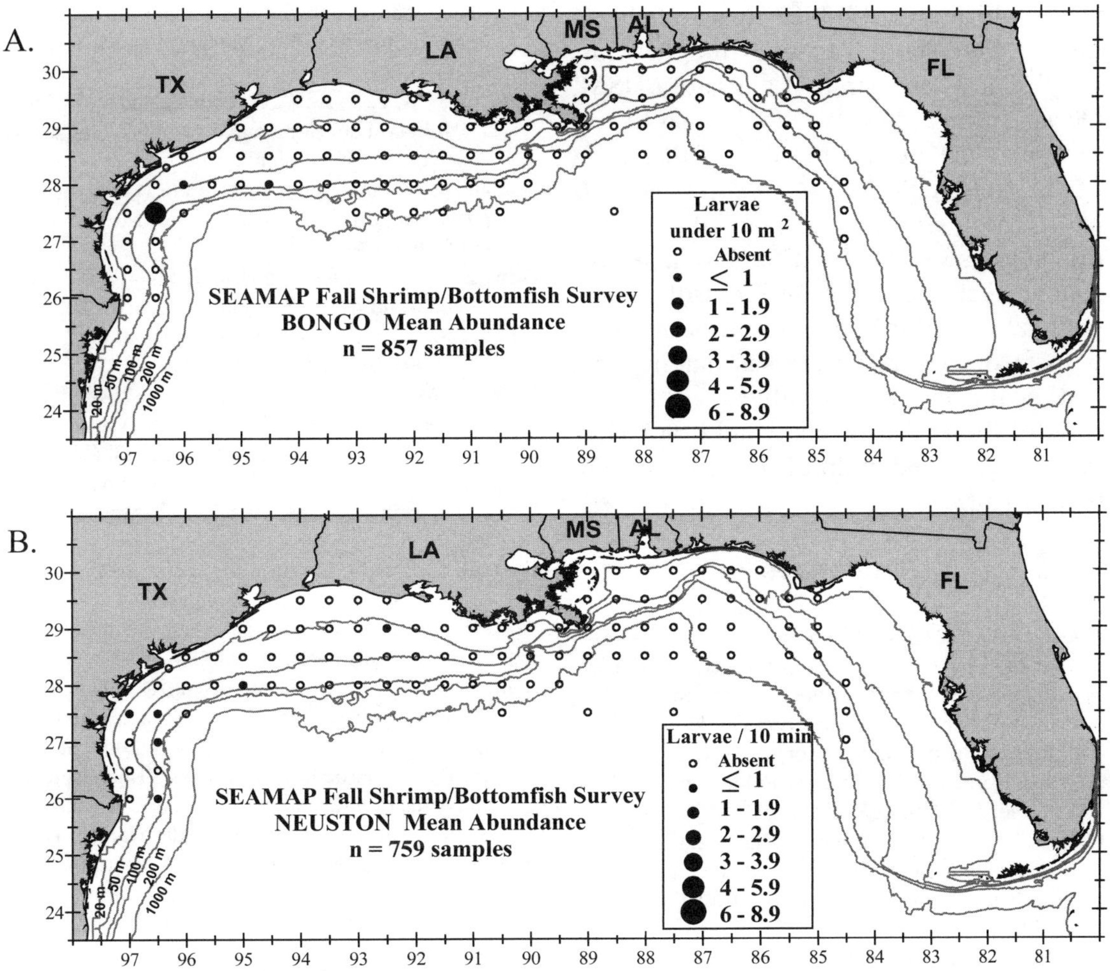

Figure 9. Mean abundance of red snapper larvae in bongo net (A) and neuston (B) net samples during SEAMAP Fall Shrimp/Bottomfish surveys, 1982–2003. Abundance in bongo samples = larvae under 10 m^2 sea surface. Abundance in neuston samples = larvae per 10 min.

ity to identify all snapper larvae in SEAMAP plankton samples, and the truncated size distribution of the catch results in underestimation of occurrence and abundance of red snapper larvae. Despite the shortcomings, these data represent the most comprehensive description of distribution and seasonal abundance of red snapper larvae available for the Gulf of Mexico. Ongoing SEAMAP resource surveys provide a reliable, albeit relative, measure of annual abundance of red snapper larvae over the size range caught, retained and identifiable in bongo and neuston nets.

Acknowledgments

The following individuals are gratefully acknowledged for their significant contributions to this work: Pam Bond, Connie Cowan, Denice Drass, Glenn Zapfe, Lanora Lang and Rosanne Brasher, NOAA/NMFS Mississippi Laboratories, Pascagoula, MS; Malgorzata Konieczna, Hanna Skolska and the Ichthyoplankton Group, Sea Fisheries Institute, Plankton Sorting and Identification Center, Szczecin and Gdynia, Poland; Kim Williams at the SEAMAP Archiving

Center, Fish and Wildlife Research Institute, St. Petersburg, FL; and Cindy Gavins, Gulf Coast Research Laboratory/USM, Ocean Springs, MS. We would also like to recognize the enduring efforts of the crews of the NOAA Ships *Chapman*, *Oregon II* and *Gordon Gunter*; and the dedication of the biologists of the NMFS, Gulf states resource agencies, Johnson Controls International and IAP World Services that participate on SEAMAP cruises making this historical data series possible.

Bill Richards, Jon Hare and Frank Hernandez provided helpful reviews of the manuscript.

References

Anderson, W. D., Jr. 2003. Lutjanidae—Snappers. Pages 1479–1504 *in* K. E. Carpenter, editor. The living marine resources of the western central Atlantic. FAO Species Identification Guide for Fishery Purposes and American Society of Ichthyologists and Herpetologists Special Publication 5. FAO, Rome. 3:1375–2127.

Collins, L. A., J. H. Finucane, and L. E. Barger. 1980. Description of larval and juvenile red snapper, *Lutjanus campechanus*. Fishery Bulletin 77:965–974.

Collins, L. A., A. G. Johnson, and C. P. Keim. 1996. Spawning and annual fecundity of the red snapper (*Lutjanus campechanus*) from the northeastern Gulf of Mexico. Pages 174–188 *in* F. Arreguin-Sanchez, J. L. Munro, M. C. Balgos, and D. Pauly, editors. Biology, fisheries and culture of tropical groupers and snappers. ICLARM Conference Proceedings 48.

Collins, L. A., G. R. Fitzhugh, L. Mourand, L. A. Lombardi, W. T. Walling, Jr., W. A. Fable, M. R. Burnett, and R. J. Allman. 2001. Preliminary results from a continuing study of spawning and fecundity in the red snapper (Lutjanidae: *Lutjanus campechanus*) from the Gulf of Mexico, 1998–1999. Proceedings of the 52nd Annual Gulf and Caribbean Fisheries Institute 52:34–47.

Ditty, J. G., and R. F. Shaw. 1992. Larval development, distribution, and ecology of cobia *Rachycentron canadum* (Family: Rachycentridae) in the northern Gulf of Mexico. Fishery Bulletin 90:688–677.

Ditty, J. G., and R. F. Shaw. 1993. Larval development of tripletail, *Lobotes surinamensis* (Pisces: Lobotidae), and their spatial and temporal distribution in the northern Gulf of Mexico. Fishery Bulletin 92:33–45.

Ditty, J. G., R. F. Shaw, C. B. Grimes, and J. S. Cope. 1994. Larval development, distribution, and abundance of common dolphin, *Coryphaena hippurus*, and pompano dolphin, *C. equiselis* (Family: Coryphaenidae), in the northern Gulf of Mexico. Fishery Bulletin 94: 275–291.

Ditty, J. G., G. G. Zieske, and R. F. Shaw. 1988. Seasonality and depth distribution of larval fishes in the northern Gulf of Mexico above latitude 26° 00' N. Fishery Bulletin 86:811–823.

Drass, D. M., K. L. Bootes, J. Lyczkowski-Shultz, B. H. Comyns, G. J. Holt, C. M. Riley, and R. P. Phelps. 2000. Larval development of red snapper, *Lutjanus campechanus*, with comparisons to co-occurring snapper species. Fishery Bulletin 98:507–527.

Gledhill, C. T. and J. Lyczkowski-Shultz. 2000. Indices of larval king mackerel, *Scomberomorus cavalla*, for use in population assessment in the Gulf of Mexico. Fishery Bulletin 98:684–691.

Hanisko, D. S., J. Lyczkowski-Shultz. 2003. Occurrence and pelagic habitat of reef fish larvae in the Gulf of Mexico. Pages 205–224 *in* D. R. Stanley and A. Scarborough-Bull, editors. Fisheries, reefs, and offshore development. American Fisheries Society, Symposium 36, Bethesda, Maryland.

Holt, S. A., and C. R. Arnold. 1982. Growth of juvenile red snapper *Lutjanus campechanus*, in the northwestern Gulf of Mexico. Fishery Bulletin 80:644–648.

Kramer, D., M. J. Kalin, E. G. Stevens, J. R. Thrailkill, and J. R. Zweifel. 1972. Collecting and processing data on fish eggs and larvae in the California Current region. NOAA Technical Report NMFS Circular 370.

Leis, J. M. and D. S. Rennis. 1983. The larvae of the Indo-Pacific coral reef fishes. University of Hawaii Press, Honolulu, Hawaii.

Lindeman, K. C., W. J. Richards, J. Lyczkowski-Shultz, D. M. Drass, C. B. Paris, J. M. Leis, M. Lara, and B. H. Comyns. 2005. Chapter 137 Lutjanidae: snappers. Pages 1549–1595 *in* W. J. Richards, editor. Early stages of Atlantic fishes: an identification guide for the western central North Atlantic. CRC Press, Boca Raton, Florida.

Morse, W. W. 1989. Catchability, growth, and mortality of larval fishes. Fishery Bulletin 87:417–446.

Parker, R. O., Jr., D. R. Colby and T. D. Willis. 1983. Estimated amount of reef habitat on a portion of the U.S. south Atlantic and Gulf of Mexico continental shelf. Bulletin of Marine Science. 33(4):935–940.

Posgay, J. A., and R. R. Marak. 1980. The MARMAP bongo zooplankton samplers. Journal of Northwest Atlantic Fishery Science 1:9–99.

Rabalais, N. N., S. C. Rabalais, and C. R. Arnold. 1980. Description of eggs and larvae of laboratory reared red snapper (*Lutjanus campechanus*). Copeia 1980/ 4:704–708.

Richards, W. J., editor. 2005. Early stages of Atlantic fishes: an identification guide for the western central North Atlantic. CRC Press/Taylor and Francis Group, Boca Raton, Florida.

Rooker, J. R., A. M. Landry, Jr., B. W. Geary, and J. A. Harper. 2004. Assessment of a shell bank and associated substrates as nursery habitat of postsettlement red snapper. Estuarine Coastal and Shelf Science 59:653–661.

Seidelmann, P. K., editor. 1992. Explanatory Supplement to the Astronomical Almanac. University Science Books, Sausalito, California.

Smith, P. E. and S. L. Richardson, editors. 1977. Standard techniques for pelagic fish egg and larva surveys. FAO Fisheries Technical Paper 175.

Scott, G. P., S. C. Turner, C. B. Grimes, W. J. Richards, and E. B. Brothers. 1993. Indices of larval bluefin tuna, *Thunnus thynnus*, abundance in the Gulf of Mexico: modeling variability in growth, mortality, and gear selectivity. Bulletin of Marine Science 53: 912–929.

Szedlmayer, S. T., and J. Conti. 1999. Nursery habitats, growth rates and seasonality of age-0 red snapper, *Lutjanus campechanus*, in the northeast Gulf of Mexico. Fishery Bulletin 97:626–635.

Woods, M. K. 2003. Demographic differences in reproductive biology of female red snapper (*Lutjanus campechanus*) in the northern Gulf of Mexico. Master's thesis. University of South Alabama, Mobile, Alabama.

American Fisheries Society Symposium 60:25–38, 2007

Patterns of Habitat Use by Newly Settled Red Snapper in the Northwestern Gulf of Mexico

Bert W. Geary, Joseph J. Mikulas, Jr., Jay R. Rooker[1], and André M. Landry, Jr.
Department of Marine Biology
Texas A&M University at Galveston
5007 Avenue U, Galveston, Texas 77551 USA

Timothy M. Dellapenna
Department of Marine Sciences
Texas A&M University at Galveston
P.O. Box 1675, Galveston, Texas 77553 USA

Abstract.—Trawl surveys were conducted to measure patterns of habitat use by newly settled red snapper *Lutjanus campechanus* at three natural banks on the inner continental shelf of Texas. Digital side-scan sonar and multibeam bathymetric data were used to define inshore (mud), ridge (shell), and offshore (mud) habitats for Freeport Rocks, Heald Bank, and Sabine Bank. Otter trawls were conducted July through September in 2003 (Heald Bank, Sabine Bank) and in 2004 (Freeport Rocks) during the settlement period of red snapper. Freeport Rocks had markedly higher densities of red snapper (91 ha^{-1}) in 2004 than Heald Bank (6 ha^{-1}) or Sabine Bank (<1 ha^{-1}) in 2003. A significant habitat effect was observed at Heald Bank and densities were higher at offshore mud habitats; no habitat effect was detected for Freeport Rocks or Sabine Bank. Growth rates varied from 0.86 $mm \cdot d^{-1}$ at Sabine Bank up to 1.12 $mm \cdot d^{-1}$ at Freeport Rocks, and rates were higher on inshore and offshore mud than ridge habitats. Otolith-based estimates of age indicated that settlers were first detected at 22–28 d and the majority of individuals were 30–60 d. Hatch dates peaked from early June to early July in both 2003 and 2004. Results from this study indicate that both shell and inshore and offshore mud habitats associated with these natural banks serve as settlement habitat of red snapper, and all three habitats have the potential to function as nursery areas of this species.

[1] Corresponding author: rookerj@tamug.edu.

Introduction

Red snapper *Lutjanus campechanus* are in decline with stocks well below long-term potential yield (GMFMC 2005). Similar to other marine fishes, fluctuations in red snapper stocks are potentially linked to variation in recruitment, which may be due to changes in environmental conditions as well as density-dependent processes during early life (Houde 1987; Leggett and Deblois 1994; Cowan and Shaw 2002). In addition, anthropogenic activities often lead to direct mortality (i.e., bycatch of juveniles from trawling) and indirect mortality due to changes in nursery habitat quantity and quality (Thrush et al. 1998; Thrush and Dayton 2002). Regardless of the cause, it is clear that recruitment variability and year-class strength of red snapper are likely determined during early life, and identifying habitats or conditions that favor survival during the nursery period is critical to the management of this species.

Several studies have indicated that juvenile red snapper use inshore low-relief habitats during early life and migrate to high-relief habitat in deeper waters as they grow (Bradley and Bryan 1975; Szedlmayer and Howe 1997; Szedlmayer and Conti 1999; Workman et al. 2002; Patterson et al. 2005). The pattern is not as clear for new settlers, with recruits often present on mud or open sand habitats with little or no structure. Szedlmayer and Conti (1999) observed newly settled red snapper on open (sand and mud) as well as low-relief shell habitat in the northeastern Gulf of Mexico (GOM); however, settlement was significantly higher at shell ridges. In the northwestern GOM, Rooker et al. (2004) collected newly settled red snapper on shell ridges, but observed large numbers of settlers in mud habitats away from low-relief shell ridges, suggesting factors other than structure may be critical during the early postsettlement period.

The aim of the present study was to determine patterns of habitat use by newly settled red snapper, and assess the quality of different banks and habitats used during early life. Using active acoustics, different substrates (inshore and offshore mud, shell ridge) associated with three prominent banks in the northwestern GOM were identified, and the resulting habitat maps were used to direct trawl surveys of red snapper during the primary settlement period. Patterns of distribution and abundance were examined at both large (banks) and small (habitats within a bank) scales. In addition, otolith-based methods were used to estimate hatch dates and growth rates of new settlers, with growth estimates providing a relative measure of habitat quality on both spatial and temporal scales.

Methods

Trawl surveys were conducted on three natural banks located along the inner to mid-continental shelf of Texas, USA (Figure 1). The banks (Freeport Rocks, Heald Bank, Sabine Bank) are dominated by shell hash and relic oyster beds with paleo-environmental evidence indicating that they originated as barrier islands (Rodriguez et al. 2000). Freeport Rocks (W 95° 18', N 28° 44') is located 22 km south of Freeport, Texas at a depth of 13–24 m, with the study area covering 80 km^2. Heald Bank (W 94° 09', N 29° 07') is 71 km southwest of Sabine Pass, Texas at 9–14 m depth, with the study area covering 20 km^2. Sabine Bank (W 93° 59', N 29° 22') is 39 km south of Sabine Pass, Texas and occurs at 8–11 m depth with our study site covering 27 km^2 of the bank. Banks were surveyed using side-scan sonar (Edge Tech 272-TD) to create habitat maps with light areas of high reflectivity representing shell substrate and dark areas of low reflectivity representing mud substrate. Habitat types were verified by weighing shell collected during trawls and by carbonate content measurement of bottom

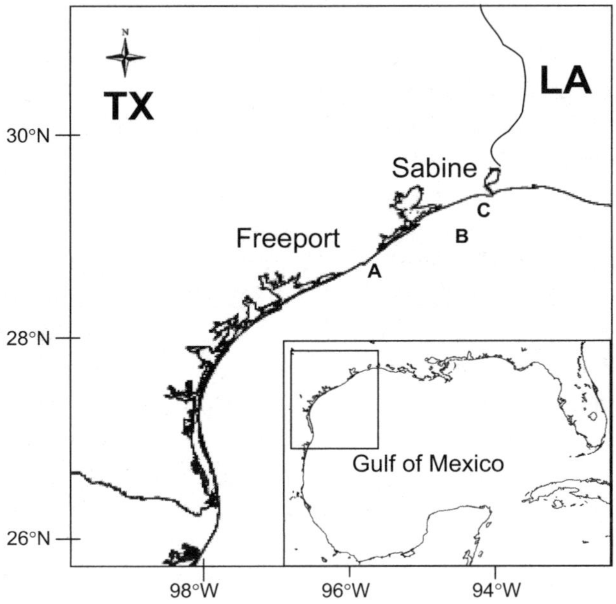

Figure 1. Location of the natural shell banks sampled for newly settled red snapper *Lutjanus campechanus* in the northwestern Gulf of Mexico; A–Freeport Rocks, B–Heald Bank, C–Sabine Bank.

sediment samples (Mikulas 2007). Ridges on all banks were comprised of sand and shell material, while areas off the ridge (both inshore and offshore) were primarily silt and mud environments. Three habitats were present at Heald Bank and Sabine Bank: inshore (mud), (shell) ridge, and offshore (mud). On Freeport Rocks, we further divided the ridge habitat into shell ridge and sand ridge (negligible shell hash present) habitats, in addition to the inshore and offshore habitats.

Bottom trawl surveys were conducted semimonthly from June through September in 2003 (Heald Bank, Sabine Bank) and 2004 (Freeport Rocks). A 6-m otter trawl with 2-cm bar mesh, 1.25-cm bar mesh liner, and 0.6-cm link tickler chain was towed at 2.5 knots for 5 min at each station. Trawl start and end coordinates were taken using GPS to calculate area sampled. A suite of environmental parameters were measured at each site (depth, dissolved oxygen, temperature, and salinity). Red snapper were picked from the sample and frozen onboard the research vessel. Samples were sorted in the laboratory, and lengths and weights were measured to the nearest 0.1 mm and 0.01 g, respectively.

Trawls with large numbers of red snapper caught (many Freeport Rocks trawls, one Heald Bank trawl) had subsamples of fish selected for otolith analysis, while all red snapper from trawls with low catch numbers (most Heald Bank trawls, all Sabine Bank trawls) were used. Sagittal otoliths were extracted and either the right or left otolith was randomly selected for age determination. Otoliths were embedded in Struer's resin and transverse sections were cut and polished following the protocol of Rooker et al. (2004). Sectioned otoliths were read (40x magnification) using an image analysis system (Olympus BX41 and Image Pro v. 4.5 software). Daily increment deposition has been validated for red snapper (Szedlmayer 1998), and growth increments were counted from the core to the otolith edge. Predicted age was based on averaging two counts for each otolith. When the mean difference of the counts was greater than 10%, a third count was taken and the mean of the two closest counts was used for age estimates. Of 216 otoliths prepared, 210 were included in analyses. Size-at-age relationships were used to generate growth estimates for each bank and each habitat within Freeport Rocks. Age-length relationships were then used to backcalculate ages and hatch dates for the entire collection of red snapper.

Seasonal patterns of red snapper abundance were apparent and numerous samples contained zero values; consequently, the assumption of normality was violated so statistical testing was restricted to samples collected during peak recruitment. This period included five trips (July 2–3, July 17–18, August 4–5, August 19–20, September 9–10) for Heald Bank and Sabine Bank in 2003 (96.3% of total catch), and three trips (July 26–27, August 10–11, September 1–2) for Freeport Rocks in 2004 (99.7% of total catch). Data were $\log_e$ transformed prior to statistical analysis to minimize heteroscedasticity. Two-way analysis of variance (ANOVA) was used to test for date and habitat effects (SPSS 2004). Tukey HSD was used to detect *a posteriori* differences among factor levels (α = 0.05). Power analysis was performed when a statistical test failed to reject the null hypothesis. Linear regression analysis was used to determine growth rates of red snapper, and analysis of covariance (ANCOVA) was used to test for differences in growth (slopes of regression) among habitats at Freeport Rocks.

Results

Water quality characteristics varied temporally, and habitat and site specific differences were detected. Water temperature increased from early July to mid September and ranged from 25.9 to 30.1°C at Freeport Rocks in 2004 and from 28.4 to 29.7°C at Heald Bank and Sabine Bank in 2003 (Figure 2). Water temperature was similar among all habitats at Freeport Rocks, Heald Bank, and Sabine Bank (P = 0.173, power = 0.365; P = 0.640, power = 0.118; P = 0.855, power = 0.073; respectively). At Freeport Rocks, salinities ranged from 31.4 to 36.3‰ with significantly higher salinities in the offshore habitat than in the other habitats (P = 0.002). Salinities were similar among habitats at Heald Bank and Sabine Bank (P = 0.927, power = 0.118; P = 0.855, power = 0.073, respectively), and increased across the sampling period (26.3–33.3‰) at both sites. At Freeport Rocks, dissolved oxygen levels (DO) ranged from 2.61 to 6.15 $mg{\cdot}L^{-1}$, with no effect of habitat on DO levels observed (P = 0.463, power = 0.179). Dissolved oxygen levels varied across survey dates at Heald Bank and Sabine Bank, and ranged from 3.69 to 7.08 $mg{\cdot}L^{-1}$, although there were no significant differences in DO by habitat (P = 0.548, power = 0.144; P = 0.224, power = 0.312, respectively).

Total red snapper catch at Freeport Rocks in 2004 (n = 1,410) was markedly higher than at either Heald Bank (n = 63) or Sabine Bank (n = 18) in 2003. Mean density of red snapper over peak recruitment at Freeport Rocks was 91.3 ± 11.32 ha^{-1} (mean ± SE), while numbers at Heald Bank and Sabine Bank were 5.8 ± 3.76 ha^{-1} and 0.3 ± 0.08 ha^{-1}, respectively (Figure 3). No habitat (P = 0.354, power = 0.285) or date (P = 0.618, power = 0.126) effect was observed at Freeport Rocks, with mean densities ranging from 76.3 ± 22.42 ha^{-1} at the shell ridge to 108.8 ± 21.79 ha^{-1} at the inshore habitat. A significant habitat effect was observed at Heald Bank (P = 0.009), with higher densities at the offshore habitat (15.5 ± 14.14 ha^{-1}) compared to the inshore (0.5 ± 0.37 ha^{-1}) and ridge (1.7 ± 0.90 ha^{-1}) habitats. A date effect on density at Heald Bank (P = 0.023), was observed, with the August 19–20 trip (when no red snapper were collected) significantly different from other trips. Red snapper densities at Sabine Bank did not differ among habitats (P = 0.259,

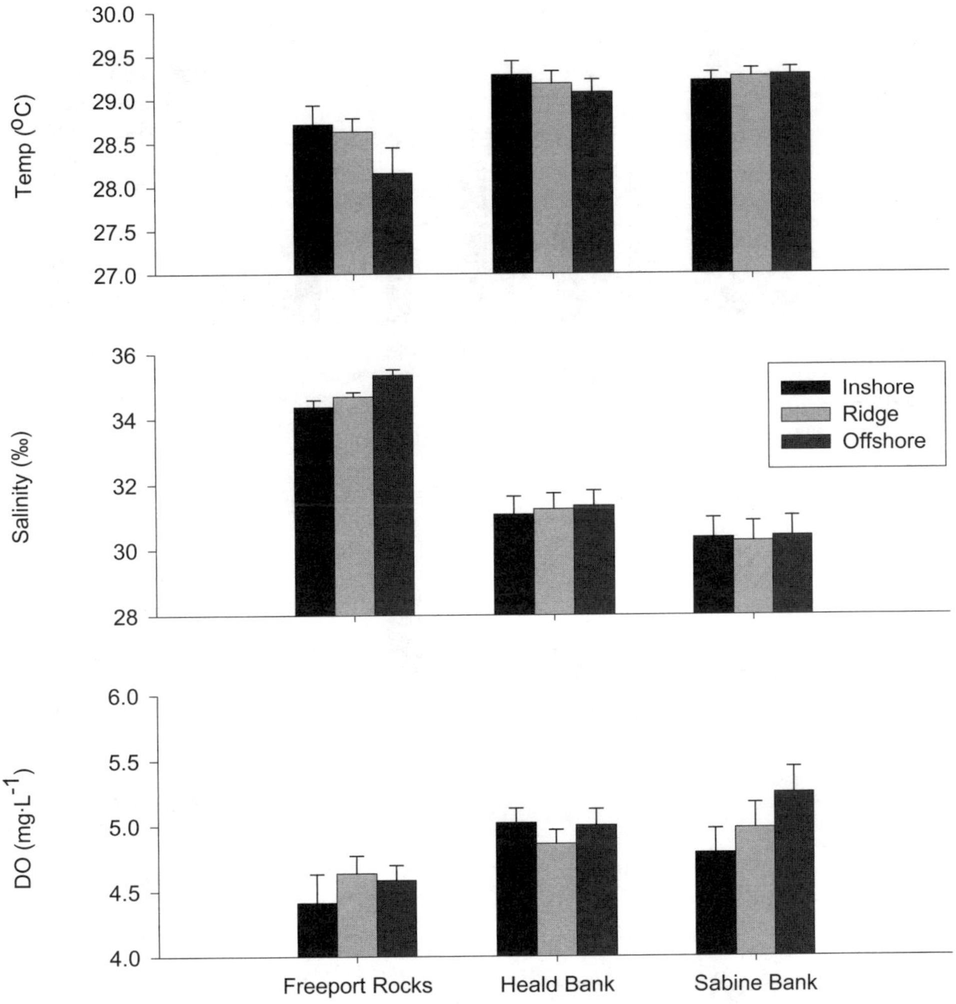

Figure 2. Mean (±1 SE) bottom water temperature (°C), salinity (‰), and dissolved oxygen (mg·L^{-1}) at natural banks in the northwestern Gulf of Mexico. Data from trips conducted on July 8–9, July 26–27, August 10–11 and September 28–29 of 2004 for Freeport Rocks, and July 17–18, August 4–5 and August 19–20 of 2003 for Heald Bank and Sabine Bank.

power = 0.288) or among sampling dates (P = 0.568, power = 0.229), ranging from 0.4 ± 0.26 ha^{-1} at the inshore habitat to 1.3 ± 0.52 ha^{-1} on the ridge.

Standard lengths (SL) of red snapper at first occurrence in our collections were 18.2 mm SL for Freeport Rocks, 14.2 mm SL for Heald Bank, and 19.3 mm SL for Sabine Bank (Figure 4). At Freeport Rocks, mean size of red snapper from shell ridge (43.9 ± 0.87 mm SL) and sand ridge (46.7 ± 0.81 mm SL) habitats were significantly larger (P < 0.001) than from inshore (40.5 ± 0.65 mm SL) or offshore habitats (35.5 ± 0.73 mm SL). Mean size of red snapper at Heald Bank (24.1 ± 6.02 mm SL) was markedly lower than that of red snapper on Sabine Bank (47.7 ± 7.47 mm SL); however, mean size among habitats did not differ significantly for either bank (P = 0.493, power = 0.165 and P = 0.868, power = 0.068, respectively). On all three banks, mean size of red snapper increased with time

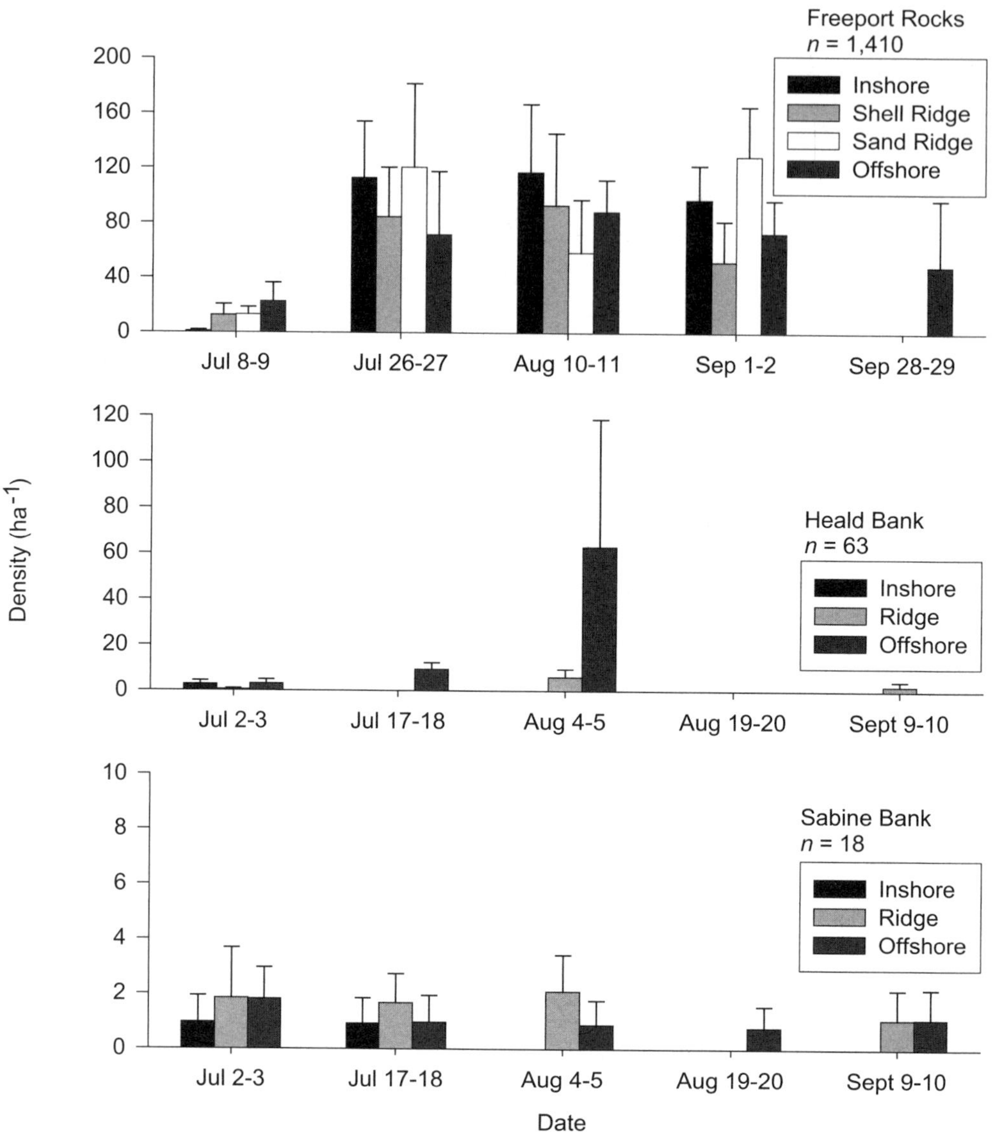

Figure 3. Mean (±1 SE) densities of red snapper *Lutjanus campechanus* collected in the northwestern Gulf of Mexico at Freeport Rocks in 2004 and in Heald and Sabine Bank in 2003. Densities are presented by sampling date and by habitat.

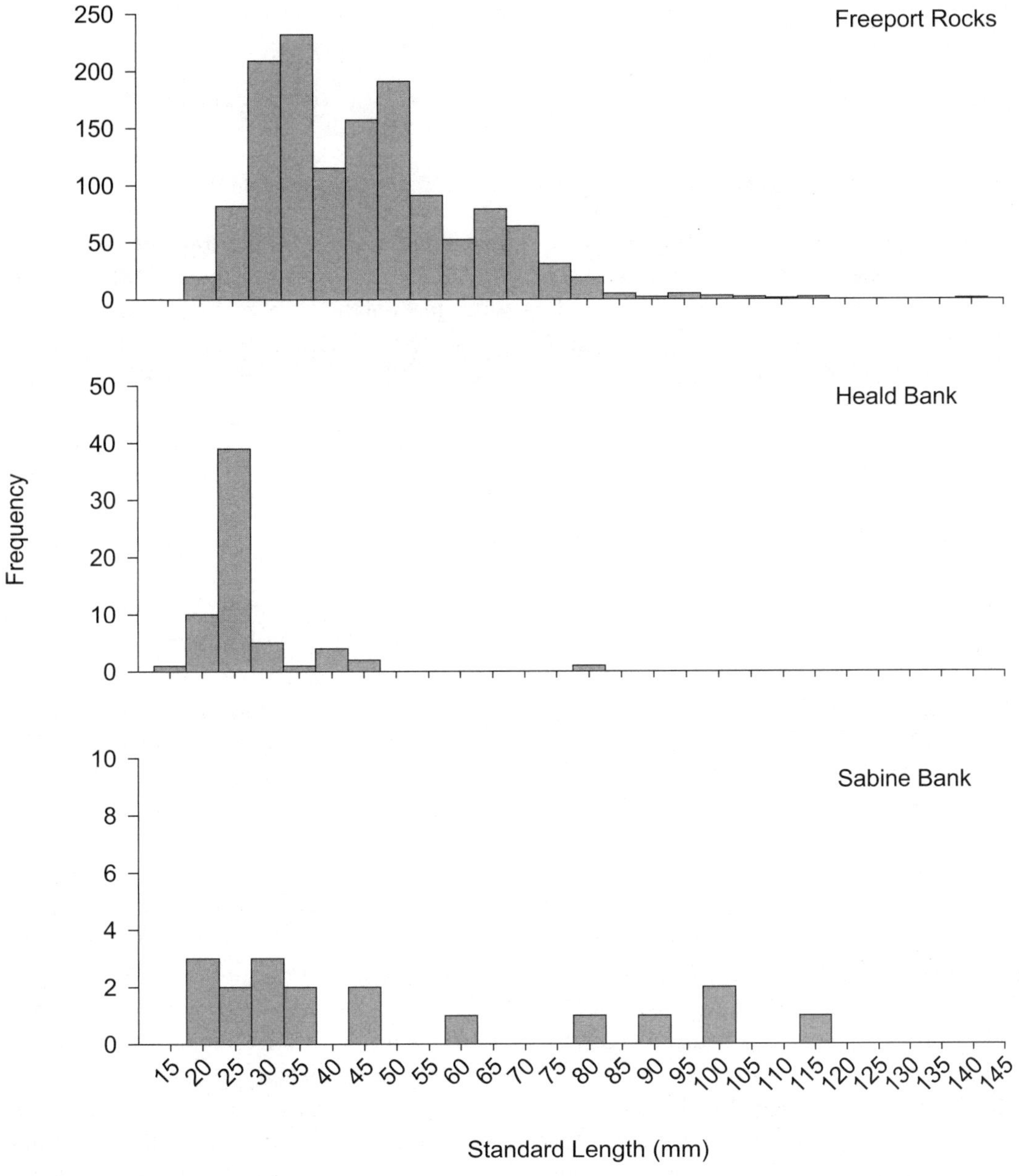

Figure 4. Length-frequency distributions of red snapper *Lutjanus campechanus* collected in the northwestern Gulf of Mexico at Freeport Rocks in 2004 and Heald Bank and Sabine Bank in 2003. Sampling period occurred during the red snapper settlement period from June–September of both years.

during the peak recruitment period. At Freeport Rocks, mean size increased from 31.6 ± 0.45 mm SL on July 26–27 to 52.0 ± 0.79 mm SL on September 1–2, while at Heald Bank mean size increased from 18.8 ± 1.34 mm SL on the July 2–3 trip to 34.0 ± 5.05 mm SL on the September 9–10 trip. Mean size also increased with time at Sabine Bank, although the pattern is not as clear due to the small sample size.

Growth rates varied among banks (Figure 5). Growth at Freeport Rocks in 2004 (1.09 $mm \cdot d^{-1}$) was higher than at either Heald Bank (1.01 $mm \cdot d^{-1}$) or Sabine Bank (0.86 $mm \cdot d^{-1}$) in 2003. Growth rates varied by bank, but were not compared statistically since banks were sampled in different years. Habitat-specific growth estimates were generated for Freeport Rocks, and growth was significantly higher ($P = 0.013$) at inshore (1.21 $mm \cdot d^{-1}$) and offshore (1.19 $mm \cdot d^{-1}$) mud habitats than at shell ridge (0.69 $mm \cdot d^{-1}$) and sand ridge (0.95 $mm \cdot d^{-1}$) habitats.

Otolith-based estimates of age indicated that early postsettlement red snapper were first detected on these banks at 22–28 d (Freeport 28 d, Heald Bank 22 d, Sabine Bank 24 d) (Figure 6). Peak abundance at Freeport Rocks in 2004 was of 30–60 d red snapper, while the majority of the catch at Heald Bank in 2003 was younger and consisted of 26–30 d red snapper; peaks in age occurred at 50–52 d and 38–40 d, respectively. Although low catch numbers were present at Sabine Bank, most red snapper were 25–35 d with a few individuals up to 130 d. Age estimates were also used to determine hatch dates, which ranged from early April to late August for all banks (Figure 7). Freeport Rocks had a unimodal hatch-date profile with a peak in June, and over 70% of the red snapper at Freeport Rocks ($n = 1000$) were produced from June spawning events. Although numbers were limited, Heald Bank exhibited a peak hatch in mid June with numerous earlier hatch dates for individual red snapper. Red snapper collected at Sabine Bank were from late May to early June hatch dates.

Discussion

Previous research suggests that structured habitats are essential nursery areas of age-0 red snapper (Szedlmayer and Conti 1999). Laboratory experiments indicate both wild and hatchery-reared juvenile red snapper show associations with structure (i.e., artificial reefs, Masuda et al. 2003) and shell habitats (Szedlmayer and Howe 1997), while field studies have found that juvenile red snapper (>60 mm SL) recruit to artificial reef structures (Workman et al. 2002; Szedlmayer and Lee 2004). In the northeastern GOM, Szedlmayer and Lee (2004) found that red snapper settled on open sand-mud habitat and later migrated to structured reef habitat by 60 mm SL. Szedlmayer and Conti (1999) found significantly higher numbers of newly settled red snapper over shell bottom than over open bottom, although red snapper were found in both habitats. Along the Texas coast, large numbers of newly settled red snapper have been reported on both shell/sand and mud habitats (Rooker et al. 2004). Similarly, findings from the current study indicate that red snapper settled throughout shell and open bottom habitats without preference for structured, shell habitats. In fact, our only significant habitat on density result was at Heald Bank, with densities of new settlers higher in the offshore mud habitat than the structured ridge habitat. Since 42 of the 63 red snapper collected came from a single trawl in the offshore mud habitat, the appearance of a habitat effect may be suspect. Findings presented here, taken together with previous research, suggest that newly settled red snapper settle to a variety of habitats, including unstructured, mud bottom habitats. Since larger red snapper were observed on ridge habitats, postsettlers likely move to more structured habitat with increasing size, which supports findings

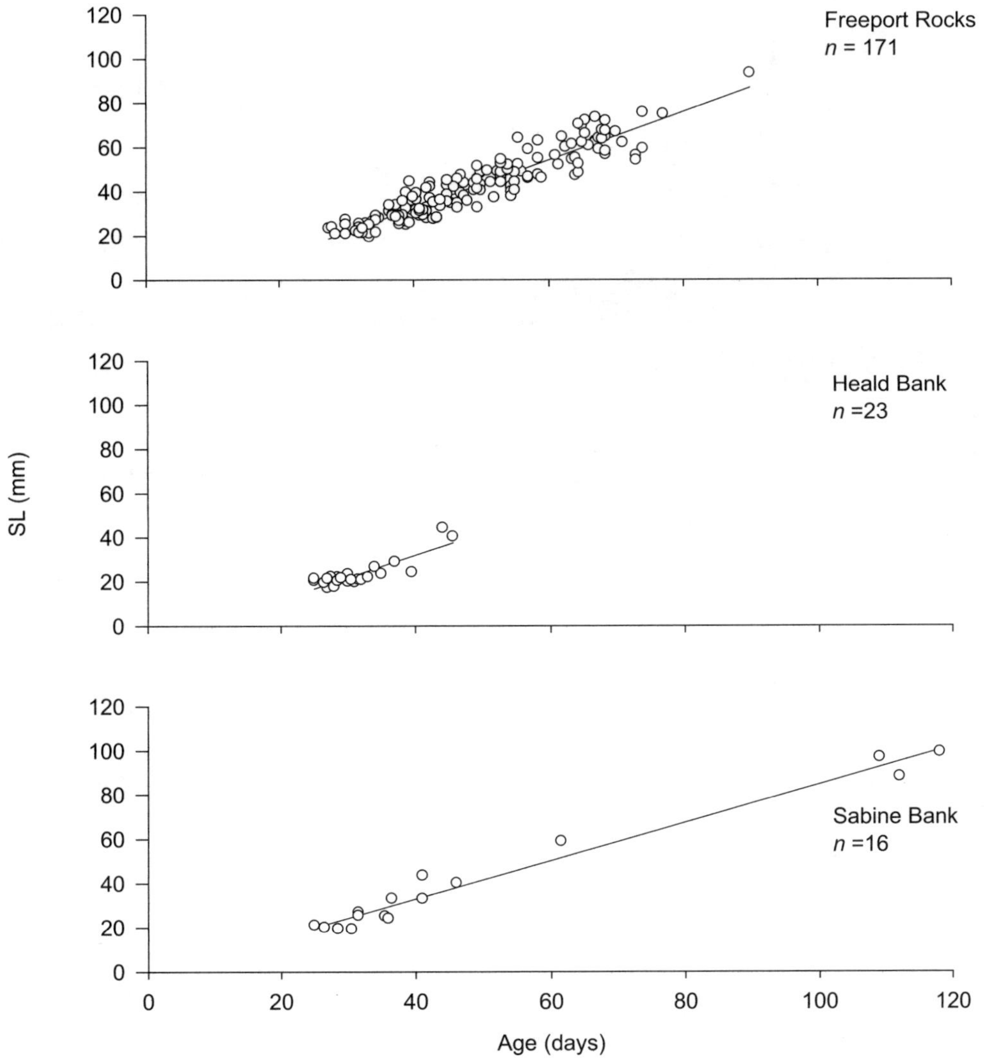

Figure 5. Otolith-based growth estimates of red snapper *Lutjanus campechanus* collected in the northwestern Gulf of Mexico at Freeport Rocks in 2004 and Heald Bank and Sabine Bank in 2003. Growth models: Freeport Rocks, SL = 1.09(age) – 11.32; Heald Bank, SL = 1.01(age) – 8.44; Sabine Bank, SL = 0.86(age) – 1.60.

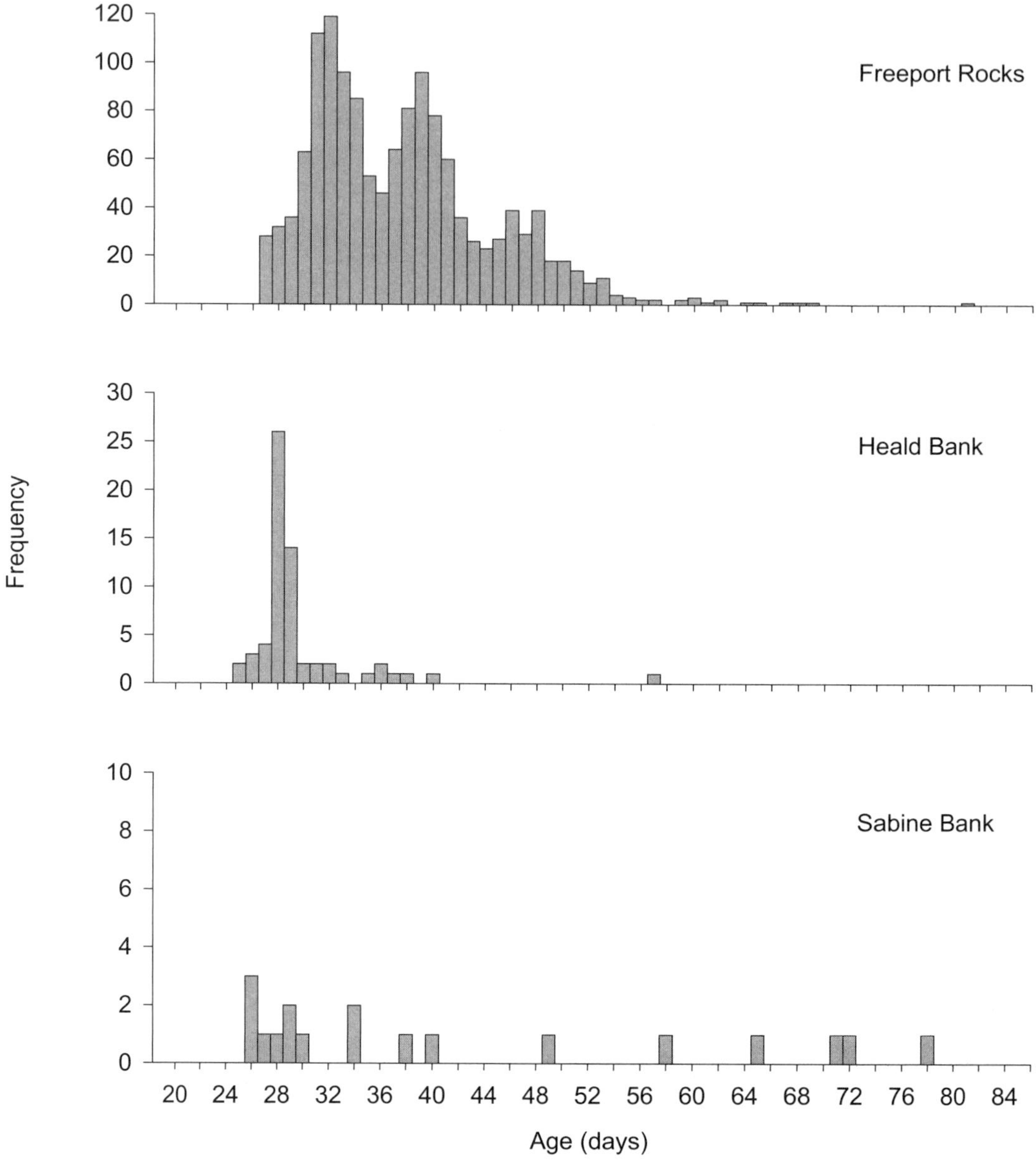

Figure 6. Age-frequency distribution of red snapper *Lutjanus campechanus* collected in the northwestern Gulf of Mexico at Freeport Rocks in 2004 and Heald Bank and Sabine Bank in 2003.

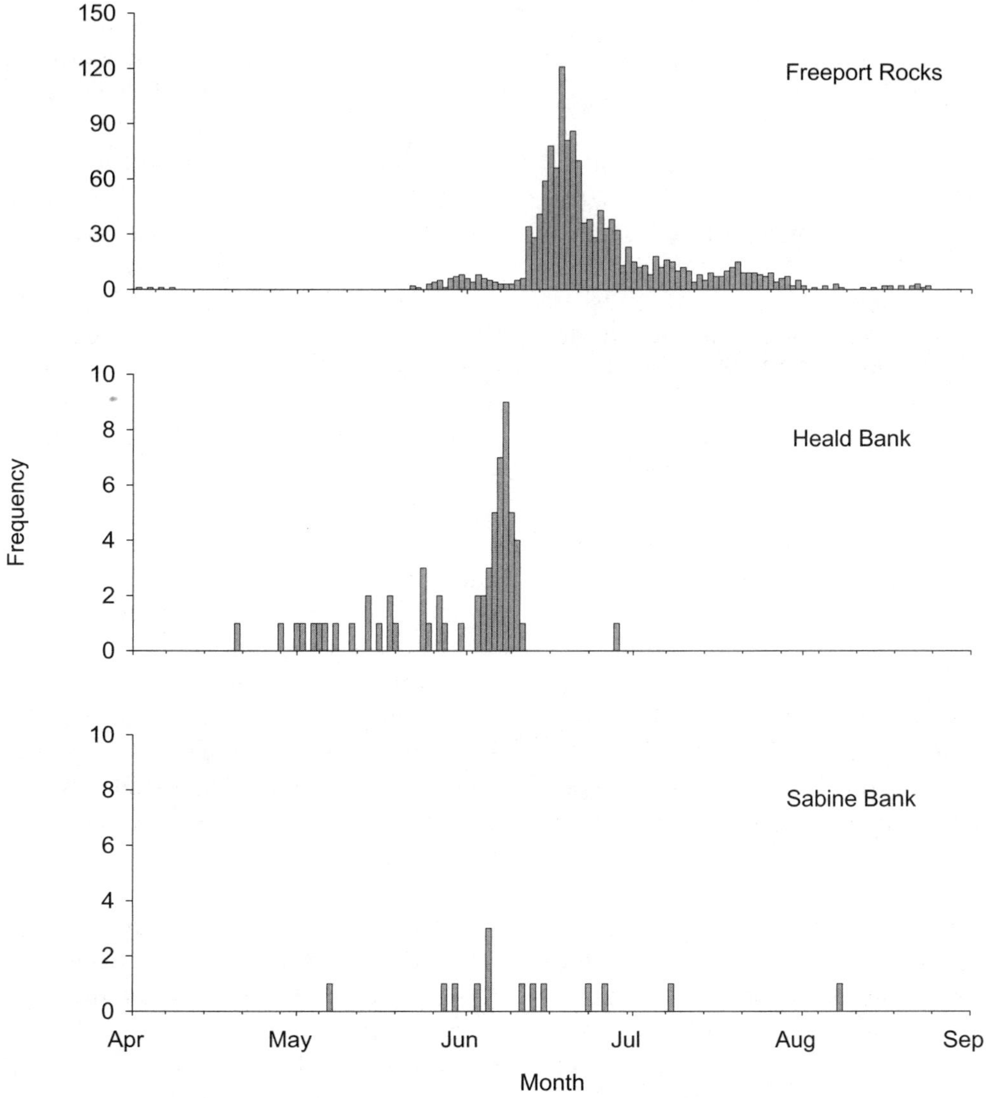

Figure 7. Hatch-date distributions of red snapper *Lutjanus campechanus* collected in the northwestern Gulf of Mexico at Freeport Rocks in 2004 and Heald Bank and Sabine Bank in 2003. Hatch dates are based on otolith-derived age estimates.

from recent studies in the north-central GOM (Wells 2007).

Red snapper recruits were first observed in early June (Sabine Bank) and early July (Freeport Rocks and Heald Bank) with peak recruitment from July to September at all three banks. Similarly, Rooker et al. (2004) reported peak recruitment of red snapper to Freeport Rocks in 2000–2001 occurred from July to August. Temporal trends are also similar to recruitment patterns reported for the northeastern GOM, where postsettlement red snapper first appeared at sampling sites in June and peaked from July to September (Szedlmayer and Conti 1999; Szedlmayer and Lee 2004).

Daily growth rates observed in this study (1.09 $mm \cdot d^{-1}$ Freeport Rocks, 1.01 $mm \cdot d^{-1}$ Heald Bank, 0.86 $mm \cdot d^{-1}$ Sabine Bank) overlap at the upper end of red snapper growth rates reported by Rooker et al. in 2004 (0.78–0.88 $mm \cdot d^{-1}$) and Szedlmayer and Conti in 1999 (0.54–0.86 $mm \cdot d^{-1}$). Similar to the previous study at Freeport Rocks (Rooker et al. 2004), growth rates were higher on the low-relief inshore and offshore mud habitats than at the high-relief shell ridge habitat. Also, the growth rate on the shell ridge measured in this study is similar to that measured on shell habitat by Szedlmayer and Conti (1999). Although faster growth over unstructured habitats has been reported for other taxa, such as flounder (Tarpgaard et al. 2005), it is by no means the rule. Other studies have found structured habitats increased growth (Tupper and Boutilier 1995; Phelan et al. 2000) or had little to no effect on growth (Tupper and Boutilier 1997; Phelan et al. 2000; Stunz et al. 2002).

Interannual differences in growth have been reported for red snapper (Szedlmayer and Conti 1999) as well as for other reef fishes (Allman and Grimes 2002; Sponaugle et al. 2006), and thus observed temporal differences are not unexpected. Cohort-specific differences in growth rates (both intra- and interannual) for reef fishes are commonly reported in the literature: red snapper 8–51% (Szedlmayer and Conti 1999; Rooker et al. 2004), gray snapper *Lutjanus griseus* 70% (Allman and Grimes 2002), bluehead wrasse *Thalassoma bifasciatum* 90% (Sponaugle et al. 2006). These differences are often linked to interannual variation in environmental parameters, including temperature (Houde 1987; Baltz et al. 1998; Werner 2002; Sponaugle et al. 2006), prey availability (Werner 2002), and dissolved oxygen (Baltz et al. 1998; Werner 2002). In the present study, mean temperature was 0.5°C higher on banks with lower growth estimates, indicating that observed growth differences among banks sampled are likely due to factors other than temperature. Precaution must be exercised when interpreting differences in growth among banks, however, because areas were sampled in different years, and therefore observed patterns may be partly explained by interannual variability rather than location. Temperatures measured in the northeastern GOM by Szedlmayer and Conti (1999) were over 4°C cooler than those in the present study while temperatures at Freeport Rocks in 2000–2001 (Rooker et al. 2004) were similar to those of the present study. Therefore, higher temperatures in the northwestern GOM may have contributed to increased growth rates relative to the northeastern GOM, and accounted for shifts in peak hatch dates measured among the three studies.

Settlement sizes and ages of red snapper reported here are similar to those of previous studies. Workman et al. (2002) observed recruits less than 20 mm SL off of Mississippi and Szedlmayer and Conti (1999) reported red snapper settled by 17 mm SL, also off of Mississippi. Similarly, postsettlement red snapper at Freeport Rocks in 2000–2001 first occurred at 16–20 mm SL (Rooker et al. 2004). Estimated red snapper settlement ages of 27 d (Texas) and 26 d (Alabama) reported previously are similar to our findings (Szedlmayer and Conti 1999; Rooker et al. 2004). Our observed planktonic larval duration of 3–4 weeks is also consistent with reports for other lutjanids in the Atlantic and Pacific (Allman and Grimes 2002; Zapata and Herron 2002; Denit and Sponaugle 2004; Mikulas 2007).

Hatch-date analysis indicated that spawning (50% hatch occurs at 25 h post-spawn (Rabalais et al. 1980)) occurred from early April through late August for all three banks. Peaks at all three banks were generally unimodal with the majority of recruits from June and July hatch dates. This is roughly two weeks later than the peak in red snapper hatch date reported by Rooker et al. (2004). Temporal variation of this nature is expected, and interannual variations in spawning periods of red snapper have been reported in other studies. Szedlmayer and Conti (1999) reported that peaks in hatch-date distributions of red snapper in the northeastern GOM vary by several weeks between consecutive years. These authors also reported that hatch-date distributions were bimodal in this region (peaks in late May-early June and July in 1994; May and June in 1995), which is different from the present study as well as from previous work in the northwestern GOM (Rooker et al. 2004). Gonadal analysis of red snapper suggests that spawning occurs in the northwestern GOM from April through October (Collins et al. 2001), indicating that a second pulse of recruits may enter the nursery grounds in the late fall. As a result, the lack of a bimodal distribution may be an artifact of our restricted sampling period.

Acknowledgments

Funding provided by the NOAA MARFIN program (grant # NA17FF2872 to J. R. Rooker). Special thanks to Captain Buddy and the crew of the R/V *Marie Hall*. Help in the field was also provided by the TAMUG students Janelle Case, Brian Fielder, Claudia Friess, Eddie Majzlik, Chris Noll, Ryan Schloesser, Harlan Hollis, and David Sanchez.

References

Allman, R. J., and C. B. Grimes. 2002. Temporal and spatial dynamics of spawning, settlement, and growth of gray snapper (*Lutjanus griseus*) from the west Florida shelf as determined from otolith microstructures. Fishery Bulletin 100:391–403.

Baltz, D. M., J. W. Fleeger, C. F. Rakocinski, and J. N. McCall. 1998. Food, density, and microhabitat: factors affecting growth and recruitment potential of juvenile saltmarsh fishes. Environmental Biology of Fishes 53:89–103.

Bradley, E., and C. E. Bryan. 1975. Life history and fishery of the red snapper (*Lutjanus campechanus*) in the northwestern Gulf of Mexico: 1970–1974. Proceedings of the 27th Gulf and Caribbean Fisheries Institute 27:77–106.

Collins, L .A., G. R. Fitzhugh, L. Mourand, L. A. Lombardi, W. T. Walling, Jr., W. A. Fable, Jr., M. R. Burnett, and R. J. Allman. 2001. Preliminary results from a continuing study of spawning and fecundity in the red snapper (Lutjanidae: *Lutjanus campechanus*) from the Gulf of Mexico, 1998–1999. Proceedings of the 52nd Gulf and Caribbean Fisheries Institute 52:34–47.

Cowan, J. H., Jr., and R. F. Shaw. 2002. Recruitment. Pages 88–111 *in* L. A. Fuiman and R. G. Werner, editors. Fishery science: the unique contributions of early life stages. Blackwell Scientific Publications, Ames, Iowa.

Denit, K., and S. Sponaugle. 2004. Growth variation, settlement, and spawning of gray snapper across a latitudinal gradient. Transactions of the American Fisheries Society 133:1339–1355.

GMFMC (Gulf of Mexico Fishery Management Council). 2005. SEDAR 7: Southeast Data, Assessment, and Review-Stock Assessment Report of SEDAR 7 Gulf of Mexico Red Snapper. Available: http://www.sefsc.noaa.gov/sedar/download/SEDAR7%20AW%20report%20main.pdf?id=DOCUMENT. (May 2007).

Houde, E. D. 1987. Fish early life dynamics and recruitment variability. Pages 17–29 *in* R. D. Hoyt, editor. American Fisheries Society, Symposium 2, Bethesda, Maryland.

Leggett, W. C., and E. Deblois. 1994. Recruitment in marine fishes: is it regulated by starvation and predation in the egg and larval stages? Netherlands Journal of Sea Research 32:119–134.

Masuda, R., K. Keller, D. A. Zieman, and J. Ogle. 2003. Association with underwater structures in hatchery-reared and wild red snapper *Lutjanus campechanus* juveniles. Journal of the World Aquaculture Society 34:140–146.

Mikulas, J. J. 2007. Habitat use, growth, and mortality of post-settlement lane snapper (*Lutjanus synagris*) on natural banks in the northwestern Gulf of Mexico. Master's thesis. Texas A&M University, College Station, Texas.

Patterson, W. F., III, C. A. Wilson, S. J. Bentley, J. H. Cowan, Jr., T. Henwood, Y. C. Allen, and T. A. Dufrene. 2005. Delineating juvenile red snapper habitat on the northern Gulf of Mexico continental shelf. Pages 277–288 *in* P. W. Barnes and J. P. Thomas, editors. Benthic habitats and the effects of fishing. American Fisheries Society, Symposium 41, Bethesda, Maryland.

Phelan, B. A., R. Goldberg, A. J. Bejda, J. Pereira, S. Hagan, P. Clark, A. L. Studholme, A. Calabrese, and K. W. Able. 2000. Estuarine and habitat-related differences in growth rates of young-of-the-year winter flounder (*Pseudopleuronectes americanus*) and tautog (*Tautoga onitis*) in three northeastern U.S. estuaries. Journal of Experimental Marine Biology and Ecology 247:1–28.

Rabalais, N. N., S. C. Rabalais, and C. R. Arnold. 1980. Description of eggs and larvae of laboratory reared red snapper (*Lutjanus campechanus*). Copeia 4:704–708.

Rodriguez, A. B., J. B. Anderson, L. A. Banfield, M. Taviani, K. Abdulah, and J. Snow. 2000. Identification of a 15-m middle Wisconsin shoreline on the Texas inner continental shelf. Palaeography, Palaeoclimatology, Palaeoecology 158:25–43.

Rooker, J. R., A. M. Landry, Jr., B. W. Geary, and J. A. Harper. 2004. Assessment of a shell bank and associated substrates as nursery habitat of postsettlement red snapper. Estuarine, Coastal and Shelf Science 59:653–661.

Sponaugle, S., K. Gorud-Colvert, and D. Pinkard. 2006. Temperature-mediated variation in early life history traits and recruitment success of the coral reef fish *Thalassoma bifasciatum* in the Florida Keys. Marine Ecology Progress Series 308:1–15.

SPSS. 2004. SPSS for Windows, release 13.0. SPSS, Chicago.

Stunz, G. W., T. J. Minello, and P. S. Levin. 2002. Growth of newly settled red drum *Sciaenops ocellatus* in different estuarine habitat types. Marine Ecology Progress Series 238:227–236.

Szedlmayer, S. T., 1998. Comparison of growth rate and formation of otolith increments in age-0 red snapper. Journal of Fish Biology 53(1):58-65.

Szedlmayer, S. T., and J. Conti. 1999. Nursery habitats, growth rates, and seasonality of age-0 red snapper, *Lutjanus campechanus*, in the northeast Gulf of Mexico. Fishery Bulletin 97:626–635.

Szedlmayer, S. T., and J. C. Howe. 1997. Substrate preference in age-0 red snapper, *Lutjanus campechanus*. Environmental Biology of Fishes 50:203–207.

Szedlmayer, S. T., and J. D. Lee. 2004. Diet shifts of juvenile red snapper (*Lutjanus campechanus*) with changes in habitat and fish size. Fishery Bulletin 102:366–375.

Tarpgaard, E., M. Mogensen, P. Gronkjaer, and J. Carl. 2005. Using short-term growth of enclosed 0-group European flounder, *Platichthys flesus*, to assess habitat quality in a Danish bay. Journal of Applied Ichthyology 21:53–63.

Thrush, S. F., and P. K. Dayton. 2002. Disturbance to marine benthic habitats by trawling and dredging: implications for marine biodiversity. Annual Review of Ecology and Systematics 33:449–473.

Thrush, S. F., J. E. Hewitt, V. J. Cummings, P. K. Dayton, M. Cryer, S. J. Turner, G. A. Funnell, R. G. Budd, C. J. Milburn, and M. R. Wilkinson. 1998. Disturbance of the marine benthic habitat by commercial fishing: impacts at the scale of the fishery. Ecological Applications 8:866–879.

Tupper, M., and R. G. Boutilier. 1995. Effects of habitat on settlement, growth, and postsettlement survival of Atlantic cod (*Gadus morhua*). Canadian Journal of Fisheries and Aquatic Sciences 52:1834–1841.

Tupper, M., and R. G. Boutilier. 1997. Effects of habitat on settlement, growth, predation risk and survival of a temperate reef fish. Marine Ecology Progress Series 151:225–236.

Werner, R. G. 2002. Habitat requirements. Pages 161–182 *in* L. A. Fuiman and R. G. Werner, editors. Fishery science: the unique contributions of early life stages. Blackwell Scientific Publications, Ames, Iowa.

Wells, R. J. D. 2007. The effects of trawling and habitat use on red snapper and the associated community. Doctoral dissertation. Louisiana State University, Baton Rouge, Louisiana.

Workman, I., A. Shah, D. Foster, and B. Hataway. 2002. Habitat preferences and site fidelity of juvenile red snapper (*Lutjanus campechanus*). ICES Journal of Marine Science 59:S43–S50.

Zapata, F. A., and P. A. Herron. 2002. Pelagic larval duration and geographic distribution of tropical eastern Pacific snappers (Pisces: Lutjanidae). Marine Ecology Progress Series 230:295–300.

American Fisheries Society Symposium 60:39–57, 2007

Video Estimates of Red Snapper and Associated Fish Assemblages on Sand, Shell, and Natural Reef Habitats in the North-Central Gulf of Mexico

R. J. David Wells[1] and James H. Cowan, Jr.
Department of Oceanography and Coastal Sciences
School of the Coast and Environment
Louisiana State University, Baton Rouge, Louisiana 70803 USA

Abstract.—Video estimation of the relative abundance of fishes is a noninvasive method commonly used to assess fish densities. This technique can be used to characterize habitat use patterns either of fish assemblages or of a particular species of interest. The objectives of this study were to quantify relative abundance of red snapper, *Lutjanus campechanus*, and to characterize with video methodology the associated fish assemblages over different habitat types. Fishes were enumerated over sand, shell, and natural hard bottom reef habitats in the north-central Gulf of Mexico (GOM) off Alabama on quarterly cruises over a two-year period with a baited stationary underwater video camera array. Red snapper showed both significantly higher abundance and larger size over the reef habitat; however, no seasonal effects were observed, indicating temporal abundance patterns were consistent among seasons. Fish assemblages differed among habitats, with significant differences between reef and shell assemblages. Efforts to identify the species that most contributed to these differences indicated that the red snapper accounted for 59% of the overall similarity within the reef fish assemblage and 20% of the total dissimilarity between the shell and reef fish assemblages. This study highlights the utility of applying video techniques to identify the importance of sand, shell, and reef habitat types both to different life stages of red snapper, and to the different fish assemblages occupying distinct habitats in the north-central GOM.

Introduction

Underwater video camera arrays have become an increasingly common tool for characterizing marine fish assemblages (Gledhill et al. 1996; Willis and Babcock 2000; Willis et al. 2000; Gledhill 2001; Rademacher and Render 2003; Cappo et al. 2004) and indexing abundances of a single species over a particular habitat type (Ellis and DeMartini 1995). This technique and other video methods are particularly desirable for estimating fish abundance when depth constraints and physical complexity of the bottom topography exist (Bortone et al. 1986; Greene and Alevizon 1989). However, difficulties associated with video censuses exist, such as biased estimates due to poor visibility, difficulty in species identification, fish movement, and

[1] Corresponding author: wellsr@tamug.edu.
Current address: Department of Marine Biology, Texas A&M University, 5007 Avenue U, Galveston, Texas 77551 USA

under-representation of small, cryptic species (Sale and Douglas 1981; Bohnsack and Bannerot 1986). Nevertheless, video methods offer unique advantages over more traditional methods (e.g., otter trawls, scuba surveys) of assessing relative fish abundance as they are nondestructive and the equipment can be deployed and retrieved rapidly from depth.

A variety of habitat types that support a diverse assemblage of fishes exists on the north-central Gulf of Mexico (GOM) continental shelf. The shelf is composed primarily of sand, mud, and silt with little or no vertical relief (Ludwick 1964; Kennicutt et al. 1995). Several studies have characterized fish assemblages over low-relief mud and sand habitats (Moore et al. 1970; Franks et al. 1972; Chittenden and McEachran 1976) while others have characterized shelf-edge bank fish assemblages from the western GOM Flower Garden Banks (Dennis and Bright 1988; Rooker et al. 1997; Gledhill 2001) to the eastern GOM Florida Middle Grounds (Smith et al. 1975; Gledhill 2001). However, extensive low-relief (cm to m) shell ridges at 20–40 m depths exist in the north-central GOM as the result of alternating periods of sea level during the Holocene transgression (Schroeder et al. 1995; McBride et al. 1999; Dufrene 2005). In addition, natural hard bottom habitats in the form of reef pinnacles, banks, and ledges exist on the shallow inner-shelf; these have been suggested as important reef habitat for red snapper and other reef fishes (Parker et al. 1983; Schroeder et al. 1988). However, little information exists regarding the function of these shell ridges and natural reefs due to long held misconceptions that little or no natural hard bottom reef habitat existed on the shallow (<40 m) north-central GOM shelf.

Red snapper, *Lutjanus campechanus*, is a demersal reef fish predominantly found along the continental shelf out to the shelf edge from North Carolina to the Yucatan Peninsula, including the GOM, but not the Caribbean Sea (Hoese and Moore 1998). Studies characterizing habitat preference of age 0 red snapper have found that they are not randomly distributed on low-relief mud and sand habitats, but that age 0 red snapper have an affinity for low-relief structure such as shell-rubble habitat (Szedlmayer and Howe 1997; Szedlmayer and Conti 1999; Patterson et al. 2005). Older sub-adult and adult red snapper are found in association with mid- to high-relief shelf features such as coral reefs, shelf-edge banks, and rock outcroppings, as well as artificial structures such as artificial reefs, petroleum platforms, and submerged wreckage (Bradley and Bryan 1975; Moseley 1966; Szedlmayer and Shipp 1994). To date, most studies investigating habitat use of red snapper have focused on mud, sand, shell, and artificial structures (Moseley 1966; Bradley and Bryan 1975; Holt and Arnold 1982; Workman and Foster 1994; Szedlmayer and Howe 1997; Szedlmayer and Conti 1999; Rooker et al. 2004; Patterson et al. 2005). However, no studies have examined habitat use patterns of juvenile and adult red snapper in conjunction with associated fish assemblages over shell ridges and natural hard bottom reefs in the shallow north-central GOM.

The objective of this study was to estimate relative abundances of red snapper and associated fish assemblages over different habitat types with underwater video methodology. Specifically, we sought to assess the efficacy of using the video methodology to investigate abundance and size-specific habitat use of red snapper among sand, shell, and natural hard bottom reef habitats. Our goal was to then delineate the relative importance of these habitats to different life stages of red snapper and the associated fish assemblage.

Methods

Study Area

Video work was conducted at sand, shell, and natural reef habitats in the north-

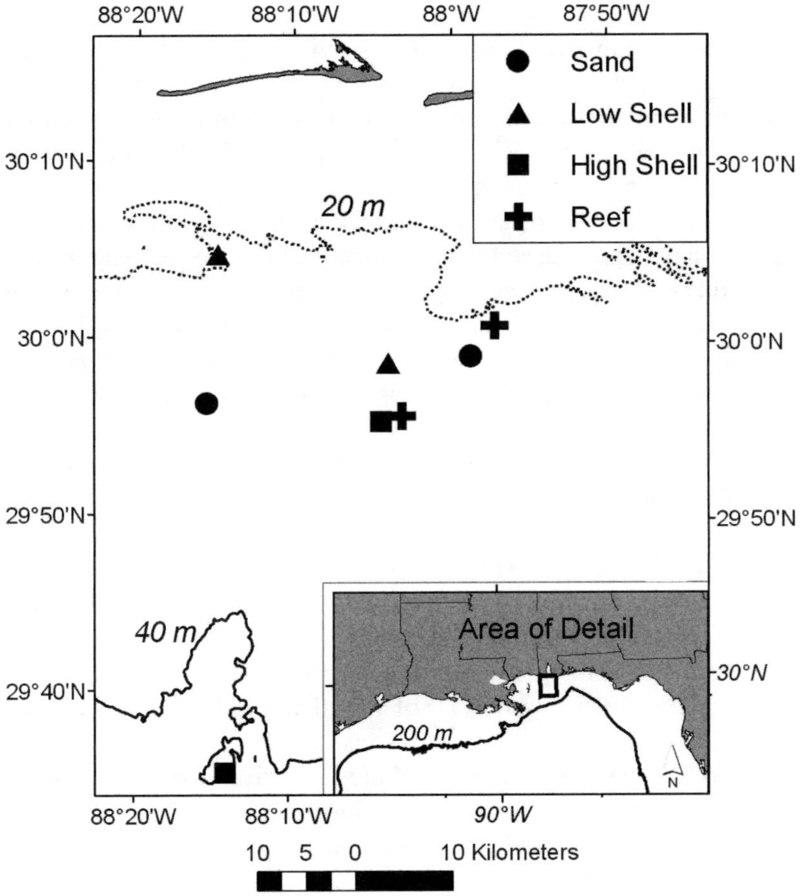

Figure 1. Map of the video study site locations in the north-central GOM. The 20 and 40 m depth contours are shown with the 200 m depth contour within the locator map representing the shelf edge.

central GOM on the Alabama and Mississippi inner continental shelf (Figure 1). Seabed characterization of the region was recently performed with digital side-scan sonar and with sediment box cores to ground truth habitat type (Dufrene 2005). Eight sampling sites were chosen for this study; these included two low-relief (cm) sand sites, four low-relief (cm to m) shell-ridge sites (2 low shell abundance and 2 high shell abundance), and two high-relief (2–4 m) natural hard bottom reef sites (Figure 1).

Video estimation

Sampling was conducted quarterly during 2004 and 2005 with a 4-camera underwater video array. The camera array consisted of four Sony DCR-VX1000 digital video camcorders housed in aluminum underwater housings. Cameras were positioned orthogonally to one another at a height of 25 cm above the bottom to provide a nearly 360° view. Each camera had a 72.5° viewing angle with an approximate viewing distance of 5 m, resulting in an estimated viewing volume of 70.4 m^3 (Rademacher and Render 2003). A series of experiments over different depths and light transmissivities was conducted by Gledhill and Lyczkowski-

Shultz (1994) to determine the accuracy of fish identification and measurements using a similar camera array. They concluded that high accuracy at 5 m from the camera was attained when the transmissivity exceeded 75%, which occurred in all of our samples, thus we were comfortable with the 5 m estimate of distance viewed. Two parallel beam lasers placed 10 cm apart were attached below each camera to aid in estimating lengths of observed fish to the nearest cm. The array was deployed for a 30-min period and was baited with a single Atlantic menhaden *Brevoortia tyrannus*, which was replaced after each deployment. All video samples were taken during daylight hours (30 min after sunrise to 30 min before sunset). Water mass characteristics were measured with a Sea-Bird SBE-25 CTD during the camera array soak period. Measurements included temperature, salinity, depth, dissolved oxygen content, and optical backscatter (or transmissivity) to gain an estimate of visibility.

Trawl sampling was also conducted adjacent to all video sites during the same seasons to obtain habitat-specific relative abundance estimates of juvenile red snapper and associated fish assemblages. Therefore, video data was compared to concomitant trawl data to obtain size-specific selectivity bias by gear type using the ratio of length-specific abundance estimates from the trawls relative to the video (Lauth et al. 2004). In addition, to investigate whether gear type biased our fish assemblage results, similar ratios correcting for the abundance-at-size by gear type and habitat type were made for those species that most contributed to the fish assemblages. Further investigation indicated that the corrected abundance-at-size did not affect our results, thus demonstrating the robustness of our fish assemblage results.

Statistical Analysis

A continuous 20 min segment of one tape was examined for fish abundances at each deployment. Tapes were chosen based upon the optimal view of the habitat of interest combined with the best visibility (i.e., in focus, good orientation relative to the current). Gledhill (2001) determined this continuous 20 min method to be optimal for reducing error in abundance estimates for sampling the taxa present, and for minimizing logistical constraints such as available time at sea. Start time began once the camera array was on the bottom and after sufficient time elapsed for the water column to clear. All fish were identified to the lowest possible taxonomic level and counted. The minimum count (*MIN*), the maximum number of a species observed at any one time on the tape, was used for all statistical comparisons. This method is commonly used for gregarious species, such as red snapper, and is analogous to the *MAXNO* of Ellis and DeMartini (1995), the *MAX* of Willis and Babcock (2000), and the *MaxN* of Cappo et al. (2004). Estimates of total length (TL) were made only at *MIN* counts to eliminate repeated measurements of the same fish. Maximum counts (*MAXIM*) were also made to obtain total counts of each fish species seen over the 20 min segment of the tape analyzed.

Video counts of red snapper were modeled with a Poisson distribution. Specifically, a log-linear fixed effects model using the GENMOD procedure in SAS was used to predict red snapper numbers, with season and habitat as factors (Willis and Babcock 2000; Willis et al. 2000) (SAS Institute, Inc. 2002). The model fit was evaluated with a maximum likelihood method and analysis of deviance. In addition, red snapper length comparisons among seasons and habitats were evaluated separately with a Kruskall-Wallis one-way analysis of variance (ANOVA) on the ranks due to the lack of normality and homogeneity of variance assumptions required by ANOVA (Systat software, Inc. 2004). Dunn's test was used to determine *a posteriori* differences among means ($\alpha = 0.05$).

Fish assemblage data were analyzed with the PRIMER (Plymouth Routines in Multivariate Ecological Research) statistical package (Clarke and Warwick 2001). This nonparametric multivariate analysis used a Bray-Curtis similarity matrix to construct similarities among samples from different habitats and seasons. Fish that were not identified to species and those with a total count of one were excluded from all statistical analyses. Therefore, twenty-five species representing 85% of the overall fish assemblage were included in the statistical analyses. A nonmetric multidimensional scaling (MDS) method was used to map the sample interrelationships in an ordination. The ANOSIM (Analysis of Similarities) permutation procedure was used to test for significant differences of fish assemblages among habitats and seasons (Warwick et al. 1990a).

To assess species-specific contributions, SIMPER (Similarity Percentages) was used as the *post hoc* analysis to indicate the contribution of a particular species to the overall fish assemblage similarity (within season or habitat) and dissimilarity (among seasons and habitats) (Clarke 1993). A cutoff percentage of 90% was used to determine those species accounting for 90% of the total similarities and dissimilarities. Additionally, a stepwise data reduction procedure, BV-STEP, was used to determine which group of species accounted for the observed patterns in the fish assemblage (Clarke and Warwick 1998). This procedure used a Spearman rank correlation coefficient of 95% as a cutoff to determine which group of species together explained most of the variability.

Patterns of species diversity among habitats and seasons were investigated with DIVERSE (Warwick et al. 1990b). This method used the Shannon diversity (H') and Pielou's evenness (J') indices. Diversity measures were estimated with the following equations:

$$H' = -\sum_i p_i \, log(p_i)$$

where p_i is the proportion of the total count from the ith species,

$$J' = H'/log\ S$$

where S (species richness) is the total number of species present in the sample. Effects of habitat and season on the diversity indices were analyzed with a two-factor ANOVA (ANOVA) (Systat software, Inc. 2004).

Species abundances and environmental correlations were investigated with canonical correspondence analysis (CCA) by using the CANOCO program (ter Braak and Smilauer 2002). This analysis is designed to maximally correlate environmental variables with fish assemblage data with a nonlinear weighted averaging method. A global permutation test with Monte Carlo permutations was used to investigate the statistical significance between the species abundances and environmental variables. Inter-set correlations of the environmental variables with the axes were used to assess the relative importance of environmental variables. These correlations are the correlation coefficients between the environmental variables and the species-derived sample scores, and are more robust to collinearity than are canonical coefficients (ter Braak and Smilauer 2002). Interset correlation coefficients with absolute values greater than or equal to 0.4 were interpreted as ecologically important (Hair et al. 1984; Rakocinski et al. 1996). Temperature, salinity, depth, and dissolved oxygen were the continuous environmental variables used and habitat types were coded as nominal variables (sand, shell, and reef). In addition, the same 25 species used for previous fish assemblage analyses were used in the CCA to reduce the bias associated with rare taxa.

Results

Forty-two (12 sand, 17 shell, and 13 reef) of the 64 possible video sampling op-

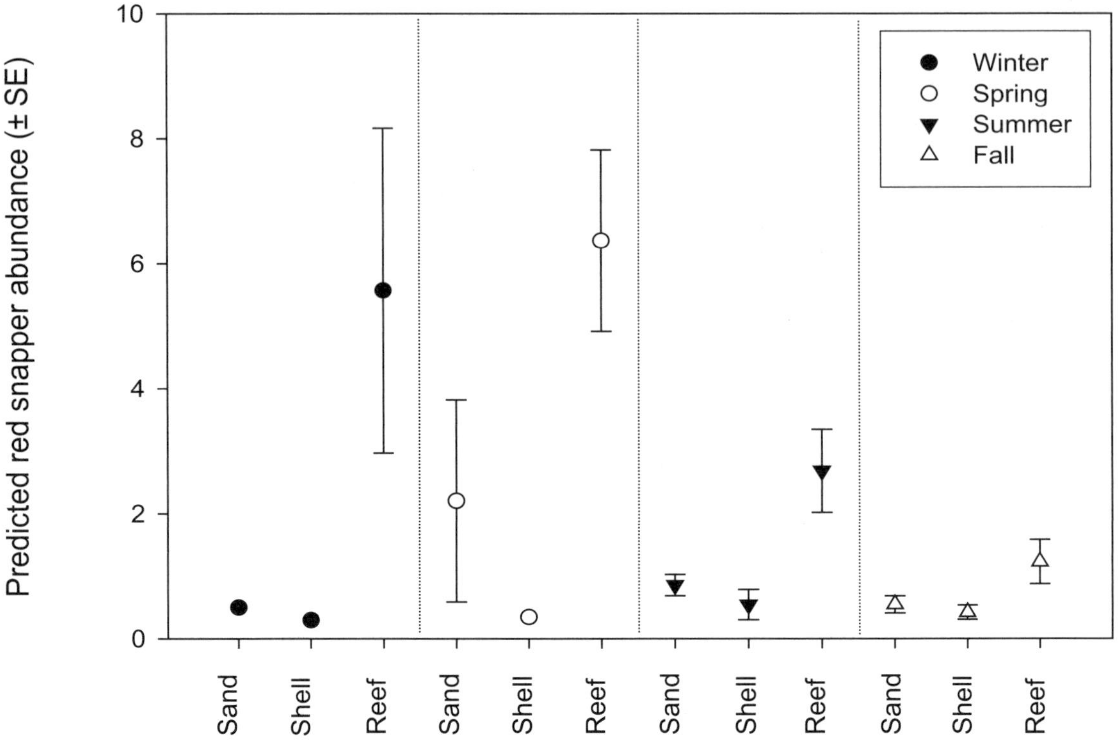

Figure 2. Relative abundance estimates of red snapper, *Lutjanus campechanus*, (±1 SE) predicted by log-linear model over sand, shell, and reef habitats by season. Relative abundance is expressed as the MIN count of red snapper 20 min^{-1} deployment.

portunities were achieved; no 2005 winter cruise was taken due to inclement weather.

Red Snapper

Relative abundance estimates of red snapper from the *MIN* index showed significant differences in abundances among habitat types (P = 0.0318) (Figure 2). Higher abundance estimates were observed over reef habitat than over either shell or sand habitats. In contrast, seasonal differences in abundance estimates were negligible (P = 0.8224), as was the interaction between habitat and season (P = 0.1260). Nevertheless, a general trend of higher abundance estimates over the reef was observed in the winter and spring followed by a decline in the summer and fall.

Size differences among red snapper were observed both spatially and temporally. Red snapper found over the reef habitats were significantly longer than conspecifics found over the sand (Dunn's Method; $P < 0.05$) (Table 1). Shell habitats supported intermediate-sized red snapper, but these showed no significant differences in length with red snapper over sand and reef habitats (Table 1). Due to insufficient numbers of red snapper observed on sand and shell habitats over all seasons, only the reef habitat was investigated for a seasonal size effect. A significant seasonal size difference existed, with the largest red snapper observed over the reef during the summer season (P = 0.002) (Table 1). Progressively smaller red snapper were seen over subsequent seasons in the fall, winter, and spring (Table 1).

Table 1. Average sizes (TL in mm ±1 SE) of red snapper observed over sand, shell, and reef habitats. Average seasonal lengths are displayed for the reef habitat. Habitats and seasons with significantly different sizes are represented by different letters and no differences are represented by similar letters ($P < 0.05$).

		Average size	Differences
Habitat	Sand	12.3 (0.30)	A
	Shell	15.0 (0.20)	AB
	Reef	25.0 (0.43)	B
Season (Reef)	Winter	25.0 (1.34)	A
	Spring	19.2 (1.67)	B
	Summer	31.5 (0.76)	A
	Fall	28.3 (6.67)	A

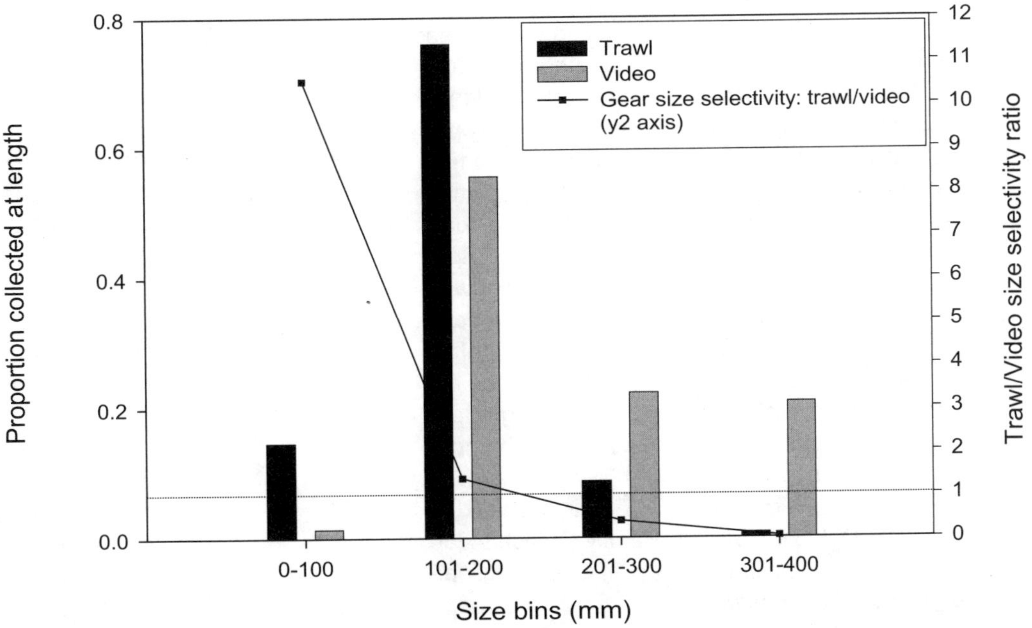

Figure 3. Size selectivity bias of red snapper, *Lutjanus campechanus*, collected with trawl and observed with underwater video methodology. Left axis represents the proportion-at-length of red snapper collected using each gear type. Right axis represents the ratio of the proportion-at-length collected from the trawl relative to the video gear using four size bins (<100 mm, 101–200 mm, 201–300 mm, and >300 mm TL). The dotted line represents a 1:1 ratio of red snapper observed in trawls relative to the video method, which would indicate no gear bias.

Table 2. Total number of fish observed from video estimates arranged in order of decreasing abundance by *MIN*. *MIN* is the maximum number observed at any one time, *MAXIM* is the total number observed over the entire tape, n is the frequency of occurrence (out of 42 camera deployments), and total length TL (± SE) is the average size of each species. Sizes were not estimated for unidentified fish, thus NA (not applicable).

Taxon	Common name	*MIN*	*MAXIM*	n	TL (±SE)
Stenotomus caprinus	Longspine porgy	141	2699	24	9.3 (0.2)
Diplectrum bivittatum	Dwarf sand perch	78	889	26	5.8 (0.3)
Lutjanus campechanus	Red snapper	72	1206	18	21.6 (1.1)
Carangidae	Family Carangidae	44	116	4	9.6 (0.6)
Haemulon aurolineatum	Tomtate	28	187	7	22.3 (0.9)
Caranx crysos	Blue runner	27	64	5	26.2 (2.2)
Micropogonias undulatus	Atlantic croaker	25	654	6	10.5 (0.5)
Trachurus lathami	Rough scad	25	132	5	7.4 (0.7)
Diplectrum formosum	Sand perch	22	83	15	8.1 (0.7)
Bothidae	Family Bothidae	16	26	14	17.9 (1.6)
Centropristis ocyurus	Bank sea bass	14	102	12	11.1 (1.0)
Syacium papillosum	Dusky flounder	13	62	8	14.8 (2.0)
Lutjanus synagris	Lane snapper	10	27	5	21.0 (2.1)
Unidentified fish	Unidentified fish	7	26	7	NA
Menticirrhus americanus	Southern kingfish	6	25	3	12.2 (1.3)
Eucinostomus gula	Silver jenny	5	28	4	10.4 (1.0)
Rhomboplites aurorubens	Vermilion snapper	5	22	1	15.0 (0.0)
Xyrichtys novacula	Pearly razorfish	4	47	3	3.3 (1.0)
Balistes capriscus	Gray triggerfish	4	14	3	23.8 (1.3)
Sciaenidae	Family Sciaenidae	4	12	4	15.0 (2.9)
Leiostomus xanthurus	Spot	4	6	2	13.3 (1.8)
Halichoeres radiatus	Puddingwife	3	13	3	15.0 (2.9)
Centropristis philadelphica	Rock sea bass	3	7	3	8.7 (4.1)
Orthopristis chrysoptera	Pigfish	3	5	3	10.7 (2.3)
Ophichthus puncticeps	Palespotted eel	3	3	2	4.7 (0.3)
Mycteroperca microlepis	Gag grouper	2	15	1	35.0 (0.0)
Chloroscombrus chrysurus	Atlantic bumper	2	6	2	15.0 (5.0)
Decapterus punctatus	Round scad	2	5	1	10.0 (0.0)
Ophidiidae	Family Ophidiidae	2	5	1	15.0 (0.0)
Ophichthidae	Family Ophichthidae	2	3	2	6.0 (4.0)
Triglidae	Family Triglidae	2	3	2	15.0 (5.0)
Archosargus probatocephalus	Sheepshead	2	2	1	33.0 (0.0)
Prionotus rubio	Blackwing searobin	1	8	1	35.0 (0.0)
Brevoortia patronus	Gulf menhaden	1	7	1	15.0 (0.0)
Sphoeroides parvus	Least puffer	1	4	1	1.0 (0.0)
Calamus leucosteus	Whitebone porgy	1	3	1	10.0 (0.0)
Serranus phoebe	Tattler	1	3	1	2.0 (0.0)
Gymnothorax nigromarginatus	Blackedge moray	1	2	1	15.0 (0.0)
Halichoeres bivittatus	Slippery dick	1	2	1	5.0 (0.0)
Carcharhinidae	Family Carcharhinidae	1	1	1	60.0 (0.0)
Sphyraena guachancho	Guaguanche	1	1	1	20.0 (0.0)

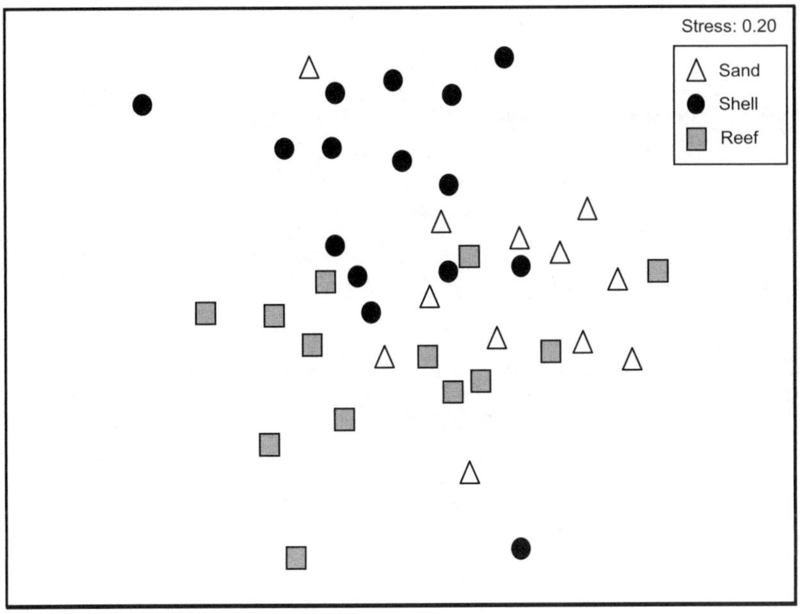

Figure 4. Multi-dimensional scaling (MDS) plot of all samples over the two-year study period. Each sample represents the 25 species analyzed for the fish assemblage.

Table 3. SIMPER results of the species that most contributed to the within-habitat similarity for each of the three habitat types: sand, shell, and reef. Mean abundance of important species within habitat type, the contribution (SIM) to the average within similarity, and the average within similarity/standard deviation (SIM/SD) ratio. A 90% cut-off was used for the cumulative % contribution of species.

Habitat	Species	Mean	SIM	SIM/SD	% contribution	% cumulative contribution
Sand	Longspine porgy	7.42	20.26	1.15	73.12	73.12
	Dwarf sand perch	0.75	3.19	0.51	11.52	84.64
	Sand perch	0.50	1.36	0.39	4.92	89.56
	Red snapper	1.17	0.81	0.31	2.93	92.49
Shell	Dwarf sand perch	3.60	11.63	1.02	58.51	58.51
	Sand perch	0.80	2.26	0.43	11.37	69.88
	Longspine porgy	1.07	2.25	0.42	11.31	81.19
	Dusky flounder	0.47	1.06	0.26	5.32	86.51
	Lane snapper	0.60	0.74	0.21	3.72	90.23
Reef	Red snapper	3.92	15.23	1.11	58.71	58.71
	Longspine porgy	2.77	5.72	0.47	22.05	80.77
	Dwarf sand perch	1.15	2.44	0.40	9.40	90.16

Table 4. SIMPER results of the species that most contributed to the dissimilarity between shell and reef habitats. Mean abundance of important species within habitat type, the contribution (DIS) to the average dissimilarity, and the average dissimilarity/standard deviation (DIS/SD) ratio. A 90% cut-off was used for the cumulative % contribution of species.

Species	$Mean_{shell}$	$Mean_{reef}$	DIS	DIS/SD	% contribution	% cumulative contribution
Red snapper	0.47	3.92	16.65	1.22	19.69	19.69
Dwarf sand perch	3.60	1.15	13.35	0.90	15.78	35.47
Longspine porgy	1.07	2.77	12.27	0.98	14.51	49.98
Tomtate	0.07	2.08	6.13	0.53	7.25	57.23
Blue runner	1.40	0.31	5.27	0.42	6.23	63.46
Rough scad	0.53	0.69	4.16	0.37	4.92	68.38
Sand perch	0.80	0.31	4.05	0.76	4.79	73.17
Dusky flounder	0.47	0.31	3.95	0.49	4.67	77.84
Atlantic croaker	0.07	1.15	3.23	0.32	3.82	81.66
Lane snapper	0.60	0.08	2.57	0.56	3.05	84.70
Bank sea bass	0.33	0.38	2.43	0.72	2.88	87.58
Vermilion snapper	0.00	0.38	1.66	0.28	1.96	89.54
Southern kingfish	0.00	0.31	1.61	0.35	1.90	91.44

Size selectivity bias of the video gear was observed for red snapper. Proportions of small red snapper were underrepresented using the video method. The gear size selectivity ratio of small red snapper collected in trawls relative to video estimates indicated that on average 10.5 (<100 mm TL) and 1.4 (101–200 mm TL) red snapper were collected in trawls relative to one red snapper observed using the video method (Figure 3). In contrast, large red snapper were more abundant in the video estimates compared to the trawled counts with 0.4 (201–300 mm TL) and 0.04 (>300 mm TL) red snapper collected in trawls relative to one red snapper seen with the video method (Figure 3).

Fish Assemblage

Thirty-three species representing 16 families were positively identified in this study (Table 2). Unidentifiable taxa were distributed among seven families and an unidentified fish category. A group of seven species best characterized the observed fish assemblage patterns shown with the BV-STEP procedure: blue runner *Caranx crysos*, bank sea bass *Centropristis ocyurus*, dwarf sand perch *Diplectrum bivitattum*, sand perch *D. formosum*, red snapper *Lutjanus campechanus*, longspine porgy *Stenotomus caprinus*, and dusky flounder *Syacium papillosum* accounted for 95.0% of the correlation among species and the observed patterns detected in the fish assemblages.

Fish assemblages showed differences among the three habitat types (Figure 4). The two-way ANOSIM indicated a significant habitat effect among fish assemblages (P = 0.008). Further examination indicated that fish assemblages over the reef and shell habitats differed significantly (P = 0.006). However, there were no seasonal differences in fish assemblages within habitats during our two-year study period (P = 0.299).

Table 3 shows the SIMPER results for species that contributed most to the overall similarity within habitat type. The longspine porgy accounted for 73.1% of the cumulative species similarity within the sand habitat, the dwarf sand perch accounted for 58.5% for the shell habitat, and the red snapper represented 58.7% of the similarity within the reef habitat. The longspine porgy, dwarf sand perch, and red snapper SIM/SD values were 1.15, 1.02, and 1.11, respectively. The SIM/SD ratios exceeding 1.0 indicates that each of the three species consistently contributed to the within habitat similarity among samples (Clarke and Warwick 2001). These three species were also important in discriminating fish assemblages between reef and shell habitats. Table 4 shows the total contribution of each species to the dissimilarity between habitat types; the three dominant species together accounted for 50% of the overall dissimilarity. The red snapper contribution was the highest at 19.7%, followed by that for dwarf sand perch (15.8%), and longspine porgy (14.5%). The red snapper most consistently contributed to these dissimilarity differences based upon the DIS/SD value of 1.22.

Habitat diversity indices varied by habitat, with highest species richness, evenness, and diversity associated with the reef fish assemblage (Table 5). Species evenness and diversity were lowest for the sand fish assemblage with increasing values over the shell habitat. However, no significant differences were detected among habitat-specific indices. Similar trends were observed by season, with the lowest values of all three indices during the winter. Species richness increased to a stable maximum over the remaining seasons, while species evenness and diversity peaked in the fall (Table 5). Similarly, no significant seasonal effects were detected among indices.

Environmental variables correlated well with species from the fish assemblages (Table 6; Figure 5). The global permutation test indicated a significant effect of CCA axis 1 (P = 0.002) and of all axes combined (P = 0.002), thereby indicating a significant relationship between species abundance and environmental variables. The first two CCA axes accounted for 62.5% of the cumulative percentage of the species-environment relationship. Shell and reef habitats correlated well with CCA axis 1, while depth, sand, and shell correlated strongly with CCA axis 2 (Table 6). Caution should be applied when interpreting nominal habitat variables in relation to CCA axes because the interset correlation coefficients are not useful (ter Braak and Smilauer 2002). Thus, nominal habitat variables were used to convey species-specific habitat use information. Species that displayed a specific habitat affinity appeared to correlate well with the corresponding nominal habitat variable (Figure 5). Red snapper and other reef-associated species were highly correlated with reef habitat type with a high negative score on axis 1. Species primarily found on shell corresponded with increasing salinity and temperature and had high positive scores on axes 1 and 2. Species that were found in association with the sand habitat showed a correlation with increased depth and high dissolved oxygen and loaded negatively on CCA axis 2.

Discussion

This study highlights the efficacy of using video methodology to assess habitat use

Table 5. Average species richness (S), Pielou evenness (J′), and Shannon diversity (H′) indices for all habitats and seasons.

		S	J′	H′
Habitat	Sand	4.0	0.608	0.941
	Shell	3.9	0.741	1.046
	Reef	4.3	0.780	1.151
Season	Winter	3.6	0.544	0.827
	Spring	4.2	0.767	1.122
	Summer	4.3	0.640	1.004
	Fall	4.2	0.888	1.231

Table 6. Canonical correspondence analysis (CCA) statistics and inter-set correlations relating environmental variables with CCA axes. Bold values denote variables with absolute value correlations ≥ 0.4.

Statistics	Axis 1	Axis 2	Axis 3	Axis 4	Total inertia
Eigenvalues	0.491	0.380	0.214	0.141	
Species-environment correlations	0.887	0.821	0.765	0.713	
Cumulative percentage variance					
of species data	9.5	16.8	21.0	23.7	
of species-environment relation	35.2	62.5	77.8	87.9	
Sum of all eigenvalues					5.18
Sum of all canonical eigenvalues					1.39
Inter-set correlations					
Environmental variables					
Depth	0.2685	**–0.5518**	0.0566	0.0588	
Temperature	0.3581	0.1259	0.0660	–0.3304	
Salinity	0.1244	0.2233	0.5389	0.0075	
Dissolved oxygen	–0.1395	–0.2579	0.0021	–0.1429	
Sand	0.0582	**–0.5508**	0.3537	0.3094	
Shell	**0.5365**	**0.5891**	0.0154	0.0291	
Reef	**–0.5609**	–0.0323	–0.3503	–0.3212	

by red snapper and associated fish assemblages in the north-central GOM. The method has its inherent biases (i.e., larger fishes were observed while smaller cryptic fishes were likely missed, effects of bait plumes on abundance estimates); however, it appears to be a practical method to characterize red snapper habitat use over a variety of substrate types. Structurally complex habitat types with high relief, such as natural and artificial reefs, rock outcrops, and petroleum platforms, require noninvasive sampling techniques. In addition, the logistical simplicity of dropping the camera array for a 30 min period makes this an appropriate method if multiple deployments over distant sites are needed, as was the case in this study.

Our study found similar habitat-specific results as others with small, intermediate, and large sized red snapper over sand, shell, and reef habitats, respectively. Juvenile red snapper were predominately collected over low-relief sand habitats, which is consistent with the findings of Rooker et al. (2004) and Patterson et al. (2005). In contrast, both sub-adult and adult red snapper were found over higher relief habitats such as the shell-rubble and natural hard bottom reef habitats. These findings are consistent with previous studies that have found adult red snapper over high relief habitats such as shelf-edge banks, mid-shelf banks, rock outcrops, coral reefs, and artificial structures (Moseley 1966; Bradley and Bryan 1975; Dennis and Bright 1988; Stanley and Wilson 2000; Gledhill 2001). It has been suggested that red snapper recruit to these high-relief habitats by about 18 months of age or 20 cm TL (Gallaway et al. 1999). Nieland and Wilson (2003), using a fishery independent survey design, found age 2 red snapper between 27.5 and 37.5 cm TL were recruited to petroleum platforms. The largest red snapper observed in the current study (mean = 25 cm TL) were associated with reef habitats, but seasonal differences in length ranged from 19.2 to 31.5 cm TL in the spring and summer, respectively. Therefore, these natural reef habitats on the shallow inner shelf may be functionally important for sub-adult and adult red snapper in the north-central GOM.

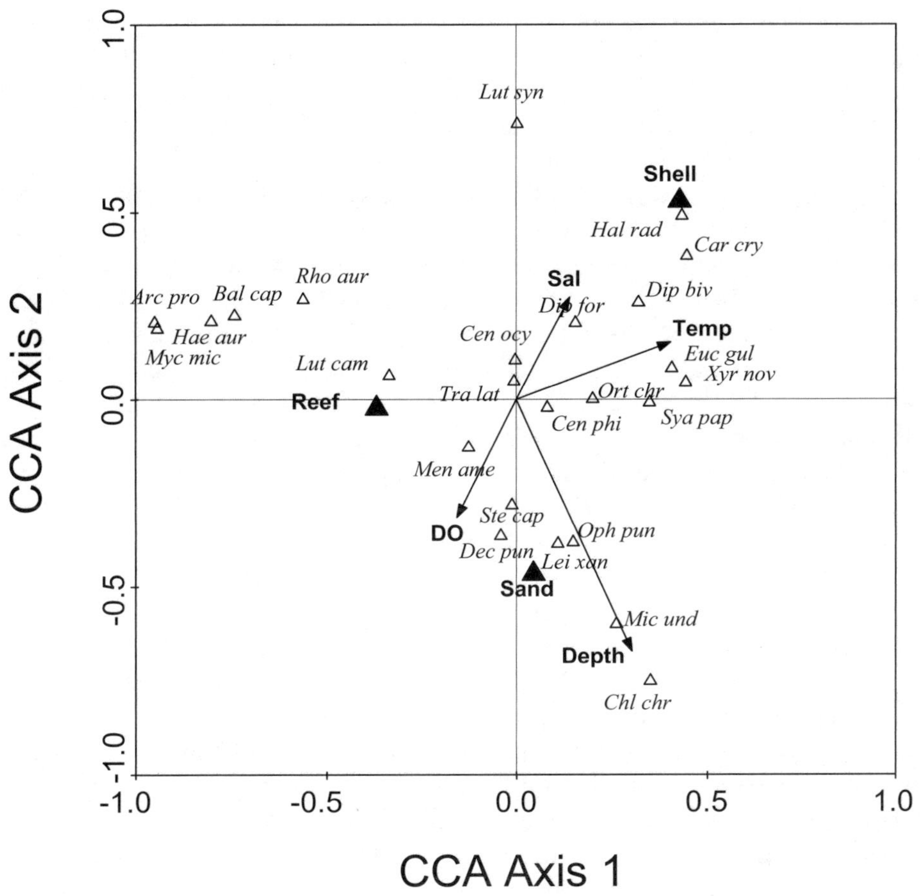

Figure 5. Biplot of axes 1 and 2 from canonical correspondence analysis of fish species and environmental variables. Environmental variables include Depth, Temp = water temperature, Sal = salinity, DO = dissolved oxygen, Sand, Shell, and Reef. Species codes: Sheepshead (*Arc pro* = *Archosargus probatocephalus*), gray triggerfish (*Bal cap* = *Balistes capriscus*), blue runner (*Car cry* = *Caranx crysos*), bank sea bass (*Cen ocy* = *Centropristis ocyurus*), rock sea bass (*Cen phi* = *Centropristis philadelphica*), Atlantic bumper (*Chl chr* = *Chloroscombrus chrysurus*), round scad (*Dec pun* = *Decapterus punctatus*), dwarf sand perch (*Dip biv* = *Diplectrum bivittatum*), sand perch (*Dip for* = *Diplectrum formosum*), silver jenny (*Euc gul* = *Eucinostomus gula*), tomtate (*Hae aur* = *Haemulon aurolineatum*), puddingwife (*Hal rad* = *Halichoeres radiatus*), pearly razorfish (*Xyr nov* = *Xyrichtys novacula*), spot (*Lei xan* = *Leiostomus xanthurus*), red snapper (*Lut cam* = *Lutjanus campechanus*), lane snapper (*Lut syn* = *Lutjanus synagris*), southern kingfish (*Men ame* = *Menticirrhus americanus*), Atlantic croaker (*Mic und* = *Micropogonias undulatus*), gag (*Myc mic* = *Mycteroperca microlepis*), palespotted eel (*Oph pun* = *Ophichthus puncticeps*), pigfish (*Ort chr* = *Orthopristis chrysoptera*),vermilion snapper (*Rho aur* = *Rhomboplites aurorubens*), longspine porgy (*Ste cap* = *Stenotomus caprinus*), dusky flounder (*Sya pap* = *Syacium papillosum*), rough scad (*Tra lat* = *Trachurus lathami*).

Seasonal size differences at the reef habitats were likely attributed to emigration and immigration of red snapper. Moseley (1966) observed changes in red snapper abundance at reefs and attributed these movement patterns to passing cold fronts, while Bradley and Bryan (1975) found similar trends and suggested the offshore movement of prey as a potential mechanism. Recent studies have found red snapper exhibit low site fidelity and moderately high dispersal from artificial reefs in the northern GOM (Patterson et al. 2001a; Patterson and Cowan 2003). However, other tagging studies have shown high site fidelity with little movement from artificial reefs in the northern GOM (Szedlmayer and Shipp 1994; Szedlmayer 1997; Szedlmayer and Schroepfer 2005). In particular, Szedlmayer and Schroepfer (2005) found no seasonal movements and long average residence times (218 d when excluding caught fish from analysis) for red snapper on artificial reefs in our study area. However, their study was conducted over a 4 year period and they interpreted long-term residence as any time greater than 117 d. Our interpretation of their Vemco telemetry data, excluding caught fish, indicated only 15.6% of tagged red snapper had residency times greater than 12 months. In addition, 65.6, 43.8, 34.4, and 21.9% of tagged red snapper in their study had residence times less than 9, 6, 3, and 1 month, respectively. Nevertheless, the large size range of red snapper in their study (43.5–84 cm TL) combined with the use of large artificial structures, such as army tanks and cars, may negate a direct comparison to our results.

In this study the largest red snapper were observed over the reef habitat during the summer season, with progressively smaller fish over the following seasons, suggesting that the larger fish were moving away. By spring, the smallest red snapper were observed over the reef, and when combined with previous size-at-age data, suggests that these fish were age 1 individuals that recruited from adjacent low-relief habitats (e.g., sand, mud, shell ridges) (Szedlmayer and Shipp 1994; Patterson et al. 2001b; Wilson and Nieland 2001). Fishing mortality may be another contributory factor to the observed trends in decreased size of red snapper over the reefs. The summer samples were the first after the recreational fishing season opened on April 21st of both years, and the reef sites sampled in this study are known to local fisherman and have been assigned the names of Southeast Banks and 17 Fathom Hole (Schroeder et al. 1988).

Red snapper abundance estimates at the reefs did not significantly change over the different seasons, suggesting that the large fish may have been replaced by smaller conspecifics. Bailey et al. (2001) performed laboratory studies on the cohabitation of juvenile and of sub-adult red snapper and found that the smaller fish were not allowed to occupy the reef when the larger fish were present. Moreover, Workman et al. (2002) found increased numbers of age 0 red snapper occupied the reefs when age 1 specimens were absent. The general trend of decreasing sizes with little change in relative abundance in this study indicates these smaller fish were likely replacing the larger conspecifics as the latter emigrated off the reef due either to natural movements or to fishing mortality. However, our results are based upon low sample sizes ranging from two reef samples in the winter to four reef samples in both the spring and summer seasons. In addition, a large amount of variability was observed in both our winter and spring estimates, which may influence the lack of any seasonal patterns. Thus, studies that aim to look at fish movement on and off a reef using underwater video methods need to incorporate more replication to gain insight into fish movement patterns.

The longspine porgy, dwarf sand perch, and red snapper were collected over all habitat types; nevertheless, they each appeared to show affinities to sand, shell, and reef habitats, respectively. The longspine porgy has been reported as one of the most abundant fish species collected over sand and mud habitats off Texas, Louisiana, and Mississippi (Moore et al. 1970; Franks et al. 1972; Chittenden and McEachran 1976; Geoghegan and Chittenden 1982). However, this species is not restricted to these habitats as Parker et al. (1979) found it to be one of the most abundant species on artificial reefs off South Carolina. The longspine porgy was the most ubiquitous species in this study because it was the dominant taxa contributing to the sand assemblage and was one of the most important contributing species to characterize the reef and shell assemblages. Little information exists on the habitat preference of the dwarf sand perch; however, limited studies have found this species' general distribution to be in association with low-relief sand and mud areas (Fraser 1971; Bortone et al. 1981). This study was the first to document the potential importance of shell habitat to the dwarf sand perch based upon its large cumulative contribution toward the shell habitat fish assemblage.

The red snapper was numerically the most dominant reef-associated species in this study. Many studies characterizing both natural and artificial reef fish assemblages have found red snapper to be abundant (Stanley and Wilson 2000; Gledhill 2001; Rademacher and Render 2003); however, no studies have shown red snapper to be the most important species to contribute to the structure of the reef fish assemblage.

The CCA technique is useful both in delineating habitat associations by species and in characterizing fish assemblages based upon habitat type. Species that correlated with the sand habitat type have been previously characterized as utilizing sand habitats. These include the Atlantic croaker *Micropogonias undulatus*, Atlantic bumper *Chloroscombrus chrysurus*, spot *Leiostomus xanthurus*, round scad *Decapterus punctatus*, southern kingfish *Menticirrhus americanus*, and palespotted eel *Ophichthus puncticeps* (Moore et al. 1970; Chittenden and McEachran 1976; Hale 1987; Hoese and Moore 1998; Pierce and Mahmoudi 2001). Additional species, other than the dwarf sand perch, that displayed an affinity for the shell habitat included the sand perch, blue runner, and puddingwife *Halichoeres radiatus*, all of which have been previously documented over a suite of habitat types (Hastings and Bortone 1976; Bortone et al. 1981; Pierce and Mahmoudi 2001). A mutualistic foraging association between the puddingwife and the bar jack, *Caranx ruber*, has been observed (Baird 1993). This social facilitation was also observed with several other species suggesting the blue runner, a close relative of the bar jack, could co-occur with the puddingwife over shell habitats for foraging purposes. All species that showed an affinity to reef habitat type in the CCA biplot were also reef associated. These species included the red snapper, vermilion snapper *Rhomboplites aurorubens*, sheepshead *Archosargus probatocephalus*, gag grouper *Mycteroperca microlepis*, gray triggerfish *Balistes capriscus*, and the tomtate *Haemulon aurolineatum* (Caldwell 1965; Smith et al. 1975; Parker et al. 1979; Sedberry and Van Dolah 1984; Moran 1988; Kellison and Sedberry 1998). Bortone et al. (1997) found the vermilion snapper to be the best indicator species for offshore artificial reef fish assemblages in the northern GOM. In addition, the tomtate has been found to be the most abundant species over rock outcrops at mid-shelf depths of 25–38 m off the South Carolina coast (Sedberry and Van Dolah 1984).

The abundance of several species correlated with environmental variables and these variables may have influenced fish assemblage structure. Depth was the only measured environmental variable that had a correlation coefficient ≥ 0.4, and both the Atlantic croaker and Atlantic bumper correlated strongly with depth. However, both species were regularly found in shallow inshore waters (<20 m) and were not limited to deeper offshore waters (Moore et al. 1970; Chittenden and McEachran 1976; Pierce and Mahmoudi 2001). In addition, several species were correlated with temperature, salinity, and dissolved oxygen content. Of notable importance was the longspine porgy with high dissolved oxygen levels and the dwarf sand perch, sand perch, and bank sea bass with high temperature and salinity. The association of the sand perch and bank sea bass with high salinity is consistent with similar analyses investigating species-environmental relationships (Bortone et al. 1997). However, the largest differences in temperature, salinity, and dissolved oxygen within any season during our video sampling were 4.75°C, 1.80 ‰, and 2.53 mg · L^{-1}, respectively. These narrow differences likely contributed to the minimal impacts that the water mass characteristics had on the species distributions and abundances.

The highest diversity indices were associated with the reef fish assemblage. Similar studies using trawl methodology found higher diversity and biomass over hard bottom habitats than over sand bottom (Wenner 1983; Sedberry and Van Dolah 1984). Higher vertical relief with more structural complexity likely provides greater microhabitat space for a suite of different species to occupy. However, our diversity indices were lower than other studies characterizing fish assemblages over natural reef habitats in the GOM (Smith et al. 1975; Dennis and Bright 1988; Rezak et al. 1990; Rooker et al. 1997). Differences in gear type likely contributed to the observed trends as these studies used SCUBA techniques that are capable of better identifying smaller cryptic species and sampling a larger area. Gledhill (2001) used a similar camera array to characterize reef fish assemblages on offshore shelf-edge banks and found much higher diversity indices; differences are likely due to the location and scale of reef habitats. Our reef sites were located on the inner continental shelf at depths between 25 and 32 m and total reef area is estimated to have ranged from m^2 to several km^2 in size. In contrast, the offshore banks Gledhill (2001) analyzed were in average water depths of 63.5 m with area sizes in the hundreds of km^2.

The underwater video methodology used in this study appeared to be size selective for larger red snapper. In comparing our gear selectivities, assumptions such as independent length-specific values between gear types, negligible variability in estimates, and similar areas sampled between gear types were not met. However, the goal of identifying the sampling bias associated with the video was achieved. Thus, this study revealed habitat use patterns of the largest red snapper utilizing these sand, shell, and reef habitats, while excluding the smallest individuals that the trawl gear was capable of collecting. Results of this size selectivity highlight the importance of using multiple gear types when quantifying fish habitat use patterns and attempting to delineate relative habitat importance.

The goal of this study was to use underwater video methods as a tool to characterize fish assemblages, with an emphasis on red snapper, over distinct habitat types. Results indicated this was a useful technique to identify fish assemblages over different habitat types in the north-central GOM. In addition, this sampling technique appeared to be a practical method for estimating relative abundance and investigating red snapper habitat use over sand, shell, and natural hard bottom reefs.

Acknowledgments

We thank the National Oceanic and Atmospheric Administration-NMFS Pascagoula, MS laboratory for providing field assistance, equipment, and boat time aboard the R/V *Caretta*. A special thanks to Drew Hopper, Kendall Falana, David Saksa, and Kevin Rademacher for their logistical support and expertise. We also thank Derek Borne, Kevin Boswell, Mark Miller, and Will Patterson for sampling assistance, Yvonne Allen for preparing the study site map, and David Nieland for constructive comments and criticism on manuscript preparations. Funding for this project was provided by National Sea Grant (grant number NA16RG2249).

References

Bailey, H. K., J. H. Cowan, Jr., and R. L. Shipp. 2001. Experimental evaluation of potential effects of habitat size and presence of conspecifics on habitat association by young-of-the-year red snapper. Gulf of Mexico Science 2:119-131.

Baird, T. A. 1993. A new heterospecific foraging association between the puddingwife wrasse, *Halichoeres radiatus*, and the bar jack, *Caranx ruber*: evaluation of the foraging consequences. Environmental Biology of Fishes 38:393–397.

Bohnsack, J. A., and S. P. Bannerot. 1986. A stationary visual census technique for quantitatively assessing community structure of coral reef fishes. NOAA Technical Report NMFS 41.

Bortone, S. A., P. Rebenack, and D. M. Siegel. 1981. A comparative study of *Diplectrum formosum* and *D. bivittatum* (Pisces: Serranidae). Florida Scientist 44:97–103.

Bortone, S. A., R. W. Hastings, and J. L. Oglesby. 1986. Quantification of reef fish assemblages: comparison of several in situ methods. Northeast Gulf Science 8:1–22.

Bortone, S. A., R. K. Turpin, R. C. Cody, C. M. Bundrick, and R. L. Hill. 1997. Factors associated with artificial-reef fish assemblages. Gulf of Mexico Science 15:17–34.

Bradley, E., and C. E. Bryan. 1975. Life history and fishery of the red snapper (*Lutjanus campechanus*) in the northwestern Gulf of Mexico: 1970–1974. Proceedings of the Gulf and Caribbean Fisheries Institute 27:77–106.

Caldwell, D. K. 1965. Systematics and variation in the sparid fish *Archosargus probatocephalus*. Bulletin of the Southern California Academy of Science 4:89–100.

Cappo, M., P. Speare, and G. De'ath. 2004. Comparison of baited remote underwater video stations (BRUVS) and prawn (shrimp) trawls for assessments of fish biodiversity in inter-reefal areas of the Great Barrier Reef Marine Park. Journal of Experimental Marine Biology and Ecology 302:123–152.

Chittenden, M. E., and J. D. McEachran. 1976. Composition, ecology, and dynamics of demersal fish communities on the Northwestern Gulf of Mexico continental shelf, with a similar synopsis for the entire Gulf. Texas A&M University Sea Grant Publication 76–208, College Station, Texas.

Clarke, K. R. 1993. Non-parametric multivariate analyses of changes in community structure. Australian Journal of Ecology 18:117–143.

Clarke, K. R., and R. M. Warwick. 1998. Quantifying structural redundancy in ecological communities. Oecologia 113:278–289.

Clarke, K. R., and R. M. Warwick. 2001. Change in marine communities: an approach to statistical analysis and interpretation, 2nd edition. PRIMER-E Ltd., Plymouth, UK.

Dennis, G. D., and T. J. Bright. 1988. Reef fish assemblages on hard banks in the northwestern Gulf of Mexico. Bulletin of Marine Science 43:280–307.

Dufrene, T. A. 2005. Geological variability and holocene sedimentary record on the northern Gulf of Mexico inner to mid-continental shelf. Master's thesis. Louisiana State University, Baton Rouge, Louisiana.

Ellis, D. M., and E. E. DeMartini. 1995. Evaluation of a video camera technique for indexing abundances of juvenile pink snapper, *Pristipomoides filamentosus*, and other Hawaiian insular shelf fishes. Fishery Bulletin 93:67–77.

Franks, J. S., J. Y. Christmas, W. L. Siler, R. Combs, R. Waller, and C. Burns. 1972. A study of the nektonic and benthic faunas of the shallow Gulf of Mexico off the state of Mississippi. Gulf Research Reports 4:1–148.

Fraser, T. H. 1971. Notes on the biology and systematics of the flatfish genus *Syacium* (Bothidae) in the Straits of Florida. Bulletin of Marine Science 21:491–509.

Gallaway, B. J., J. G. Cole, R. Meyer, and P. Roscigno. 1999. Reduction of juvenile red snapper bycatch in the U.S. Gulf of Mexico shrimp trawl fishery. North American Journal of Fisheries Management 19:342–355.

Geoghegan, P., and M. E. Chittenden, Jr. 1982. Reproduction, movements, and population dynamics of the longspine porgy, *Stenotomus caprinus*. Fishery Bulletin 80:523–540.

Gledhill, C. T., and J. Lyczkowski-Shultz. 1994. Final Report to MARFIN: integration of video, acoustic and larval index methods to assess reef fish populations. NOAA/NMFS MS Laboratories, Pascagoula, Mississippi.

Gledhill, C. T., J. Lyczkowski-Shultz, K. Rademacher, E. Kargard, G. Crist, and M. A. Grace. 1996. Evaluation of video and acoustic index methods for assessing reef-fish populations. ICES Journal of Marine Science 53:483–485.

Gledhill, C. T. 2001. Reef fish assemblages on Gulf of Mexico shelf-edge banks. Doctoral dissertation. University of South Alabama, Mobile, Alabama.

Greene, L. E., and W. S. Alevizon. 1989. Comparative accuracies of visual assessment methods for coral reef fishes. Bulletin of Marine Science 44:899–912.

Hair, J. F., R. E. Anderson, L. Tatham, and B. J. Grablowsky. 1984. Multivariate Data Analysis. Macmillan, New York.

Hale, L. S. 1987. Distribution, abundance, reproduction, food habitats, age, and growth of round scad, *Decapterus punctatus*, in the south Atlantic Bight. Fishery Bulletin 85:251–268.

Hastings, P. A., and S. A. Bortone. 1976. Additional notes on tropical marine fishes in the northern Gulf of Mexico. Florida Scientist 39:123–125.

Hoese, H. D., and R. H. Moore. 1998. Fishes of the Gulf of Mexico, 2nd edition. Texas A&M Press, College Station, Texas.

Holt, S. A. and C. R. Arnold. 1982. Growth of juvenile red snapper, *Lutjanus campechanus*, in the northwestern Gulf of Mexico. Fishery Bulletin 80:644–648.

Kellison, G. T., and G. R. Sedberry. 1998. The effects of artificial reef vertical profile and hole diameter on fishes off South Carolina. Bulletin of Marine Science 62:763–780.

Kennicutt, M. C., W. W. Schroeder, and J. M. Brooks. 1995. Temporal and spatial variations in sediment characteristics on the Mississippi-Alabama continental shelf. Continental Shelf Research 15:1–18.

Lauth, R. R., J. Ianelli, and W. W. Wakefield. 2004. Estimating the size selectivity and catching efficiency of a survey bottom trawl for thornyheads, *Sebastolobus* spp. using a towed video camera sled. Fisheries Research 70:27–37.

Ludwick, J. C. 1964. Sediments in the northeastern Gulf of Mexico. Pages 204–240 *in* R. L. Miller, editor. Papers in Geology: Shepard Commemorative Volume. MacMillan, New York.

McBride, R. A., L. C. Anderson, A. Tudoran, and H. H. Roberts. 1999. Holocene stratigraphic architecture of a sand-rich shelf and the origin of linear shoals: northeastern Gulf of Mexico. Society for Sedimentary Geology, Special Publication 64:95–126.

Moore, D., H. A. Brusher, and L. Trent. 1970. Relative abundance, seasonal distribution, and species composition of demersal fishes off Louisiana and Texas, 1962–1964. Contributions in Marine Science 15:45–70.

Moran, D. 1988. Species Profiles: Life histories and environmental requirements of coastal fishes and invertebrates (Gulf of Mexico)-red snapper. U.S. Fish and Wildlife Service Biological Report 82(11.83).

Moseley, F. N. 1966. Biology of the red snapper, *Lutjanus aya Bloch*, of the northwestern Gulf of Mexico. Publications of the Institute of Marine Science, University of Texas 11:90–101.

Nieland, D. L., and C. A. Wilson. 2003. Red snapper recruitment to and disappearance from oil and gas platforms in the northern Gulf of Mexico. Pages 73–81 *in* D. R. Stanley and A. Scarborough-Bull, editors. American Fisheries Society, Symposium 36, Bethesda, Maryland.

Parker, R. O., Jr., R. B. Stone, and C. C. Buchanan. 1979. Artificial reefs off Murrells Inlet, South Carolina. Marine Fisheries Review 41:12–24.

Parker, R. O., D. R. Colby, and T. D. Willis. 1983. Estimated amount of reef habitat on a portion of the U.S. South Atlantic and Gulf of Mexico continental shelf. Bulletin of Marine Science 33:935–940.

Patterson, W. F., III, J. C. Watterson, R. L. Shipp, and J. H. Cowan, Jr. 2001a. Movement of tagged red snapper in the northern Gulf of Mexico. Transactions of the American Fisheries Society 130:533–545.

Patterson, W. F., III, J. H. Cowan, Jr., C. A. Wilson, and R. L. Shipp. 2001b. Age and growth of red snapper, *Lutjanus campechanus*, from an artificial reef area off Alabama in the northern Gulf of Mexico. Fishery Bulletin 99:617–627.

Patterson, W. F., III, and J. H. Cowan, Jr. 2003. Site fidelity and dispersion of red snapper associated with artificial reefs in the northern Gulf of Mexico. Pages 181–193 *in* D. R. Stanley and A. Scarborough-Bull, editors. American Fisheries Society, Symposium 36, Bethesda, Maryland.

Patterson, W. F., III, C. A. Wilson, S. J. Bentley, J. H. Cowan, Jr., T. Henwood, Y. C. Allen, and T. A. Dufrene. 2005. Delineating juvenile red snapper habitat on the northern Gulf of Mexico continental shelf. Pages 277–288 *in* P. W. Barnes and J. P. Thomas, editors. Benthic habitats and the effects of fishing. American Fisheries Society, Symposium 41, Bethesda, Maryland.

Pierce, D. J., and B. Mahmoudi. 2001. Nearshore fish assemblages along the central west coast of Florida. Bulletin of Marine Science 68:243–270.

Rademacher, K. R., and J. H. Render. 2003. Fish assemblages around oil and gas platforms in the Northeastern Gulf of Mexico: developing a survey design. Pages 101–122 *in* D. R. Stanley and A. Scarborough-Bull, editors. American Fisheries Society, Symposium 36, Bethesda, Maryland.

Rakocinski, C. F., J. Lyczkowski-Shultz, and S. L. Richardson. 1996. Icthyoplankton assemblage structure in Mississippi Sound as revealed by canonical correspondence analysis. Estuarine, Coastal and Shelf Science 43:237–257.

Rezak, R., S. R. Gittings, and T. J. Bright. 1990. Biotic assemblages and ecological controls on reefs and banks of the northwest Gulf of Mexico. American Zoologist 30:23–25.

Rooker, J. R., Q. R. Dokken, C. V. Pattengill, and G. J. Holt. 1997. Fish assemblages on artificial and natural reefs in the Flower Garden Banks National Marine Sanctuary, USA. Coral Reefs 16:83–92.

Rooker, J. R., A. M. Landry, B. W. Geary, and J. A. Harper. 2004. Assessment of a shell bank and associated substrates as nursery habitat of postsettlement red snapper. Estuarine, Coastal and Shelf Science 59:653–661.

Sale, P. F., and W. A. Douglas. 1981. Precision and accuracy of visual census techniques for fish assemblages on coral patch reefs. Environmental Biology of Fishes 6:333–339.

SAS Institute, Inc. 2002. Statistics, version 9.0. SAS Institute, Inc., Cary, North Carolina.

Schroeder, W. W., A. W. Shultz, and J. J. Dindo. 1988. Inner-shelf hardbottom areas, northeastern Gulf of Mexico. Gulf Coast Association of Geological Societies Transaction 38:535–541.

Schroeder, W. W., A. W. Shultz, and O. H. Pilkey. 1995. Late Quaternary oyster shells and sea-level history, inner shelf, northeast Gulf of Mexico. Journal of Coastal Research 11:664–674.

Sedberry, G. R., and R. F. Van Dolah. 1984. Demersal fish assemblages associated with hard bottom habitat in the South Atlantic Bight of the USA. Environmental Biology of Fishes 11:241–258.

Smith, G. B., H. M. Austin, S. A. Bortone, R. W. Hastings, and L. H. Ogren. 1975. Fishes of the Florida Middle Grounds with comments on ecology and zoogeography. Florida Marine Research Publications 9, St. Petersburg, Florida.

Stanley, D. R., and C. A. Wilson. 2000. Variation in the density and species composition of fishes associated with three petroleum platforms using dual beam hydroacoustics. Fisheries Research 47:161–172.

Systat software, Inc. 2004. SigmaStat, version 3.1. Systat software, Inc., Point Richmond, California.

Szedlmayer, S. T., and R. L. Shipp. 1994. Movement and growth of red snapper, *Lutjanus campechanus*, from an artificial reef area in the northeastern Gulf of Mexico. Bulletin of Marine Science 55:887–896.

Szedlmayer, S. T. 1997. Ultrasonic telemetry of red snapper, *Lutjanus campechanus*, at artificial reef sites in the northeast Gulf of Mexico. Copeia 4:846–850.

Szedlmayer, S. T., and J. C. Howe. 1997. Substrate preference in age-0 red snapper, *Lutjanus campechanus*. Environmental Biology of Fishes 50:203–207.

Szedlmayer, S. T., and J. Conti. 1999. Nursery habitats, growth rates, and seasonality of age-0 red snapper, *Lutjanus campechanus*, in the northeast Gulf of Mexico. Fishery Bulletin 97:626–635.

Szedlmayer, S. T., and R. L. Schroepfer. 2005. Long-term residence of red snapper on artificial reefs in the northeastern Gulf of Mexico. Transactions of the American Fisheries Society 134:315–325.

ter Braak, C. J. F., and P. Smilauer. 2002. CANOCO Reference manual and CanoDraw for windows user's guide: software for canonical community ordination, version 4.5. Microcomputer Power, Ithaca, New York.

Warwick, R. M., K. R. Clarke, and J. M. Gee. 1990a. The effects of disturbance by soldier crabs, *Mictyris platycheles*. H. Milne Edwards, on meiobenthic community structure. Journal of Experimental Marine Biology and Ecology 135:19–33.

Warwick, R. M., H. M. Platt, K. R. Clarke, J. Agard, and J. Gobin. 1990b. Analysis of macrobenthic and meiobenthic community structure in relation to pollution and disturbance in Hamilton Harbour, Bermuda Journal of Experimental Marine Biology and Ecology 138:119–142.

Wenner, C. A. 1983. Species associations and day-night variability of trawl caught fishes from the inshore sponge-coral habitat, South Atlantic Bight. Fishery Bulletin 81:532–552.

Willis, T. J., and R. C. Babcock. 2000. A baited underwater video system for the determination of relative density of carnivorous reef fish. Marine and Freshwater Research 51:755–763.

Willis, T. J., R. B. Millar, and R. C. Babcock. 2000. Detection of spatial variability in relative density of fishes: comparison of visual census, angling, and baited underwater video. Marine Ecology Progress Series 198:249–260.

Wilson, C. A. and D. L. Nieland. 2001. Age and growth of red snapper, *Lutjanus campechanus*, from the northern Gulf of Mexico off Louisiana. Fishery Bulletin 99:653–664.

Workman, I. K., and D. G. Foster. 1994. Occurrence and behavior of juvenile red snapper, *Lutjanus campechanus*, on commercial shrimp fishing grounds in the northeastern Gulf of Mexico. Marine Fisheries Review 56:9–11.

Workman, I., A. Shah, D. Foster, and B. Hataway. 2002. Habitat preferences and site fidelity of juvenile red snapper (*Lutjanus campechanus*). ICES Journal of Marine Science 59:S43–S50.

American Fisheries Society Symposium 60:59–75, 2007

Swimming Performance and Behavior of Red Snapper: Their Application to Bycatch Reduction

Glenn R. Parsons[1]
Department of Biology
The University of Mississippi
University, Mississippi 38677 USA

Daniel G. Foster
National Marine Fisheries Service
3209 Frederic Street, Pascagoula, Mississippi 39567 USA

Abstract.—We examined the swimming performance and behavior of red snapper, *Lutjanus campechanus*. Our intention was to use this information toward developing a more efficient bycatch reduction device for use in the Gulf of Mexico shrimp trawl fishery. Using a Brett type swim tunnel, we found a significant effect of fish size and season on red snapper critical swimming speeds. For fish ranging between 6 and 17 cm standard length, critical swimming speeds ranged from about 35–70 cm/s, depending upon season. However, critical swimming speeds did not differ between day and night. This was an important observation since the majority of shrimp trawling in the Gulf occurs at night. We constructed and tested in the laboratory, a Vortex Generating Bycatch Reduction Device (VGBRD) that may prove useful in the shrimp trawl fishery. During behavioral tests during daylight, 79.2% of red snapper exited the VGBRD in an average of 4.1 min. However, during night-time tests, only 17.6% of red snapper exited the VGBRD in an average of 5.0 min. Behavioral tests revealed a strong negative phototactic response in dark adapted red snapper. We found that, during night-time tests when the VGBRD was illuminated with LED lights placed downstream of the exit, 96% of red snapper exited the device in 7.1 min. In color/contrast choice experiments, red snapper unerringly associated with the dark colored (black or dark green) panel placed on the bottom of the experimental tank. In another set of experiments, we found that snapper displayed a strong optomotor response, i.e. the tendency to following and match speeds with a moving pattern. Illumination, color/contrast, and/or the optomotor response may improve bycatch reduction device performance.

Introduction

Designing an effective bycatch reduction device (BRD) for use in the shrimp trawl fishery should take into consideration the behavior and swimming ability of both the target and nontarget species. An understanding of fish behavior under a variety of conditions is important in developing effective bycatch re-

[1] Corresponding author: bygrp@olemiss.edu.

duction technology and Engaas et al. (1999) reported that "efficient separation of fish from shrimp will require taking advantage of behavioral differences between these species." Unfortunately, for many commercially important species, there is little information regarding behavior and even less concerning swimming ability. The paucity of information regarding fish swimming is particularly unacceptable upon recognizing that accurate swimming performance data can be valuable for making management decisions regarding trawl fisheries. Knowledge of fish swimming ability can be valuable in trawl design and in determining how and when a particular design is used, thereby improving trawl selectivity and reducing unwanted bycatch, i.e. nontargeted species. By understanding fish swimming performance and behavior, the potential for effective bycatch reduction in trawl fisheries is greatly increased. In a world of declining fish stocks, it is imperative that managers use every weapon in their arsenal to reduce or eliminate the catch of unwanted species.

The shrimp trawl fishery is one of the most valuable commercial fisheries in the Gulf of Mexico. The importance of this fishery to local and state economies cannot be over-stated. Additionally, the red snapper *Lutjanus campechanus* is arguably one of the most sought-after species in the Gulf of Mexico recreational fishery and is likewise a valuable component of the commercial catch. Unfortunately, the snapper fishery has been declining for some time. Regulations aimed at stemming this decline have been implemented and a target date for restoring snapper stocks to a 30% spawning potential ratio has been set at 2032. Although the cause of declining snapper stocks are likely manifold, it is well documented that juvenile snapper are a significant portion of shrimp trawl bycatch with some 20 million estimated to have been taken in shrimp trawling in 1989 (Nichols et al. 1990). Warren (1994) estimated that the capture of juvenile red snapper in shrimp trawl bycatch was 34 million fish per year. More recent estimates of juvenile snapper bycatch in the shrimp trawl fishery vary from 25 to 45 million per year (SEDAR 2005). Although a portion of these fish may survive capture, the vast majority do not and are discarded (Gutherz and Pellegrin 1988). Goodyear and Phares (1990) estimate that 90% of the fishing mortality of age 0 and 1 red snapper results from shrimp trawl bycatch. A reduction in juvenile red snapper mortality by reducing their capture during trawling may contribute to increasing snapper stocks. The Gulf of Mexico and South Atlantic Fishery Management Councils consider a 44% reduction in juvenile red snapper shrimp trawl mortality necessary for recovery of the snapper fishery.

In this paper we report on how red snapper critical swimming speeds change with season, size, and time of day (day versus night). Additionally, we examined the optomotor response, color preference, and the response of dark adapted red snapper to illumination. Finally, we examined the behavioral response of red snapper to changing water flow patterns using a Vortex Generating Bycatch Reduction Device (VGBRD) designed for this study.

Methods

Collection and Maintenance of Fish

Red snapper juveniles ranging in size from 30 to 250 mm Standard Length (SL) were collected from the northern Gulf of Mexico shrimp grounds by the National Marine Fisheries Service (NMFS), Pascagoula Laboratory. Fish were trawled from artificial webbing reefs designed by the NMFS, Pascagoula Laboratory (Workman and Foster 2002). Fish were immediately placed in an aerated holding tank on the deck of the research vessel and were allowed to acclimate for at least 12 h, overnight, in the holding tank before transport back to the University of Mississippi. We observed lower mortality rates when fish were allowed a short acclimation time, as opposed to being immediately transported after collection. In some cases, particularly during winter months, it was necessary to deflate the over-inflated swim bladder after trawl collections. Using a tuberculin needle and syringe with the plunger removed, the swim bladder was punctured and relieved of excess pressure. This allowed fish to immediately right themselves and descend to the bottom of the tank. Few fish died from this treatment and all survivors typically began actively feeding within 24 h of their arrival at the University of Mississippi. To assess seasonal effects, red snapper were collected during February, April, July, October and December of 2002 and 2003. Approximately 50–100 fish were collected during each of these months. Additionally, because all size groups are not available during any particular month, this allowed the complete size range to be examined. Fish were immediately returned to the laboratory and held in twin 1300 L re-circulating seawater systems (all projects using vertebrates must be reviewed and approved by the University of Mississippi, Institutional Animal Care and Use Committee). These systems use biological filtration, mechanical filtration, and UV sterilizers for maintaining water quality. Prior to swimming performance evaluation, fish were maintained under the same environmental conditions (temperature, salinity, photoperiod) from which they were collected. Before beginning swimming and/or behavioral testing, fish were allowed a minimum one week period of acclimation to recover from collecting stress. Fish were fed *ad-libitum* daily a diet of squid, shrimp, and fish. Only those fish that appeared to be healthy and were actively feeding were used for study. Fish were not fed for 24 h prior to swimming or behavioral testing such that the fish were in a postabsorptive state.

Critical Swimming Speed Determination

A swim tunnel (Figure 1) located at The University of Mississippi, Department of Biology, was used for the study. The swim tunnel was 1000 L in volume. Water current was propeller generated and the working section was approximately 100 cm long and 38 cm wide. Flow filters placed within the tunnel were used to promote rectilinear flow (flow that is microturbulent). Water was continuously forced through the working section of the tunnel by a propeller connected to a variable speed motor. Water velocity in the tunnel was measured using a Marsh-McBirney electronic flowmeter. Swim tunnel conditions of temperature, salinity and dissolved oxygen were adjusted to match the conditions of the holding tank prior to swimming fish. During swimming trials, dissolved oxygen was maintained at or near saturation.

During a typical swimming test, an individual was transferred from the holding tank into the working section of the flume one day before the swimming test. In this manner, the fish was allowed 12–24 h to acclimate prior to testing. The swimming trial was begun by subjecting fish to a current of 10 cm/s for 30 min. If fish swam the entire 30 min bout, the current speed was increased to 15 cm/s. This process was repeated until all fish fatigued. As an individual fatigued, it was removed and the current speed, the amount of time the fish swam at that speed, and the standard length (SL) were recorded. Because fish actually swim better with successive swimming trials (known as the training effect), an individual fish was only used once for the swimming performance study. However, these fish were used later for the behavioral study. Thirty minute critical swimming speeds were calculated from the above information using the equation:

$$U\text{crit} = u1 + (u2)(t1/t2)$$

Figure 1. The Brett swim tunnel used for red snapper critical swimming speed determinations.

where $u1$ was the highest speed at which the fish swam the prescribed time period, $t1$ was the time the fish swam at the fatigue velocity, $t2$ was the prescribed time period, and $u2$ was the velocity increment.

Nighttime swimming trials were conducted under total darkness using an infrared video system. Fish were acclimated to the tunnel for a minimum of 12–24 h. A typical trial began at 2000 hours and would generally end about 6 h later.

Flow Dynamics of the Vortex Generating BRD

In this portion of the study we examined the changing patterns of water flow in the Vortex Generating Bycatch Reduction Device (VGBRD). Because bycatch reduction devices may attempt to exploit the response of red snapper to changing patterns of water flow, the information gathered in this portion of the study was essential to predicting the effectiveness of BRD design. To evaluate this response, the VGBRD was constructed (Figure 2) employing plates arranged in such a manner that water flow was interrupted and a vortex was generated just behind and under the plates (Figure 2). The plates were constructed of black Plexiglas. The front plate was 70 cm long and 10 cm wide. The top plate was 70 cm long and 70 cm wide. The device was designed such that the front plate angle, relative to the top plate, could be varied between 45, 60, and 75°.

To evaluate the effect of water speed and front plate angle on vortex (or vortices) generation and water speed in the BRD, we altered the speed of the water entering the BRD (position #1) as well as the angle of the front plate (45, 60 and 75°) and then mea-

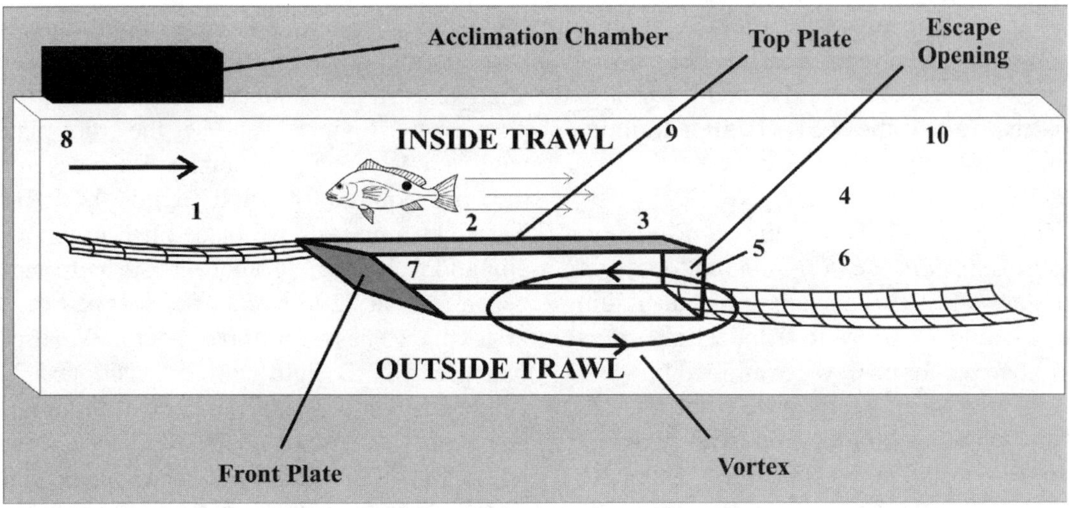

Figure 2. The behavior test tank containing the Vortex Generating Bycatch Reduction Device used for red snapper behavioral testing. Numbers 1–6 indicate positions where current speeds were measured. Numbers 7–10 indicate positions for placement of artificial lighting.

sured the water speed at each of positions 2–6 (Figure 2). We also examined the appearance of the vortex (vortices) that were generated in the BRD using video. Flow visualization was accomplished using a fused glass air stone, placed in or near the escape opening of the BRD. Very fine bubbles produced by this air stone were entrained in the vortices, and were easily recorded on video for later evaluation.

Response of Red Snapper to the VGBRD

To examine red snapper response to the VGBRD, one to 10 fish ranging in size from 7.5 to 15.3 cm SL were removed from the holding tank and placed into the acclimation chamber (Figure 2) of the behavior test tank. For all VGBRD tests, salinity varied from 30 to 33 ppt and temperature from18–22°C. Fish were placed in the acclimation chamber 2–12 h prior to testing. The acclimation chamber was painted black to reduce excitement caused by external activities in and around the test tank. Additionally, this effectively eliminated all light exposure during nighttime swimming tests. To begin a test, the current speed was set to 50 cm/s at position #1 (Figure 2) and a false bottom in the acclimation chamber was removed. This resulted in fish being drawn into the flowing water, whereupon they immediately began swimming. The time from when the fish were first exposed to flow, until the fish exited the BRD was recorded. In many cases, tests were videotaped for later analysis.

Day versus night.—We examined how red snapper would respond to the VGBRD during daylight and night swimming. For daylight tests, fish were transferred from the holding tank to the acclimation chamber of the behavior test tank at about 0900 hours and were tested approximately 2 h later as described above. We tested the response of red snapper to a single current speed (ca 50 cm/s measured at position #1, Figure 2) and front plate angle (45°) by recording the time it took for fish to exit the VGBRD as well as the percentage of fish that exited. For nighttime observations, we began the behavioral test at 2100 hours in total darkness after 2.0 h dark acclimation. Fish behavior was recorded until

2400 hours whereupon the room lights came on. We continued recording fish behavior for the next several hours and timed how long it took fish to exit the VGBRD after illumination.

Response of Dark Adapted Red Snapper to Artificial Lighting.—After observing that dark adapted red snapper in our holding tanks would move away from a light source (negative phototaxis), we examined how dark adapted red snapper responded to the VGBRD when artificial illumination was provided. Fish were tested in the behavioral test tank as described above. However, either green Cyalume light sticks or blue-green LED (Digi Key Co., part # 404–1094-ND) lights were placed at various positions (Figure 2) around the behavioral test tank and the time for red snapper to exit was recorded.

Front plate angle and artificial illumination.—We examined the response of red snapper to three front plate angles (45, 60 and 75°) and day versus night. For day-time tests, we transferred fish from the holding tank at about 0900 hours and tested them approximately two hours later as described above. We used a single current speed, 50 cm/s measured at position #1, Figure 2, and set the front plate angle to one of the three angles. We measured the time required for fish to exit the VGBRD. For night-time tests we placed fish in the acclimation chamber at about 1600 hours and began tests at 2200 hours after fish had been in total darkness for three hours. We used blue-green LED light sources positioned 30 cm downstream of the VGBRD exit (Figure 2, position #10). We video taped dark tests using an "Aqua-Vue" infra-red camera system. The infra-red camera was positioned just above the VGBRD exit such that all fish escaping could be detected. In some cases video taping lasted as long as 6 h For these tests, fish that did not exit after two hours were not used in the analysis.

Red Snapper Color/Contrast Preference Experiment

In this experiment we investigated the preference of red snapper for certain colors/contrasts. We placed five, 25 × 25 × 0.25 cm

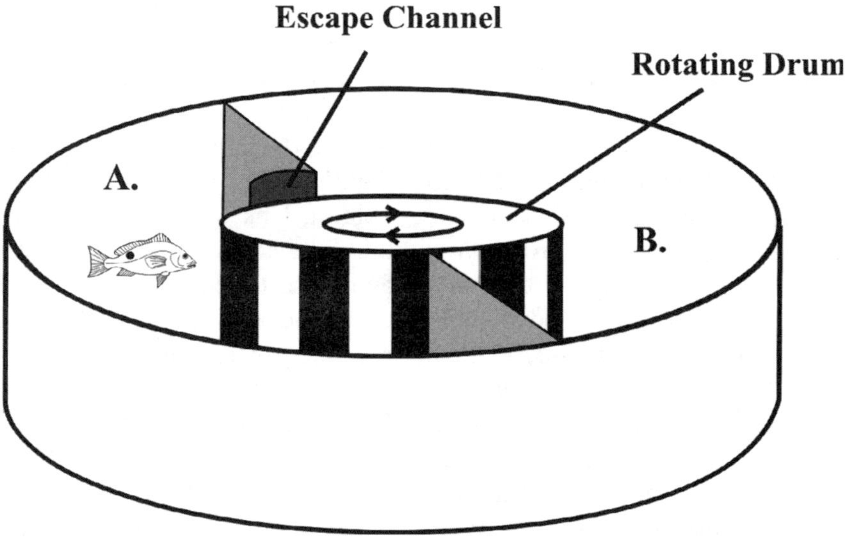

Figure 3. The optomotor response test tank. Fish were placed on side A and by following the rotating drum, were encouraged to swim through the escape channel to the opposite side of the tank (side B).

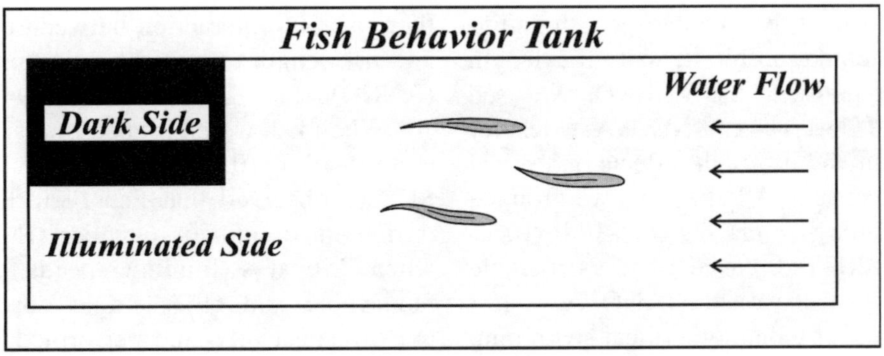

Figure 4. The behavior test tank modified for the dark/light choice experiment.

colored panels (red, white, green, black and yellow) on the bottom of a 2 m × 2 m ×1 m (water depth) holding tank illuminated with fluorescent lighting. The bottom of the holding tank was light blue in color, the salinity was 33 ppt, and the temperature was 21°C. All experiments were conducted between 1200 and 1530 hours. One panel was placed in the center of the tank and the others were placed in each corner, 50 cm from the tank walls. A single, light adapted, red snapper was introduced into the holding tank at the center, surface of the water, and the time required for the fish to associate with a particular panel was recorded. To eliminate any possible preference for a particular area of the tank, or any bias due to the position of the panels in the tank, the panels were systematically rotated through all possible positions around the tank and the experiments repeated. Likewise, the researcher releasing the fish stood in a position such that the fish could not observe the scientist during tests.

Optomotor Response Experiment

In the experiments described here, we tested whether we could exploit the optomotor response to induce fish to move from one area (inside the trawl), through a narrow channel, to another area (outside the trawl). We used a circular holding tank (117 cm in diameter, 25.5 cm depth) with a drum (38 cm in diameter) with vertical black and white stripes, positioned in the center of the holding tank (Figure 3). The entire tank was then divided into two areas via a partition. Fish were placed on one side of the partition and the drum was rotated at about 30 rpm. An escape channel allowed snapper to move to the opposite side of the partition by following the pattern on the rotating drum. All tests were conducted during daylight hours at 33 ppt salinity and 17°C temperature.

Dark/Light Choice Experiment

In this experiment the fish behavior test tank was divided into dark and light sections (Figure 4). We placed a partition near the rear of the test chamber and illuminated one side while the other was dark. Illumination was provided by blue-green LED light sources. Temperature in the test tank was 17°C and salinity was 33 ppt. After red snapper were dark adapted beginning at 2000 hours (at night), fish were released into flowing water at the head of the test chamber. As fish failed to maintain position against current and moved toward the rear of the chamber, we recorded the number of fish choosing the illuminated side versus the dark side.

Statistical Analyses

We used simple linear regression to describe the relationship between critical swimming speeds and fish standard length and to examine the relationship between fish length and the response to the VGBRD. We used Analysis of Covariance (ANCOVA) to test for size, season and day/night effects on critical swimming speeds. Additionally, we grouped fish into three size classes (8.0–10.0, 10.1–12.0 and 12.1–14.0 cm SL) and used single factor analysis of variance (ANOVA) to test for the effect of month on critical swimming speed for each size-class. We also used ANOVA to test for the effect of VGBRD front plate angle and the effect of day versus night on time for red snapper to exit the VGBRD. For significance testing, *P*-values less than or equal to 0.05 were considered significant. Relationships were considered approaching significance when *P*-values were greater than 0.05 and less than 0.10.

Results

Critical Swimming Speed Determination

We obtained one hour critical swimming speeds over all seasons of the year, for 193 red snapper ranging in size from 6 to 17.7 cm SL. Critical swimming speeds ranged from 28.8 to 81.5 cm/s. There was a highly significant effect (ANCOVA; $F_{1, 181} = 54.3$; $P < 0.0001$) of size on critical swimming speed across all months of the study (Figures 5–9). Critical swimming speeds for the smallest red snapper were generally in the range of about 30–40 cm/s while the largest was about 60–75 cm/s. We also observed a highly significant effect of month (ANCOVA; $F_{1, 187} = 19.3$; $P < 0.0001$) on critical swimming speeds of red snapper and a significant interaction was detected between month and standard length on critical swimming speed (ANCOVA; $F_{1, 187} = 6.66$; $P = 0.0106$). We observed no significant effect of time of day (day versus night) on critical swimming speed during any season (ANCOVA; $F_{1, 187} = 0.04$; $P = 0.836$) and likewise there was no interaction between day/night and fish length on critical swimming speed (ANCOVA; $F_{1, 187} = 0.005$; $P = 0.941$).

When fish were grouped into size classes (8.0–10.0, 10.1–12.0 and 12.1–14.0 cm SL) we observed that fish performed best during the month of October (Figure 10) when critical swimming speeds averaged 56.4, 60.4 and 69.7 (single observation) cm/s, respectively, and performed worst in February when speeds averaged 44.9, 50.0 and 58.4 cm/s, respectively. There was a highly significant effect of month on critical swimming speeds across all size classes (ANOVA; $F_{1, 132} = 934$; $P < 0.0001$; $F_{1, 230} = 2277$; $P < 0.0001$; $F_{1, 123} = 1392$; $P < 0.0001$; respectively).

Flow Dynamics of the VGBRD

We examined flow dynamics of the VGBRD at three different front plate angles (45, 60 and 75°). We were particularly interested in the flow dynamics at position #5 (Figure 2), the entrance to the escape opening. We found that the highest front plate angle actually produced less effective negative water flow (3–9 cm/s opposite to free stream flow) at the escape opening. At the lower angles, the flow of water into the escape opening was about 4–11 cm/s in opposition to free stream flow (Table 1). We characterized the water flow qualitatively at each angle. At a front plate angle of 45° the vortices were typically uniform and consistently generated. At 60 and 75°, the vortices became more irregular and appeared simply as turbulent/chaotic water flow at the highest angle.

Response of Red Snapper to the VGBRD

Day versus night— We tested fish 8.4–15.3 cm SL during the day and fish 7.5–14.9 cm SL at night (in darkness). We found that 79.2% of fish exited the VGBRD in an aver-

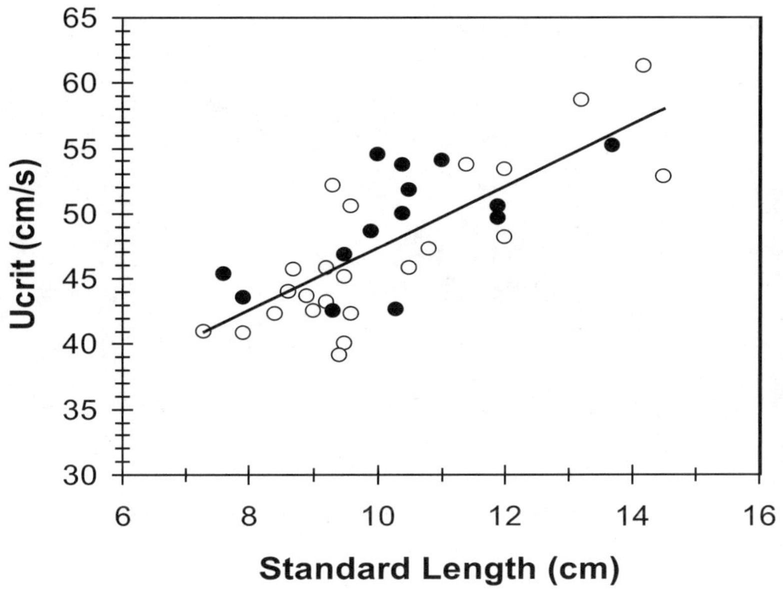

Figure 5. Critical swimming speeds for red snapper *Lutjanus campechanus* collected in February. Filled data points are night swims (in the dark under infra-red light) and unfilled points are day swims. For regression analysis day and night swims were pooled. ($Y = 23.9 + 2.35X$; $N = 38$; $R^2 = 0.60$, $F = 53.4$; $P < 0.0001$).

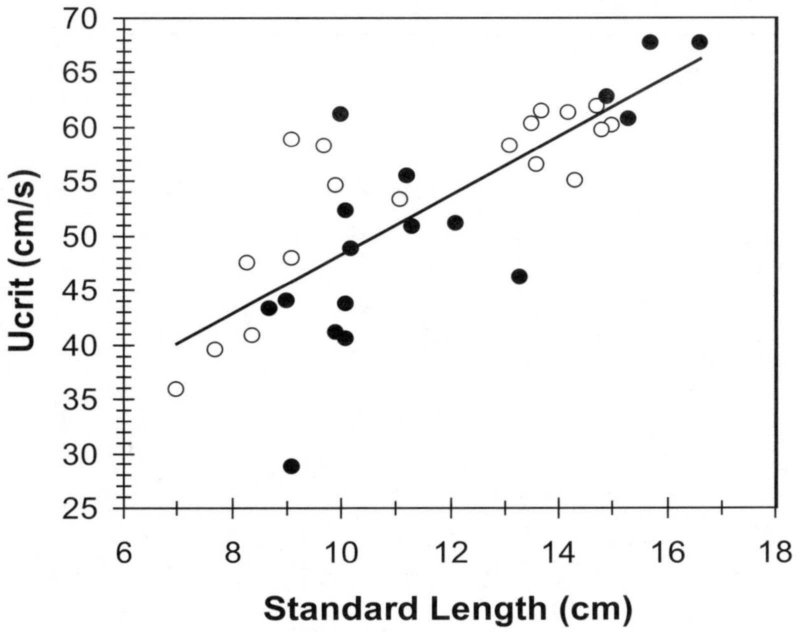

Figure 6. Critical swimming speeds for red snapper *Lutjanus campechanus* collected in April. Filled data points are night swims (in the dark under infra-red light) and unfilled points are day swims. For regression analysis day and night swims were pooled. ($Y = 21.0 + 2.7X$; $N = 35$; $R^2 = 0.60$; $F = 48.6$; $P < 0.0001$).

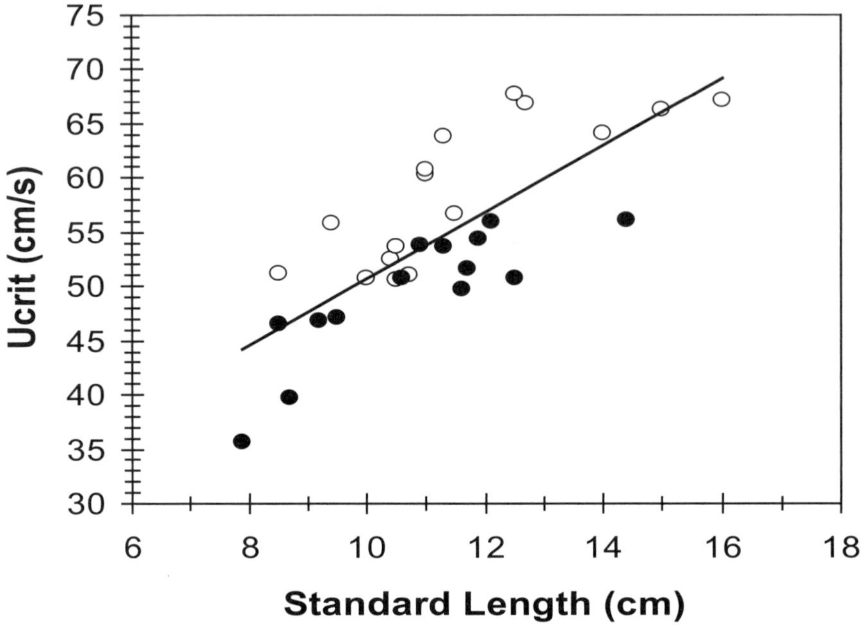

Figure 7. Critical swimming speeds for red snapper *Lutjanus campechanus* collected in July. Filled data points are night swims (in the dark under infra-red light) and unfilled points are day swims. For regression analysis day and night swims were pooled. ($Y = 20.0 + 3.1X$; $N = 30$; $R^2 = 0.59$, $F = 39.6$; $P < 0.0001$).

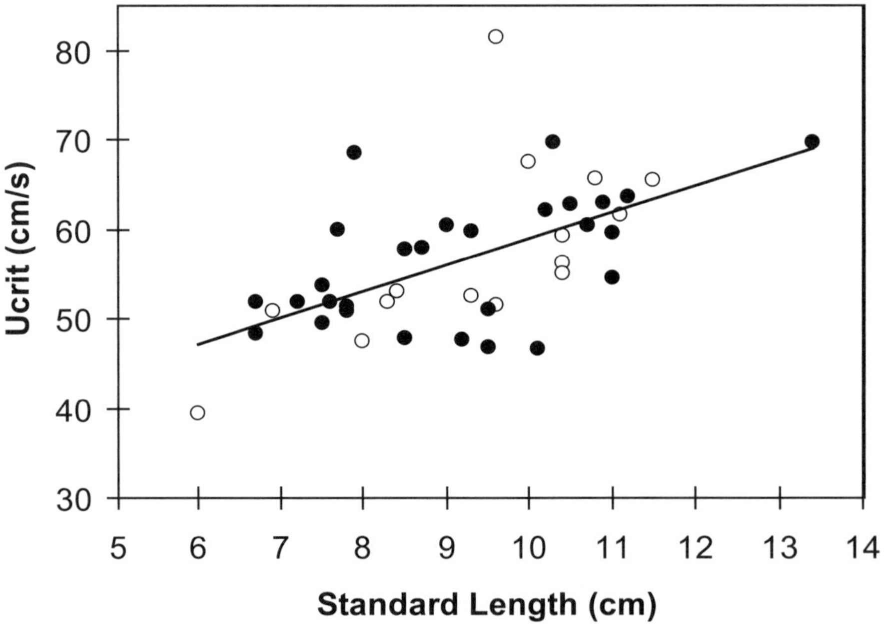

Figure 8. Critical swimming speeds for red snapper *Lutjanus campechanus* collected in October. Filled data points are night swims (in the dark under infra-red light) and unfilled points are day swims. For regression analysis day and night swims were pooled. ($Y = 29.6 + 2.9X$; $R^2 = 0.33$, $N = 43$; $F = 20.6$; $P < 0.0001$).

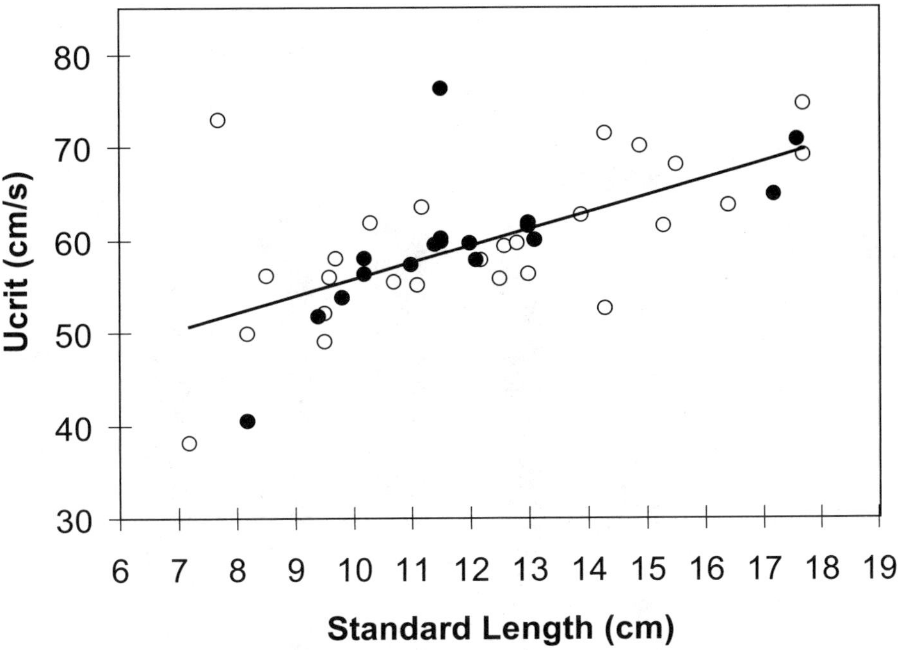

Figure 9. Critical swimming speeds for red snapper *Lutjanus campechanus* collected in December. Filled data points are night swims (in the dark under infra-red light) and unfilled points are day swims. For regression analysis day and night swims were pooled. ($Y = 37.6 + 1.8X$; $N = 42$; $R^2 = 0.41$, $F = 29.6$; $P < 0.0001$).

age of 4.1 min (±1.2 SE, $N = 45$) during the day and 17.6% of fish exited in an average of 5.0 min (±1.1 SE, $N = 27$) at night (Figure 11). When dark adapted fish (8.4–13.1 cm SL) were suddenly exposed to light (room lights came on at 2400 hours), we observed 100% of fish exiting the VGBRD and escape was observed on average 3.98 (±0.97 SE, $N = 23$) min after illumination (Figure 11). We observed a significant effect of illumination (day, night/darkness, night/illumination) on time to exit the BRD (ANOVA; $F_{2, 94} = 4.37$; $P = 0.0154$). We found no significant effect of fish length on their response to the VGBRD ($R^2 = 0.012$, $P = 0.5$, $F = 0.47$, $N = 95$).

Response of Dark Adapted Red Snapper to Artificial Illumination.—We examined the response of dark adapted red snapper to artificial illumination. When a single Cyalume light stick was placed in the holding tank, after fish were dark adapted, we observed a strong, negative phototactic response, (i.e., snapper were repelled by the light). We then examined the response of red snapper to artificial illumination in the VGBRD by placing Cyalume light sticks in various positions (Figure 2, positions 7–10) around the behavior test tank. When Cyalume lights were placed under the top plate of the VGBRD (Figure 2, position 1) the fish were reluctant to approach the escape opening and out of 10 fish tested, none were observed to exit. We likewise placed Cyalume lights just under the acclimation chamber (Figure 2, position #8) and found that fish were reluctant to leave the acclimation chamber and enter free stream flow. Once entering the test chamber, snapper positioned themselves at the very rear of the test chamber and, out of 10 fish, none were observed to exit. Another group of experiments was conducted with Cyalume lights positioned at the rear of the test chamber (Figure 2, position #10). In this case we found that fish were encouraged to exit and 53% of snap-

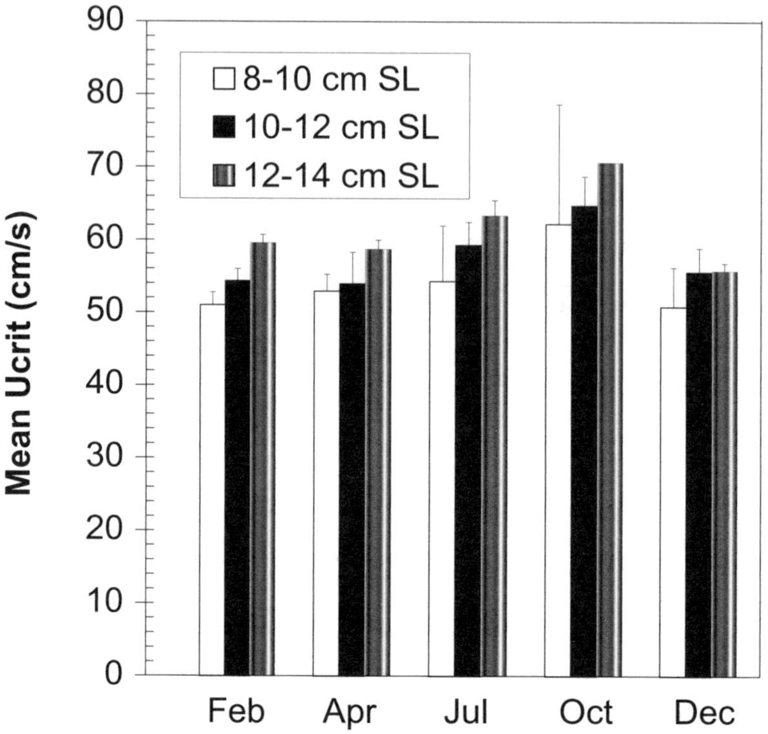

Figure 10. Change in red snapper *Lutjanus campechanus* critical swimming speed with season for three size classes of fish. Means and standard errors are shown except for the October, 12.1–14.0 cm SL size-class which was a single observation.

per exited in an average of 10.9 min after the test began.

In another set of experiments, we used LED lights in place of Cyalume lights. We placed the lights such that they shone directly onto the escape opening (Figure 2, position #9). Snapper positioned themselves as far away from the light source as possible and did not exit from the VGBRD. We then placed LED lights downstream of the escape opening (Figure 2, position #10). In this case, the fish stayed ahead of the light source which meant they tended to stay in the region of the reduced flow just down stream of the escape opening. In this arrangement we observed 96% of fish exiting the test tank in an average of 7.1 min.

Front plate angle— We tested the effect of three VGBRD front plate angles (45, 60 and 75°) on red snapper time to exit during both day and night. The effect of day versus night on time to exit approached significance (ANOVA; $F_{1, 61} = 3.30$; $P = 0.074$), there was no significant effect of front plate angle on time to exit (ANOVA; $F_{2, 61} = 0.048$; $P = 0.95$), and there was no significant interaction between front plate angle and time of day (ANOVA; $F_{2, 61} = 0.96$; $P = 0.39$). We therefore pooled the results for all angles and tested day versus night and found a significant effect (ANOVA; $F_{1, 65} = 7.63$; $P = 0.008$). Snapper exited the VGBRD on average 1.8 min (±0.39 SE) after entering the test chamber in daylight tests and 4.5 min (±1.1 SE) in nighttime tests. However, in daylight tests 92% of fish exited the VGBRD whereas in nighttime tests 74% of fish exited.

Color/Contrast Preference Experiment

In 61 trials, red snapper associated with the black panel 54 times (88.5%) in an average of 7.3 s (range = 2–22 s). On seven occasions (11.5%), snapper associated with the dark green panel in an average of 12.6 s (range = 7–22 s). Fish were never observed to associate with any other colored panels. In a typical trial, the fish would pause for a few seconds at the surface, apparently to orient itself, and would then make a direct descent toward the black panel. After association with the panel, the fish would frequently swim in a head down fashion with its nose against the panel.

Optomotor Response Experiment

We observed that red snapper would readily follow a revolving pattern (optomotor response) into a narrow escape channel. In tests involving 18 snapper varying in size from 8.5 to 17 cm SL, fish were observed to follow the rotating drum from one side of the partition to the other in every case. When a group of six fish were introduced into the system, all fish exited at the same time, within 15 s of the drum beginning rotation. Some fish did not respond as readily and required several minutes to exit (1–9 min). Other fish exited in seconds (15–56 s). Average time to exit for all fish was 2.6 min.

Dark/Light Choice Experiment

We tested 21 fish ranging in size from 7.1 to 17.5 cm SL. In all cases (100%), as fish swam against current and were unable to maintain station, they chose to position themselves on the dark side of the test tank.

Table 1. The effect of water speed entering the VGBRD and front plate angle, on water speed (cm/s) within the VGBRD. The positions in the VGBRD correspond to those numbers that appear in Figure 2. Water speed at position #1 is defined as the speed entering the BRD.

	Positions in VGBRD					
	1	2	3	4	5	6
Front plate angle 45°.	21.0	17.5	19.5	14.0	–4.0	6.5
	22.0	19.5	22.5	17.5	–4.0	9.5
	32.0	24.0	26.0	26.5	–4.0	12.5
	33.5	32.5	31.5	27.5	–5.0	17.5
	53.0	45.0	46.0	42.0	–9.0	24.0
Front plate angle 60°.	"	"	"	20.0	–3.0	4.0
	"	"	"	21.0	–4.0	5.0
	"	"	"	30.0	–4.0	10.0
	"	"	"	39.0	–7.0	11.0
	"	"	"	48.0	–11.0	19.0
Front plate angle 75°.	"	"	"	15.0	–3.0	9.0
	"	"	"	16.0	–3.0	12.0
	"	"	"	20.0	–4.0	18.0
	"	"	"	21.0	–4.0	20.0
	"	"	"	33.0	–9.0	29.0

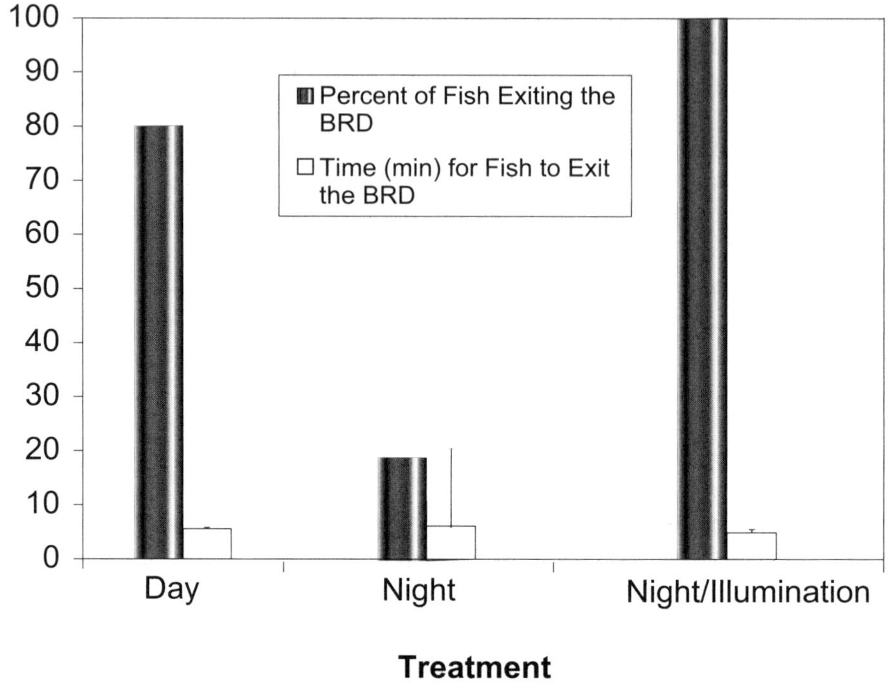

Figure 11. The response of red snapper *Lutjanus campechanus* in the VGBRD to various light treatments. Fish were tested during the day in light, during the night in dark, and then in both dark and light conditions. This plot shows the percentage of fish that exited the BRD under each condition and the mean time required to exit.

Discussion

Critical Swimming Speed Determinations

In examining critical swimming speeds we observed a significant effect of fish size. This was an expected finding and similar results have been observed across many taxa. Small organisms typically demonstrate reduced absolute speeds when compared to larger organisms of the same species. There was little difference in swimming ability measured during the day when compared with night performance. This was an important observation for the development of a bycatch reduction device that might be effective for eliminating or reducing red snapper bycatch. The senior author has observed white crappie, *Pomoxis annularis* in an apparent state of torpor and unable (or unwilling) to swim during night. If a similar situation had been observed in red snapper, the development of a successful bycatch reduction device effective for this species might have been in jeopardy.

The critical swimming speeds observed suggest that red snapper are not weak swimmers. The smallest fish (6–7 cm SL) were observed to swim at 35–40 cm/s and the largest (16–17 cm SL) swam at 65–70 cm/s. While these speeds are considerably slower than typical absolute trawl speeds (125–160 cm/s), the speed inside the trawl in the vicinity of the bycatch reduction device is likely slower due to the reduction in flow caused by the trawl netting itself. Additionally, these speeds are those that can be maintained for one hour of continuous swimming. If a fish can maintain station for an hour in the vicinity of the bycatch reduction device then the chance of escape may be greatly increased. These results suggest that water flow speeds in and around the BRD escape openings should not be considerably greater than 35–40 cm/s. Results in this study are consistent with an *in situ* study where scuba divers measured the optimal flow ranges in association with BRD's to be between 20 and 50 cm/s (Watson et al. 1993). It is also important to recognize that critical swimming speeds may be a conservative estimate of the actual swimming ability of a fish. When a fish is removed from its natural environment, transported to the laboratory, held under artificial conditions, and then subjected to the rigors of swimming trials, it may be reasonably anticipated that there would be a negative effect on performance. This means that our critical swimming speeds may be an underestimate of the actual performance of the fish. However, if a management recommendation is made based on laboratory estimates of critical swimming speeds, any error would be in favor of the fish.

Response of Red Snapper to the VGBRD

In this portion of the study, we examined red snapper behavior in the VGBRD under various conditions. Our intent was to identify any red snapper behavioral characteristics that might result in the development of new, or the modification of existing, bycatch reduction devices. During daylight tests, red snapper were observed to locate the exit in a short period of time and leave the VGBRD in large numbers. It appeared that fish responded to the vortex that was generated at the escape opening and used the reversed flow to exit the device. However, another possibility was that the fish were drawn to the escape opening because they perceived the test environment as a threat and sought out the opening as a place of refuge. The test of the VGBRD in complete darkness at night provided very different results and may provide information regarding this "refuge hypothesis." 0Red snapper exited the VGBRD in much lower numbers at night, a disturbing result because the vast majority of shrimp trawling in the Gulf of Mexico takes place at night. However, the poor nighttime results suggested that the response to the change in flow may not be as important as the refuge response. Further support for this came when dark adapted red snapper were suddenly exposed to artificial illumination in the VGBRD and all fish were observed to exit quickly. In fact, the stimulus was so intense that some snapper were actually observed to exit through gaps in the netting that we did not think a fish could pass through. This negative phototactic response has been observed in various fish species (Guthrie and Muntz 1986).

The above results suggested that while flow quality and speed may be important, illumination during nighttime trawling may increase the efficiency of the VGBRD. However, the position as well as the intensity of the light source in the trawl may alter the effectiveness of the VGBRD. Lights improperly placed may discourage fish escape. For example, when lights were shone directly on the exit, fish were prevented from remaining in the reduced flow region near the escape channel and was counter-productive to escape. However, when lights were placed downstream of the VGBRD exit, fish were observed to stay ahead of the light source in the vicinity of the exit and snapper escape was improved. These results suggest that the best position for lights was downstream of the VGBRD exit. In this position, red snapper preferred to stay ahead of the lights and therefore spent more time in the vicinity of the VGBRD. Additionally, illumination allowed fish to see the escape opening and may have encouraged some fish to take refuge there.

We evaluated the effect of VGBRD front plate angle on snapper escape during both day and night. There was no effect of front plate angle on time to escape during day or night. We did, however, observe an effect of day versus night. Snapper did not escape as readily during nighttime tests, despite the fact that we used artificial lighting (LEDs) in the VGBRD. We never obtained the kind of results (100% escape) that was observed when room lights were used for illumination. These results suggest that the intensity of lighting may be an important factor in snapper escape.

Color/Contrast Preference

In the color/contrast preference experiments, red snapper consistently chose the darkest panels. We suggest that when stressed, such as during trawl capture, snapper will seek refuge and a dark panel against a light colored background may be perceived as an escape opening. The behavior of red snapper, i.e. swimming in head down fashion across the surface of the panel, suggests that the fish may perceive these dark panels as a place of refuge. It is not clear if a dark adapted red snapper, suddenly exposed to artificial lighting in a trawl, would exhibit the same type of behavioral responses that were observed in this experiment. However, further research on exploiting this "refuging response" in BRD development to facilitate red snapper escape is warranted.

Optomotor Response

Engaas et al. (1999), when testing the Jones/Davis BRD, observed that red snapper turned forward and swam in the direction of water flow when entering the trawl mouth and suggested that this was in response to the visual stimulus provided by the turtle excluder device (TED) grid. However, after passing through the TED the fish typically took up positions in low flow areas. Apparently, when fish were near the escape openings they did not exit because of the strong optomotor response to the moving net.

Wardle (1983, 1986) used underwater observations to describe the behavior of "roundfish" when encountering a trawl. They observed larger fish holding station within the trawl mouth for very long periods of time whereas smaller ones maintained station for short periods and then turned and swam toward the cod end. Some fish have been shown to match their swimming speed to the speed of the moving background, i.e. the trawl netting. This reaction to a moving background is termed the optomotor response and was described by Harden-Jones (1963). This response is so strong that Hemmings (1973) was able to induce haddock to maintain station in the mouth of a moving trawl even when the netting behind the mouth of the net had been completely removed.

In this study, we examined the optomotor response in red snapper in an attempt to exploit this response for bycatch reduction device design. Snapper demonstrate a strong optomotor response. In every case, snapper were observed to follow the moving pattern and were induced to swim from one side of the trawl to the other. These results suggest that a rotating drum may be useful for directing movement of snapper toward an escape opening. We have already incorporated the optomotor response into a BRD design that we hope to test soon.

Dark/Light Choice

When snapper were given a choice between dark and light areas in a simulated trawl situation, they consistently chose the dark area. This behavior was another manifestation of the negative phototactic response already described. This suggests that illumination of a down-stream area of the trawl may be used to encourage snapper to move to darkened areas in a trawl, even when fish are in a flow field. If we can move snapper to a particular area in a trawl, this could be a first step in facilitating their escape.

Acknowledgments

The authors would like to thank W. Bet-Sayad, B. Cage, J. Watson, and L. Saxon. This work was supported by grant #NA17FF2031 to GRP from the National Oceanic and Atmospheric Administration, MARFIN program.

References

Engaas, A., D. Foster, B. D. Hataway, J. Watson, and I. Workman. 1999. The behavioral responses of juvenile red snapper (*Lutjanus campechanus*) to shrimp trawls that utilize water flow modifications to induce escapement. Marine Technology Society Journal 33:43–50.

Goodyear, C. P. and P. Phares. 1990. Status of red snapper stocks of the Gulf of Mexico: Report of National Marine Fisheries Service/Southeast Fisheries Science Center, Miami Laboratory Contribution CDR 89/90–05, Miami, Florida.

Gutherz, E. J., and G. J. Pellegrin. 1988. Estimate of the catch of red snapper, *Lutjanus campechanus*, by shrimp trawlers in the U.S. Gulf of Mexico. Marine Fisheries Review 50:17–25.

Guthrie, D. M. and W. R. A. Muntz. 1986. The optical propertis of natural waters and their effects on underwater visibility. Pages 89–128 *in* T. J. Pitcher, editor. Behaviour in teleost fishes. Chapman and Hall, London.

Harden-Jones, F. R. 1963. The reaction of fish to moving backgrounds. Journal of Experimental Biology 40:437–446.s

Hemmings, C. C. 1973. Direct observation of the behaviour of fish in relation to fishing gear. Helgolander Wissenschaftliche Meeresuntersuchungen 24:348–360.

Nichols, S., A. Shah, G. J. Pellegrin Jr., and K. Mullin. 1990. Updated estimates of shrimp fleet bycatch in the offshore waters of the U.S. Gulf of Mexico. Report to the Gulf of Mexico Fishery Management Council. National Marine Fisheries Service, Southeast Fisheries Science Center, Mississippi Laboratories, Pascagoula, Mississippi.

Southeast Data, Assessment, and Review. 2005. Stock assessment report of SEDAR 7, Gulf of Mexico Red Snapper, Assessment Report 1. SEDAR, Charleston, South Carolina.

Wardle, C. S. 1983. Fish reaction to towed fishing gear. Pages 167–196 *in* A. G. McDonald and I. G. Priede, editors. Experimental biology at sea. Academic Press, London.

Wardle, C. S. 1986. Fish behaviour and fishing gear. Pages 609–643 *in* T. J. Pitcher, editor. Behaviour of teleost fishes. Chapman and Hall, London.

Warren, B. 1994. Win-Win bycatch solutions. National Fisheries Conservation Center, Seattle, Washington.

Watson, J., I. Workman, D. Foster, C. Taylor, A. Shah, J. Barbour, and D. Hataway. 1993. Status report on the potential of gear modifications to reduce finfish bycatch in shrimp trawls in the southeastern United States. NOAA Technical Memorandum NMFS-SEFSC-327.

Workman, I. and D. Foster. 2002. The Webbing Reef: A tool used in the study of juvenile red snapper *Lutjanus campechanus*. OCEANS 2002 IEEE/MTS Proceedings, 146–150.

American Fisheries Society Symposium 60:77–104, 2007

Seasonal and Size Specific Diet and Prey Demand of Red Snapper on Alabama Artificial Reefs

JESSICA R. MCCAWLEY[1]
Florida Fish and Wildlife Conservation Commission
Division of Marine Fisheries Management
620 South Meridian Street, Tallahassee, Florida 32399 USA

JAMES H. COWAN, JR.
Coastal Fisheries Institute and
Department of Oceanography and Coastal Sciences
School of the Coast and Environment
Louisiana State University
Baton Rouge, Louisiana 70803 USA

Abstract.—Red snapper, *Lutjanus campechanus*, is a heavily exploited reef fish that is harvested in the Gulf of Mexico from both natural and artificial reefs. Since regulations were imposed in 1990, the stock has begun to recover; this recovery has also been attributed in part to an increase in artificial habitats/reefs. However, little is known about the role artificial reefs play in the trophic dynamics of the species. To this end, the seasonal and size-specific diet of red snapper was examined through stomach content analysis of individuals collected from artificial reefs in the north-central Gulf of Mexico off Alabama between May 1999 and April 2000. Diet information was subsequently combined with data from the literature to obtain a first-order estimate of prey demand of the red snapper population on artificial reefs off Alabama. Results indicate both that diet varied with season and red snapper size and that the overall diet was comprised primarily of demersal crustaceans, fish, and pelagic zooplankton. Annual prey demand of the red snapper population on Alabama artificial reefs was estimated to be over 31 million kg. Red snapper derived most of their nutrition from sand/mud- and water column-associated organisms, not from organisms associated with reefs. Based on these results, and the results from other red snapper studies off Alabama, Alabama artificial reefs may be attracting, not producing red snapper.

Introduction

Red snapper *Lutjanus campechanus* (Poey, 1860) supports the most important recreational and commercial offshore finfish fishery in the northern Gulf of Mexico (Fischer et al. 2004). Knowledge of the role artificial habitats play in the life history of this species is crucial to making informed management decisions about this fishery. Since 1990, regulations imposed by National Marine Fisheries Service, including size and bag limits and total allowable catches, have

[1] Corresponding author: jessica.mccawley@myfwc.com.

helped the overexploited stock begin to recover (Schirripa and Legault 1997; Patterson 1999); despite these actions, Gulf of Mexico red snapper continue to be overfished and are undergoing overfishing (Goodyear 1995; Schirripa and Legault 1999; SEDAR 2005). The recovery of the stock has also been attributed in part to an increase in artificial habitats in the form of oil and gas platforms and artificial reefs (Bailey et al. 2001). However, some scientists question whether these structures are a positive influence because of doubts about whether they produce or attract fish (see *Fisheries* Vol. 22, April 1997).

The Alabama shelf, one of the centers of red snapper abundance in the northern Gulf of Mexico (Goodyear 1995; Schirripa and Legault 1997; Patterson 1999; Patterson et al. 2001), has over 4000 km^2 of artificial reef permit area (Shipp 1999), where anywhere from 8,000 (Minton and Heath 1998) to 20,000 (Patterson 1999; Bailey et al. 2001) artificial reefs have been deployed to enhance fishing. According to Strelcheck (2001), red snapper are the most abundant finfish on artificial reefs in the Hugh Swingle permit area off Alabama. By examining factors that might be enhanced by artificial reefs, such as food availability, the production versus attraction debate can be addressed (Vose et al. 1997). The type and amount of food necessary to support the most abundant species on the reefs must first be quantified (Meier and Steimle 1997), after which prey supply versus prey demand can be addressed empirically. Estimating the diet composition and prey demand of the red snapper population is a first step toward the resolution of the production versus attraction debate off Alabama.

Little is known about the foraging habits of reef fishes on temperate artificial reefs (Bohnsack and Sutherland 1985; Howe 2001; Appendix 1 for review), especially off Alabama. Of diet studies on red snapper in Alabama waters, two had small sample sizes (Siegel 1983; Bailey 1995), one sampled primarily juveniles (Szedlmayer and Lee 2004), another focused on red snapper from deep, natural reefs (Pinnacles Reef Tract) (Weaver et al. 2001), and the other looked at diel feeding periodicity (Ouzts and Szedlmayer 2003). Although Siegel (1983) collected seasonal data, he found no significant seasonal trends, possibly due to small sample size.

Some researchers suggest that reef associated fishes such as red snapper might not be feeding at the reef, but rather on mud/sand-associated organisms that surround the reef (Bohnsack 1989; Bohnsack et al. 1991; Sedberry and Cuellar 1993; Nelson and Bortone 1996; Bohnsack et al. 1997b). This behavior may create an important energetic link between artificial reefs and the surrounding habitat (Parrish 1989) implying that the fish biomass on artificial reefs is dependent upon trophic subsidies from the surrounding environment. The importance of the reef itself, versus the water column or adjacent sediments as feeding grounds, is still poorly understood (Sedberry and Cuellar 1993) and may vary by location and by species. If reef fishes are not feeding on reef-associated organisms, then they may only be attracted to artificial reefs as a result of a behavioral preference (Bohnsack 1989). If true, this may reveal important management implications concerning the relationship between artificial reefs and reef fishes, as well as for the use of artificial reefs to enhance fishing.

Additional information about red snapper diet on Alabama artificial reefs can provide insight into the role that artificial reefs play in red snapper feeding ecology. Off Alabama a quantitative estimate of the prey demand of the red snapper population living on artificial reefs is lacking. Moreover, the role of artificial reefs in the life history of the species is inconclusive. To this end, we examined the affects of artificial reefs on adult red snapper trophic dynamics, focusing on seasonal and size-specific changes in their diet over a year-long period. Diet data then were used, along with data on abundance and size-distributions from Strelcheck et al. (2005) and Szedlmayer and Furman (2000), and an empirically-derived estimate of time-specific consumption rate (Palomares and Pauly 1989) to obtain first-order estimates of the annual and seasonal prey demand of the red snapper population on Alabama artificial reefs.

Methods

Monthly sampling.—Red snapper were collected with hook and line from artificial reefs in the northern Gulf of Mexico (Gulf) off the coast of Alabama between May 1999 and April 2000. Most fish were caught by recreational fishermen in the Hugh Swingle General Permit Area (see map of study area in Strelcheck et al. 2007, this volume). However, some larger red snapper were opportunistically collected from local spearfishing and angling fishing tournaments. Knowing that red snapper are prone to regurgitation (Adams and Kendall 1891; Camber 1955; Moseley 1966; Bradley and Bryan 1975; Parrish 1987), we collected 39–86 fish per month to ensure that we obtained a significant number of fish with prey in their stomachs.

At capture, all red snapper were weighed to the nearest 0.01 kg, their total length (TL) and fork length (FL) measured to the nearest mm, and their sex determined before the stomach was removed. Stomachs were severed at the esophagus and duodenum below the pyloric sphincter, slit to allow complete preservation, and then preserved in 10% formalin for at least 48 h. They then were transferred to 70% isopropyl alcohol until they could be sorted. Stomachs were dissected, prey items removed and identified to the lowest possible taxonomic level, and each taxon weighed to the nearest 0.01 g after being blotted dry. Empty stomachs were either labeled as 'genuinely empty' or 'distended' according to the description of Treasurer (1988).

Enumeration of Stomach Contents.—The relative contribution of each of several prey categories was determined by using four methods: (1) percent composition by weight (%W); (2) percent composition by number (%N); (3) percent frequency of occurrence (%FO); and, (4) percent index of relative importance (%IRI). Percent frequency of occurrence was calculated as in Bowen (1996): %FO = number of stomachs containing one particular prey category/number of stomachs with any prey (excluding bait). The index of relative importance was calculated as (Pinkas et al. 1971; modified by Hacunda 1981): IRI = (%N +%W) × %FO. Percent IRI was calculated by dividing the IRI value for each prey category by the sum of the IRI values and multiplying by 100. These descriptive indices were used to describe the overall diet, as well as to evaluate the diet by size-class of red snapper (size classes = 200–299 mm FL, 300–399, 400–499, 500–599, and >600) and on a seasonal basis (summer = June, July, and August; fall = September, October, November; winter = December, January, February; and spring = March, April, and May). However, %W was the primary index used to describe the diet, as Bowen (1996) suggested it to be the best descriptive index if the ultimate goal is to measure the contribution of the prey to the predator's nutrition (Bowen 1996).

The identifiable contents of all red snapper stomachs combined were divided into seven major prey categories (Table 1): fish; adult mantis shrimp, *Squilla empusa*; crabs; penaeid shrimp; squid, *Loligo* sp.; pelagic zooplankton; and miscellaneous benthic-associated species (hereafter, miscellaneous benthic species). Larval fish were not included in the pelagic zooplankton category; they were grouped with fish because it was often difficult to determine if they had flexed. In addition, not all demersal species were grouped in the miscellaneous benthic species category. This category contained only those taxa that did not fall within one of the other major categories. The diet also consisted of an unidentified material category, which was defined as that having no recog-

Table 1. Prey contained in all red snapper *Lutjanus campechanus* stomachs and the 300-499 mm FL subset of stomachs collected on Alabama artificial reefs based upon four descriptive indices for seven prey categories ranked in decreasing order of importance for each index. %W = percent weight, %N = percent number, %FO = percent frequency of occurrence, %IRI = percent index of relative importance, and Misc. benthic sp. = Miscellaneous benthic species.

Category	Prey Type	%W[a] (rank)	%W[b] (rank)	%N (rank)	%FO (rank)	%IRI (rank)
All stomachs	Unidentified material	35.91 (1)				
	Fish	19.48 (3)	28.70 (1)	25.88 (2)	38.81 (2)	31.14 (2)
	Adult *Squilla empusa*	12.59 (4)	16.08 (4)	14.02 (4)	21.27 (4)	9.41 (4)
	Crab	20.25 (2)	26.79 (2)	24.35 (3)	35.82 (3)	26.93 (3)
	Shrimp	1.70 (6)	2.19 (6)	2.04 (6)	3.73 (5.5)	0.23 (6)
	Loligo sp.	0.43 (8)	0.54 (7)	0.31 (7)	1.49 (7)	0.02 (7)
	Pelagic zooplankton	7.97 (5)	23.51 (3)	31.02 (1)	39.93 (1)	32.01 (1)
	Misc. benthic sp.	1.67 (7)	2.20 (5)	2.38 (5)	3.73 (5.5)	0.25 (5)
300-499	Unidentified material	37.82 (1)				
	Fish	18.37 (2)	27.33 (2)	24.78 (2)	36.63 (2)	28.54 (2)
	Adult *S. empusa*	13.78 (4)	17.69 (4)	15.95 (4)	21.51 (4)	10.82 (4)
	Crab	15.43 (3)	20.80 (3)	18.37 (3)	27.33 (3)	16.01 (3)
	Shrimp	2.49 (6)	3.20 (5)	2.88 (5)	4.65 (5)	0.42 (5)
	Loligo sp.	0.43 (8)	0.52 (7)	0.32 (7)	1.16 (7)	0.01 (7)
	Pelagic zooplankton	10.16 (5)	28.42 (1)	35.60 (1)	45.93 (1)	43.97 (1)
	Misc. benthic sp.	1.52 (7)	2.06 (6)	2.10 (6)	3.49 (6)	0.22 (6)

[a] Represents the % weight including the unidentified material category
[b] Represents the % weight excluding the unidentified material category

nizable bones or hard parts, thus preventing classification into any of the categories listed above. Unidentified material was not included in all of the analyses because %N, %FO, and IRI cannot be determined for this category.

Diet studies can introduce bias depending on how prey items identified with different taxonomic resolution are grouped together (Hansson 1998). To eliminate bias and to provide more detailed diet information, four of the seven major prey categories (fish, pelagic zooplankton, crabs, and miscellaneous benthic species) were further subdivided and combined with the other three prey types that consisted of a single species or genus, e.g., adult *S. empusa*, penaeid shrimp, and *Loligo* sp., for a total of 47 groups (see Table 2 for a complete listing). This more detailed breakdown of prey was also examined by season and size-class using the same descriptive indices listed above.

To further analyze the diet data, the PRIMER statistical package (Clarke and Warwick 1994) was used. Because this study possessed groups defined *a priori*, such as month, season, and size-class, the nonparametric permutation procedure ANOSIM (Analysis of Similarities, PRIMER) (Clarke and Warwick 1994) was used to test for significant differences among seasons and size classes. Prey categories that most contributed to the observed differences among season and size-class were elucidated with BVSTEP. SIMPER (Similarity Percentages, PRIMER), a multivariate multiple permutations test, was used to examine the contribution that a prey type made to the average within-group (season or size-class) similarity and between-group dissimilarity (Clarke 1993).

Caloric Density.—The caloric density (calories/g) of the major diet items either was estimated directly with bomb calorimetry (McCawley 2003) or was borrowed from the literature. An index of caloric importance (ICI) was calculated for each prey type for the overall diet as well as by season and by size-class with the formula: ICI = (%W × C) × %FO, where C = calories/g wet weight. Percent ICI (% ICI) was calculated with the formula: %ICI = (ICI for each prey category/ sum of the ICI values) × 100. A prey importance index (PI) also was calculated for each prey category for the overall diet as well as by season and by size-class according to the equation from Pope et al. (2001):

$$PI_i = \frac{1}{P}\sum_{j=1}^{P}\left(\frac{W_{ij}X_i}{\sum_{i=1}^{Q} W_{ij}X_i}\right)$$

where i = prey type;
  j = fish with prey (here red snapper stomachs);
  P = number of fish with food in their stomachs;
  W_i = weight (g) of food category i;
  X_i = caloric value (cal/g wet weight) of food category i; and,
  Q = number of food categories.

Percent PI (% PI) was calculated by multiplying each PI value by 100. The %ICI and %PI indices were compared to %W and %IRI to determine which index best described the diet.

Prey Habitat Preference.—A habitat preference was specified from the literature for each of the 47 red snapper prey categories. Five major habitat types were identified: sargassum-associated (*SA*); sand/mud-associated (*SM*); reef- or structure-associated (*R*); water column associated (*WC*); and, those prey found on a variety of habitats (*V*). A *SA* organism was defined as that which lives among floating sargassum. Sand/mud-associated organisms were defined as those organisms that live on the sand or mud bottom, as well as those that spend most of their time burrowed in the mud (such as a shrimp eel, Ophichthidae or mantis shrimp). A *R* organism (e.g., sea horses family Syngnathidae) was liberally defined as an organism that would not otherwise be found in a particular habitat unless a reef (artificial or natural) or some type of structure was present. Water column organisms were either mostly planktonic organisms or those swimming in the water column, such as *Loligo* sp. An organism that was not characteristic of any one habitat type was classified as being found on a variety of habitats. These habitat types were paired with each prey's %W contribution to the diet and then summed by habitat type to determine the cumulative contribution made to the diet by prey from each habitat. The cumulative habitat contribution was examined for the overall diet as well as by season and by size-class. The average caloric density for prey from each habitat type also was determined.

Annual Prey Demand.—To obtain an estimate of population consumption for red snapper on Alabama artificial reefs, an estimate of Q/B = 1.44% per day for *Lutjanus campechanus* was taken from Palomares and Pauly (1989). Q represents the amount of food consumed, B represents biomass, and thus Q/B is a time-specific ratio of the food consumed to the weight of the consumer. To estimate an annual weight-specific prey demand for red snapper, Q/B was multiplied by the number of days in a year (365) and then multiplied by the mean (±SE) biomass of red snapper on 14 experimental artificial reefs in the Hugh Swingle Permit Area determined by Strelcheck et al. (2005) from catch-per-unit-effort (CPUE) data. An annual prey demand estimate also was obtained in a similar manner with biomass data collected via visual census in Strelcheck et al. (2005). However, in this prey demand estimate, the mean (±SE)

Table 2. Most detailed taxonomic breakdown of stomach contents (47 prey categories) for all red snapper *Lutjanus campechanus* collected on Alabama artificial reefs based upon four descriptive indices. Prey categories are ranked for each index in decreasing order of importance. %W = percent weight, %N = percent number, %FO = percent frequency of occurrence, and %IRI = percent index of relative importance and Misc. benthic species = Miscellaneous benthic species.

Prey Type	%W[a] (rank)	%W[b] (rank)	%N (rank)	%FO (rank)	%IRI (rank)
Unidentified material	35.91 (1)				
Fish					
Unidentified fish	9.97 (3)	13.40 (2)	11.03 (3)	17.16 (4)	15.35 (3)
Family Ophichthidae	3.21 (7)	4.09 (8)	3.94 (8)	6.34 (8)	1.86 (8)
Family Triglidae	1.75 (9)	2.35 (10)	2.06 (10)	2.61 (12)	0.42 (11)
Family Haemulidae	0.35	0.46	0.56	0.75	0.03
Family Syngnathidae	0.07	0.12	0.15	0.75	0.01
Ophidion sp.	0.50	0.62	0.47	0.75	0.03
Decapterus sp.	0.25	0.30	0.30	0.37	0.01
Lagodon rhomboides	0.27	0.37	0.37	0.37	0.01
Anchoa hepsetus	0.19	0.23	0.07	0.37	<0.01
Fish larvae	2.93 (8)	6.75 (6)	6.92 (7)	11.94 (6)	5.97 (7)
Crabs					
Unidentified crabs	8.09 (4)	11.08 (4)	9.84 (5)	17.54 (3)	13.43 (4)
Family Portunidae	0.49	0.64	0.91	2.24	0.13
Portunus gibbesii	5.77 (5)	7.41 (5)	7.78 (6)	11.19 (7)	6.22 (6)
P. sayi	1.08	1.34	1.41	2.99 (11)	0.30 (12)
P. spinimantus	0.45	0.55	0.53	1.12	0.04
P. spinicarpus	0.13	0.16	0.29	0.37	0.01
P. ordiwayi	0.11	0.14	0.09	0.37	<0.01
P. sebae	<0.01	<0.01	0.05	0.37	<0.01
Callinectes sp.	0.25	0.31	0.19	0.37	0.01
C. sapidus	0.89	1.11	0.70	1.87	0.12
C. exasperatus	0.21	0.26	0.19	0.37	0.01
C. danae	0.01	0.01	0.05	0.37	<0.01
Ovalipes floridanus	1.20	1.65 (12)	0.77	2.61	0.23
Calappa flammea	0.60	0.75	0.50	0.75	0.03
C. agusta	0.28	0.35	0.19	0.37	0.01
Hepatus epheliticus	0.17	0.31	0.37	0.75	0.02
Parthenope granulata	0.52	0.73	0.50	0.75	0.03
Pelagic zooplankton					
Larval *Squilla empusa*	4.52 (6)	11.78 (3)	13.94 (2)	21.64 (1)	20.37 (2)
Unidentified mollusk larvae	0.84	1.44	1.20	1.87	0.18
Crab megalopa and zoea	0.31	0.50	0.77	1.87	0.09
Order Amphipoda	0.91	5.76 (7)	10.25 (4)	16.79 (5)	9.84 (5)
Order Euphausicaea	<0.01	0.75	0.75	0.75	0.04
Order Isopoda	0.01	0.38	0.50	0.75	0.02
Order Mysidacea	<0.01	0.03	0.20	0.75	0.01
Order Calanoida	<0.01	0.01	0.01	0.37	<0.01
Family Sergestidae	0.03	0.13	0.30	1.49	0.02
Family Palaemonidae	<0.01	0.05	0.06	0.37	<0.01
Cavolinia sp.	1.35	2.62 (9)	2.98 (9)	4.10 (9)	0.84 (9)
Sagita sp.	<0.01	0.06	0.06	0.75	<0.01

Table 2. (Continued)

Prey Type	%W[a] (rank)	%W[b] (rank)	%N (rank)	%FO (rank)	%IRI (rank)
Adult *Squilla empusa*	12.59 (2)	16.08 (1)	14.02 (1)	21.27 (2)	23.43 (1)
Penaeid shrimp	1.70 (10)	2.19 (11)	2.04 (11)	3.73 (10)	0.58 (10)
Squid	0.43	0.54	0.31	1.49	0.05
Miscellaneous Benthic species					
Phylum Arthropoda	<0.01	0.08	0.09	0.37	<0.01
Pagurus sp.	0.30	0.37	0.37	0.37	0.01
Sicyonia sp.	0.10	0.13	0.28	0.75	0.01
Glycera sp.	<0.01	0.05	0.01	0.37	<0.01
Albunea paretii	1.27	1.57	1.62 (12)	1.87	0.22

[a] Represents the % weight including the unidentified material category
[b] Represents the % weight excluding the unidentified material category

biomass of all reef fish was multiplied by the mean percent abundance of red snapper on all reefs (to determine the % biomass of red snapper) before being multiplied by the Q/B ratio. For visual census data from Strelcheck et al. (2005), mean biomass was calculated only from reefs where visibility was greater than or equal to 4 m. The resulting weight (kg) of food needed to support the snapper population on one artificial reef then was used to determine the amount of food needed to support a population of red snapper on 14; 8,000; and 20,000 artificial reefs. The latter two numbers were taken from Minton and Heath (1998) and Patterson (1999) and were used to bound the estimate of prey demand for the entire snapper population on artificial reefs off Alabama. This extrapolation was done assuming that red snapper biomass on all artificial reefs is similar to that found at the experimental reefs in Strelcheck et al. (2005). Estimates of consumption based upon CPUE and visual census data were further partitioned by percent weight (%W) among the seven major prey categories found in the diet.

To provide additional insight, annual prey demand also was estimated with visual census data from Szedlmayer and Furman (2000), who observed a mean size of red snapper of 279 ± 59 mm standard length (SL) and a mean abundance (±SD) of 86.3 ± 69.4 on 28 artificial reefs off Alabama. A distribution of SL of red snapper ($N = 5000$) was simulated with the Monte-Carlo method (Sokal and Rohlf 1981). A normal probability function was constructed based on mean length and standard deviation specified by Szedlmayer and Furman (2000). To estimate SL for an individual fish, a probability value ranging from 0.0001 to 0.9999 was randomly drawn (with replacement) and the corresponding SL was assigned to the fish; SL then was converted to fork length (FL) with the regression equation $FL = 1.669 \times SL + 5.911$ taken from measurements ($n = 50$) from red snapper collected off Alabama. The assigned FLs then were used to obtain an estimated weight for each red snapper with a length-weight regression: Log weight (kg) = 3.014(*log* FL) – 4.7799 obtained from all snapper collected in this study. A mean weight (±SD) then was determined from the estimated weights. Biomass of red snapper at a reef was calculated by multiplying the mean abundance of red snapper observed in Szedlmayer and Furman (2000) by the mean weight (±SD) of red snapper estimated here from their data. Annual prey demand of red snapper was determined from these data by multiplying annual Q/B estimates by the estimated biomass of red snapper on the reefs they observed. Assuming that red snapper biomass on all artificial reefs was similar to the reefs they studied, the prey demand of a red snapper population on one reef was used to determine the amount of food needed to support a red snapper population on 14; 8,000; and 20,000 artificial reefs and then partitioned into the seven major prey categories as described above. These annual prey demand estimates based on data from Szedlmayer and Furman (2000) were compared to annual estimates based on CPUE and visual census data from Strelcheck et al. (2005).

Results

Enumeration of Overall Diet

Stomach contents of 656 red snapper ranging from 207 to 913 mm FL were examined. Of these, 268 (40.9%) stomachs contained identifiable prey, 262 (39.9%) were empty, 63 (9.6%) contained only bait, and 63 (9.6%) contained only unidentified material. Of the 262 stomachs classified as empty, 169 (64.5%) were considered 'truly' empty and 93 (35.5%) were considered distended or empty due to regurgitation. The empty and bait only stomachs were excluded from further analyses. All red snapper containing prey were staged as adults (Jackson et al. 2007, this volume) and ranged in length from 240 to 913 mm FL with a mean of 463 mm, a median of 426 mm, and a mode of 410 mm FL.

Seven prey categories.—The unidentified material category contributed the largest proportion to red snapper diet by %W (35.9%), followed by crab (20.2%), fish (19.5%), adult *S. empusa* (12.6%) and pelagic zooplankton (8.0%) (Table 1). After exclusion of the unidentified material category (listed by descending %W), fish, pelagic zooplankton, crab, and adult *S. empusa* were the principal components of red snapper diet when all stomachs were combined (Table 1). However, no single group was largest by all indices. Pelagic zooplankton was the largest category by %N, %FO, and %IRI, whereas fish was the largest category by %W.

Forty-seven prey categories.—When the stomach contents for all the nonempty red snapper collected were divided into their highest taxonomic resolution (47 prey categories) unidentified material was again the largest category (Table 2). After excluding unidentified material, the ten most abundant taxa in the diet (contributing over 81%, listed by descending %W) were adult mantis shrimp *S. empusa*, unidentified fish, larval *S. empusa*, unidentified crabs, iridescent swimming crab *Portunus gibbesii*, larval fish, amphipods, family Ophichthidae, *Cavolinia* sp., and family Triglidae. Adult *S. empusa* was the largest category in all indices except %FO, wherein larval *S. empusa* was the largest contributor. In general, smaller prey (e.g., amphipods and larval *S. empusa*) were more important by %N than by %W. The ranking of prey by %FO and %IRI found amphipods, penaeid shrimp, and larval *S. empusa* to be more important than in %W. The largest categories by %W revealed that demersal crustaceans were important contributors to the diet in the form of adult *S. empusa*, unidentified crabs, and *P. gibbesii*. Demersal fishes (family Ophichthidae and family Triglidae), as well as unidentified fishes and larval fish, were also important contributors to the overall diet.

Enumeration of Seasonal Diet

All nonempty red snapper collected were divided into 5 size classes; however, not every size-class of red snapper was collected in every month. Before testing for seasonal differences in red snapper diet, we chose to include only those size classes that were collected in all seasons. As such, red snapper in the 300–399 and 400–499 mm FL size classes (N = 452) were collected in every season. Fish within this size range are indicative of the predominant size classes of red snapper inhabiting several experimental reefs in the Hugh Swingle reef permit area (Strelcheck 2001) and represent the dominant sizes of recreationally harvested red snapper off Alabama and Louisiana (375–425 mm FL) (Fischer et al. 2004). Thus, these two size classes of red snapper were combined (300–499 mm) for the statistical evaluation of red snapper diet by season.

The descriptive indices for the 300–499 mm size-group were similar to results pooled over all sizes (Table 1). For this subset, the diet as a whole was composed primarily of pelagic zooplankton by all four indices. The next most important diet items were fish, then crabs, followed closely by adult *S. empusa*. Compared to all stomachs combined, pelagic zooplankton and adult *S. empusa* made up a larger portion of this subset diet, while fish and crab made up smaller portions. The contributions made by penaeid shrimp, *Loligo sp.*, and miscellaneous benthic species were largely unchanged.

Seven prey categories.—Gut content examination by %W by season for the 300–499 mm subset indicated that unidentified material was the largest contributor to the diet in every season (Table 3). After excluding the unidentified material from further analysis, examination of seasonal diet by %W (Table 3) revealed fish were present in red snapper diets in all seasons (comprising between 24.5 and 31.8% W), but they did not contribute the greatest amount by %W in any season. The diet in summer and winter was composed predominately of adult *S. empusa* (33.0% W and 34.4% W respectively), in fall of crabs (35.1% W), and spring of pelagic zooplankton (46.5% W).

Demersal crustaceans (crabs and adult *S. empusa*) were present in all seasons, comprising between 41.9% and 54.4% by %W of the diet in summer, fall, and winter. During the spring, red snapper fed on high numbers of pelagic zooplankton. However, the amount of pelagic zooplankton consumed appeared to be inversely related to the amount of demersal crustaceans eaten by red snapper, a pattern that was especially evidenced by the shift in spring from consuming primarily demersal crustaceans to feeding on pelagic zooplankton. When the seasonal diet was examined by %N, %FO, and %IRI (Table 3), the same trends were present; however pelagic zooplankton comprised a larger portion of the diet in all seasons.

Percent weight data by season for the 300–499 mm red snapper for seven prey categories were included in the test for significance. ANOSIM found a highly significant difference among the %W data by season (p = 0.001) despite a low (0.089) R-value. Even though ANOSIM found an overall significant difference among seasons, low R-values (with significant P-values) from the ANOSIM pairwise comparisons between seasons indicate significance, but high overlap, revealing that red snapper were feeding on nearly the same kinds of organisms year-round.

BVSTEP results revealed that differences among season were attributable to four influential prey types: fish, crab, adult *S. empusa*, and pelagic zooplankton. There was a 99.5% correlation between these prey types and the overall pattern seen in the samples. The SIMPER results revealed that red snapper collected in winter had the largest number of prey types contributing to within-season similarity, with all four influential prey types contributing. Spring fish had the fewest prey types with only two prey types contributing. Demersal crustaceans were the largest contributors (45.0–48.0%) to every season's within-season similarity except

Table 3. Taxonomic breakdown of stomach contents from 300-499 mm FL red snapper *Lutjanus campechanus* collected on Alabama artificial reefs by season based upon four descriptive indices for seven prey categories ranked in decreasing order of importance for each index. %W = percent weight, %N = percent number, %FO = percent frequency of occurrence, and %IRI = percent index of relative importance, and Misc. benthic sp. = Miscellaneous benthic species.

Season	Prey Type	%W[a] (rank)	%W[b] (rank)	%N (rank)	%FO (rank)	%IRI (rank)
Summer	Unidentified material	37.84 (1)				
	Fish	15.47 (3)	24.48 (2)	21.39 (3)	38.10 (3)	24.20 (3)
	Adult *Squilla empusa*	21.85 (2)	32.96 (1)	26.55 (2)	40.48 (2)	33.36 (1)
	Crab	9.98 (4)	14.46 (4)	15.67 (4)	23.81 (4)	9.94 (4)
	Shrimp	3.59 (6)	5.64 (5)	4.64 (5)	9.52 (5)	1.36 (5)
	Loligo sp.	0.00 (8)	0.00 (7)	0.00 (7)	0.00 (7)	0.00 (7)
	Pelagic zooplankton	9.60 (5)	20.09 (3)	29.37 (1)	45.24 (1)	30.99 (2)
	Misc. benthic sp.	1.67 (7)	2.38 (6)	2.38 (6)	2.38 (6)	0.16 (6)
Fall	Unidentified material	32.05 (1)				
	Fish	16.11 (3)	24.84 (3)	23.37 (3)	29.55 (3)	23.37 (3)
	Adult *S. empusa*	5.66 (5)	6.82 (4)	7.95 (4)	9.09 (4.5)	2.20 (4)
	Crab	27.18 (2)	35.09 (1)	31.06 (1)	36.36 (1.5)	39.46 (1)
	Shrimp	1.89 (7)	2.27 (6)	2.27 (6)	2.27 (6)	0.17 (6)
	Loligo sp.	0.00 (8)	0.00 (7)	0.00 (7)	0.00 (7)	0.00 (7)
	Pelagic zooplankton	12.94 (4)	25.68 (2)	29.97 (2)	36.36 (1.5)	33.20 (2)
	Misc. benthic sp.	4.17 (6)	5.30 (5)	5.37 (5)	9.09 (4.5)	1.59 (5)
Winter	Unidentified material	27.41 (1)				
	Fish	18.80 (3)	27.55 (2)	26.80 (2)	35.14 (2)	29.74 (2)
	Adult *S. empusa*	29.62 (2)	34.43 (1)	34.23 (1)	40.54 (1)	43.35 (1)
	Crab	16.69 (4)	19.99 (3)	19.37 (3)	27.03 (3)	16.57 (3)
	Shrimp	0.31 (6)	0.36 (6)	1.35 (5)	2.70 (5.5)	0.07 (5)
	Loligo sp.	0.00 (8)	0.00 (7)	0.00 (7)	0.00 (7)	0.00 (7)
	Pelagic zooplankton	7.15 (5)	17.11 (4)	17.57 (4)	18.92 (4)	10.22 (4)
	Misc. benthic sp.	0.02 (7)	0.56 (5)	0.68 (6)	2.70 (5.5)	0.05 (6)
Spring	Unidentified material	51.26 (1)				
	Fish	23.28 (2)	31.84 (2)	27.43 (2)	42.86 (2)	22.87 (2)
	Adult *S. empusa*	0.67 (7)	1.71 (6)	0.26 (6)	2.04 (6)	0.04 (6)
	Crab	9.19 (4)	14.02 (3)	8.52 (3)	22.45 (3)	4.55 (3)
	Shrimp	3.57 (5)	4.08 (4)	3.06 (4)	4.08 (4.5)	0.26 (4)
	Loligo sp.	1.58 (6)	1.81 (5)	1.12 (5)	4.08 (4.5)	0.11 (5)
	Pelagic zooplankton	10.45 (3)	46.54 (1)	56.92 (1)	75.51 (1)	72.17 (1)
	Misc. benthic sp.	0.00 (8)	0.00 (7)	0.00 (7)	0.00 (7)	0.00 (7)

[a] Represents the % weight including the unidentified material category
[b] Represents the % weight excluding the unidentified material category

spring when pelagic zooplankton was the largest contributor. Fish also was characteristic of every season and was the second largest contributor (27.6–30.2% contribution) in all seasons except fall. However, low values of average similarly (<34.1) for each season indicate that within-season diet composition was variable and not dominated by the influence of a single prey type.

Forty-seven prey categories.—Unidentified material was the largest component of the diet in all seasons contributing between 27.4 and 51.3% W to the forty-seven prey category breakdown. After exclusion of this category, adult *S. empusa* were the largest contributors to the diet in summer and winter by %W. *P. gibbesii* was the largest contributor in the fall and amphipods were the largest contributor in the spring. On the whole, the largest contributors by %W were larger organisms, such as unidentified fish, unidentified crabs, and fishes from the families Triglidae and Ophichthidae. In contrast, smaller organisms, were more important by %N, %FO, and %IRI, especially in spring and fall. Overall, red snapper diets in summer and winter were comprised mainly of demersal crustaceans and spring and fall had more pelagic zooplankton and larval fish.

When evaluating the 300–499 mm red snapper diet by 47 prey categories, using the %W data, ANOSIM again found that red snapper diet varied significantly by season (p = 0.001) despite high diet overlap (R = 0.106). BVSTEP results revealed that these differences were attributable to a combination of six prey categories having a 95.1% correlation with the overall pattern of the samples. The six prey categories consisted of unidentified fish, larval fish, unidentified crab, adult *S. empusa*, larval *S. empusa*, and amphipods. SIMPER results revealed that the prey categories that contributed to within-season similarity were generally some of the five largest categories by %W. Overall, average within-season similarity values decreased when compared to the SIMPER analysis using seven prey categories. Moreover, SIMPER again showed demersal crustaceans contributed to every season's within-season similarity, making the smallest contribution in spring and the largest contribution in summer. Likewise pelagic zooplankton, made a contribution to within-season similarity, and the largest contribution in spring.

Enumeration of Size-Specific Diet

Seven prey categories.—All nonempty stomachs were pooled for examination of size-class differences in red snapper diet. Red snapper were divided into five size classes: 200–299 mm, 300–399 mm, 400–499 mm, 500–599 mm, and over 600 mm. However, because all size classes were not collected during all seasons, our data do not permit either a direct statistical comparison of diet by size-class or a size by season interaction, thus only descriptive results are given.

By %W, unidentified material was the largest diet component of every size-class except for red snapper over 600 mm, where crab was the largest category. However, after excluding unidentified material, fish were present in the diets of all size classes and were the largest contributor by %W to the diets of 200–299 mm and 500–599 mm red snapper (Figure 1). The diet of 300–399 mm fish was almost equally proportioned among the seven prey categories, however crab was the largest contributor. Crab also dominated the diet of >600 mm red snapper. The diet of the 400–499 mm red snapper was dominated by pelagic zooplankton. There was an apparent shift in diet by size-class. As red snapper got larger they ate more demersal crustaceans and less pelagic zooplankton. Demersal crustaceans made up 18.2% W of the diet in the 200–299 mm fish and contributed over 68.9% W to the diet of the >600 mm fish, whereas pelagic zooplankton made up 27.3%W in the 200–299 mm fish and only 6.5% W in the red snapper over 600 mm.

The trends in diet by size-class remain approximately the same for the other diet indices. However, smaller organisms made a larger contribution in the other indices; pelagic zooplankton became the largest category of the 300–399 mm snapper by %N, %FO, and %IRI. Similarly in the 500–599 mm red snapper, fish was the largest group by %W and %IRI, but pelagic zooplankton was the largest group by %N and %FO.

Because of data inadequacies a test for significance was not run; however, BVSTEP found that differences did exist among the diets of different red snapper size classes and these differences resulted from six influential prey types: fish, adult *S. empusa*, crab, penaeid shrimp, pelagic zooplankton, and miscellaneous benthic species. There

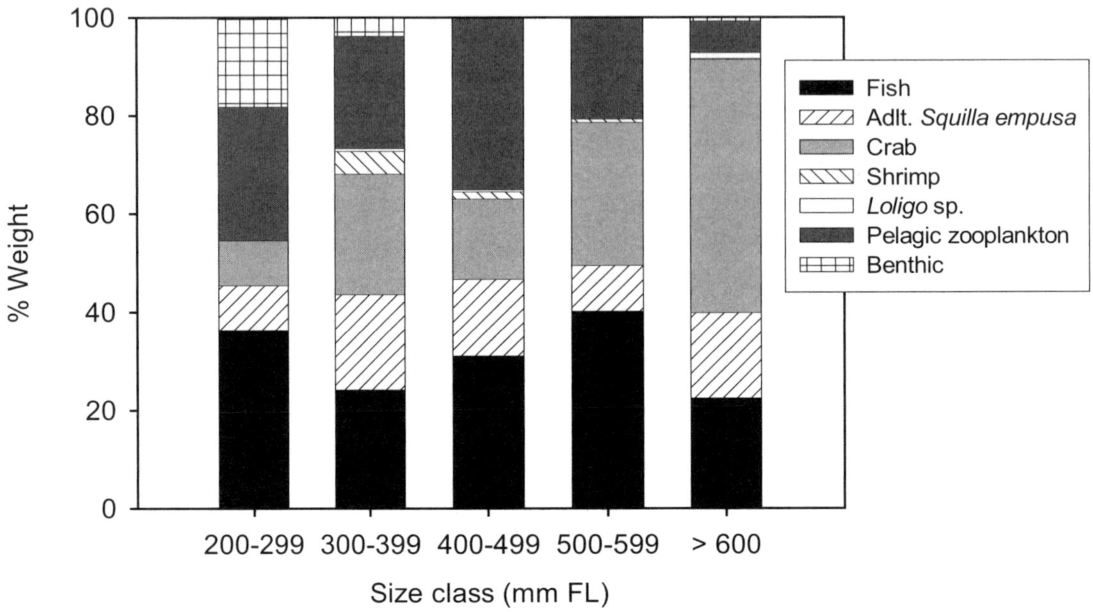

Figure 1. Prey contained in all red snapper *Lutjanus campechanus* stomachs collected on Alabama artificial reefs broken down by size class by % weight for seven prey categories. Benthic = Miscellaneous benthic species.

was an 82.2% correlation between these prey types and the overall pattern found in the samples. The SIMPER results revealed that the 300–399 mm size-class had the largest number of contributing prey types (fish, crab, pelagic zooplankton, and adult *S. empusa*) to within-size-class similarity. The 200–299 mm size-class had the smallest number of contributing species, with fish and pelagic zooplankton accounting for over 90% of the within-size-class similarity. Fish was the largest contributor to within-size-class similarity in the 200–299, 300–399, and 500–599 mm red snapper diets. Pelagic zooplankton was the largest contributor to within-size-class similarity of 400–499 mm red snapper and crab the largest contributor to within-size-class similarity of fish over 600 mm. Red snapper over 600 mm had the highest average similarity (37.3), meaning that stomach contents of red snapper in this size-class were more similar to each other than they were in any other size-class. The aforementioned trend of larger red snapper eating more demersal crustaceans, and less pelagic zooplankton was again evident as the 200–299 mm red snapper did consume pelagic zooplankton, but did not have a demersal crustacean as a contributor to within-season similarity. Snapper over 600 mm had demersal crustaceans present in their diet, but no pelagic zooplankton.

Forty-seven prey categories.—Size-class differences also were examined when the diet was divided into the 47 different prey categories mentioned previously. Unidentified material was the largest contributing category to every size-class by %W ranging from 16.6% W in the 200–299 mm fish to 47.8% W in the fish over 600 mm. After excluding the unidentified material category, the largest contributing prey category varied by size-class. The largest category for the 200–299 mm size-class was larval fish. For the 300–399 mm fish and fish over 600 mm, adult *S. empusa* was the largest prey category. Larval *S. empusa* was the largest category for the 400–499 mm red snapper, and unidentified crab was the largest category for the 500–599 mm red snapper.

Overall, larger organisms, such as *P. gibbesii*, ophichthid fishes, adult *S. empusa*, unidentified fish, and unidentified crabs, made a greater contribution by %W. By %N, %FO, and %IRI, some smaller organisms, such as larval fish, larval *S. empusa*, and amphipods, made a greater contribution to the diet. In general, as snapper got larger they ate more demersal crustaceans and less pelagic zooplankton; this trend was identified by all of the indices.

Descriptive calculations with BVSTEP indicated that differences among size classes were the result of seven influential prey types: unidentified fish, larval fish, unidentified crab, *P. gibbesii*, adult *S. empusa*, larval *S. empusa*, and amphipods. There was a 97.0% correlation among these prey types and the patterns detected in the samples.

Caloric Density

Caloric density (calories/g) was determined for each of the 47 major red snapper prey categories, either directly by bomb calorimetry or taken from literature values. These values were used with %W and %FO to determine the %ICI for each of the 47 prey types (Appendix 2). The results indicated that adult *S. empusa* was the most important prey category with a %ICI of 35.9%, followed by larval *S. empusa*, *P. gibbesii*, and larval fish. ICI could not be determined for unidentified fish and crabs because a caloric value cannot be assigned to these categories. Caloric density values were also used to determine the %PI for each of the seven and 47 prey categories. A comparison was made among %W, %IRI, %ICI and %PI for the seven major prey categories (Figure 2). The %PI values were very similar to the %W values; however, the %ICI values were higher for some of the prey types with higher caloric densities, such as fish and crab. Thus, we believe %ICI to be more informative than %IRI or %PI because it better takes into account the effects of caloric density when describing diet contribution. When the diet was examined by season, %PI gave similar results to %W, but %ICI revealed a slightly different picture (Table 4). For example, in the fall %IRI ranked pelagic zooplankton as second in importance and %ICI ranked fish second, the difference due to the caloric differences between these prey items. Similar results occurred when the data were examined by size-class.

Prey Habitat Preference

After assigning a habitat type to each of the 47 prey categories, the %W values were summed for each habitat type (Appendix 3). Sand- or mud-associated organisms made up the largest portion of the overall diet, followed by water column organisms. Reef-associated organisms only made a 1.3%W contribution to the diet of red snapper off Alabama. When the habitat preference of the prey was examined by season (Table 5), sand- or mud-associated organisms not associated with reefs dominated every season except spring, when water column organisms made up half of the diet. Prey derived from the water column was the second most important category by %W in other seasons. Reef-associated fauna were not present in summer and winter diets and contributed less than 1% in fall and spring.

When prey item habitat preference was examined by red snapper size-class (Table

5), water column species were the largest contributors by %W to the diet of the 200–299 and 400–499 mm size classes. Sand- or mud-associated organisms were the largest contributors to the 300–399, 500–599, and >600 mm size classes. Reef-associated prey species made up less than 1%W in the 200–299, 300–399, and 400–499 mm size classes, and only 2.5%–4.8% in the 500–599 and over 600 mm size classes. There does not appear to be a size related trend in the habitat over which red snapper feed based upon these results. Red snapper of all sizes appear to be feeding either in the water column or in the sand/mud areas surrounding the reef. Despite the fact that few reef-associated species were consumed, reef-associated prey had the highest average caloric density of all the prey types, with water column and sand/mud-associated prey being the next highest.

Annual Prey Demand

The estimates of annual mean prey demand of the red snapper population on artificial reefs off Alabama indicate that, depending on the data source and collection method, between 6.4×10^5 and 31.1×10^6 kg of food are required annually to support the red snapper population found on Alabama artificial reefs (Table 6A). Estimates based on Szedlmayer and Furman (2000) biomass data resulted in the highest prey demand and Strelcheck et al. (2005) visual census data gave the lowest. When these estimates were partitioned into the seven major red snapper diet categories, red snapper off Alabama could require up to 5.6×10^6 kg of fish, 8.4×10^6 kg of demersal crustaceans (crabs and adult *Squilla empusa*) and 4.6×10^6 kg of pelagic zooplankton annually (Table 6B).

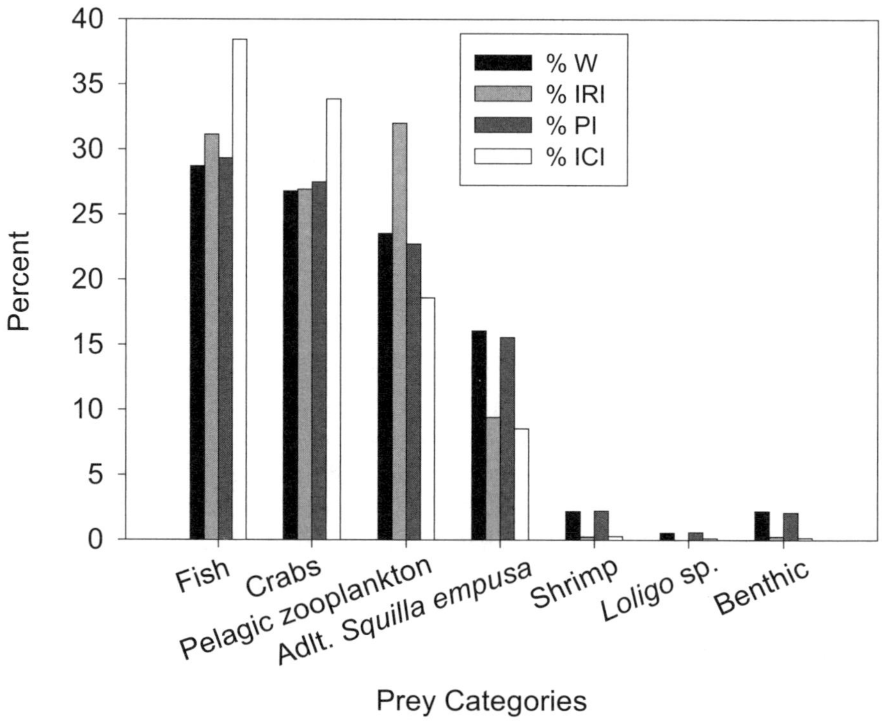

Figure 2. Comparison of prey importance in red snapper *Lutjanus campechanus* diet on Alabama artificial reefs by four indices of prey importance for seven prey types. %W = percent weight, %IRI = percent index of relative importance, %PI = percent prey importance index, %ICI = percent index of caloric importance, and Benthic = Miscellaneous benthic species.

Table 4. Comparison of four prey importance indices by season for all red snapper *Lutjanus campechanus* collected on Alabama artificial reefs based upon seven prey categories. Prey categories are ranked for each index in decreasing order of importance. %W = percent weight, %IRI = percent index of relative importance, %ICI = percent index of caloric importance, %PI = percent prey importance index, and Misc. benthic sp. = Miscellaneous benthic species.

Season	Prey Type	%W (rank)	%IRI (rank)	%ICI (rank)	%PI (rank)
Summer	Fish	28.08 (2)	29.72 (2)	33.70 (2)	28.63 (2)
	Adult *Squilla empusa*	23.23 (3)	18.12 (3)	16.05 (3)	22.25 (3)
	Crab	31.48 (1)	38.01 (1)	43.02 (1)	32.57 (1)
	Shrimp	2.26 (5)	0.26 (5)	0.30 (5)	2.33 (5)
	Loligo sp.	0.50 (7)	0.02 (7)	0.03 (7)	0.50 (7)
	Pelagic zooplankton	13.23 (4)	13.77 (4)	6.84 (4)	12.60 (4)
	Misc. benthic sp.	1.22 (6)	0.11 (6)	0.06 (6)	1.11 (6)
Fall	Fish	24.85 (3)	23.80 (3)	28.20 (2)	25.44 (2)
	Adult *S. empusa*	6.25 (5)	1.90 (5)	1.47 (5)	6.25 (5)
	Crab	34.25 (1)	38.72 (1)	48.26 (1)	34.51 (1)
	Shrimp	2.08 (6)	0.15 (6)	0.16 (6)	2.08 (6)
	Loligo sp.	0.00 (7)	0.00 (7)	0.00 (7)	0.00 (7)
	Pelagic zooplankton	25.62 (2)	32.98 (2)	20.27 (3)	24.99 (3)
	Misc. benthic sp.	6.94 (4)	2.45 (4)	1.64 (4)	6.71 (4)
Winter	Fish	25.48 (2)	25.93 (2)	31.70 (2)	25.90 (2)
	Adult *S. empusa*	33.35 (1)	42.06 (1)	36.96 (1)	32.90 (1)
	Crab	19.49 (4)	16.87 (3)	20.98 (3)	19.87 (4)
	Shrimp	0.33 (6)	0.06 (5)	0.03 (5.5)	0.33 (6)
	Loligo sp.	0.00 (7)	0.00 (7)	0.00 (7)	0.00 (7)
	Pelagic zooplankton	20.83 (3)	15.02 (4)	10.30 (4)	20.61 (3)
	Misc. benthic sp.	0.51 (5)	0.05 (6)	0.03 (5.5)	0.39 (5)
Spring	Fish	34.16 (2)	28.40 (2)	43.47 (1)	35.02 (2)
	Adult *S. empusa*	1.81 (5)	0.08 (5)	0.10 (5.5)	1.64 (5)
	Crab	18.53 (3)	8.29 (3)	13.88 (3)	19.13 (3)
	Shrimp	3.22 (4)	0.24 (4)	0.36 (4)	3.21 (4)
	Loligo sp.	1.27 (7)	0.06 (6)	0.10 (5.5)	1.47 (6)
	Pelagic zooplankton	39.58 (1)	62.90 (1)	42.05 (2)	38.10 (1)
	Misc. benthic sp.	1.43 (6)	0.04 (7)	0.03 (7)	1.43 (7)

Discussion

Diet Comparison with Other Studies.— This is the most comprehensive study of adult red snapper diet on Alabama artificial reefs to date. As in other studies (see Appendix 1 for review), red snapper in this study were found to feed opportunistically on a variety of organisms, the proportions of which changed seasonally and with size. Red snapper off Alabama ate primarily demersal crustaceans (crabs and adult *S. empusa*), fish, and pelagic zooplankton. Parrish (1987), in a literature review of lutjanids, reported that the principal food groups in most studies are fish and decapod crustaceans, and that anguilliform fishes, like the ophichthids we found, were common in the snapper diets he reviewed. Parrish (1987) also found that crabs, specifically portunid and calappid crabs, and shrimps and other crustaceans (especially stomatopods) were frequently consumed.

The proportions of the major red snapper diet categories changed significantly with season; the diet in summer and winter was dominated by adult *S. empusa*, crabs were the largest category in the fall, and pelagic zooplankton the largest category in the spring. These findings also are relatively similar to other studies of red snapper diet (Appendix 1). However, we recognize that the short, one year duration of our study, and likely changes in local prey availability necessitates that conclusions based upon seasonal data from

Table 5. Habitat association of prey consumed by red snapper *Lutjanus campechanus* given as percent weight by season for 300–499 mm FL snapper and by size class (mm FL). SM = sand- and/or mud-associated; R = reef-associated; WC = water-column-associated; SA = sargassum-associated; and V = a variety of habitats.

Summary of Prey Affiliation by Season for 300–499 mm FL Red Snapper

	Percent Weight Contribution by Season			
Habitat Type	summer	fall	winter	spring
SM	47.3	47.9	55.3	13.2
R	0	0.55	0	0.31
WC	25.9	31.8	21.1	60.4
SA	0	0	0	0
V	0	0	5.3	0

Summary of Prey Affiliation by Size Class (mm FL)

	Percent Weight Contribution by Size Class				
Habitat Type	200-299	300-399	400-499	500-599	>600
SM	27.3	46.4	31.1	30.3	59.8
R	0	0.3	0.2	4.8	2.5
WC	54.6	28.1	45.9	29.1	7.7
SA	0	0	0	0	7.8
V	0	1.0	1.3	0	0

this and several of the other studies (Appendix 1) should be interpreted with caution.

Our research provided a size-class (200 to >600 mm) comparison of adult red snapper diet off Alabama, which has been lacking in other studies. A shift in diet with size was indicated: as red snapper grew, the amount of pelagic zooplankton in their diet decreased and the amount of demersal crustaceans increased, with fish remaining a component in the diet of all size classes. However, unlike most of the other studies summarized in Appendix 1, we observed patterns with size that are somewhat atypical. Surprisingly, we observed a small decrease in the amount of fish in the diet of the largest red snapper we examined. However, our data should be interpreted with caution because not all size classes of red snapper were collected at all times of the year. Most red snapper diet studies have found that as individuals grow, there is a shift in their diet from invertebrates and zooplankton to a dependency on fish (Moseley 1966; Bradley and Bryan 1975; Szedlmayer and Lee 2004). Few studies (e.g., Camber 1955) have reported that large red snapper continue to eat demersal crustaceans. Also, pelagic zooplankton contributed to the diet of all size classes in our study, decreasing in importance only for snapper larger than 500 mm FL. Moseley (1966) found a decline in the occurrence of zooplankton in red snapper diets after 170 mm SL. Moseley (1966) concluded that it is inefficient for red snapper larger than 110 mm SL to exist on zooplankton. However, Siegel (1983) found that adult red snapper up to 550 mm SL off Alabama continued to consume significant amounts of pelagic zooplankton, especially larval decapods and stomatopods.

Foraging in the water column on zooplankton has been reported previously for red snapper and other lutjanids. Parrish (1987) found numerous studies of lutjanid diets that reported large zooplankton as an important part of the diet. Haight et al. (1993), in a study of deepwater lutjanids at Penguin Bank, Hawaii, found a diet of primarily zooplankton.

Likewise, Weaver and Sulak (2001) estimated that 1.5 million zooplankters would be available to the reef fish in a meter wide swath in the water column above the deep reef community of the Pinnacles Reef Tract off Alabama. Zooplankton is the primary source of prey for that deep-water reef fish community. Results of our study, combined with results of previous work, strongly support the conclusion that red snapper are opportunistic feeders, and apparently do not consume significant amounts of prey derived directly from reef habitat.

Prey Preference.—Red snapper in this study fed opportunistically on nonreef-associated benthic fauna and pelagic zooplankton; however, it is unclear whether their feeding habits vary due to preference for certain prey, or due to the differences in the availability of food in the environment (Cailliet 1976). Bradley and Bryan (1975) believed

Table 6. A = Estimates of annual mean prey demand of a red snapper *Lutjanus campechanus* population on artificial reefs off Alabama. B= Estimates of annual mean prey demand of a red snapper *Lutjanus campechanus* population on artificial reefs off Alabama broken down by seven prey categories. Misc. benthic sp. = Miscellaneous benthic species.

A.

Biomass Data Source	Number of Reefs	Minimum Prey Demand (kg)	Mean Prey Demand (kg)	Maximum Prey Demand (kg)
Strelcheck 2001 CPUE	1	76	82	88
	14[a]	1,069	1,154	1,239
	8,000[b]	611,021	659,346	707,751
	20,000[c]	1,527,553	1,648,365	1,769,377
Strelcheck 2001 Visual census	1	32	38	43
	14	450	525	600
	8,000	257,386	300,270	342,754
	20,000	643,464	750,675	856,886
Szedlmayer and Furman 2000 Visual census	1	395	975	1556
	14	5,532	13,655	21,777
	8,000	3,161,036	7,802,780	12,444,124
	20,000	7,902,590	19,506,951	31,110,311

B.

Biomass Data Source	Prey Type	Mean Prey Demand (kg) Per Number of Reefs			
		1	14[a]	8,000[b]	20,000[c]
Strelcheck 2001 CPUE	Fish	24	331	189,213	473,033
	Adult *Squilla empusa*	13	186	106,012	265,031
	Crabs	22	309	176,621	441,553
	Penaeid shrimp	2	25	14,438	36,096
	Loligo sp.	0	6	3560	8900
	Pelagic zooplankton	19	271	154,997	387,492
	Misc. benthic sp.	2	25	14,504	36,260
Strelcheck 2001 Visual census	Fish	11	151	86,169	215,422
	Adult *Squilla empusa*	6	84	48,279	120,696
	Crabs	10	141	80,434	201,086
	Penaeid shrimp	1	12	6575	16,438
	Loligo sp.	0	3	1621	4053
	Pelagic zooplankton	9	124	70,586	176,466
	Misc. benthic sp.	1	12	6605	16,513
Szedlmayer and Furman 2000 Visual census	Fish	280	3919	2,239,174	5,597,935
	Adult *Squilla empusa*	157	2195	1,254,562	3,136,404
	Crabs	261	3658	2,090,156	5,225,390
	Penaeid shrimp	21	299	170,864	427,160
	Loligo sp.	5	74	42,131	105,327
	Pelagic zooplankton	229	3210	1,834,250	4,585,626
	Misc. benthic sp.	21	300	171,644	429,110

[a] 14 = Number of experimental reefs from Strelcheck et al. (2005)
[b] 8,000 = Estimated minimum number of artificial reefs off Alabama from Minton and Heath (1998)
[c] 20,000 = Estimated maximum number of artificial reefs off Alabama

that young red snapper were eating a wider variety of prey items (such as more zooplankton) in winter due to the scarcity of more preferred prey. Thus, red snapper could be eating whatever is available around the reef, which may not be the most preferred items with respect to caloric content. Results of our study combined with our review of the available literature (Appendix 1) also suggest the need to compare in more detail the diets of red snapper collected on natural versus artificial habitats, as almost all of the recent studies, including ours, are based upon fishes collected at artificial reefs.

Assuming that fish is a preferred prey because of its caloric content, our results may have been biased by differential digestion rates of organisms, causing less fish to be found in the stomachs than the amount that was actually eaten. Longley and Hildebrand (1941) studied the rate of digestion of lutjanids in the Dry Tortugas and found that after 3.5 h, prey fish in the stomachs were almost completely disintegrated, while digestion of crabs had barely occurred. However, Popova and Sierra (1985) report that the digestion rate of lane snapper, *Lutjanus synagris*, and gray snapper, *Lutjanus griseus*, to be about 40 h in winter and about 20 h in summer since digestion is temperature dependent. Thus, depending on the digestion rate of red snapper, some fish prey could have been underrepresented in the diet. In contrast, when examining gastric evacuation in Atlantic horse mackerel *Trachurus trachurus*, Temming and Herrmann (2001) concluded that the greater the energy density of an organism, the slower it is evacuated. As such, we found numerous organisms of varying caloric content in the stomachs of red snapper including fish having a high caloric content. Thus, we are unsure if fish was underrepresented in the diet.

Prey Demand.—To date, no other estimates of red snapper prey demand on artificial reefs exist. We found that over 31 million kg of prey could be required annually to sustain the red snapper population on artificial reefs off Alabama. However, whether or not these prey demands are being met remains to be answered. The distribution and abundance of red snapper off Alabama over mud bottom, which once supported much lower densities than at present, has been altered (Cowan et al. 1999; Shipp 1999). The permit areas that were once habitat for juvenile reef fish are now home to as many as 20,000 artificial reefs (Cowan et al. 1999; Shipp 1999), where large numbers of red snapper may now be overexploiting their prey resources. However, to quantitatively address this issue, prey production and turnover rates need to be measured and then compared with prey demand estimates.

Differences in annual prey demand estimates based on data from Szedlmayer and Furman (2000) and Strelcheck et al. (2005) are due to the differences in the size of the red snapper on the reefs each sampled, as well as the size of the experimental reefs each examined. The red snapper on the reefs sampled by Strelcheck et al. (2005) were smaller than the red snapper found on the reefs examined by Szedlmayer and Furman (2000), and the artificial reefs that Strelcheck et al. (2005) sampled were smaller than those examined by Szedlmayer and Furman (2000). Moreover, the estimate of annual prey demand based on data from Strelcheck et al. (2005) assumed all artificial reefs off Alabama were of the size he studied, whereas the estimate based on data from Szedlmayer and Furman (2000) assumed all artificial reefs off Alabama were larger. Because artificial reefs off Alabama vary in size, thus causing variability in the number and size of the fish they hold, the annual prey demand for the red snapper population on Alabama artificial reefs probably lies somewhere between these two estimates. Our result may be conservative because we did not take into account the red snapper population over natural hard-bottoms off Alabama.

Contribution of Reef versus Off-Reef Prey.—Foraging by red snapper off Alabama does not appear to be associated with the reef structure *per se*, and thus they may be gaining little nutritional support from reef-associated fauna. Red snapper diet was composed primarily of benthic organisms, such as portunid crabs, adult *S. empusa*, ophichthid fishes, triglid fishes, and *Ophidion* sp.; these organisms typically are associated with mud or sand substrates. Several other studies of red snapper diet describe foraging habits that support this contention (Moseley 1966; Beaumariage and Bullock 1976; Futch and Bruger 1976; Siegel 1983, Parrish 1989). Bohnsack et al. (1991 and 1997b) suggest that feeding on encrusting organisms is not a major attractant for fishes to artificial reefs because most reef fishes depend on pelagic prey and surrounding benthos for food. Similarly, studies of a variety of reef fishes also indicate that snappers, groupers, grunts and other species often eat benthic organisms not associated with artificial or natural reefs (lane snapper, Duarte and Garcia 1999; gray snapper, Croker 1962; Nassau grouper, *Epinephelus striatus*, Eggleston et al. 1998; pigfish, *Orthopristis chrysoptera*, Howe 2001). In a review of lutjanid feeding ecology, Parrish (1987) reports that most snapper probably remain within a few meters of the bottom because most of their prey must be captured from the substrate. He classifies red snapper as an intermediate depth feeder, which means they forage anywhere from relatively shallow water up to 100 m depth. He contends that such feeders either forage widely from shelter over soft bottom to gain food or forage by "patrolling up to several meters off the bottom for nektonic prey" as well as "periodically foraging on substrate for fully benthic forms." Our study found that red snapper consumed some organisms, such as larval *S. empusa*, fish larvae, amphipods, and *Cavolinia* sp., which reside higher in the water column, and other assorted pelagic zooplankton as well as mud-associated species.

Red snapper foraging primarily on benthic invertebrates and pelagic zooplankton suggests that they may be a vital link between the reef community and surrounding habitats if they are translocating energy from the water column and the adjacent sand or mud bottom to the reef through defecation (Parrish 1989). Sedberry and Cuellar (1993) suggested that vermilion snapper off South Carolina and Georgia are important in transferring energy from benthic sand habitats and the water column to the reef because they feed on crustaceans that are not reef-associated. Lindquist et al. (1994), in a study off North Carolina, determined that the sand-associated organisms around the reefs are an important source of energy for artificial reef fish. Davis and Birdsong (1973) describe coral reefs and other habitat interfaces as rich in diversity, suggesting they "represent 'cross roads' between foraging and refuge areas." Artificial reefs can be seen in much the same context, as red snapper seem to fill their energetic demands from habitats other than reefs. It is still unclear if red snapper are simply leaving reefs to forage on nearby mud bottoms or if they are feeding during their transit between reefs, or both. In contrast, Szedlmayer and Lee (2004) found that small red snapper over artificial reefs ate reef-associated prey, such as fish (*Halichoeres* sp., *Serranus* sp. and *Centropristis* sp.) and shrimp (*Lysmata* sp. and *Synalpheus* sp.), although they examined primarily juvenile red snapper which were smaller (10 mm–230 mm SL) than those in our study. Perhaps smaller red snapper are more reef dependent than their larger counterparts or less likely to forage away from structure.

Patterns of Foraging.—It is also possible that red snapper diet may change with distance from shore. Weaver et al. (2001) found red snapper on deep, natural reefs off Alabama (Pinnacles Reef Tract) to be eating similar organisms to red snapper found on the shallow artificial reefs where our red snapper were collected, however the red snapper they examined ate considerably more fish (56% N) (in the form of planktivores, such as red barbier, *Hemanthias vivanus*; striped codlet, *Bregmaceros cantori*; and pike-conger eels, *Hoplunnis sp.*). Weaver and Sulak (2001) found that 90–99% of the fish on the Pinnacles Reef Tract on the Mississippi-Alabama shelf (50–110 m) were small planktivores. These fish served as prey for larger reef fish, and were found in the stomachs of the red snapper they sampled.

In contrast, similarly sized red snapper on inshore artificial reefs ate fewer fish (26%N), but more demersal crustaceans and pelagic zooplankton. As such, red snapper appear to be occupying a different trophic position than similar sized red snapper on deeper reefs further offshore. This suggests that there is a potential for density-dependant food limitation in inshore waters due to elevated adult red snapper densities that are atypical of the inshore habitat, calling into question the role of artificial reefs with respect to the production of new fish biomass.

Succession of Artificial Reefs.—Hueckel and Buckley (1987) discussed the succession of colonization of artificial reef communities in Puget Sound. They believed that in the first stage of development, predators that are aggregated at a reef feed primarily on organisms from the surrounding benthic community. In the second stage, piscivores colonize the reef and feed on fishes that are feeding on the benthic community. In the latter stages of succession, the reef itself can produce an adequate amount of prey to support fish that feed on reef-attached organisms. In their study, 70% of reef fish were feeding on reef-attached organisms. According to their theory of reef colonization, the prey that we found red snapper eating suggests that the artificial reefs off Alabama are indicative of immature communities with low amounts of reef-associated prey. Thus, something may be occurring to keep these reefs "young." Perhaps the passage of tropical storms, which redistribute red snapper (Watterson et al. 1998; Patterson et al. 2001) and reset the benthic community via scouring from wave activity, never allow the artificial reefs to accumulate a mature community of these encrusting organisms, thus producing reef ecosystems that never fully mature.

Attraction versus Production.—Examining a theoretical idea about the attraction versus production debate, we can draw conclusions about the role of artificial reefs in the life history of red snapper. Bohnsack (1989) stated that attraction and production are not mutually exclusive, but rather opposite extremes along a gradient. He stated that five criteria are important for determining whether attraction or production was occurring within a reef system: reef availability, fishing intensity, population controls, reef dependency, and behavior of the target species. Increased production would be likely at locations where reef availability was low, fishing intensity was low, the population of interest was habitat limited, the species of interest was more obligately reef dependent, and the behavior of that species was demersal or territorial. Increased attraction would be likely at locations where reef availability was high, fishing intensity was high, the population of interest was recruitment limited, the species of interest was only partially or opportunistically reef dependent, and the behavior of that species was migratory (Bohnsack 1989). A broad look at the Alabama system using the criteria developed by Bohnsack (1989) provides insight about where the reefs off Alabama fall within the attraction versus production continuum for red snapper. From the current study, we know that red snapper rarely feed on reef dependent prey species. We also know that reef availability is high off Alabama, and that the area is known to have large amounts of low-relief natural hard bottom habitat (Strelcheck et al. 2005). Fishing intensity on these reefs is high as indicated by the large fraction of the total recreational catch that is landed off Alabama (37% of the total recreational catch in the northern Gulf of Mexico, MRFSS 1993–2003); this is corroborated by high estimates of fishing mortality in Watterson (1998) and Schirripa and Legault (1999). There is no evidence that the availability of natural or artificial habitat limits the current population size of red snapper (Cowan et al. 1999), especially given the high mortality of prerecruits attributable to bycatch in the shrimp fishery. Many studies incorrectly conclude that higher densities of organisms around artificial reefs is evidence for increased production (Bohnsack et al. 1997a) by assuming that the amount of hard-bottom habitat is limiting reef fish populations (Bohnsack 1989; Bohnsack et al. 1991; Polovina 1991; Bohnsack et al. 1997a; Bohnsack et al. 1997b; Lindberg 1997). Finally, Patterson et al. (2001) and Patterson and Cowan (2003), showed that red snapper move more frequently and travel greater distances than previously thought, indicating only moderate site fidelity. When all of the above information is interpreted in light of Bohnsack's (1989) criteria, it seems likely that artificial reefs off Alabama are functioning more to attract red snapper than to produce them.

We conclude that red snapper are attracted to artificial reefs off Alabama as a result of a behavioral preference rather than for increased foraging opportunities. Instinctual behavioral responses may explain why red snapper congregate around artificial structures, even when by doing so they may decrease success in foraging (Strelcheck et al. 2005). Shipley and Cowan (University of South Alabama, unpublished data) used Ecopath with Ecosim and Ecospace, based upon a synthesis of much of the data reported here and elsewhere from studies off Alabama, to show that red snapper are capable of creating foraging halos of depleted prey resources around individual reefs. In these simulations, reefs spaced too closely together caused the halos to overlap, negatively affecting the fitness of red snapper occupying the reefs. Bohnsack (1989) suggests that evolutionary experience of fishes elicits responses that are not necessarily adaptive. If red snapper are attracted to Alabama reefs from other areas in the Gulf (Patterson et al. 2001), then limited or over-exploited prey resources coupled with high fishing mortality could make Alabama's artificial reefs a net sink for red snapper production in the northern Gulf. Strelcheck et al. (this volume) concludes this explicitly based upon *G/Z* ratios (Houde 1989) calculated for this region.

Acknowledgments

This research was funded by the Alabama Department of Conservation Marine Resources Division and a Mississippi-Alabama Sea Grant Fellowship. The authors would like to thank W. Patterson, W. Ingram, M. Woods Jackson, A. Strelcheck, and J. Jackson for their red snapper sampling as well as their discussions about this work. We would also like to thank J. Prerost for help with invertebrate identification. The authors thank the Thierry family for allowing us both to sample their catches and to conduct this research aboard their vessel, the *Lady Ann*. This research was completed in partial fulfillment of the master's thesis titled "Diet and Prey Demand of Red Snapper, *Lutjanus campechanus*, on Alabama Artificial Reefs" at the University of South Alabama under the direction of R. Shipp and J. Cowan. Thanks also to Dauphin Island Sea Laboratory technical support, K. Kirsch, A. Gunter, M. Dardeau, J. Cowan, G. Chaplin, and Y. Hintz, as well as the entire maintenance crew. We would also like to thank the following colleagues: S. Baker, A. Fisher, A. Gonzales, B. Jones, S. Keenan, M. Lingo, D. Nieland, and C. Wood. The authors would also like to thank the anonymous reviewers and the editor who provided constructive comments that greatly improved this manuscript.

References

Adams, A. C., and W. C. Kendall. 1891. Report upon an investigation of the fishing grounds off the West Coast of Florida. Bulletin of the United States Fish Commission 9:289–312.

Bailey, H. K., IV. 1995. Potential interactive effects of habitat complexity and sub-adults on young of the year red snapper (*Lutjanus campechanus*) behavior. Master's thesis. University of South Alabama, Mobile, Alabama.

Bailey, H. K., J. H. Cowan, Jr., and R. L. Shipp. 2001. Experimental evaluation of potential effects of habitat size and presence of conspecifics on habitat association by young-of-the-year red snapper. Gulf of Mexico Science 19:119–131.

Beaumariage, D. S., and L. H. Bullock. 1976. Biological research on snappers and groupers as related to fishery management requirements. Pages 86–94 *in* H. R. Bullis, Jr. and A. C. Jones, editors. Proceedings: colloquium on snapper-grouper fishery resources of the western central Atlantic Ocean. Florida Sea Grant Colloquium Report 17.

Bohnsack, J. A. 1989. Are high densities of fishes at artificial reefs the result of habitat limitation or behavioral preference? Bulletin of Marine Science 44:631–645.

Bohnsack, J. A., and D. L. Sutherland. 1985. Artificial reef research: a review with recommendations for future priorities. Bulletin of Marine Science 37:11–39.

Bohnsack, J. A., A. E. Ecklund, and A. M. Szmant. 1997a. Artificial reef research: is there more than the attraction-production issue? Fisheries 22(4):14–16.

Bohnsack, J. A., A. E. Ecklund, and A. M. Szmant. 1997b. Productivity of experimental modular artificial reefs off southeastern Florida. Technical working papers from a symposium on artificial reef development. Atlantic States Marine Fisheries Commission Special Report 64. Washington, D.C.

Bohnsack, J. A., D. L. Johnson, and R. F. Ambrose. 1991. Ecology of artificial reef habitats and fishes. Pages 61–107 *in* W. Seaman, Jr. and L. M. Sprague, editors. Artificial Habitats for Marine and Freshwater Fisheries. Academic Press, San Diego, California.

Bowen, S. H. 1996. Quantitative description of the diet. Pages 513–532 *in* B. R. Murphy and D. W. Willis, editors. Fisheries techniques, 2nd edition. American Fisheries Society, Bethesda, Maryland.

Bradley, E., and C. E. Bryan. 1975. Life history and fishery of the red snapper (*Lutjanus campechanus*) in the northwestern Gulf of Mexico: 1970–1974. Proceedings of the Gulf and Caribbean Fisheries Institute 27:77–106.

Britton, J. C., and B. Morton. 1989. Shore Ecology of the Gulf of Mexico. University of Texas Press, Austin, Texas

Cailliet, G. M. 1976. Several approaches to the feeding ecology of fishes. Pages 1–13 *in* C. A. Simenstad and S. J. Lipovsky, editors. Fish food habitat studies 1st Pacific northwest technical workshop. Washington Sea Grant, Seattle, Washington.

Camber, C. I. 1955. A survey of the red snapper fishery of the Gulf of Mexico, with special reference to the Campeche Banks. Florida Board of Conservation Technical Series 12:1–64.

Casanova, J. P. 1999. Chaetognatha. Pages 1353–1374 *in* D. Boltovskoy, editor. South Atlantic zooplankton. Backhuys Publishers, Leiden, Netherlands.

Clarke. K. R. 1993. Non-parametric multivariate analyses of changes in community structure. Australian Journal of Ecology 18:117–143.

Clarke, K. R., and R. M. Warwick. 1994. Change in marine communities: an approach to statistical analysis and interpretation. Natural Environmental Research Council, Plymouth, Devon, UK.

Collins, J. W. 1885. The red snapper grounds in the Gulf of Mexico. Bulletin of the United States Fish Commission 5:145–146.

Cowan, J. H., Jr., W. Ingram, J. McCawley, B. Sauls, A. Strelcheck, and M. Woods. 1999. The attraction vs. production debate: does it really matter from the management perspective? A response to the commentary by Shipp, R. L., 1999, Gulf of Mexico Science XVII: 51-55. Gulf of Mexico Science 17:137–138.

Croker, R. A. 1962. Growth and food of the gray snapper, *Lutjanus griseus,* in Everglades National Park. Transactions of the American Fisheries Society 91:379–383.

Cummins, K. W., and J. C. Wuycheck. 1971. Caloric equivalents for investigations in ecological energetics. Mitteilungen. Internationale Vereiningung fuer Theoretische und Angewandte Limnologie 18:158.

Davis, W. P., and R. S. Birdsong. 1973. Coral reef fishes which forage in the water column. Helgolaender Wissenschaftliche Meeresuntersuchungen 24:292–306.

Duarte, L. O., and C. B. Garcia. 1999. Diet of the lane snapper, *Lutjanus synagris* (Lutjanidae), in the Gulf of Salamance, Colombia. Caribbean Journal of Science 35:54–63.

Eggleston, D. B., J. J. Grover, and R. N. Lipcius. 1998. Ontogenetic diet shifts in Nassau grouper: trophic linkages and predatory impact. Bulletin of Marine Science 63:111–126.

Fauchald, K. 1977. The Polychaete worms: definitions and keys to the orders, families, and genera. Natural History Museum of Los Angeles County, Los Angeles.

Fischer, A. J., M. S. Baker, C. A. Wilson. 2004. Red snapper (*Lutjanus campechanus*) demographic structure in the northern Gulf of Mexico based on the spatial patterns in growth rates and morphometrics. Fishery Bulletin 102:593–603.

Futch, R. B., and G. E. Bruger. 1976. Age, growth, and reproduction of red snapper in Florida waters. Pages 165–184 *in* H. R. Bullis, Jr. and A. C. Jones, editors. Proceedings: Colloquium on snapper-grouper fishery resources of the western central Atlantic Ocean. Texas A&M University Sea Grant College and Mississippi-Alabama Sea Grant Consortium. Florida Sea Grant College Rep. 17.

Gallaway, B. J. 1980. Pelagic, reef and demersal fishes and macrocrustacean/biofouling communities. *In* W. B. Jackson and E. F. Wilkens, editors. Volume 2. Environmental assessment of Buccaneer gas and oil field in northwestern Gulf of Mexico, 1975–1980. NOAA Technical Memorandum NMFS-SEFC 48.

Gibbons, M. J., V. A. Spiridonov, and G. A. Tarling. 1999. Euphausiacea. Pages 1241–1279 *in* D. Boltovskoy, editor. South Atlantic Zooplankton. Backhuys Publishers, Leiden, Netherlands.

Goodyear, C. P. 1995. Red snapper in U.S. waters of the Gulf of Mexico. National Marine Fisheries Service, Southeast Fisheries Science Center, Miami Laboratory, Miami MIA-95. /96–05.

Hacunda, J. S. 1981. Trophic relationships among demersal fishes in a coastal area of the Gulf of Maine. Fishery Bulletin 79:775–788.

Haight, W. R., J. D. Parrish, and T. A. Wayes. 1993. Feeding ecology of deepwater lutjanid snappers at Penguin Bank, Hawaii. Transactions of the American Fisheries Society 122:328–347.

Hansson, S. 1998. Methods of studying fish feeding: a comment. Canadian Journal of Fisheries and Aquatic Sciences 55:2706–2707.

Harris, R. K., T. Nishiyama, and A. J. Paul. 1986. Carbon, nitrogen, and caloric content of eggs, larvae, and juveniles of the walleye pollock, *Theragra chalcogramma.* Journal of Fish Biology 29:87–98.

Hoese, H. D., and R. H. Moore. 1998. Fishes of the Gulf of Mexico, 2nd edition. Texas A&M University Press, College Station, Texas.

Houde, E. D. 1989. Subtleties and episodes in the early life of fishes. Journal of Fish Biology 35 (Supplement A):29–38.

Howe, J. C. 2001. Diet composition of juvenile pigfish, *Orthopristis chrysoptera* (Perciformes: Haemulidae), from the northern Gulf of Mexico. Gulf of Mexico Science 1:55–60.

Hueckel, G. S., and R. M. Buckley. 1987. The influence of prey communities on fish species assemblages on artificial reefs in Puget Sound, Washington. Environmental Biology of Fishes 19:195–214.

Jackson, M. W., J. H. Cowan, Jr., and D. L. Nieland. 2007. Demographic differences in northern Gulf of Mexico red snapper reproductive maturation: implications for the unit stock hypothesis. Pages 217–227 *in* W. F. Patterson, III, J. H. Cowan, Jr., G. R. Fitzhugh, and D. L. Nieland, editors. Red snapper ecology and fisheries in the U.S. Gulf of Mexico. American Fisheries Society, Symposium 60, Bethesda, Maryland.

Kaplan, E. H. 1988. A field guide to southeastern and Caribbean seashores Cape Hatteras to the Gulf Coast, Florida, and the Caribbean. Houghton Mifflin, Boston, Massachusetts.

Lindberg, W. J. 1997. Can science resolve the attraction-production issue? Fisheries 22(4):10–13.

Lindquist, D. G., L. B. Cahoon, I. E. Clavijo, M. H. Posey, S. K. Bolden, L. A. Pike, S. W. Burk, and P. A. Cardullo. 1994. Reef fish stomach contents and prey abundance on reef and sand substrata associated with adjacent artificial and natural reefs in Onslow Bay, North Carolina. Bulletin of Marine Science 55:308–318.

Longley, W. H. and S. F. Hildebrand. 1941. Systematic catalogue of the fishes of Tortugas, Florida. Carnegie Institute of Washington, Publication 535(34):1–331.

Matsuura, Y., and M. P. Olivar. 1999. Fish larvae. Pages 1445–1496 *in* D. Boltovskoy, editor. South Atlantic Zooplankton. Backhuys Publishers, Leiden, Netherlands.

McCawley, J. R. 2003. Diet and prey demand of red snapper, *Lutjanus campechanus*, on Alabama artificial reefs. Master's thesis. University of South Alabama, Mobile, Alabama.

McCawley, J. R., J. H. Cowan, Jr., and R. L. Shipp. 2006. Feeding periodicity and prey habitat preference of red snapper, *Lutjanus campechanus* (Poey, 1860), on Alabama artificial reefs. Gulf of Mexico Science 24:14–27.

Meier, M., and F. Steimle. 1997. Management priorities for research on artificial reefs. Technical Working Papers from a Symposium on artificial reef development. Atlantic States Marine Fisheries Commission. Special Report 64, Washington, D.C.

Minton, R. V., and S. R. Heath. 1998. Alabama's Artificial Reef Program: Building oases in the desert.

Gulf of Mexico Science 1:105–106.
Morgan, S. G., and A. J. Provenzano, Jr. 1979. Development of pelagic larvae and postlarvae of *Squilla empusa* (Crustacea, Stomatopoda), with an assessment of larval characters within the Squillidae. Fishery Bulletin 77:61–90.
Morris, M. J., and T. L. Hopkins. 1983. Biochemical composition of crustacean zooplankton from the eastern Gulf of Mexico. Journal of Experimental Marine Biology and Ecology 69:1–19.
Moseley, F. N. 1966. Biology of the red snapper, *Lutjanus aya* Bloch, of the northwestern Gulf of Mexico. Publication of the Institute of Marine Science, Texas 11:90–101.
Murano, M. l999. Mysidacea. Pages 1099–1140 *In* D. Boltovskoy, editor. South Atlantic Zooplankton. Backhuys Publishers, Leiden, Netherlands.
Nelson, B. D., and S. A. Bortone. 1996. Feeding guilds among artificial reef fishes in the northern Gulf of Mexico. Gulf of Mexico Science 14(2):66–80.
Ouzts, C. O., and S. T. Szedlmayer. 2003. Diel feeding patterns of red snapper on artificial reefs in the north-central Gulf of Mexico. Transactions of the American Fisheries Society 132:1186–1193.
Palomares, M. L., and D. Pauly. 1989. A multiple regression model for predicting the food consumption of marine fish populations. Australian Journal of Marine and Freshwater Research 40:259–273.
Parrish, J. D. 1987. The trophic biology of snappers and groupers. Pages 405–463 *in* J. J. Polovina and S. Ralston, editors. Tropical Snappers and Groupers, Biology and Fisheries Management. Westview Press, Boulder, Colorado.
Parrish, J. D. 1989. Fish communities of interacting shallow-water habitats in tropical oceanic regions. Marine Ecology Progress Series 58:143–160.
Patterson, W. F., III. 1999. Aspects of the population ecology of red snapper *Lutjanus campechanus* in an artificial reef area off Alabama. Doctoral dissertation. University of South Alabama, Mobile, Alabama.
Patterson, W. F., III, and J. H. Cowan, Jr. 2003. Site fidelity and dispersion of red snapper associated with artificial reefs in the northern Gulf of Mexico. Pages 181–193 *in* D. R. Stanley and A. Scarborough-Bull, editors. Fisheries, reefs, and offshore development. American Fisheries Society, Symposium 36, Bethesda, Maryland.
Patterson, W. F., III, J. C. Watterson, R. L. Shipp, and J. H. Cowan, Jr. 2001. Movement of tagged red snapper in the northern Gulf of Mexico. Transactions of the American Fisheries Society 130:533–545.
Pechenik, J. A. 1996. Biology of the Invertebrates. 3rd Edition. Wm. C. Brown Publishers, Dubuque, Iowa.
Pinkas, L., M. S. Oliphant, I. L. K. Iverson. 1971. Food habits of albacore, bluefin tuna, and bonito in California waters. California Fish and Game 152:1–105.
Pohle, G., F. L. Mantelatto, M. L. Negreiros-Fransozo, A. Fransozo and F. L. Mantellatto. 1999. Larval Decapoda (*Brachyura*). Pages 1281–1351 *in* D. Boltovskoy, editor. South Atlantic Zooplankton. Backhuys Publishers, Leiden, Netherlands.
Polovina, J. J. 1991. Fisheries applications and biological impacts of artificial habitats. Pages 153–176 *in* W. Seaman, Jr. and L. M. Sprague, editors. Artificial habitats for marine and freshwater fisheries. Academic Press, San Diego, California.
Pope, K. L., M. L. Brown, W. G. Duffy, and P. H. Michaletz. 2001. A caloric-based evaluation of diet indices for largemouth bass. Environmental Biology of Fishes 61:329–339.
Popova, O. A., and L. M. Sierra. 1985. Feeding of black jack, *Caranx ruber*, in the Gulf of Batabano on the Cuban Shelf. Digestion Rate and Daily Rations. Journal of Ichthyology 25:105–118.
SEDAR (Southeast Data, Assessment, and Review) 2005. SEDAR 7: Gulf of Mexico red snapper complete stock assessment report. SEDAR, Charleston, South Carolina.
Schirripa, M. J., and C. M. Legualt. 1997. Status of the red snapper in U.S. waters of the Gulf of Mexico. NOAA, NMFS, Southeast Fisheries Science Center, Miami Laboratory Contribution MIA-97/98–05.
Schirripa, M. J., and C. M. Legault. 1999. Status of the red snapper in the U.S. waters of the Gulf of Mexico: Updated through 1998. NOAA, NMFS, Southeast Fisheries Science Center, Miami Laboratory. Contribution SFD-99/00–75.
Sedberry, G. R., and N. Cuellar. 1993. Planktonic and benthic feeding by the reef-associated vermilion snapper, *Rhomboplites aurorubens* (Teleostei, Lutjanidae). Fishery Bulletin 91:699–709.
Shipp, R. L. 1999. The artificial reef debate: Are we asking the wrong questions? Gulf of Mexico Science 17:51–55.
Siegel, D. M. 1983. Aspects of the feeding ecology of snapper fishes (Lutjanidae). Master's thesis. University of West Florida, Pensacola, Florida.
Smith, D. L., and K. B. Johnson. 1996. A Guide to Marine Coastal Plankton and Marine Invertebrate Larvae, 2nd edition. Kendall/Hunt Publishing, Dubuque, Iowa.
Sokal, R. R., and R. J. Rohlf. 1981. Biometry, 2nd edition. Freeman, New York.
Starck, W. A. 1968. A list of fishes of Alligator Reef, Florida with comments on the nature of the Florida reef fish fauna. Undersea Biology 1:5–40.

Stearns, S. 1884. On the position and character of the fishing grounds of the Gulf of Mexico. Bulletin of the United States Fish Commission 4:289–290.

Strelcheck, A. 2001. The influence of reef design and nearest-neighbor dynamics on artificial-reef fish assemblages. Master's thesis. University of South Alabama, Mobile, Alabama.

Strelcheck, A. J., J. H. Cowan, Jr., and W. F. Patterson, III. 2007. Site fidelity, movement, and growth of red snapper: Implications for artificial reef management. Pages 147–162 *in* W. F. Patterson, III, J. H. Cowan, Jr., G. R. Fitzhugh, and D. L. Nieland, editors. Red snapper ecology and fisheries in the U.S. Gulf of Mexico. American Fisheries Society, Symposium 60, Bethesda, Maryland.

Strelcheck, A. J., J. H. Cowan, Jr., and A. Shah. 2005. Influence of reef location on artificial-reef fish assemblages in the northcentral Gulf of Mexico. Bulletin of Marine Science 77(3):425–440.

Stuck, K. C. 1978. The pelagic amphipods of the Mississippi Gulf Coast and associated offshore areas. Master's thesis. University of Southern Mississippi, Hattiesburg, Mississippi.

Szedlmayer, S. T., and C. Furman. 2000. Estimation of abundance, mortality, fecundity, age frequency, and growth rates of red snapper, *Lutjanus campechanus*, from a fishery independent stratified random survey. Gulf and South Atlantic Fisheries Foundation Final Report #70–01-126091/21165.

Szedlmayer, S. T., and J. D. Lee. 2004. Diet shifts of juvenile red snapper (*Lutjanus campechanus*) with changes in habitat and fish size. Fishery Bulletin 102:366–375.

Temming, A., and J. P. Herrmann. 2001. Gastric evacuation of horse mackerel. II. The effects of different prey types on the evacuation model. Journal of Fish Biology 58:1246–1256.

Todd, C. D., M. S. Laverack, and G. A. Boxshall. 1996. Coastal marine zooplankton: a practical manual for students, 2nd edition. Cambridge University Press, Cambridge, UK.

Treasurer, J. W. 1988. Measurement of regurgitation in feeding studies of predatory fishes. Journal of Fish Biology 33:267–271.

Van der Spoel, S., and J. R. Dadon. 1999. Pteropoda. Pages 649–706 *in* D. Boltovskoy, editor. South Atlantic Zooplankton. Backhuys Publishers, Leiden, Netherlands.

Vose, F. E., M. H. Posey, and W. J. Lindberg. 1997. How reasonable is it to assume enhanced productivity by artificial reefs, based on experiments in the eastern Gulf of Mexico? Technical Working Papers from a symposium on artificial reef development. Atlantic States Marine Fisheries Commission. Special Report 64, January 1997. Washington, D.C.

Watterson, J. C. 1998. Estimates of site fidelity and short term movements of red snapper (*Lutjanus campechanus*) based upon mark/recapture on north central Gulf of Mexico artificial reefs. Master's thesis. University of South Alabama, Mobile, Alabama.

Watterson, J. C., W. F. Patterson, III, R. L. Shipp, and J. H. Cowan, Jr. 1998. Movement of red snapper, *Lutjanus campechanus*, in the north central Gulf of Mexico: potential influence of hurricanes. Gulf of Mexico Science 16:92–104.

Weaver, D. C., G. D. Dennis, and K. J. Sulak. 2001. Northeastern Gulf of Mexico coastal and marine ecosystem program: community structure and trophic ecology of demersal fishes on the Pinnacles Reef Tract. Final Synthesis Report. United States Department of the Interior, United States Geological Survey, USGS BSR-2001–0008.

Weaver, D. C., and K. J. Sulak. 2001. Trophic subsidies in the twilight zone: zooplankton availability and food web structure of deep reef fishes along the Mississippi-Alabama outer continental shelf. *In* M. McKay, J. Nides, and D.Vigil, editors. Proceedings: Twentieth Annual Gulf of Mexico Information Transfer Meeting, December 2000. United States Department of the Interior, Minerals Management Service, Gulf of Mexico, OCS Region. OCS Study MMS 2001–082, New Orleans, Louisiana.

Williams, A. B. 1984. Shrimps, lobsters, and crabs of the Atlantic Coast of the eastern United States, Miami to Florida. Smithsonian Institute Press, Washington, D.C.

Wissing, T. E., R. M. Darrnell, M. A. Ibrahim, and L. Berner, Jr. 1973. Caloric values of marine animals from the Gulf of Mexico. Contributions in Marine Science 17:1–7.

Appendix 1. Literature review of *Lutjanus campechanus* diet studies in the northern Gulf of Mexico. FO = frequency of occurrence.

Source	Location of study	Fish size	# stomachs examined (# w/ food)	Description of Results
Stearns 1884	Northern Gulf of Mexico	?	450 (1)	[a] "fine sand worms and sponge like stuff" were reported. (Camber 1955 stated that find sand worms were probably *Cavolinia longirostris*)
Collins 1885	Gulf of Mexico	?	?	[a b] Fish were the principal prey items and some pelagic plankton (urochordates)
Adams and Kendall 1891	Florida	?	Small #	[a] Flying fish, eels, crabs (*Calappa* sp.), and mantis shrimp (*Squilla* sp.)
Camber 1955	Campeche Banks	Juvenile	15 (14)	Penaeid shrimp
		Adult	100 (24)	Penaeid shrimp, crabs, octopus, conch, *Sicyonia* sp. and fish associated with coral reefs
Moseley 1966			712 (187)	
	Louisiana	Juvenile	(28)	39%FO of crustaceans in one sample, another sample had 60% *Squilla* sp. and 27% unidentified fish
		Adult	(46)	80% fish and 20% crustaceans in one sample, another sample had 44% fish and 8% crustaceans
	Texas	Juvenile	(45)	41% crustaceans in one sample and 89% crustaceans in another
		Adult	(68)	Fish dominated the diets in all samples (40-59%) and crustaceans made up 32% in one sample
Bradley and Bryan 1975	Texas	Juvenile and sub-adult	575 (258)	* Mostly dependent on shrimp throughout the year, but crabs and mantis shrimp also were important. Primary food items by season: summer-squid and fish; fall- octopods; winter and spring-shrimp and other crustaceans. Winter had the most varied diet. Zooplankton were not present in the diet after 150 mm FL.
		Adult	1139 (190)	* Primarily ate fish throughout the year, but in summer ate more crustaceans. Primary food items by season: summer-unid. fish, *Callinectes danae* and *Sicyonia dorsalis*; fall-fish, *S. dorsalis* and *C. danae*; winter-unid. fish and eels; spring-unid. fish, eels, mantis shrimp, and *Sicyonia* sp. Summer had the largest variety of organisms and winter the smallest variety.
Beaumariage and Bullock 1976	Florida	Juvenile	?	Invertebrates (shrimps, crabs, squids, and mud burrowing shrimp) made up a considerable portion of the diet
Futch and Bruger 1976	Florida, west of Clearwater	?	213 (56)	Invertebrates were represented more than fish. The inverts were associated with sand-shell bottom, such as Stomatopods and Decapods. Some larvae found in the stomachs also suggested some water column feeding.
Gallaway 1980	Buccaneer oil and gas platform in NW Gulf	70-450 mm FL	?	* *Squilla* sp. a major contributor in summer and spring; fish also important in summer; fall comprised of fish, shrimp and swimming crabs; winter most of diet was bait and a few fish
Siegel 1983	Primarily Alabama but also some samples from Louisiana	1-250 mm SL (Juveniles)	582 (289)	* Fish were most prominent throughout the year, followed by crabs (mostly portunids and albunids) and shrimp (Sicyonidae, Sergestidae, Penaeidae). Less frequent occurrence of zooplankton (such as amphipods). High occurrence of decapod and stomatopod larvae in June. Overall lack of significant seasonal trends. Consumed many different prey types.
	and Florida	251-550 mm SL (Adults)	? (34)	* Fish and crabs (portunids and albunids) were the largest part of the diet. All sizes of adults consumed crabs, sicyonid and penaeid shrimp, and decapod and stomatopod larvae. Squid had minimal importance. Diversity of prey was highest in summer and lowest in winter. Summer and winter had larval *S. empusa* in abundance. Overall lack of significant seasonal trends. Consumed few different types of prey.
Bailey 1995	Alabama	330-691 mm TL	98 (45)	Principal prey items in summer were rock shrimp and crabs, also ate some eels and unidentified fish
Weaver et al. 2001	Alabama (Pinnacles Reef Tract)	192-465 mm SL	26 (?)	Characterized red snapper as a generalized carnivore. Found them eating 56% fish (all of which were deep water species) and 25.7% pelagic zooplankton. Also found 5% crab, 5% mantis shrimp, and 2.6% squid.
Ouzts and Szedlmayer 2003	Alabama Hugh Swingle Permit Area	185-590 mm SL	432 (164)	Fish prey dominated red snapper diet for entire diel cycle. Sand prey dominated dusk and dawn and reef prey dominated day and night. Pelagic prey were unimportant for all periods.
Szedlmayer and Lee 2004	Alabama	10-280 mm SL	1639 (789)	Red snapper on open habitat (most <70 mm SL) ate non-reef-associated species such as mysid shrimp, chaetognaths, squid, and copepods. Reef-associated red snapper (70-280 mm SL) ate reef associated species such as fish (*Halichoeres* sp., Blennidae, and Serranidae), squid, portunid crabs, and shrimp (*Sicyonia* sp. and Squillidae)
McCawley et al. 2006	Alabama Hugh Swingle Permit Area	267-590 mm FL	268 (166)	Red snapper fed continuously throughout the day and night. Fed above the reef on water-column associated organisms during the day and away from the reef on sand/mud associated organisms at night.
This study	Alabama Hugh Swingle Permit Area	207-913 mm FL (Adult)	656 (331)	* Overall ate primarily demersal crustaceans (crab and adult *Squilla empusa*), fish, and pelagic zooplankton. Summer and winter dominated by adult *Squilla empusa*, fall dominated by crabs, and spring dominated by pelagic zooplankton. Diversity of prey was highest in summer and lowest in winter. As snapper got larger pelagic zooplankton decreased and demersal crustaceans increased, while fish remained a constant component. Primarily ate species not associated with reefs.

* Seasonal study
[a] Cited in Camber (1955)
[b] Cited in Parrish (1987)

Appendix 2. Percent index of caloric importance (%ICI) of red snapper *Lutjanus campechanus* diet items and the three values that compromise the index for 47 prey categories. Prey items are ranked in decreasing order of importance by the %ICI index. %W = percent weight, Cal/g wet wt = Calories/gram wet weight, %FO = percent frequency of occurrence, and %ICI = percent index of caloric importance.

Prey Type	%W	Cal/g wet wt	Source	%FO	%ICI (rank)
Fish					
Unidentified fish	13.40			17.16	
Family Ophichthidae	4.09	1278.67	McCawley 2003	6.34	4.03 (6)
Family Triglidae	2.35	882.25	McCawley 2003	2.61	0.66
Family Haemulidae	0.46	1234.13	McCawley 2003	0.75	0.05
Family Syngnathidae	0.12	1009.22	McCawley 2003	0.75	0.01
Ophidion sp.	0.62	1011.11	McCawley 2003	0.75	0.06
Decapterus sp.	0.30	1235.95	McCawley 2003	0.37	0.02
Lagodon rhomboides	0.37	1846.77	McCawley 2003	0.37	0.03
Anchoa hepsetus	0.23	1073.70	McCawley 2003	0.37	0.01
Fish larvae	6.75	1190.25	Harris et al. 1986	11.94	11.65 (4)
Crabs					
Unidentified crabs	11.08			17.54	
Family Portunidae	0.64	1146.58	McCawley 2003	2.24	0.20
Portunus gibbesii	7.41	1331.04	McCawley 2003	11.19	13.40 (3)
P. sayi	1.34	1146.58	McCawley 2003	2.99	0.56
P. spinimantus	0.55	939.02	McCawley 2003	1.12	0.07
P. spinicarpus	0.16	1146.58	McCawley 2003	0.37	0.01
P. ordwayii	0.14	1146.58	McCawley 2003	0.37	0.01
P. sebae	<0.01	1146.58	McCawley 2003	0.37	0.00
Callinectes sp.	0.31	1519.65	McCawley 2003	0.37	0.02
C. sapidus	1.11	1483.58	McCawley 2003	1.87	0.37
C. exasperatus	0.26	1519.65	McCawley 2003	0.37	0.02
C. danae	0.01	1519.65	McCawley 2003	0.37	0.00
Ovalipes floridanus	1.65	1059.01	McCawley 2003	2.61	0.55
Calappa flammea	0.75	1053.39	McCawley 2003	0.75	0.07
C. augusta	0.35	1053.39	McCawley 2003	0.37	0.02
Hepatus epheliticus	0.31	1053.39	McCawley 2003	0.75	0.03
Parthenope granulata	0.73	1295.81	McCawley 2003	0.75	0.09
Pelagic zooplankton					
Larval *Squilla empusa*	11.78	686.40[a]	Wissing et al. 1973	21.64	21.25 (2)
Unidentified mollusk larvae	1.44	686.40[a]	[a]	1.87	0.22
Crab megalopa and zoea	0.50	686.40[a]	[a]	1.87	0.08
Order Amphipoda	5.76	686.40[a]	Wissing et al. 1973	16.79	8.05 (5)
Order Euphausicaea	0.75	708.70	Morris and Hopkins 1983	0.75	0.05
				0.75	0.02
Order Isopoda	0.38	686.40[a]	Wissing et al. 1973	0.75	0.00
Order Mysidacea	0.03	714.00	Morris and Hopkins 1983	0.37	0.00
Order Calanoida	0.01	663.90	Morris and Hopkins 1983	1.49	0.02
Family Sergestidae	0.13	659.00	Morris and Hopkins 1983	0.37	0.00
Family Palaemonidae	0.05	686.40[a]	[a]	4.10	0.89
Cavolinia sp.	2.62	686.40[a]	[a]	0.75	0.00
Sagita sp.	0.06	686.40[a]	[a]		
Adult *Squilla empusa*	16.08	865.41	McCawley 2003	21.27	35.93 (1)
Penaeid shrimp	2.19	1166.86	McCawley 2003	3.73	1.16 (7)
Squid	0.54	1222.06	McCawley 2003	1.49	0.12
Miscellaneous Benthic species					
Phyllum Arthropoda	0.08	792.00	Cummins and Wuycheck 1971	0.37	0.00
Pagurus sp.	0.37	695.30[b]	[b]	0.37	0.01
Sicyonia sp.	0.13	654.89	McCawley 2003	0.75	0.01
Glycera sp.	0.05	639.00	Cummins and Wuycheck 1971	0.37	0.00
Albunea paretii	1.57	695.30[b]	[b]	1.87	0.25

[a] Estimated from other pelagic zooplankton values in this study.
[b] Estimated from other miscellaneous demersal species in this study.

Appendix 3. Percent weight (%W) of stomach contents and habitat association of prey consumed by red snapper *Lutjanus campechanus* for all stomachs collected from Alabama artificial reefs. SM = sand- and/or mud-associated; R = reef-associated; WC = water-column-associated; SA = sargassum-associated; and V = a variety of habitats.

Prey Type	%W	Habitat	Source
Fish			
Unidentified fish	13.40		
Family Ophichthidae	4.09	SM	Hoese and Moore 1998
Family Triglidae	2.35	SM	Hoese and Moore 1998
Family Haemulidae	0.46	R	Hoese and Moore 1998
Family Syngnathidae	0.12	R	Starck 1968
Ophidion sp.	0.62	SM	Hoese and Moore 1998
Decapterus sp.	0.30	R	Starck 1968
Lagodon rhomboides	0.37	R	Hoese and Moore 1998
Anchoa hepsetus	0.23	WC	Hoese and Moore 1998
Fish larvae	6.75	WC	Matsuura and Olivar 1999
Crabs			
Unidentified crabs	11.08		
Family Portunidae	0.64	SM	Britton and Morton 1989
Portunus gibbesii	7.41	SM	Britton and Morton 1989
P. sayi	1.34	SA	Williams 1984
P. spinimantus	0.55	SM	Williams 1984
P. spinicarpus	0.16	SM	Williams 1984
P. ordwayii	0.14	SM	Britton and Morton 1989
P. sebae	<0.01	SM	Kaplan 1988
Callinectes sp.	0.31	SM	Britton and Morton 1989
C. sapidus	1.11	SM	Britton and Morton 1989
C. exasperatus	0.26	SM	Britton and Morton 1989
C. danae	0.01	SM	Williams 1984
Ovalipes floridanus	1.65	SM	Kaplan 1988
Calappa flammea	0.75	SM	Williams 1984
C. augusta	0.35	SM	Britton and Morton 1989
Hepatus epheliticus	0.31	SM	Williams 1984
Parthenope granulata	0.73	V	Williams 1984
Pelagic zooplankton			
Larval *Squilla empusa*	11.78	WC	Morgan and Provenzano 1979
Unidentified mollusk larvae	1.44	WC	Todd et al. 1996
Crab megalopa and zoea	0.50	WC	Pohle et al. 1999
Order Amphipoda	5.76	WC	Stuck 1978
Order Euphausicaea	0.75	WC	Gibbons et al. 1999
Order Isopoda	0.38	WC	Smith and Johnson 1996
Order Mysidacea	0.03	WC	Murano 1999
Order Calanoida	0.01	WC	Pechenik 1996
Family Sergestidae	0.13	WC	Williams 1984
Family Palaemonidae	0.05	WC	Pechenik 1996
Cavolinia sp.	2.62	WC	Van der Spoel and Dadon 1999
Sagita sp.	0.06	WC	Casanova 1999
Adult *Squilla empusa*	16.08	SM	Britton and Morton 1989
Penaeid shrimp	2.19	SM	Britton and Morton 1989
Squid	0.54	WC	Britton and Morton 1989
Miscellaneous Benthic species			
Phyllum Arthropoda	0.08	SM	Pechenik 1996
Pagurus sp.	0.37	SM	Britton and Morton 1989
Sicyonia sp.	0.13	SM	Williams 1984
Glycera sp.	0.05	SM	Fauchald 1977
Albunea paretii	1.57	SM	Williams 1984

Summary of Habitat Types

Habitat Type	% Weight
SM	41.2
R	1.3
WC	31.0
SA	1.3
V	0.7

American Fisheries Society Symposium 60:105–121, 2007

Fidelity of Red Snapper to Petroleum Platforms in the Northern Gulf of Mexico

MEGAN P. WESTMEYER[1], CHARLES A. WILSON, III[2], AND DAVID L. NIELAND[2]
Department of Oceanography and Coastal Sciences, Louisiana State University, Baton Rouge, Louisiana 70803 USA

Abstract.—The habitat value of petroleum platforms for red snapper, *Lutjanus campechanus*, is poorly understood. However, it is widely recognized that the presence of thousands of platforms in the northern Gulf of Mexico (GOM) has affected the distribution of red snapper by the addition of hard substrate habitat. We evaluated the habitat value of petroleum platforms by monitoring the fidelity of red snapper to these structures with acoustic telemetry. In May 2003, 125 red snapper were captured with hook and line at several platforms in a 35-km^2 portion of the South Timbalier oil and gas lease blocks, 50 km south of Port Fourchon, Louisiana. Following anaesthetization with MS-222, an individually coded acoustic pinger was surgically implanted into the peritoneal cavity of each specimen. After a short recovery period fish were released at five platforms in the study area. Presence of individual snapper was recorded with omnidirectional acoustic receivers attached to eight platforms. Red snapper exhibited little movement among platforms in the study area; however, logistic regression showed a high initial fidelity to release location which subsequently decreased over time. Therefore, site fidelity was estimated to be high in the short-term, but much lower in the long-term. However, study results were confounded by tag detection issues that may have resulted in long-term site fidelity being underestimated. Overall, estimates provided of red snapper fidelity to petroleum platforms should lead to more effective management of this species by adding to the knowledge of the function of platforms as habitat and their importance to the GOM red snapper population.

Introduction

The first petroleum platform on the Outer Continental Shelf (OCS) of the Gulf or Mexico (GOM) was installed in 1942 (Pulsipher et al. 2001). By 1997 almost 5,600 platforms had been erected on the GOM OCS, mostly off the coasts of Louisiana and Texas (Pulsipher et al. 2001). These platforms account for a significant percentage of U.S. domestic petroleum production, but they also function as artificial reefs by providing settling habitat for algae and encrusting invertebrates and shelter and/or food for many species of reef fishes and predatory pelagic fishes (Hasting et al. 1976; Dugas et al. 1979; Gallaway et

[1] Corresponding author: mwestmeyer@scaquarium.org. Current address: Sustainable Seafood Initiative, South Carolina Aquarium, 100 Aquarium Wharf, Charleston, South Carolina 29401 USA
[2] Current address: Louisiana Sea Grant College Program, Louisiana State University, Baton Rouge, Lousiana 70803 USA

al. 1981; Bohnsack and Sutherland 1985; Bull and Kendall 1994; Render 1995 Fabi et al. 2002; Stachowitsch et al. 2002). The reef function of platforms was most significant on the shallow shelf of the northwest GOM where the substrate was primarily muddy and sandy with limited natural hard bottom (Parker et al. 1983; Render 1995). Pulsipher et al. (2001) estimated platforms have increased the total amount of reef habitat by as much as 10–25% in the northwest GOM. Platforms west of the Mississippi River may be especially important habitat because they rise above the nepheloid layer and bottom waters that may become seasonally hypoxic (Render 1995).

Red snapper, *Lutjanus campechanus,* are known to be a predominant member of the reef fish community found around platforms in the northern GOM (Hasting et al. 1976; Stanley and Wilson 1989; Stanley and Wilson 1990, Render 1995). Multiple studies conducted through several decades have consistently reported red snapper to be one of the most abundant fishes encountered at oil platforms, with large schools of subadult and adults fish comprising between 37 and 80% of total fish abundance (Continental Shelf Associates 1982; Gallaway et al. 1981 and Nieland and Wilson 2002; Putt 1982; Rademacher and Render 2003; Render 1995). Despite the large number of red snapper found associated with platforms, little information exists about the roll platforms play in the ecology of the species or the threat that decommissioning and removal of older platforms may pose to the red snapper population in the northwest GOM. Therefore, it is necessary to study and document the extent of red snapper residency at rigs to determine the potential effects of platform removal on this already depleted population.

Remote acoustic telemetry may provide an ideal tool to examine red snapper site fidelity to and movement among petroleum platforms. Acoustic telemetry enables continuous acoustic tracking, thus offering several benefits over direct observation and conventional tag and recapture techniques, such as the ability to investigate diel movements, habitat utilization, and site residence times (Eklund and Schull 2001). Traditional mark and recapture studies, where fish are usually harvested permanently after recapture, are limited to collection of two data points: release location and capture location (Klimley et al. 1998). Most such studies are also dependent upon the cooperation and accuracy of fishermen for data collection. In some cases, however, researchers perform recapture and subjects are released and recaptured multiple times (Watterson et al. 1998; Patterson et al. 2001). Even this type of mark and recapture study usually provides only vague descriptions of movement of a mobile subject (Hart and Summerfelt 1975).

The objective of this study was to estimate red snapper site fidelity to and movement among petroleum platforms with acoustic telemetry. We investigated both the short-term (days to weeks) site fidelity and the long-term (weeks to months) site fidelity of red snapper to petroleum platforms on the OCS of the GOM off the coast of Louisiana. Results of this study should provide managers with useful information regarding the artificial reef function of petroleum platforms for red snapper, and provide guidance relative to removal versus reefing of obsolete platforms.

Methods

Study Site Description

The study was conducted on a cluster of petroleum platforms, locally called "the circle," located approximately 50 kilometers south of Port Fourchon, Louisiana (Figure 1). The area is part of the Miner-

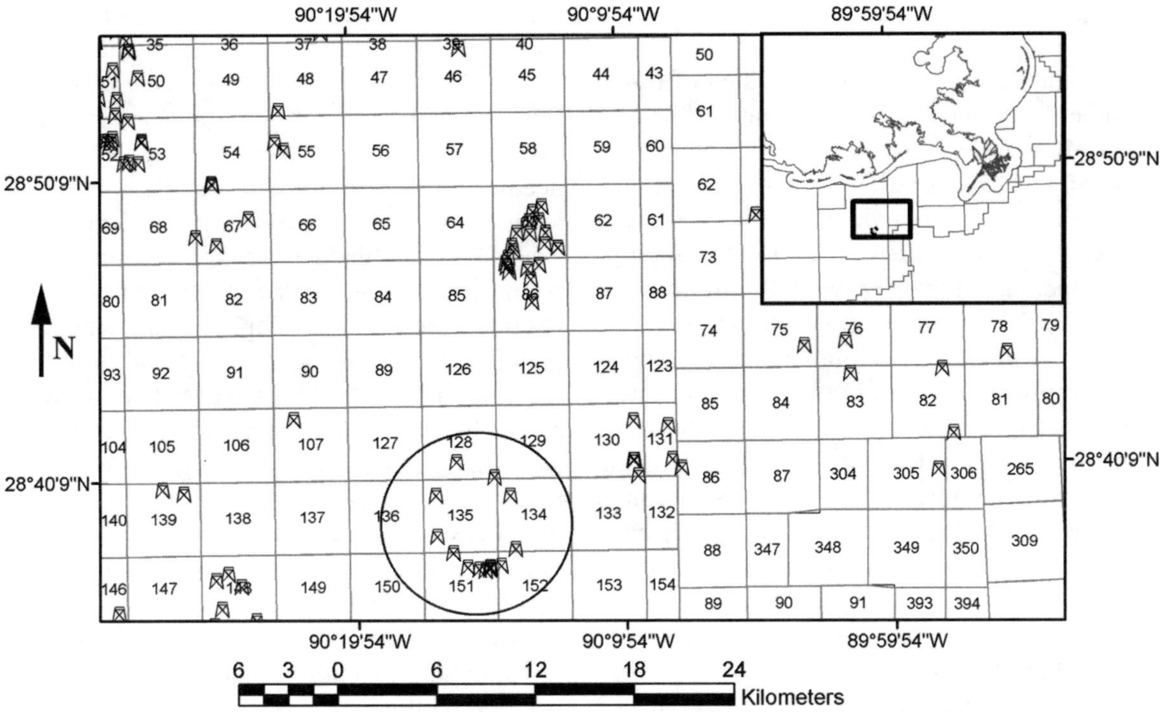

Figure 1. Map of the study area in the Minerals Management Service's South Timbalier 128, 134, 135, 151, and 152 oil and gas lease blocks. The numbers denote the lease block areas and the remaining symbols denote individual petroleum platform locations. All tagged red snapper *Lutjanus campechanus* were released at platforms within the circled area. The inset in the upper right corner shows the location of the study area with respect to the coast of Louisiana.

als Management Service's oil and gas lease blocks South Timbalier 128, 134, 135, 151, and 152. These leases are held primarily by ChevronTexaco Corporation, which first erected platforms there in the 1960s and then again in the 1980s. The area was chosen due to its unique circular configuration, the close proximity of 12 standing platforms, the existence of numerous artificial reefs within the 35 km^2 area, and the high frequency with which commercial and recreational fishermen visit the area. The closest platforms outside this area are in the South Timbalier 130 lease block approximately eight km to the east.

The majority of the platforms in the circle are 6- and 8-pile structures with the exception of the ST151-Y complex, which consists of four individual platforms connected by catwalks. The water depth in this area ranges from 30 to 42 m and the prevailing current direction in this region is westerly (L. Rouse, Louisiana State University, personal communication). The study area is farther offshore than the normal extent of the hypoxic zone formed during the summer months by nutrient and freshwater input from the Mississippi River (Rabalais et al. 2002). The substrate in this region is muddy with the exception of the litter and shell hash that has accumulated within the immediate vicinity of platforms (Hasting et al. 1976; Parker et al. 1983; Love et al. 1999).

Equipment

The telemetry system consisted of transmitters and hydrophones manufactured by VEMCO LTD. The transmitters were model V8SC-2H 4K acoustic tags (pingers) that had a random pulse train delay of 150–300 s and predicted battery life of over 400 d. All pingers operated at 69 kHz, were 9 mm in diameter by 30 mm in length, and weighed 5 g in air. Each emitted a unique acoustic signal determined by the timing between the seven pulses in the train. In addition to the internal pinger, all red snapper were fitted with a Floy internal anchor tag (FM-95W). Printed on the internal and external sections of each tag were the tag number, an offer of a reward, and a 1–800 telephone number. Fishermen who returned the tags and pingers were given a nominal reward.

A multipart receiver system, VEMCO's VR2 model, consisted of an omnidirectional hydrophone, an acoustic signal receiver, an ID detector, a data logging memory, and a battery, all of which were housed in a submersible cylindrical case measuring 34 cm in length by 6 cm in diameter and weighing approximately 1.2 kg. The receiver memory was capable of storing 300,000 detections and its replaceable batteries had an estimated life of six to eight months. Data were uploaded to a laptop computer via a magnetic probe, which allowed easy upload in the field. Data from each receiver were imported into a spreadsheet, cleaned and reorganized, then combined and archived in a database. Data were exported from the database as text documents for use in SAS (SAS Institute Inc., 2003).

Pinger Implantation

Acoustic pingers were surgically implanted in the peritoneal cavities of red snapper on 21, 22, 28, and 29 May 2003. Fish were captured with hook and line from among several capture locations that were distributed across the study area, but most red snapper were caught in the western part of the circle. Following capture, fish were held in an aerated, flow through tank and monitored for any ill effects of handling. A 16-gauge hypodermic needle sterilized in a dilute povidone iodine solution was used to puncture and deflate each fish's air bladder through the sidewall of the body a few scales below and behind the tip of the opercular flap. Fish showing no major trauma were moved to a tank with a solution of MS-222 at a concentration of 80 mg/L. Individual fish then were moved to a foam-lined plastic box for surgical implantation of an acoustic pinger. During surgery the gills were irrigated with a 50 mg/L solution of MS-222 pumped through tubes by a small submersible bilge pump. Pingers were inserted through 2–3 cm incisions made in the abdominal cavity behind and below the pectoral fin. Pingers then were pushed slightly forward, away from the internal organs. The anchor end of the Floy tag was inserted at the posterior end of the incision and the wound was sutured with Ethicon 3–0 absorbable chromic gut thread. The area was patted dry and the wound was sealed with Nexaband S/C Topical Skin Closure, a veterinary glue. The glue was allowed to dry for one minute and then fish were moved to an oxygenated recovery tank. After a 15–30 min recovery period, tagged individuals were released into a 29 ft open-ended hoop net, the bottom of which was angled towards the platform. Video from a remotely operated vehicle (ROV) was recorded as a means to confirm fish exited the hoop net once released.

Seventy-two percent of tagged fish were caught and released at the same location (hereafter native), while 28% were caught and subsequently relocated to nearby platforms prior to release (hereafter relocated). This was done to estimate if homing occurred and to test if relocated snapper were more likely to leave their release location. A result of this process was release locations were slightly more geographically balanced across the study area than capture locations were.

Receiver Deployment

Acoustic receivers were deployed at seven platforms in the study area on 21 and 22 May, 2006 (Table 1; Figure 2). One of these receivers was lost in late June during a tropical storm after which additional receivers were deployed at an artificial reef and another platform (Table 1).

Data Analyses

Short- and long-term site fidelity was estimated from acoustic tagging data. Short-term site fidelity was examined with plots of the daily location of each tagged red snapper to determine if specimens had moved among receiver locations. Fish recapture data (location and date) were included in these plots. Long-term site fidelity was examined with a logistic regression model computed with Proc Logistic in SAS (SAS Institute Inc., 2003). All red snapper removed from the population by fishermen were omitted from the logistic regression analysis because removal created a forced absence. Red snapper from ST151-I and ST135-Q were removed from the regression dataset because of missing data due to receiver malfunction and late deployment (see below). Time was divided into ten-day periods beginning with the day of release for each fish. If a snapper was detected (i.e., perceived present) within a given period, it was coded as category one; if a snapper was not detected during the period, it was coded category zero. Native and relocated snapper were treated as two different populations in the logistic regression. A potentially confounding factor was introduced in late July 2003 when a thermocline developed below the depth of the receivers (see Results). Therefore, the presence of the thermocline was treated as a dummy variable ($z = 1$ when present, $z = 0$ when absent) in the logistic regression model.

Results

During this study 125 red snapper were tagged and implanted with acoustic tags. Tagged fish ranged from 28 to 47 cm TL, with a mean length of 36 cm (Figure 3). The majority of individuals were smaller than the legal size for both recreational and commercial harvest. Age estimated from fish length indicate these fish were predominately in the 2–4 year old age classes, with the possibility of some 5 year old fish present (Wilson and Nieland 2001). ROV video footage revealed that released red snapper

Table1. Receiver deployment locations, dates, and number of specimens released at each site.

Platform name	Latitude	Longitude	Date of deployment	Number of specimens released
ST135-M	28° 38.102’	90° 16.914’	05/21/2003	63
ST151-I	28° 37.540’	90° 16.334’	05/21/2003	17
ST128-R	28° 40.041’	90° 14.724’	05/22/2003	22
ST134-S	28° 39.409’	90° 14.132’	05/22/2003	0
ST151-Y	28° 36.970’	90° 14.974’	05/22/2003	5
ST134-W	28° 37.630’	90° 13.979’	05/22/2003	0
ST151-O	28° 36.926’	90° 15.146’	05/22/2003	0
Artificial Reef	28° 38.215’	90° 16.020’	06/16/2003	0
ST135-Q	28° 39.466’	90° 16.929’	07/29/2003	19

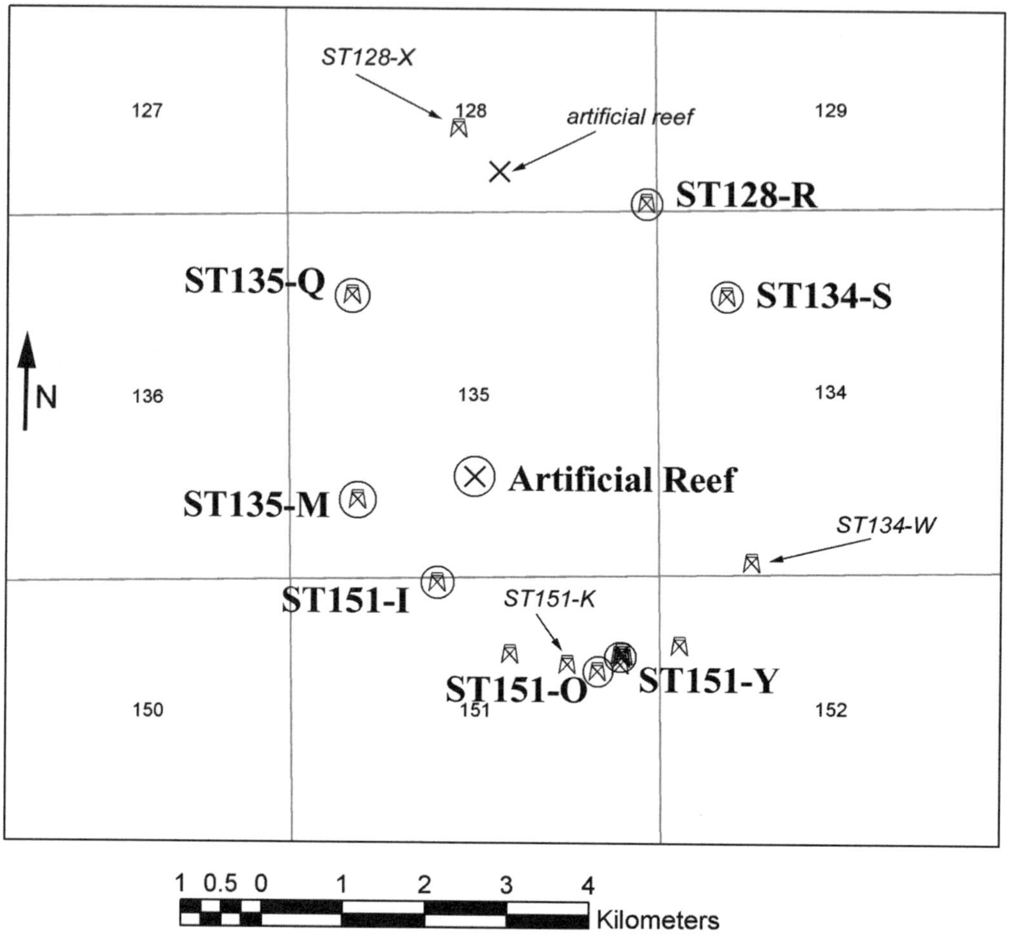

Figure 2. An enlargement of the area circled in Figure 1, displaying the final locations of acoustic receivers at seven platforms and one artificial reef in the study area. Small numbers denote Minerals Management Service (MMS) lease blocks, receiver locations are circled and labeled with designations assigned by the MMS; other structures of interest are also labeled.

had little difficulty orienting towards the platform.

Storm Damage

In late June 2003, Tropical Storm Bill moved through the study area and affected some of the deployed receivers. Receivers at ST134-W and ST134-S were either lost or damaged. No red snapper had been released at either location and the salvaged data file from the latter location contained no detections. Receivers and mounting apparatuses at ST128-R, ST151-Y, and ST151-O were unharmed. Some of the cables at ST151-I and ST135-M became detached during the storm, so the receivers were temporarily suspended from the platform cross beams on 11 July until the cabling system could be repaired. The receiver from ST151-I was also flooded, but examination of the data file revealed that the receiver had actually ceased to collect data six hours after it was originally deployed on 21 May due to equipment malfunction. On 29 July the cabling systems at all platforms were repaired. On this date a receiver was also deployed at ST135-Q, replacing the receiver at ST134-W, because 19 red snapper had been released at

ST135-Q and no specimens had been released at ST134-W; therefore, red snapper released at ST135-Q were only detected if they were still present at that site after 29 July.

Thermocline

While repairing the cabling at ST135-M on 29 July, a diver sensed the presence of a thermocline at approximately the same depth as the cross member used to temporarily deploy the receiver. When the receiver was positioned below the cross member, before repairs on the cabling system, it was below the thermocline. After the cabling system was repaired and the receiver was remounted as in the original design, it was moved above the thermocline. At the exact time and date the receiver was moved above the thermocline a decline in ping detection rate is evident in the data. Around the same date other receivers also experienced decreases in ping detection rate. VEMCO reported that a thermocline could dampen and deflect the acoustic signals (G. Coady, VEMCO Ltd., personal communication). We conducted a test at a platform by concurrently hanging one receiver above the thermocline and one below the thermocline for 30 min. The receiver below the thermocline detected 20 times as many pings as the receiver above the thermocline.

Unfortunately, environmental data were not collected continuously throughout the study because of low frequency of visits to the study site and constraints on equipment transport to the oil field during trips to upload data. Historic cruise data (1994–2003) collected within 15 mi of the study area in the month of July show evidence that thermoclines are common in this region at depths between 15 and 25 m with temperature differences across the cline of up to 8° C (N. Rabalais, Louisiana Universities Marine Consortium, personal communication, 11 February 2004). A 29 July press release from the Louisiana Universities Marine Consortium (LUMCON 2003) stated that two tropical storms in late June and mid July disrupted the large hypoxic zone which typ-

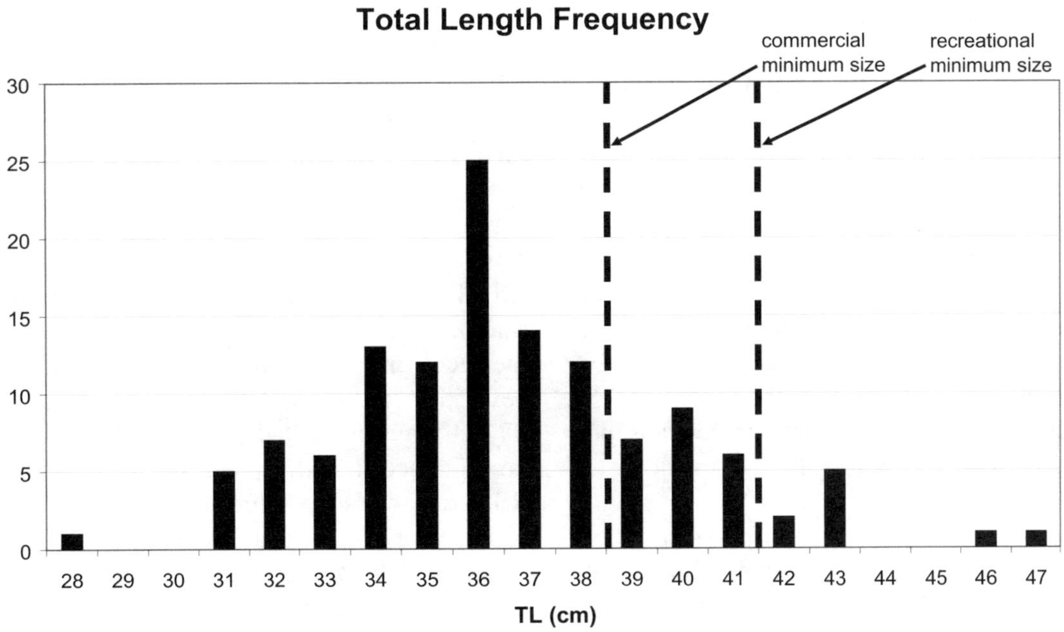

Figure 3. Histogram of the total length frequency of tagged red snapper *Lutjanus campechanus*.

ically forms in the northern GOM, mixing oxygenated surface waters with low oxygen bottom waters, and indicated a small zone of hypoxia (with concentrations of dissolved oxygen less than 2.0 mg/l) to the west of the study area which could have expanded in size as the summer progressed. This hypoxic zone did contain a strong thermocline in late July (Nancy Rabalais, Louisiana Universities Marine Consortium, personal communication, 11 February 2004) and could have been a contributing factor to the thermocline detected in the study area. Subsequent CTD casts from charter vessels during this study revealed that the water column in the study area was moderately mixed with no severe thermoclines by mid October 2003.

Short-term site fidelity

Seventy-eight percent ($n = 97$) of tagged fish were detected between 1 and 202 d following release (mean days at large = 64) (Figure 4). The remaining 22% ($n = 28$) were never detected during the study and were omitted from further analysis. The initial lack of or interrupted receiver coverage at ST135-Q and ST151-I may account for 24 of the 28 undetected tags. The other four undetected tags could have been caused by pinger malfunctions. Fish detected by receivers were estimated to be within 75 m of the platform based on VEMCO estimates of tag performance and ambient conditions in the study environment (Peabody and Wilson 2006). In general, detection versus nondetection varied between individual fish and with time. Most fish were detected within the vicinity of one platform on a daily basis with few absences longer than 2 d before final acoustic detection.

Red snapper movement between receivers appeared to be infrequent and erratic (i.e., no overall directionality or homing behavior), with movement among platforms documented for only seven of the 97 fish detected (Figure 4). Approximately three weeks after release, four individuals from ST151-I moved to ST135-M, the next platform to the north, stayed at that platform for two to three weeks, then were not detected again (tag numbers RS 16, 17, 19, and 21). One of these fish (RS 19) was eventually caught at ST151-K (28° 36.977' N, 90° 15.374' W), a platform 2 km to the south of ST151-I. A fifth fish (RS 56) was released at ST151-Y and detected at ST151-O (only 300 m to the west) the next day, then was never detected again. A sixth red snapper (RS 67) was released at ST135-M, stayed there for approximately 45 d, and then moved to an artificial reef 1.5 km to the east where it remained for two months before the final acoustic detection of this tag. Last, RS 81 was released at platform ST135-M where it remained for over two months. Just before that fish left the area, it was periodically detected at the artificial reef over a few days.

Homing behavior was not exhibited by any of the 35 relocated snapper. Of the seven snapper that showed movement between receivers, six were relocated fish (RS 16, 17, 19, 21, 56, and 67). Of these, four were caught at ST135-Q and relocated to ST151-I (RS 16, 17, 19, and 21). The northward movement shown by these four fish to ST135-M and then subsequent disappearance could have been homing behavior to ST135-Q, which was north of ST135-M. However, RS 19 was caught at ST151-K, farther to the south, and RS 21, which was the only one of these four snapper remaining in the population when the receiver was deployed at ST135-Q, was never detected there. RS 56 was captured at ST151-I and relocated to ST151-Y; there is no evidence that it homed toward ST151-I. RS 67 was also a relocated fish originally captured at ST134-S and released at ST135-M. Although this snapper did move east to the artificial reef receiver, it was never detected at ST134-S. The only location from where red snapper were relocated that was

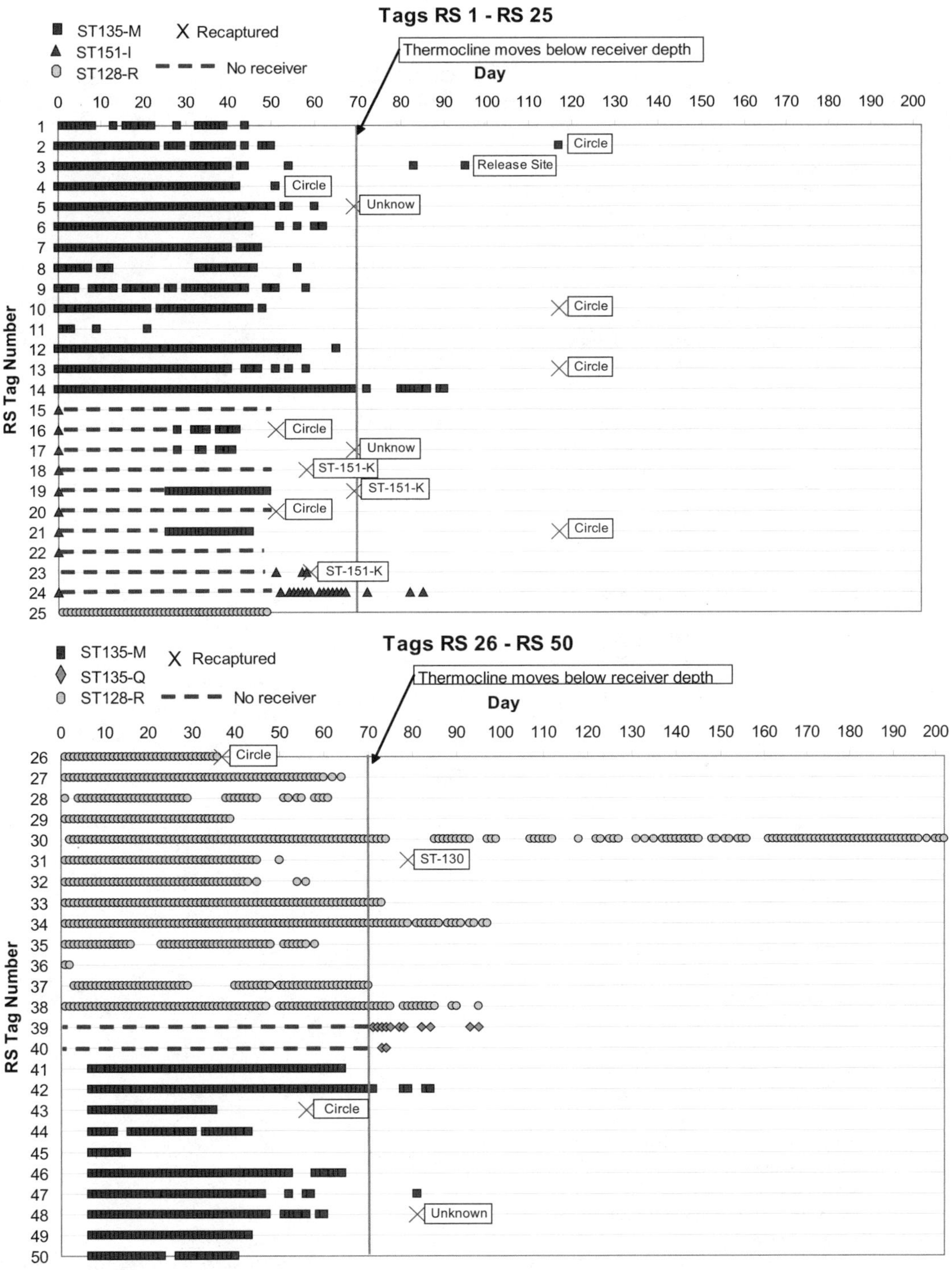

Figure 4. Daily locations of each tagged red snapper *Lutjanus campechanus* a) RS 1–25, b) RS 26–50 c) RS 51–75 and d) RS 76–97. Different symbols represent different locations, dashed lines denote a period during which the receiver was missing or not functioning, Xs and labels denote approximate date and location of recapture if known, and the vertical line at day 70 denotes the suspected interference of the thermocline with ping detection rate.

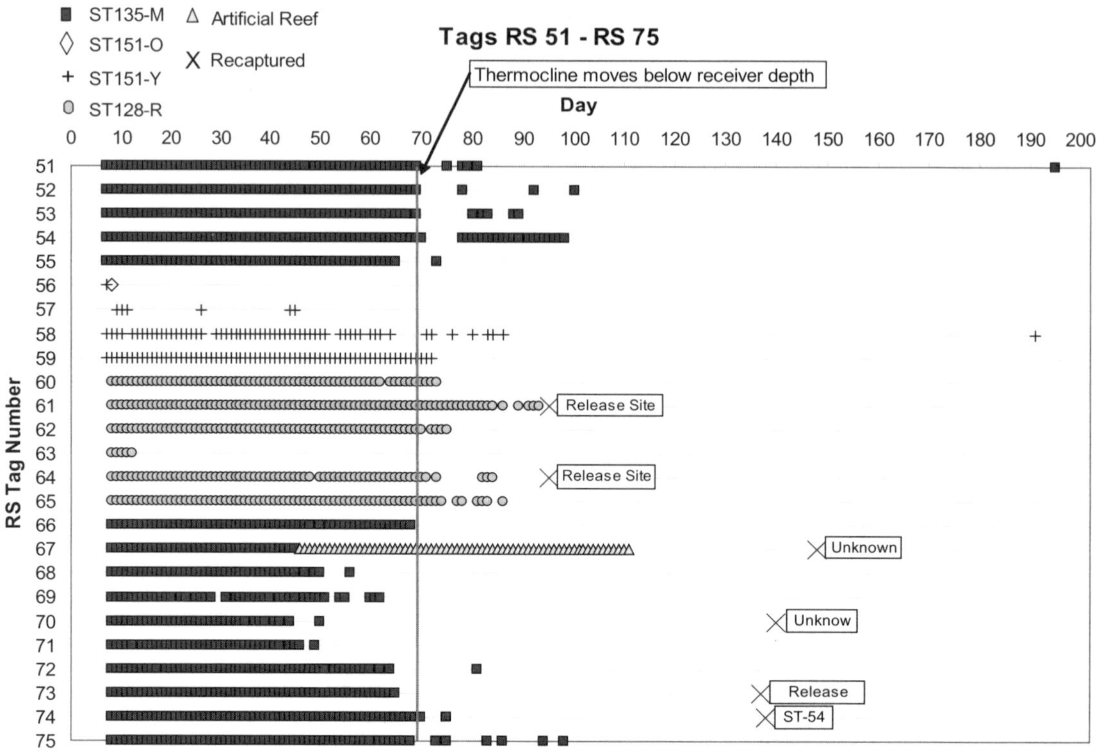

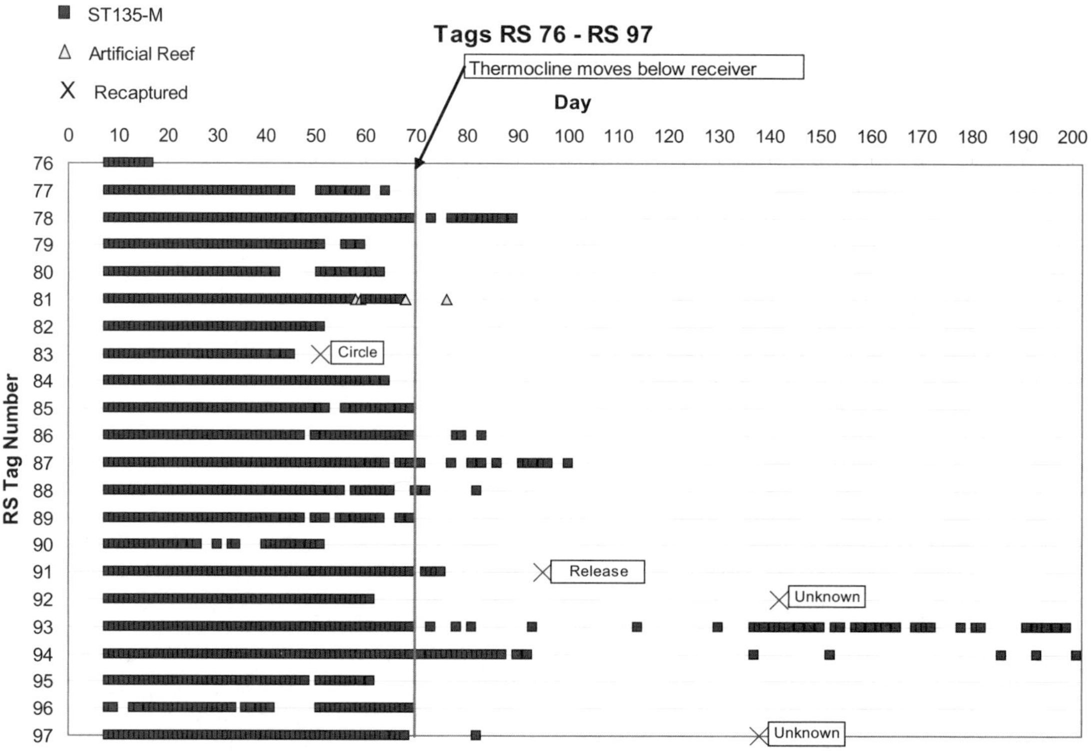

Figure 4. (Continued)

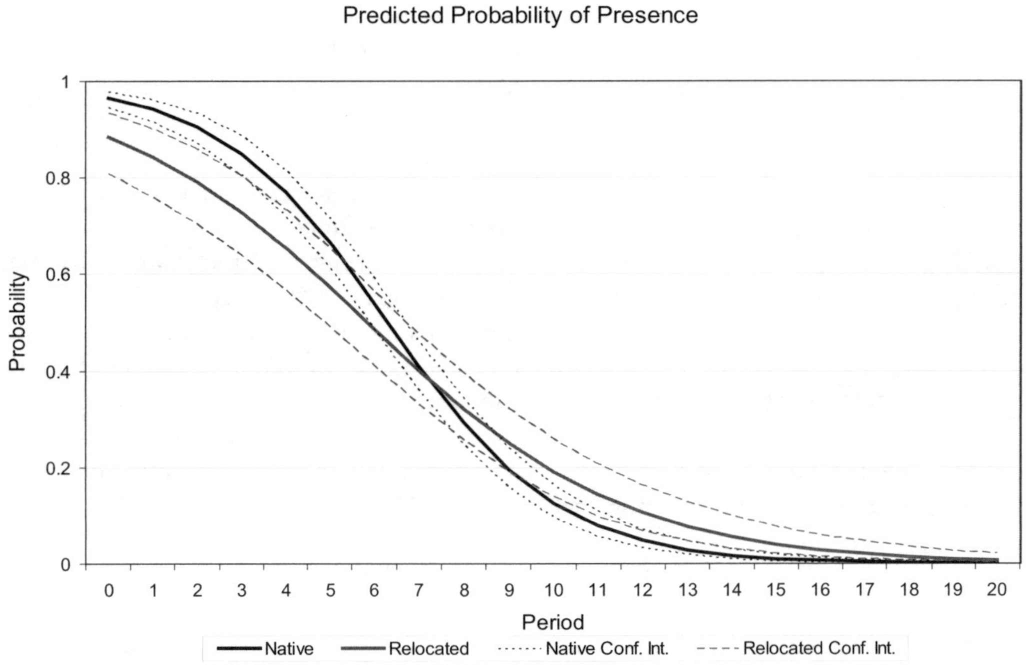

Figure 5. Predicted probability of presence at release location for both native and relocated snapper. Dashed lines are 95% confidence intervals.

not covered by a receiver was an artificial reef located between platforms ST128-X (28° 40.564' N, 90° 16.121' W) and ST128-R (Figure 2). Some snapper also were relocated from ST135-Q, which was not covered by a receiver until late July.

Long-term site fidelity

The results of the logistic regression model indicated time ($\chi^2_{df=1} = 259.33$; $p < 0.001$) , relocating fish ($\chi^2_{df=1} = 8.21$; $p = 0.004$), and thermocline presence ($\chi^2_{df=1} = 50.12$; $p < 0.001$) all significantly affected fish detection. The probability of detection decreased with time, with the 50% probability estimated to occur at approximately day 70 (Figure 5). In the earlier periods, relocated fish had a lower probability of detection at study receivers than did fish released at their capture location. During the later periods the relocated population had a higher probability of detection, but the 95% confidence intervals of the two functions overlapped during the latter time periods.

Pinger Malfunction

After pingers were returned by fishermen, each was tested in the laboratory and most were pinging as expected, including some tags returned as late as October 2003. In January 2004 we discovered that a great number of the pingers stored in the laboratory were no longer functioning. These pingers had been activated for only eight months, which is well below the predicted a battery life of over 400 d. VEMCO tested the tags and found that the battery power was expended, which resulted from an internal malfunction. Further testing revealed that some pingers were still transmitting an acoustic signal, though the output was very low, indicating that the battery capacity had declined within days of testing (G. Coady, VEMCO Ltd., personal communication, 26 February

2003). Because most pingers received from fishermen during the first few months of the study were operating correctly, we assume that most pingers still in the field were operating correctly during these initial months in which the majority of the acoustic data were collected. Thus, we completed the analyses as planned despite the likelihood some fish present at platforms were not detected due to pinger malfunction.

Discussion

Acoustic tagging studies of fish movement rely on several assumptions being met in order to infer fish site fidelity from detection of acoustic tags. Obviously, detection is contingent upon tags and receivers functioning properly. Acoustic signals propagate well in saltwater, but many obstacles to sounds waves or their detection may be found in the marine environment. Acoustic signals, or pings, can be deflected, obscured, or absorbed by solutes, suspended matter, plankton, fish, air bubbles, thermoclines, water turbulence, raindrops, wind, wave action, boat motors, submerged structures, and even biological noise like snapping shrimp (Winter 1983; Wolcott 1995). The range of detection for an acoustic telemetry system depends on the size, strength, and frequency of the acoustic transmitter (lower frequencies propagate farther), the level of ambient noise in the environment, and the detectability of the receiver (Klimley et al. 1998).

This study potentially was affected by issues of both tag and receiver functionality. Though the tag batteries did expire earlier than expected, we believe they functioned through the first six months of the study, thus analysis of site fidelity only extended to 200 d. We estimated that acoustic tags used in this study, given ambient conditions at platforms, could be detected within approximately 75 m of the receiver (Peabody and Wilson 2006). While this limited our ability to detect a fish's presence in areas immediately adjacent to platforms, our detection radius was sufficient to detect the presence of fish within a given platform without fear of detecting acoustic signals fish released at nearby platforms. Two of our receivers were damaged by Tropical Storm Bill in June 2003, which affected our ability to continuously track fish at two of our seven sites. Data from replaced receivers indicated some fish released at those sites were detected after several weeks of having nonfunctional receiver systems. A strong thermocline that persisted for several weeks in summer 2003 also may have affected tag detection. Although total detections per tag did decline, nearly half (46%) of tagged fish were detected when the thermocline was present.

Potential tag detection problems caused by tag battery life and the presence and effect of the thermocline in midsummer force us to divide discussion of red snapper site fidelity to platforms into two time stanzas: less than 70 d post tagging (approximate onset of the thermocline) and between 70 and 200 d post tagging. In the first stanza, daily location of each fish was used to examine site fidelity on a scale of days to weeks. In general, tagged fish did not regularly move from platform to platform, but remained at a site for a few weeks to even a month or two. Acoustic data suggested high short-term site fidelity and logistic regression models predicted approximately 50% of tagged fish would be present at platforms 70 d post tagging. Relocated fish showed no evidence of homing behavior to original capture locations. The majority of tagged snapper did not exhibit movement among receiver locations on a daily or weekly, or in most cases, even a monthly basis and none of the 35 relocated snapper in this study were detected or recaptured at their capture sites. Although six of the seven red snapper that did move between receiver locations were relocated snapper, none showed evidence of homing back to their original capture location. Patterson et al. (2001) also reported relocated snapper were more likely to move than fish not transported to other artificial reef sites prior to release.

Limits to carrying capacity of platforms or red snapper behavior may explain increased movement of relocated fish observed in our study. Any given platform can likely only support a finite biomass of red snapper, although some locations may have greater carrying capacity than others. Some platforms may provide better habitat than others because of either quantity or quality of food, competition with other species for resources, or lower risk of predation. Alternatively, if the environment at a platform is already ecologically balanced or saturated with red snapper or other species, interspecific or intraspecific competition for resources may effect the emigration of a certain number of red snapper to regain the ecological balance (Mueller et al. 1994; Overholtzer and Motta 1999; Munday et al. 2001).

Based on fish movements shown in acoustic data, we conclude that on a short time scale (days to weeks) red snapper exhibited high site fidelity to platforms. However, as the time scale increased (weeks to months) red snapper appeared to display lower site fidelity. Overall, the probability that tagged fish were present at release locations decreased as time increased. Some of the decrease in presence with time could be due to battery failure in acoustic tags or the onset of a thermocline. We attempted to control for the confounding effect of battery failure by restricting our analysis to the time period for which we estimated batteries were reliable (<200 d). The significant decline in detection of tagged fish when the thermocline was present remains troubling, but while the total number of detections per fish diminished, nearly half the tagged fish were detected at release sites during the second (post-thermocline onset) stanza of the study. Therefore, despite the caveats listed above, we perceive the decline in tag detection in the second stanza mostly reflects movement of fish away from release sites. If true, then our findings suggest lower long-term site fidelity than reported in previous conventional and ultrasonic tagging studies.

Though our short-term site fidelity estimates are mostly consistent with previous studies, our estimates of long-term site fidelity are lower than others have reported. We report the time to 50% of fish presence at release sites was approximately 70 d regardless if fish were moved prior to release or not. In a tag-recapture study of the coast of Florida, fishermen returned 26% of tagged fish and >90% had been recaptured at their site of release after being at liberty for an average of 113 d (Beaumariage 1969). Beaumariage and Bullock (1976) reported that red snapper show definite specific reef residency in shallow water and that the only extensive movement occurred at reefs in water deeper than 15 fathoms, which may have been forage-motivated or in response to reproductive stimuli. Results of two tagging studies of red snapper off the coast of Texas (Fable 1980; Gallaway et al. 1981) suggested high site fidelity to petroleum platforms with extremely low percentages of tag returns or re-sightings showing any movement from release locations. Szedlmayer and Shipp (1994) reported high site fidelity of red snapper to artificial reefs off Alabama from tag-recapture data, as 57% percent of fish were recaptured at their site of release and 76% of recaptures made within 2 km of release sites. Patterson and Cowan (2003) estimated a site fidelity rate from conventional tagging data of approximately 25% year^{-1}; thus, their model predicted 50% of fish were present at release sites after 180 d. Strelcheck et al. (2007, this volume) used similar methodology as Patterson and Cowan (2003) and estimated annual site fidelity rates of approximately 50% year^{-1}. Schroepfer and Szedlmayer (2006) estimated the probability that fish implanted with ultrasonic tags were detected within a 1.1 km^2 area around artificial reef sites off Alabama was approximately 50% after one year.

No previous study has attempted to estimate red snapper site fidelity to petroleum platforms off Louisiana, although Gallaway et al. (1981) and Fable (1980) performed conventional tagging experiments on platforms off the Texas coast when many fewer platforms existed on the GOM OCS (Pulsipher et al. 2001). It is possible that estimated site fidelity of red snapper off the coast of Louisiana was lower than reported in other regions due to the proximity and plethora of available habitat in the form of platforms. Frazer and Lindberg (1994) reported stone crabs (genus *Menippe*) exhibited lower site fidelity at more closely spaced artificial reefs. This relates to the 'resource mosaic hypothesis' proposed by Lindberg et al. (1990), which states that as reef spacing decreases, so does access to prey on the soft-bottom around each reef. The foraging halos, or areas of depleted prey, at closely spaced reefs may overlap and cause a disproportional depletion of resources. At closely spaced reefs, foraging fish may be forced to travel greater distances to locate a suitable amount of prey, possibly encountering new habitat. Platforms are sometimes closely spaced, within a few hundred meters of each other. Therefore, red snapper may encounter numerous other suitable habitats while foraging away from platforms. Because platforms offer such high-relief, vertical habitat, they may support higher densities of fish than natural reefs (Wilson et al. 2003). Thus, foraging halos at platforms may be even larger than would be found at natural reefs, forcing platform red snapper to forage at greater distances where they may lose sensory orientation to the platform and lose the ability to find their way back to that same platform. Last, the presence of hypoxia in the vicinity of our study sites in summer 2003 also may have further diminished prey resources and may explain the lower site fidelity estimates we report.

Overall, low site fidelity estimates reported herein have implications for our understanding of how red snapper utilize petroleum platforms as artificial reefs. It is clear from our acoustic data that fish do move between structures and closely spaced (<5 km) platforms. Thus, platforms, pipelines, and artificial reefs may form a functionally interconnected network of manmade reef fish habitats on the shelf. It is important to note that most fish recaptured by fishermen were reported as being caught at nearby platforms. Therefore, while site fidelity estimates to individual platforms was low in this study, no evidence exists of extensive movement of fish such as reported in some conventional tagging studies (e.g., Beaumariage 1969; Patterson et al. 2001; Szedlmayer and Shipp 1994). Restricted movement of red snapper off Louisiana is consistent with metapopulation structure inferred from genetic analysis (Gold and Sallaint 2007, this volume) and from differences in population dynamics among GOM regions (Fischer et al. 2004; Jackson et al. 2007, this volume). Alternatively, fish tagged in this study may have been smaller than sizes at which distance of movement might be expected to increase (Patterson et al. 2001).

Acknowledgments

We would like to extend our sincere appreciation for the advice and insightful comments of James Cowan, LSU Department of Oceanography and Coastal Sciences, and James Geaghan, LSU Department of Experimental Statistics. We also extend our thanks to the research associates, especially Mark Miller, and graduate students in the LSU Department of Oceanography and Coastal Sciences for assisting in tagging trips and field data collection and equipment maintenance. Thanks are also due to David Stanley of Golder Associates, Ed Chesney of the Louisiana Universities Marine Consortium and Mark Benfield and Larry Rouse of the LSU Department of Oceanography and Coastal Sciences for their help with project planning or data analysis. We greatly appreciate the help of ChevronTexaco Corporation, especially the pilots, crews, and staffs of South Timbalier 151 and the Leeville heliport and Dan Allen, senior ecologist. Thanks are due to the owner, captains and crews of Steve Tomeny Charters (Steve, Ed, Darryl, TC, Gabe, Travis, Roman, Matt, and Chad). This research was supported by the Minerals Management Service, U.S. Department of the Interior through the Coastal Marine Institute of Louisiana State University under a cooperative agreement with Louisiana State University.

References

Beaumariage, D. S. 1969. Returns from the 1965 Schlitz tagging program including a cumulative analysis of previous results. Florida Department of Natural Resources Technical Series 59:1–38.

Beaumariage, D. S., and L. H. Bullock. 1976. Biological research on snappers and groupers as related to fishery management requirements. Pages 86–94 *in* H. R. Bullis, Jr. and A. C. Jones, editors. Colloquium on snapper-grouper fishery resources of the western central Atlantic ocean. Texas A&M University Sea Grant College and Mississippi-Alabama Sea Grant Consortium.

Bohnsack, J. A., and D. L. Sutherland. 1985. Artificial reef research: a review for recommendations for future priorities. Bulletin of Marine Science 37(1):11–39.

Bull, A. S., and J. J. Kendall, Jr. 1994. An indication of the process: offshore platforms as artificial reefs in the Gulf of Mexico. Bulletin of Marine Science 55(2–3):1086–1098.

Continental Shelf Associates. 1982. Study of the effect of oil and gas activities on reef fish populations in the Gulf of Mexico OCS area. U.S. Department of Interior, Minerals Management Service, Gulf of Mexico OCS Region, MMS 82–10, New Orleans, Louisiana.

Dugas, R., V. Guillory, and M. Fischer. 1979. Oil rigs and offshore sport fishing in Louisiana. Fisheries 4(6):2–10.

Eklund, A. -M., and J. Schull. 2001. A stepwise approach to investigating the movement patterns and habitat utilization of goliath grouper, *Epinephelus itajara*, using conventional tagging, acoustic telemetry and satellite tracking. In *Electronic Tagging and Tracking in Marine Fisheries*, editor. J. R. Sibert and J. L. Nielsen, 189–216. Kluwer Academic Publishers, Netherlands.

Fabi, G., F. Grati, A. Lucchetti, and L. Trovarelli. 2002. Evolution of the fish assemblage around a gas platform in the northern Adriatic Sea. ICES Journal of Marine Science 59(Supplement S):S309–S315.

Fable Jr., W. A. 1980. Tagging studies of red snapper (*Lutjanus campechanus*) and vermilion snapper (*Rhomboplites aurorubens*) off the South Texas coast. Contributions in Marine Science 23:115–121.

Frazer, T. K., and W. J. Lindberg. 1994. Refuge spacing similarly affects reef-associated species from three phyla. Bulletin of Marine Science 55(2–3):388–400.

Gallaway, B. J., L. R. Martin, R. L. Howard, G. S. Boland, and G. D. Dennis. 1981. Effects on artificial reef and demersal fish and macrocrustacean communities. Pages 237–299 *in* B. S. Middleditch, editors. Environmental effects of offshore oil production: the Buccaneer Gas and Oil Field study. Plenum Press, New York.

Gold, J. R., and E. Saillant. 2007. Population structure of red snapper in the northern Gulf of Mexico. Pages 201–216 *in* W. F. Patterson, III, J. H. Cowan, Jr., G. R. Fitzhugh, and D. L. Nieland, editors. Red snapper ecology and fisheries in the U.S. Gulf of Mexico. American Fisheries Society Symposium 60, Bethesda, Maryland.

Hart, L. G., and R. C. Summerfelt. 1975. Surgical procedures for implanting ultrasonic transmitters into flathead catfish (*Pylodictis olivaris*). Transactions of the American Fisheries Society 104(1):56–59.

Hasting, R. W., L. H. Ogren, and M. T. Mabry. 1976. Observations on the fish fauna associated with offshore platforms in the northeastern Gulf of Mexico. Fishery Bulletin 74:387–402.

Jackson, M. W., J. H. Cowan, Jr., D. A. Nieland. 2007. Demographic differences in northern Gulf of Mexico red snapper reproductive maturation: implications for the unit stock hypothesis. Pages 217–227 *in* W. F. Patterson, III, J. H. Cowan, Jr., G. R. Fitzhugh, and D. L. Nieland, editors. Red snapper ecology and fisheries in the U.S. Gulf of Mexico. American Fisheries Society, Symposium 60, Bethesda, Maryland.

Klimley, A. P., F. Voegeli, S. C. Beavers, and B. J. Le Boeuf. 1998. Automated listening stations for tagged marine fishes. Marine Technology Society Journal 32(1):94–101.

Lindberg, W. J., T. K. Frazer, and G. R. Stanton. 1990. Populations effects of refuge dispersion for adult stone crabs (Xanthidae, Menippe). Marine Ecology Progress Series 66:239–249.

Love, M. S., J. Caselle, and L. Snook. 1999. Fish assemblages on mussel mounds surrounding seven oil platforms in the Santa Barbara Channel and Santa Maria Basin. Bulletin of Marine Science 65(2):497–513.

LUMCON (Louisiana Universities Marine Consortium). 2003. Tropical storms put the oxygen back into the "Dead Zone." Available: http://www.lumcon.edu/Information/News/default.asp?XMLFilename=20030731Hypoxia.xml. (August, 2006).

Mueller, K. W., G. D. Dennis, D. B. Eggleston, and R. I. Wicklund. 1994. Size-specific social interactions and foraging styles in a shallow water population of mutton snapper, *Lutjanus analis* (Pisces: Lutjanidae), in the central Bahamas Environmental Biology of Fishes 40:175–188.

Munday, P. L., G. P. Jones, and M. J. Caley. 2001. Interspecific competition and coexistence in a guild of coral-dwelling fishes. Ecology 82(8):2177–2189.

Nieland, D. L., and C. A. Wilson. 2002. Red snapper recruitment to and disappearance from oil and gas platforms in the northern Gulf of Mexico. Pages 270–279 *in* J. McKay, J. Nides, and D. Vigil, editors. Gulf of Mexico fish and fisheries: bringing together new and recent research. U.S. Department of the Interior, Minerals Managment Service, Gulf of Mexico OCS Region, New Orleans, Louisiana.

Overholtzer, K. L., and P. J. Motta. 1999. Comparative resource use by juvenile parrotfishes in the Florida Keys. Marine Ecology Progress Series 177:177–187.

Parker, R. O. Jr., D. R. Colby, and T. D. Willis. 1983. Estimated amount of reef habitat on a portion of the U.S. south Atlantic and Gulf of Mexico continental shelf. Bulletin of Marine Science 33(4):935–940.

Patterson, W. F., III and J. H. Cowan, Jr. 2003. Site fidelity and dispersion of red snapper associated with artificial reefs in the northern Gulf of Mexico. Pages 181–193 *in* D. R. Stanley and A. Scarborough-Bull, editors. Fisheries, reefs, and offshore development. American Fisheries Society, Symposium 36, Bethesda, Maryland.

Patterson, W. F., III, J. C. Watterson, R. L. Shipp, and J. H. Cowan, Jr. 2001. Movement of tagged red snapper in the northern Gulf of Mexico. Transactions of the American Fisheries Society 130:533–545.

Peabody, M. B., and C. A. Wilson. 2006. Fidelity of red snapper (*Lutjanus campechanus*) to petroleum platforms and artificial reefs in the northern Gulf of Mexico. U.S. Department of the Interior, Minerals Management Service, Gulf of Mexico OCS Region, Louisiana. OCS Study 2006–005. New Orleans, Louisiana.

Pulsipher, A. G., O. O. Iledare, D. V. Mesyanzhinov, A. Dupont, and Q. L. Zhu. 2001. Forecasting the number of offshore platforms in the Gulf of Mexico OCS to the year 2023. Prepared by the Center for Energy Studies, Louisiana State University, Baton Rouge, Louisiana. OCS Study MMS 2001–013. U.S. Department of the Interior, Minerals Management Service, Gulf of Mexico OCS Region, New Orleans, Louisiana.

Putt, R. E. J. 1982. A quantitative study of fish populations associated with a platform within Buccaneer Oil Field, northwestern Gulf of Mexico. Master's thesis. Texas A&M University, College Station, Texas.

Rabalais, N. N., R. E. Turner, and W. J. Wiseman, Jr. 2002. Gulf of Mexico Hypoxia, A.K.A. "The Dead Zone." Annual Review of Ecology and Systematics 33:235–263.

Rademacher, K. R., and J. H. Render. 2003. Fish assemblages around oil and gas platforms in the northeastern Gulf of Mexico: developing a survey design. Pages 101–122 *in* D. R. Stanley and A. Scarborough-Bull, editors. Fisheries, reefs, and offshore development. American Fisheries Society, Symposium 36, Bethesda, Maryland.

Render, J. H. 1995. The life history (age, growth and reproduction) of red snapper (*Lutjanus campechanus*) and its affinity for oil and gas platforms. Doctoral dissertation. Louisiana State University, Baton Rouge, Louisiana.

SAS Institute Inc. 2003. Statistics Version 9.0. SAS Institute Inc. Cary, North Carolina.

Schroepfer, R. L., and S. T. Szedlmayer. 2006. Estimates of residence and site fidelity for red snapper Lutjanus campechanus on artificial reefs in the northeastern Gulf of Mexico. Bulletin of Marine Science 78(1):93–101.

Stachowitsch, M., R. Kikinger, J. Herler, P. Zolda, and E. Geutebruck. 2002. Offshore oil platforms and fouling communities in the southern Arabian Gulf (Abu Dhabi). Marine Pollution Bulletin 44(9):853–860.

Stanley, D. R., and C. A. Wilson. 1989. Utilization of offshore platforms by recreational fishermen and SCUBA divers off the Louisiana coast. Bulletin of Marine Science 44(2):767–775.

Stanley, D. R., and C. A. Wilson. 1990. A fishery-dependent based study of fish species composition and associated catch rates around oil and gas structures off Louisiana. Fishery Bulletin 88:719–730.

Strelcheck, A. J., J. H. Cowan, Jr., and W. F. Patterson, III. 2007. Site fidelity, movement, and growth of red snapper: implications for artificial reef management. Pages 147–162 *in* W. F. Patterson, III, J. H. Cowan, Jr., G. R. Fitzhugh, and D. L. Nieland, editors. Red snapper ecology and fisheries in the

U.S. Gulf of Mexico. American Fisheries Society, Symposium 60, Bethesda, Maryland.

Szedlmayer, S. T., and R. L. Shipp. 1994. Movement and growth of red snapper, *Lutjanus campechanus*, from an artificial reef area in the northeastern Gulf of Mexico. Bulletin of Marine Science 55(2–3):887–896.

Watterson, J. C., W. F. Patterson, III, R. L. Shipp, and J. H. Cowan, Jr. 1998. Movement of red snapper, *Lutjanus campechanus*, in the north central Gulf of Mexico: potential effects of hurricanes. Gulf of Mexico Science 1998(1):92–104.

Wilson, C. A. and D. L. Nieland. 2001. Age and growth of red snapper, *Lutjanus campechanus*, from the northern Gulf of Mexico off Louisiana. Fishery Bulletin 99(4):653–664.

Wilson, C. A., A. Pierce, and M. W. Miller. 2003. Rigs and reefs: a comparison of the fish communities at two artificial reefs, a production platform, and a natural reef in the northern Gulf of Mexico. Prepared by the Coastal Fisheries Institute, School of the Coast and Environment. Louisiana State University. U.S. Department of the Interior, Minerals Mgmt. Service, Gulf of Mexico OCS Region. OCS Study MMS 2003–009. New Orleans, Louisiana.

Winter, J. D. 1983. Underwater biotelemetry. Pages 371–395 *in* L. A. Nielsen and D. L. Johnson, editors. Fisheries techniques. American Fisheries Society, Bethesda, Maryland.

Wolcott, T. G. 1995. New options in physiological and behavioural ecology through multichannel telemetry. Journal of Experimental Marine Biology and Ecology 193:257–275.

American Fisheries Society Symposium 60:123–144, 2007

Factors Affecting Catch and Release (CAR) Mortality in Fish: Insight into CAR Mortality in Red Snapper and the Influence of Catastrophic Decompression

JODIE L. RUMMER[1]
Department of Zoology
University of British Columbia
6270 University Boulevard, Vancouver, British Columbia V6T 1Z4 Canada

Abstract.—The red snapper *Lutjanus campechanus* fishery is arguably one of the most important in the Gulf of Mexico, but habitat destruction, climate change, and serial overfishing has resulted in significant population declines in red snapper and other high-profile fisheries species. The red snapper fishery may be one of the best examples where management strategies that promote catch and release (CAR) have failed. Populations have not recovered despite CAR management strategies, likely because CAR mortality is high; however, the basis for CAR mortality is unclear. Numerous studies associated with fishing-induced mortality were reviewed in an attempt to make generalizations as to how red snapper and other high-profile fisheries species respond to CAR. A framework for understanding CAR mortality in red snapper and other species was constructed based on four pillars: retrieval conditions, species and size relationships, handling, and release conditions. Each of these fishing factors was examined as to relative impact toward CAR. A predictive model was generated from all available data on CAR mortality. For a deep-water fish like red snapper, the underlying problem is directly related to capture depth, particularly injuries related to rapid swim bladder (SB) overinflation and catastrophic decompression syndrome (CDS). If not immediately lethal, depth-related injuries may have long term effects on growth and immune function that could go unnoticed and are unaccounted for in traditional field studies; all other fishing factors will only intensify this baseline impairment. Management plans are typically built under the assumption that CAR mortality is below 20%, but it is widely accepted that this is a gross underestimate. Modeling from this review suggests that, in red snapper, mortality may be as low as 20% but only if fish are caught between 0 and 20 m depths. This is not the case, and CAR mortality may reach 100% if fish are retrieved from deeper than 110 m. Current CAR management strategies are ineffective, and not enough information exists to impose maximum fishing depths. Given these limitations, a logical approach would be to restrict particular areas such that fish populations can be protected from all fishing and CAR activity, therefore protecting age, size, and sex classes and ratios. For fish species like red snapper, where overfishing is widespread and CAR mortality is high, or other species where CAR is unclear and a thorough investigation as to depth-related CAR mortality has not been performed, strategies based on space (i.e., marine protected areas and no-take reserves), rather than time or numbers (i.e., season closures, size limits, bag limits, etc.), have the greatest potential for overall conservation and sustainability and should be strongly considered.

[1] Corresponding author: rummer@zoology.ubc.ca.

Introduction

The Gulf of Mexico red snapper *Lutjanus campechanus* fishery developed in the mid 1800s and advanced alongside technology over almost two centuries; the red snapper has since become an icon species of the Gulf of Mexico (Moran 1988; Schirripa and Legault 1999). Recreational catch often surpasses the 2,000 metric ton annual commercial catch in the U.S., and the fishery as a whole is worth over $40 million U.S. annually (Schirripa and Legault 1999; Stevens 2004; SEDAR 7 2005). However, red snapper, as well as many other reef species, are overfished relative to established benchmarks for resource sustainability and have been declared "severely overfished" and appear on "species to avoid" lists (Helm and Smullen 1997; Stevens 2004; Ault et al. 2005). Historically, a fishery could be sustainable when populations have a spatial refuge from the fishery (Pauly et al. 2002). Although red snapper are traditionally caught from deep waters, which may offer some protection from the fishery, they are also caught near shore and exercise high site fidelity, making them an easy target for exploitation and necessitating management strategies for adequate protection (Moran 1988; Schirripa and Legault 1999; Coleman et al. 2000; Pritchard 2005). Stringent regulations such as season closures, bag limits, and size limits, implemented since 1990, have increased the proportion of red snapper caught that are subsequently released in the recreational fishery to over 50%, a 10-fold increase since the early 1980s (Figure 1). However, these strategies have failed to reverse the decline in red snapper populations, likely because not enough red snapper survive the catch and release process (Schirripa and Legault 1999; GMFMC 2000; Stevens 2004; Rummer and Bennett 2005).

Catch-and-release (CAR) fishing has been historically viewed as an approach toward conservation and ethical sustainability, but fisheries management strategies such as season closures, quotas, bag limits, and size limits that have been implemented in the recreational and commercial red snapper fisheries for almost two decades also promote CAR activity (Schirripa and Legault 1999; Casselman 2005; Cooke et al. 2005). Time closures and quotas are ineffective strategies, as they temporarily shift pressure to other species, and red snapper may still be caught and suffer release mortality in those fisheries (Stevens 2004; Coleman et al. 2000; GMFMC 2000). Bag limits can be problematic if fishing trips are not limited or the number of anglers increases (Coleman et al. 2000). Size limits and species prohibition strategies are only practical for hardy species in shallow-water systems, and few management plans are in place to preserve size, age, social structure, or the natural sex ratios of reef fish like red snapper (Coleman et al. 2000). The key to a successful CAR management strategy is that fish actually survive the CAR experience, which may be unlikely (Casselman 2005; Rummer and Bennett 2005). CAR-related mortality is reported to range from 16 to 30% depending on species, gear, depth, and season, but may be even higher (Figure 2) (Davis 2002; Stevens 2004; Casselman 2005).

According to Muoneke and Childress' 1994 review, CAR accounts for less than 15% mortality in lake trout and pikes occasionally exceeds 30% among drums, basses, trouts, and catfishes, but averages 68% among brown trout *Salmo trutta*, bluegill *Lepomis macrochirus*, crappie *Pomoxis annularis*, striped bass *Morone saxatilis*, and coho salmon *Oncorhynchus kisutch* (Muoneke and Childress 1994). Data from the 118 CAR studies involving over 120,000 fish in Casselman's (2005) review displayed average CAR mortality at 16.2% (Casselman 2005). Bartholomew and Bohnsack reported a comparable 18% average after reviewing 53 CAR studies; although averages for release mortality were described

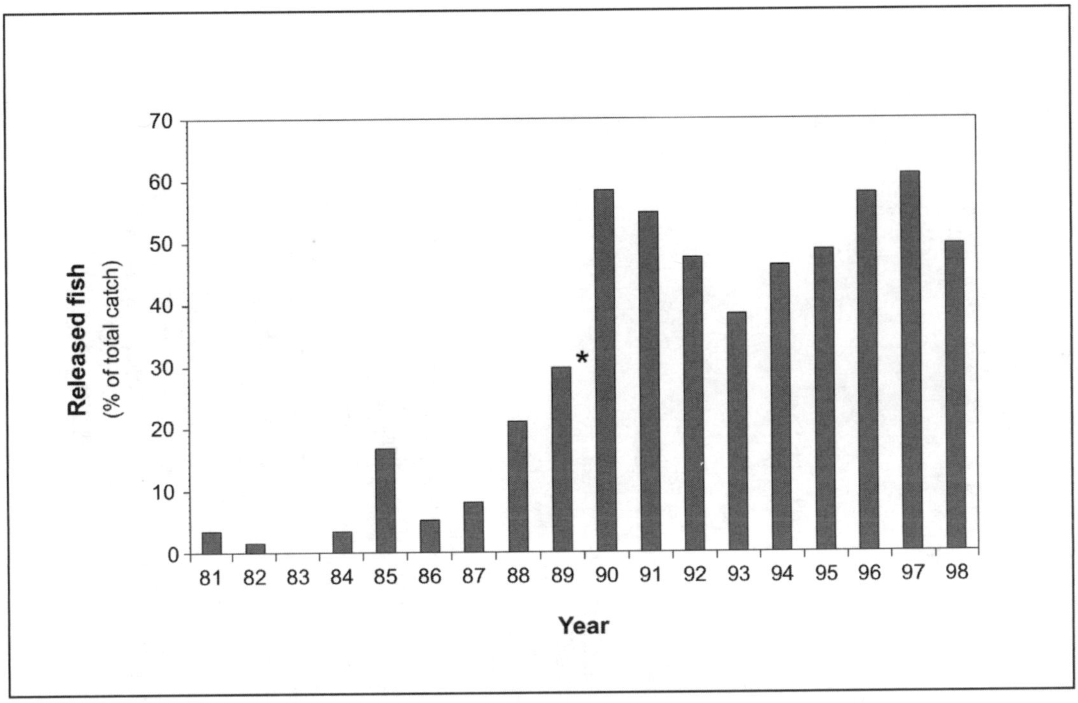

Figure 1. Estimated fractions of red snapper *Lutjanus campechanus* caught and released by recreational anglers between 1981 and 1998 in the Gulf of Mexico. The asterisk demarcates the onset of regulatory practices. Modified from Shirripa and Legault 1999.

as heavily skewed, varying substantially by species and within species, and ranging from 0 to 95% (Bartholomew and Bohnsack 2005). For red snapper, specifically, Porch and colleagues estimate release mortality to average approximately 46%, ranging 15–88% in both recreational and commercial fisheries combined (reviewed in SEDAR 7 2005). A framework for understanding the relative importance of factors responsible for CAR mortality and the underlying causes does not yet exist, hence the wide range in mortality rates reported (Figure 2) and the multitude of factors influencing overall CAR mortality. The purpose of this work was first to build a framework for understanding the factors that ultimately lead to CAR mortality in fish in general, and second, to integrate information into a predictive model to be used in assessing the interactions between factors that further affect CAR mortality. Not only will this approach give insight as to how red snapper respond to CAR but also how CAR mortality can be understood in other species as well.

Methods

I. Qualitative Approach

To approach CAR mortality qualitatively, a framework was built starting with the four general fishing factors that attribute to CAR mortality: retrieval conditions, species and size, handling, and release conditions (Figure 3). Based on data from over 200 studies investigating at least 40 species (Figure 2), the framework was expanded to include sub-elements under each of the four fishing factors (Figure 3). Some of the sub-elements outlined have been the focus of other studies (e.g., hook type and location, retrieval time and depth, swim bladder (SB) physiology and

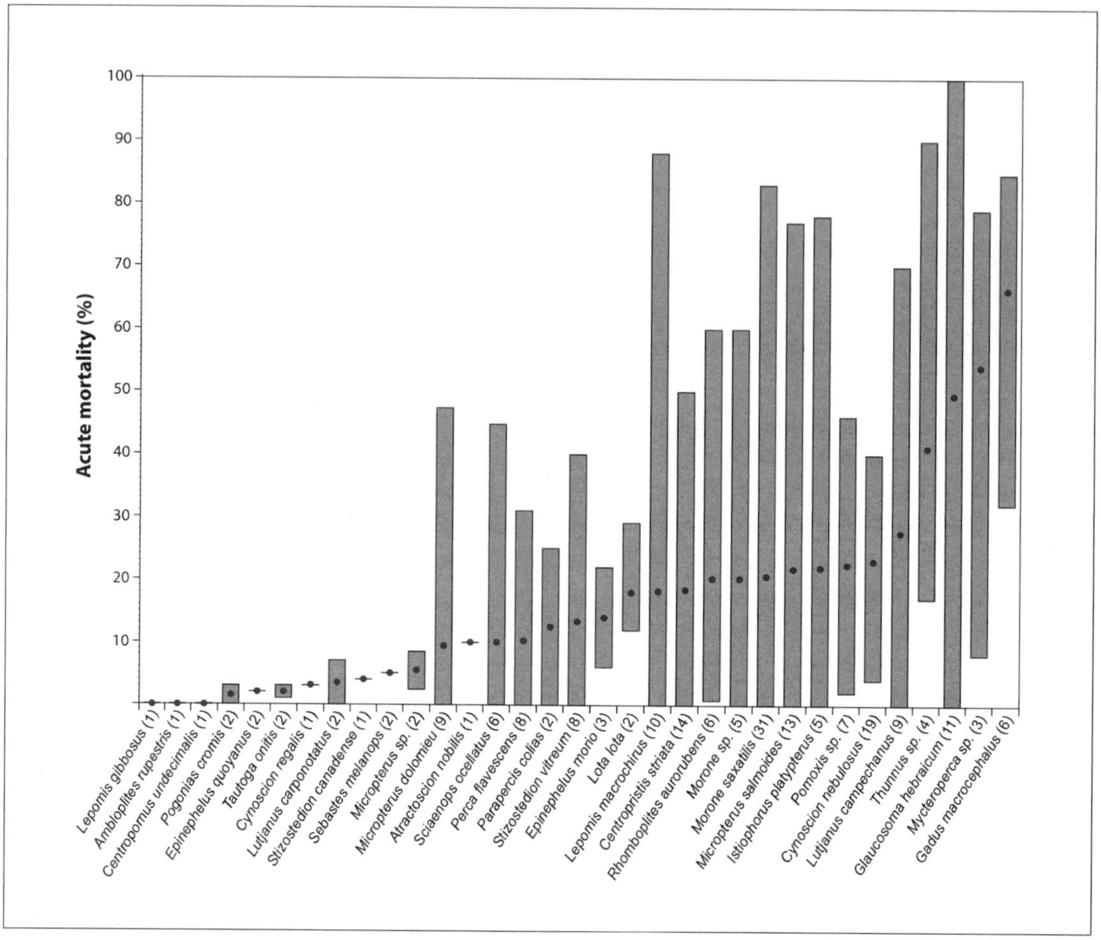

Figure 2. Summary of CAR mortality rates (%) in various physoclistous fish species. Means are derived from the average mortality reported from each study considered, but the bar extends to include the highest and lowest rates published for each species. The number in parentheses, following the species name, represents the number of studies considered. Data were compiled from reviews by (Muoneke and Childress 1994; Bartholomew and Bohnsack 2005; Casselman 2005) and studies by (Beggs et al. 1980; Fable 1980; Low 1981; Bugley and Shepherd 1991; Lee 1992; Bruesewitz et al. 1993; Gitschlag and Renaud 1994; Render and Wilson 1994; Murphy et al. 1995; Keniry et al. 1996; Render and Wilson 1996; Wilson and Burns 1996; Shasteen and Sheehan 1997; Bettoli and Osborne 1998; Nelson 1998; Collins et al. 1999; Bettoli et al. 2000;Cooke et al. 2001; Cooke et al. 2002a; Cooke et al. 2002b; Cooke et al. 2003a; Cooke et al. 2003b; Aalbers et al. 2004; Neufeld and Spence 2004; Cooke et al. 2005; Millard and Mohler 2005; Bettinger et al. 2005; Rummer and Bennett 2005; St. John and Syers 2005; Nichol and Chilton 2006)

morphology, air exposure, tactile protocol, temperature, and predation); others, like life history, health, capture history, reproductive stage, and diet and prandial status have not been as easily addressed (Figure 3). Acute effects, when identifiable, were documented for each factor but demarcated with a question mark when data were unavailable. The latent effects of CAR (including mortality) that have been investigated thus far were listed as the final component of the framework.

II. Quantitative Approach

The framework devised to describe the various factors responsible for CAR mortality was integrated into a predictive model that was generated to assess the influence of and potential interactions between factors on CAR mortality as a whole. Ideally, a suite of data from one comprehensive study where CAR mortality in red snapper was investigated and all parameters were examined equally would be used. A statistical model could then be developed to predict mortality in red snapper caught from a specific depth under various conditions. Unfortunately, a comprehensive study on red snapper does not yet exist. CAR mortality data in red snapper were used when available, but when necessary, average mortality rates from studies on other species were used as well. Assuming that each of the factors examined significantly affects mortality, a predictive model that can be validated and tested statistically was generated.

While it is possible to assume that certain factors will result in increased mortality, it is difficult to predict whether relationships are linear and which factors interact with others, resulting in multiplicative, rather than additive effects. However, it is known that stressors, in general, are often additive or multiplicative but rarely subtractive (Wedemeyer et al. 1990). The seven parameters that have been most heavily investigated in the literature were considered for the model and regarded as having additive effects on mortality. The seven parameters used include capture depth, venting, retrieval rate, hook type, surface to depth temperature differential, presence of surface predators, and handling and hook location. Only studies where one parameter was explicitly monitored or factors were investigated independently were utilized. Handling and hook location were integrated into the model as one factor because, aside from studies where deeply embedded hooks were not removed prior to fish release (i.e., line was cut instead of removing hook), it is assumed that handling time would increase if the hook was embedded deeply.

The following model describes the effects of the seven factors on general CAR mortality. Y_i is a binary variable measuring whether fish i is dead (1) or alive (0) when inspected after exposure to one of the seven parameters investigated. The first parameter, capture depth$_i$, is the depth (measured in meters) from which fish i was retrieved; percentages (0–100%) were assigned to each depth to represent corresponding mortality rates for each retrieval depth investigated. The second parameter, venting$_i$, is a binary variable recording whether fish i was vented (1) or not (0); percentages were assigned to vented (1) and unvented (0) fish to represent corresponding mortality rates. Thirdly, retrieval rate$_i$ is a binary variable and records the relative rate at which fish i was brought to the surface, either slow enough for acclimation to neutral buoyancy, which depends on species and was typically only observed in research-based collections where commercial and recreational fishing gear was not used, or fast, similar to commercial fishery retrieval rates; percentages were assigned to slow (0) and fast (1) rates to represent corresponding mortality rates. Fourth, hook type$_i$ is a binary variable recording if fish i was captured with

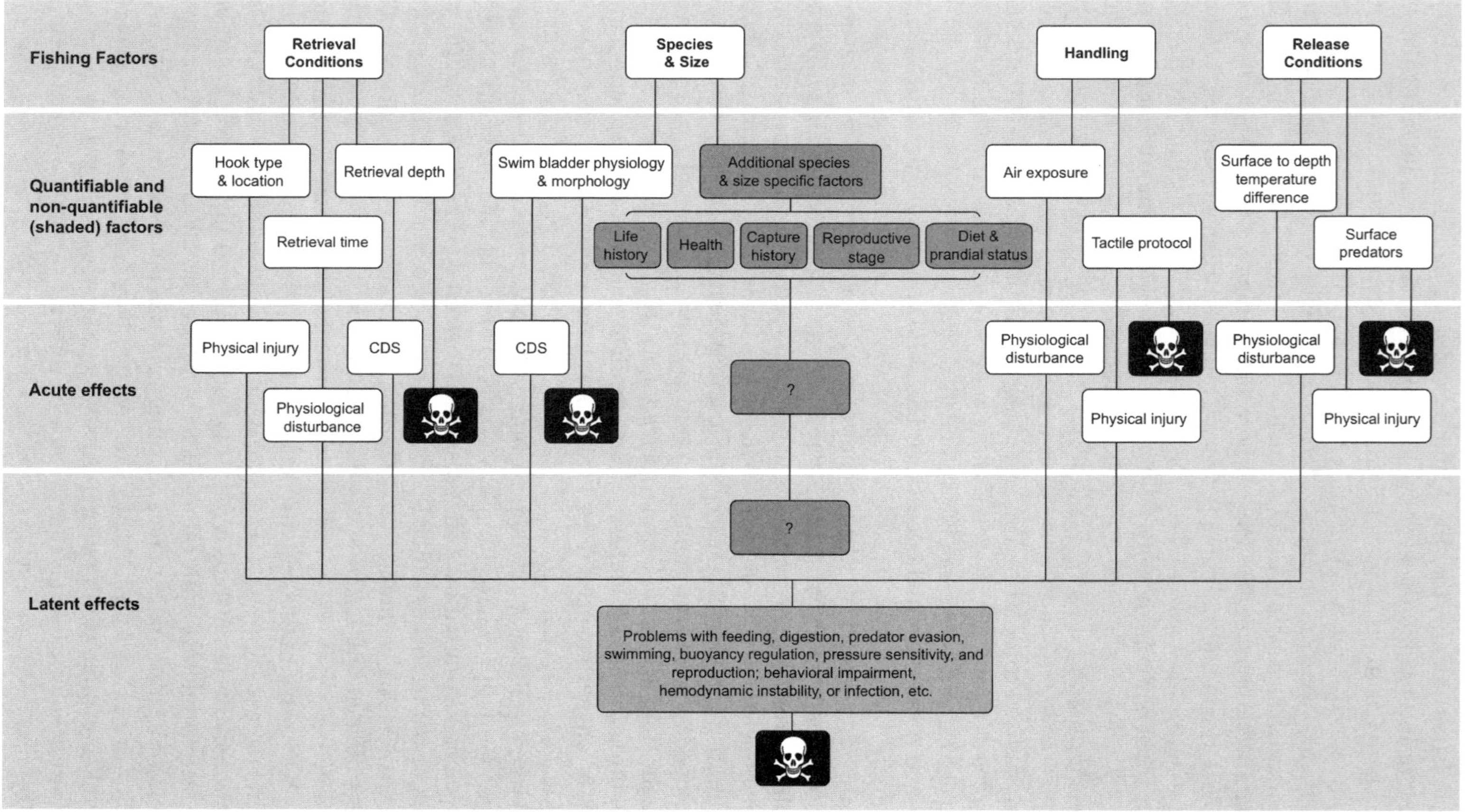

Figure 3. Framework for understanding catch and release (CAR) mortality based on four main fishing factors (top panel), eight parameters most commonly investigated as well as five factors that are not easily identifiable or quantifiable (second panel), and the acute and latent effects of CAR (third and fourth panels). Boxes shaded in gray represent uncertainties in the overall understanding of each factor's contribution to CAR, and black boxes with skull and crossbones represent probable mortality from the respective aspect of CAR or acute effects related to CAR.

a J hook (1) or circle hook (0); perecentages were assigned to each hook type to represent each corresponding mortality rate. Fifth, water temperature$_i$ is a binary variable recording the relative temperature of surface waters either being warm or outside the species' optimal temperature range (1), or cool or optimal for that particular species upon release (0); percentages were assigned to each temperature differential to represent mortality rates. The sixth parameter, predation$_i$, is a binary variable recording if there were surface predators present (1) or not (0) when fish *i* was released; percentages were assigned to the presence or absence of surface preadators to represent corresponding mortality rates. Finally, the seventh parameter, handling time & hook location$_i$, is a binary variable that combines both parameters and is (1) for long handling time and visceral hook location and (0) for short handling time and superficial hook location; percentages were assigned to each category to represent corresponding mortality rates. The model, which predicts Y_i, whether fish *i* is dead or alive, using *a-h* as constants and ε_i as the error term, was calculated using SigmaStat statistical program Version 3.0 (Systat Software Corp, Richmond, California, USA) and assumes that mortality increases with depth. The power of the test for each parameter (i.e., the probability of accepting an incorrect H_0 [coefficient is 0] when H_1 [coefficient is the estimated value] is true) was calculated at $\alpha = 0.05$. When power is low (<0.40) a nonsignificant result ($P > 0.05$) is inconclusive. The predictive model is as follows:

Given that, (1) data from an array of species were utilized for this model, (2) no two studies were executed identically, and (3) no study incorporated every factor of interest, the approach to modeling utilized for this contribution is not statistically concrete. However, it is critical that, even if only a theoretical predictive tool at this point, we have a starting point for future studies investigating CAR mortality in red snapper and other species.

Results and Discussion

I. Qualitative Approach

Fishing Factors: Retrieval conditions

Hook type/location.—Hook type and location have been the most thoroughly examined factors in CAR mortality studies and the leading causes for CAR mortality in shallow water species, including red snapper retrieved shallower than 30 m (Muoneke and Childress 1994; Watterson et al. 1998; GMFMC 2000; Burns et al. 2004). Fishing style, fish size, species feeding mode, and species mouth morphology necessitate an array of hook types (shape and barb presence/numbers) and sizes (Muoneke and Childress 1994; Bartholomew and Bohnsack 2005; Casselman 2005; Cooke et al. 2005). Circle hooks generally result in lower mortality rates (0–34%) than other hook shapes because they are rarely swallowed (Cooke et al. 2003b; Cooke et al. 2005; Millard and Mohler 2005). For this reason, circle hooks are recommended for red snapper as well as many other species. However, circle

$$Y_i = \frac{\exp\begin{pmatrix} a + b*\textbf{Capture depth}_i + c*\textbf{Venting}_i + d*\textbf{Retrieval rate}_i + e*\textbf{Hook type}_i \\ + f*\textbf{Water temperature}_i + g*\textbf{Predation}_i + h*\textbf{Handling \& hook location}_i \end{pmatrix}}{1+\exp\begin{pmatrix} a + b*\textbf{Capture depth}_i + c*\textbf{Venting}_i + d*\textbf{Retrieval rate}_i + e*\textbf{Hook type}_i \\ + f*\textbf{Water temperature}_i + g*\textbf{Predation}_i + h*\textbf{Handling \& hook location}_i \end{pmatrix}} + \varepsilon_i$$

hooks require more time to remove and eye hooking, which may permanently impair vision, is common (Cooke et al. 2003b; Cooke et al. 2005; Millard and Mohler 2005). J-hooks are easier to set and remove compared to circle hooks but result in higher mortality rates (0–46%) because they are more prone to embed deeply, resulting in damage to heart, liver, gill arches, kidneys, and intestines (Cooke et al. 2005). Barbless hooks will not embed in a fish easily, and for that reason, are less desirable to anglers (Cooke et al. 2001). However, barbless hooks are easier to remove than barbed counterparts and therefore reduce handling time, tissue damage, and ultimately mortality rates (Cooke et al. 2001). Bait type may also result in differences in hooking mortality. Lunging behavior, common in carnivorous species like the red snapper, regularly results in esophageal hooking and therefore increased hook removal and air exposure time and chance of additional injury (Muoneke and Childress 1994; Wilde et al. 2000; Burns et al. 2004; Bartholomew and Bohnsack 2005; Casselman 2005). Some investigations, however find data on hooking-induced CAR mortality inconclusive (Aalbers et al. 2004; Bartholomew and Bohnsack 2005; St. John and Syers 2005). While hooking may rarely result in immediate mortality, latent effects of hooking and multiple hooking events should be considered, especially for red snapper and other long-lived and high site fidelity species that may encounter angling often (Bartholomew and Bohnsack 2005). Clearly, mortality related to hook type is related to how and where the hook penetrates the fish, bait type, and how difficult it is to remove the hook once the fish is retrieved from the water (Pelzman 1978; Murphy et al. 1995; Nelson 1998; Wilde et al. 2000; Aalbers et al. 2004; Burns et al. 2004; Lindsay et al. 2004; Bartholomew and Bohnsack 2005).

Retrieval time.—Acute and latent effects of retrieving a fish rely not only on specific details of hooking, as outlined above, but also on the degree and duration of struggle as the fish is brought to the surface (Gustaveson et al. 1991; Tufts et al. 1991; Ferguson and Tufts 1992; Cooke et al. 2001; Stephens et al. 2002; Cooke et al. 2003a; Suski et al. 2003; Bartholomew and Bohnsack 2005; Bettinger et al. 2005; Casselman 2005; Morrissey et al. 2005; Lupes et al. 2006). The physiological effects of play (how long it takes to retrieve a fish) have been well studied. Acute effects include changes in heart rate, cardiac output, blood pressure, ventilation rate, plasma parameters (e.g., catecholamines, corticosteroids, glucose, lactate, chloride, and osmolarity), respiratory and metabolic acid-base balance, and reductions in muscle energy stores. The acute physiological effects of retrieval may take several hours to return to baseline levels, potentially resulting in cellular and tissue damage, immune suppression, changes in behavior, and ultimately increased mortality (Beggs et al. 1980; Wood et al. 1983; Tufts et al. 1991; Muoneke and Childress 1994; Wells 1996; Davis 2002; Cooke et al. 2001; Manire et al. 2001; Cooke et al. 2003a; Cooke et al. 2003b; Bartholomew and Bohnsack 2005; Casselman 2005; Cooke et al. 2005; Cooke and Suski 2005). Despite the fact that the overall physiological response to retrieval and play has been clearly outlined, monitoring changes in physiological parameters upon capture, prior to release, and post release has not been found adequate for predicting mortality (Davis 2002).

Retrieval depth.—Countless field studies have considered gear types and retrieval times responsible for CAR mortality in shallow water species, but for a deep-water fish like red snapper, the fundamental concern with CAR is capture depth (Rummer and Bennett 2005). It is well known that, if a fish moves above the level at which it is in hydrostatic equilibrium with its environment, the decrease in hydrostatic pressure (1 atm for every 10 m of water) leads to an expansion of the SB

(Harden-Jones 1952). If ambient pressure is rapidly reduced, catastrophic decompression (CD) may result in SB overexpansion (Figure 4) or rupture and internal injuries, collectively referred to as catastrophic decompression syndrome (CDS) (Harden-Jones 1952; Keniry et al. 1996; Schmidt-Nielsen 1997; Collins et al. 1999; Neufeld and Spence 2004; Rummer and Bennett 2005; St. John and Syers 2005). Injuries associated with CDS can often be observed superficially, immediately as a fish is brought to the surface. In the field, researchers have observed that up to 50% of red snapper retrieved from depth possess superficial injuries, most of which were related to SB over-expansion (e.g., stomach eversion from the mouth and intestinal protrusion) (Gitschlag and Renaud 1994; Rummer and Bennett 2005). However, external symptoms of CDS in red snapper have not been found to be accurate predictors of mortality (Gitschlag and Renaud 1994; Rummer and Bennett 2005). The lack of correlation between external injuries and mortality is problematic in many other deep-water species as well. For example, Neufeld and Spence retrieved burbot (*Lota lota*) directly from depths ranging 13–31 m and found 22% of the fish died within 10 min, all of which exhibited varying degrees of CDS, but no trend could be established from superficial observations (Neufeld and Spence 2004). Necropsy results revealed severe internal injuries, and mortality was likely due to ruptured blood vessels, hemorrhaging, and hematomas in the pericardial region (Neufeld and Spence 2004). Information on CDS in red snapper has been uncovered largely through systematic laboratory experiments and thorough necropsy immediately following CD (Rummer and Bennett 2005). Cardiac injuries, including hemorrhaging and hematomas, that would likely be fatal in nature, were sustained by 18% of red snapper decompressed from pressures equivalent to 30 m and in 90% of red snapper decompressed from pressures corresponding to 110 m depth (Rummer and Bennett 2005). It is certain that external symptoms of CD account for only a slight proportion of the overall detriment sustained by the fish when retrieved from depth; therefore, the underlying causes for mortality can probably only be uncovered via thorough necropsy (Gitschlag and Renaud 1994; reviewed in Muoneke and Childress 1994; Keniry et al. 1996; Cooke and Suski 2005; Morrissey et al. 2005; Rummer and Bennett 2005; Nichol and Chilton 2006). Capture depth may be the most important factor influencing CAR mortality in deepwater species, but the response likely varies by species and fish size; this area of study has not yet received ample attention (Gitschlag and Renaud 1994; Keniry et al. 1996; Rummer and Bennett 2005; St. John and Syers 2005).

Fishing Factors: Species and size

Swim bladder physiology and morphology.—Studies on SB organization and physiology date back to the early 1800s. However, linking a species' SB physiology and morphology to CD-mediated overinflation and expansion patterns and ultimately the type and degree of injuries a fish sustains upon retrieval from depth, is a new area of study. It is known that species with closed (physoclistous) SBs are more prone to CD-related CAR mortality than species with open (physostomous) SBs, lacking functional SBs, or from surface or shallow (<5 m) waters (Davis 2002; Neufeld and Spence 2004; Bartholomew and Bohnsack 2005; Morrissey et al. 2005). Physostomes can rapidly remove excess gas by using the pneumatic duct as an escape valve and are therefore less likely to experience rapid SB overexpansion (Harden-Jones 1952). For the remainder of this review, however, the focus will be on physoclists, like red snapper, that must utilize the slow process of gas resorption into the blood to empty the SB. It may take several hours for the oval window to sufficiently resorb (re-

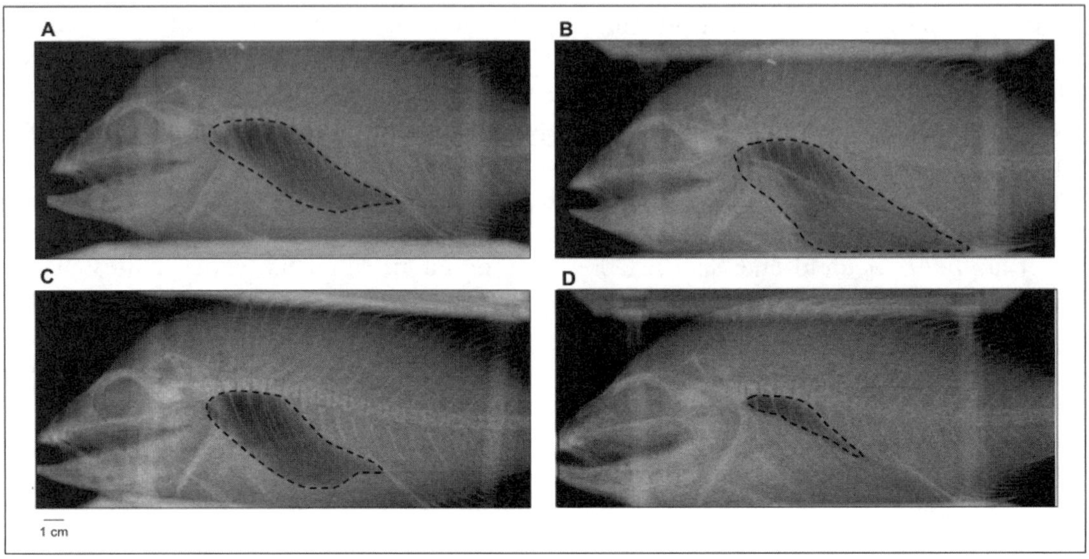

Figure 4. Lateral aspect X-ray images taken of red snapper acclimated to ambient pressure of 1atm (panels A and C). Panel B is following decompression of the fish from panel A. The fish from panel A was decompressed from a pressure of 6atm, a simulated depth of 50 m, at a rate of 0.1atm/s. Panel D represents the fish from panel C following decompression from a pressure of 12atm. An acclimation depth of 110 m was simulated, and decompression was executed at a rate of 0.1atm/s, during which time the fish's swim bladder ruptured. The broken line in each image demarcates the swim bladder boundary. Panels A and B were modified from Rummer and Bennett (2005), and panels C and D are images compiled from unpublished data from Rummer.

move) gas from the SB to maintain neutral buoyancy, and if pressure rapidly decreases, SB overexpansion will occur (Harden-Jones 1952; Keniry et al. 1996). Resorption rates are not as well characterized as SB secretion (filling) rates, and so most of our understanding is based on secretion data. Red snapper can acclimate to changes in depth (by filling the SB), while maintaining neutral buoyancy, at a rate no faster than 0.52 m per hour, a rate comparable to averages reported for other species, ranging 0.21–2.5 m per hour (Alexander 1972; Wittenberg and Wittenberg 1974; Ribbink and Hill 1979; Harden-Jones and Scholes 1985; Rummer and Bennett 2005). Secretion and resorption are mechanistically different and vary by species, but secretion rates are generally faster than if not equal to resorption rates, meaning 0.52 m per hour is a conservative estimate for resorption. This indicates that a red snapper would have to be retrieved from 50 m over a period of at least four days in order to avoid SB overexpansion upon retrieval. Healthy fish devoid of CDS have been retrieved from depth for research purposes when divers cage fish at depth and subsequently initiate a step-wise ascent over hours or days to bring fish to surface pressures without the risk of CD, but the protocol is clearly not feasible in recreational or commercial fisheries (Haight et al. 1993; Neufeld and Spence 2004).

The most obvious differences among SBs are at the morphological level. Differences in volume, shape, and elasticity vary by species and are also most pronounced between freshwater (FW) and seawater (SW) teleosts. If calculated relative to water density, the SB of a FW teleost occupies approximately 7% of the body volume, and the SB of a SW teleost occupies slightly less, approximately 5% of the body volume (Harden-Jones 1952; Alexander 1972; Alexander 1993). In line with this concept, red snapper SBs occupy an average 4.86% (n = 64) of the total body volume (Rummer, unpublished data). Expansion patterns depend on volume but also differ with SB shape. SB shape is typically ellipsoidal, as is seen in the red snapper, but several species defy this trend and possess multi-lobed or even heart-shaped SBs (Barimo 1998; Davenport 1999; Carpenter 2004; Rummer and Bennett 2005; Strand et al. 2005). Change in SB shape during CD is also influenced by the passive resistance generated from the SB wall and surrounding tissues (Harden-Jones 1952). A thick-walled SB, lacking substantial elastic properties, and therefore resisting expansion (e.g., *Gadus* spp.), may be more likely to tear or rupture than a thin-walled, less resistant SB (e.g. *Perca* spp.) (Harden-Jones 1952; Rogers et al. 1986; Nichol and Chilton 2006). SB rupture is rare in red snapper and has been observed in only 3% of fish in laboratory studies (Figure 4), which may give insight into the elasticity of red snapper SBs (Rummer and Bennett 2005; Rummer, unpublished data).

If SB rupture is common for a species, however, repair time is crucial. SB rupture has been observed in 90% of cod investigated and found to occur after pressure reductions greater than 50% or if fish are retrieved from deeper than 30 m (Harden-Jones 1952; Wilson Jr. and Smith Jr. 1985; Nichol and Chilton 2006). However, tears in cod SBs are repaired quickly, within 1–2 d; whereas, red snapper average 14 d, and other species may take longer than 4–8 weeks (Rankin et al. unpublished data; Rummer, unpublished data; Bruesewitz et al. 1993). Loss of SB function via rupture or overexpansion affects maneuverability that, in some species, can result in compensatory fin movements and a 20% increase in energy expenditure to maintain position in the water column (Harden-Jones 1952; Alexander 1972; Alexander 1993; Gitschlag and Renaud 1994). Reduced vertical migration rates and erratic recuperation behavior are common and probably related to SB volume leakage (Strand et al. 2005; Nichol and Chilton 2006). The SB is nature's solution to the buoyancy problem in teleosts and an anatomical and physiological feature potentially responsible for the extensive adaptive radiation of modern teleost fishes. However, the SB may also be the basis for the ultimate demise of red snapper and other high profile fisheries species in a CAR-based fishery. All things considered, capture depth is the underlying factor responsible for the greater part of CAR mortality, and so it is reasonable to begin species-specific CAR mortality investigations with an extensive understanding of SB physiology and morphology.

Additional species and size specific factors influencing CAR mortality.—Life history and reproductive stage, diet and prandial status, health, and capture history are aspects that can affect how a fish responds to the initial hooking and retrieval processes, SB expansion, as well as recovery post release. However, data are limited, contradictory, or only anecdotal. Depending on species, size may attribute to post release survival; large fish appear to descend faster, but small fish ultimately recuperate faster from the initial stress (Muoneke and Childress 1994; Nelson 1998; Wilde 1998; reviewed in Bartholomew and Bohnsack 2005; reviewed in Casselman 2005; Millard and Mohler 2005; Bettinger et al. 2005; Nichol and Chilton 2006). Other studies have found the opposite or no trend at all, further implying species dependence on size and survival relationships (Gitschlag and Renaud 1994; Bettoli and Osborne 1998; Wilde et al. 2000; Davis 2002).

Slight differences within a species attributed life history stage or reproductive status may affect post release survival both directly, by affecting hormone levels and the magnitude of the stress response, and indirectly via SB expansion patterns and internal organ displacement (Pankhurst and Dedual 1994; Machias and Tsimenides 1996). Ripe gonads in both male and female fish occupy a substantial portion of the body cavity and may alter SB expansion patterns and consequently, internal organ displacement. Intraspecific variations may be due to differences in gonad shape and size (large and tubular in females versus flat and thin in males). Fish with substantial body fat present in the abdominal cavity may also respond differently upon SB overexpansion; fat may insulate internal organs thus preventing or alleviating compaction injuries. However, excess abdominal body fat decreases available body cavity space for the SB to expand and may result in a lower threshold for when SB expansion-mediated displacement injuries shift to compaction injuries (Rummer and Bennett 2005). Postprandial physiological parameters could magnify the physiological stress associated with exhaustive exercise experienced by the fish upon retrieval as well, but little information exists to expand on this point (Busk et al. 2000; Hicks and Bennett 2004). Finally, multiple CAR events increase the probability of severe injuries (Nichol and Chilton 2006). Whether this increase is due to unhealed physical injuries or the chronic effects of a previous physiological disturbance has yet to be investigated (Nichol and Chilton 2006). A thorough understanding of species and size relationships relative to CAR and factors affecting mortality is needed.

Fishing Factors: Handling

Air exposure.—The time interval during which a fish is brought to the surface and returned to the water is crucial to post release survival in red snapper as well as many other species. Burns et al. (2002) suggest a direct relationship between short surface intervals and increased likelihood for post release survival. Air exposure can be detrimental to many species but is a necessary component of the de-hooking and release process. Rock bass, *Ambloplites rupestris*, exposed to air for less than one minute require up to two hours to fully recover (Cooke et al. 2001). However, Bettoli and Osborne (1998) found air exposure to be unrelated to CAR mortality in striped bass, *Morone saxatilis*. Air temperature, rather than exposure time, more strongly influences CAR mortality in striped bass and sablefish, *Anoplopoma fimbria* (Bettoli and Osborne 1998; Lupes et al. 2006). Stress parameters (plasma cortisol and glucose) were significantly elevated, and the immunological response was suppressed in sablefish exposed to elevated air temperatures (Lupes et al. 2006). Direct cause for release mortality cannot be easily defined and generalizations cannot yet be made, but it is clear that air exposure (time and temperature) negatively affects post release survival.

Tactile protocol.—Excessive handling and use of landing nets when fish are retrieved and released can cause physical injury and physiological stress (reviewed in Casselman 2005). Many protocols however, recommend handling a fish long enough to vent the over-expanded SB with a cannula or hypodermic needle prior to releasing the fish (Keniry et al. 1996; Wilson and Burns 1996; Burns et al. 2002). The technique alleviates compression on internal organs and allows an otherwise positively buoyant fish to quickly return to depth (Keniry et al. 1996; Wilson and Burns 1996; Burns et al. 2002). Results from some venting studies remain inconclusive, and some data suggest the process is detrimental (Gotschall 1964; Bruesewitz et al. 1993; Render and Wilson 1994 & 1996). The venting process will undoubtedly increase handling time and air exposure, and many investigators recommend avoiding venting for those reasons (reviewed in Casselman 2005). Properly venting a fish requires knowledge of the species' internal anatomy and, if done improperly, can result in increased mortality due to infection or damage to vital organs (Parrish and Moffitt 1993). By and large, if a fish exhibits a noticeably expanded SB or external symptoms of SB overexpansion that would necessitate venting, the fish has already sustained displacement and compaction injuries, and all other fishing factors will only amplify this baseline level of insult.

Fishing Factors: Release conditions

Difference between depth and surface water temperature.—CAR mortality in many species positively correlates with the temperature differential between conditions at depth of capture and the surface temperatures (Muoneke and Childress 1994; Murphy et al. 1995; Nelson 1998; Wilde 1998; Wilde et al. 2000; Bartholomew and Bohnsack 2005; Bettinger et al. 2005; Campbell et al. unpublished data). Warm surface waters can account for up to an additional 7–31% increase in mortality in Lutjanids, Percids, and Serranids (Fable 1980; Low 1981; Bugley and Shepherd 1991; Gitschlag and Renaud 1994; Muoneke and Childress 1994; Keniry et al. 1996; Wilson and Burns 1996; Shasteen and Sheehan 1997; Collins et al. 1999; Bartholomew and Bohnsack 2005; Casselman 2005; Rummer and Bennett 2005). Mortality is likely due to a suppressed immunological response, as signified by elevations in plasma lactate and glucose, indicative of cortisol release (Gustaveson et al. 1991; Lupes et al. 2006). Elevated surface water temperatures also correlate with low dissolved oxygen, which could be detrimental to released fish when at a high respiratory demand (reviewed in Bartholomew and Bohnsack 2005). Delayed mortality may be reduced by holding fish in recovery tanks prior to release or releasing fish into cages that are subsequently lowered to an intermediate depth (Matteson and Hannah, unpublished data; Gitschlag and Renaud 1994; Shasteen et al. 1997; Bettinger et al. 2005; St. John and Syers 2005). However, recovery tanks and recompression cages may only allow time for fish to recover from physiological stress. If fish sustain internal organ damage, although it may be to a lesser extent if fish are immediately recompressed, the latent effects may manifest themselves after physiological parameters have returned to baseline levels.

Surface predators.—Birds, large fish, and marine mammals commonly prey on injured, released red snapper and may account for 20% CAR mortality when fish are retrieved from 20 to 30 m depths (Parker 1985; Rummer and Bennett 2005). Although perceived high, predation has not been thoroughly investigated because experimental protocol either determine predation via surface observations or release fish into cages where predation is not a risk (Parker 1985; Gitschlag and Renaud 1994; Davis 2002; Burns et al. 2004; Bartholomew and Bohnsack 2005). In fact, Diamond and colleagues still measured 71% CAR mortality in cage-released red snapper retrieved from 45 m depths (Diamond and Campbell, unpublished data). Fish that are vented prior to release are able to avoid surface predators directly, as they are no longer positively buoyant and therefore able to freely swim back to depth (Parker 1985; Davis 2002; Burns et al. 2004; Bartholomew and Bohnsack 2005). Although Burns et al. (2002) retrieved small red snapper from 55 m, vented them, and then released them into cages, they still recorded 70% mortality. The venting procedure will also alleviate predation indirectly by decreasing the fish's target strength, making the fish a less obvious target for echo-locating predators than if the SB was overinflated and target strength high (Love 1969, 1971, and 1977; Keniry et al. 1996; Collins et al. 1999). Caged release and sinker-mediated release techniques have been recommended for releasing CD fish so that venting is not necessary and the risk of predation and detrimental surface water conditions can be assuaged, but the reality of CDS as a result of retrieval depth remains an issue and will compromise survival over the long term.

II. Quantitative Approach

The ultimate goal is to be able to utilize the framework outlined in Figure 3 for designing future experiments to understand the basis for CAR mortality in red snapper and other fish species. The full theoretical logistic regression model (Table 1) included all parameters as significant contributions toward mortality, even though some were less significant than others (e.g., venting; $p = 0.023$, predation; $p = 0.049$, and handling time and hook location; $p = 0.035$). A likelihood ratio test (test statistic 77.680) of the theoretical model showed that the model was an adequate fit ($P < 0.001$) of the response variable. The ability of the theoretical model to correctly predict Y_i (whether the fish was alive or dead) was 98% (with 1.5% of live fish predicted to be dead while 22.9% of dead fish were predicted as alive). Given the limitations in data available, mortality rates from an array of species were used in conjunction with red snapper specific data to generate this predictive model as a starting point for future investigations.

To most effectively communicate CAR mortality data used to generate the predictive model for this review, the baseline mortality rate (%) as a function of capture depth was plotted (Figure 5) (Rummer and Bennett 2005). The remaining six parameters were factored in, one by one until all were integrated into the mortality curve, represented by the curve furthest to the left on the graph (Figure 5). For comparison, a capture depth dependent mortality curve, according to a comprehensive set of field data collected by

Table 1. Parameter estimates for the predictive model built to describe whether a fish would be alive or dead as a result of capture depth, venting, retrieval rate, hook type, surface to depth temperature differential, surface predators, and handling time and hook location (first column). The constant and coefficients (± standard error) for the model are represented in the first and second columns respectively. The Wald statistic, which is the coefficient divided by the standard error and describes how significantly each independent variable predicts the dependent variable (mortality), is represented in the third column. The odds ratio (95% confidence interval), which represent the lower and upper ends of the confidence interval in which the true odds ratio lies, is reported in the fourth column. *P* values (fifth column) were calculated from the Wald statistic and based on chi-square distribution with one degree of freedom. When *P* values are small ($P < 0.05$), this indicates high probability that the independent variables affect the dependent variable (mortality).

Variable	Coefficient (S.E.)	Wald statistic	Odds ratio (95% C.I.)	P
% Mortality	(*a*) -3.133 (0.385)	66.252	0.044 (0.021-0.093)	<0.001
Capture depth	(*b*) 0.062 (0.012)	29.023	1.064 (1.040-1.088)	<0.001
Venting	(*c*) 1.777 (0.780)	5.184	5.911 (1.287-27.279)	0.023
Retrieval rate	(*d*) 2.710 (0.693)	15.299	15.027 (3.865-58.427)	<0.001
Hook type	(*e*) 2.965 (0.684)	18.803	19.399 (5.078-74.104)	<0.001
Water temperature	(*f*) 3.815 (0.691)	30.462	45.391 (11.710-175.946)	<0.001
Predation	(*g*) 1.532 (0.783)	3.833	4.628 (0.998-21.458)	0.049
Handling time and hook location	(*h*) 1.677 (0.795)	0.795	5.347 (1.125-25.424)	0.035

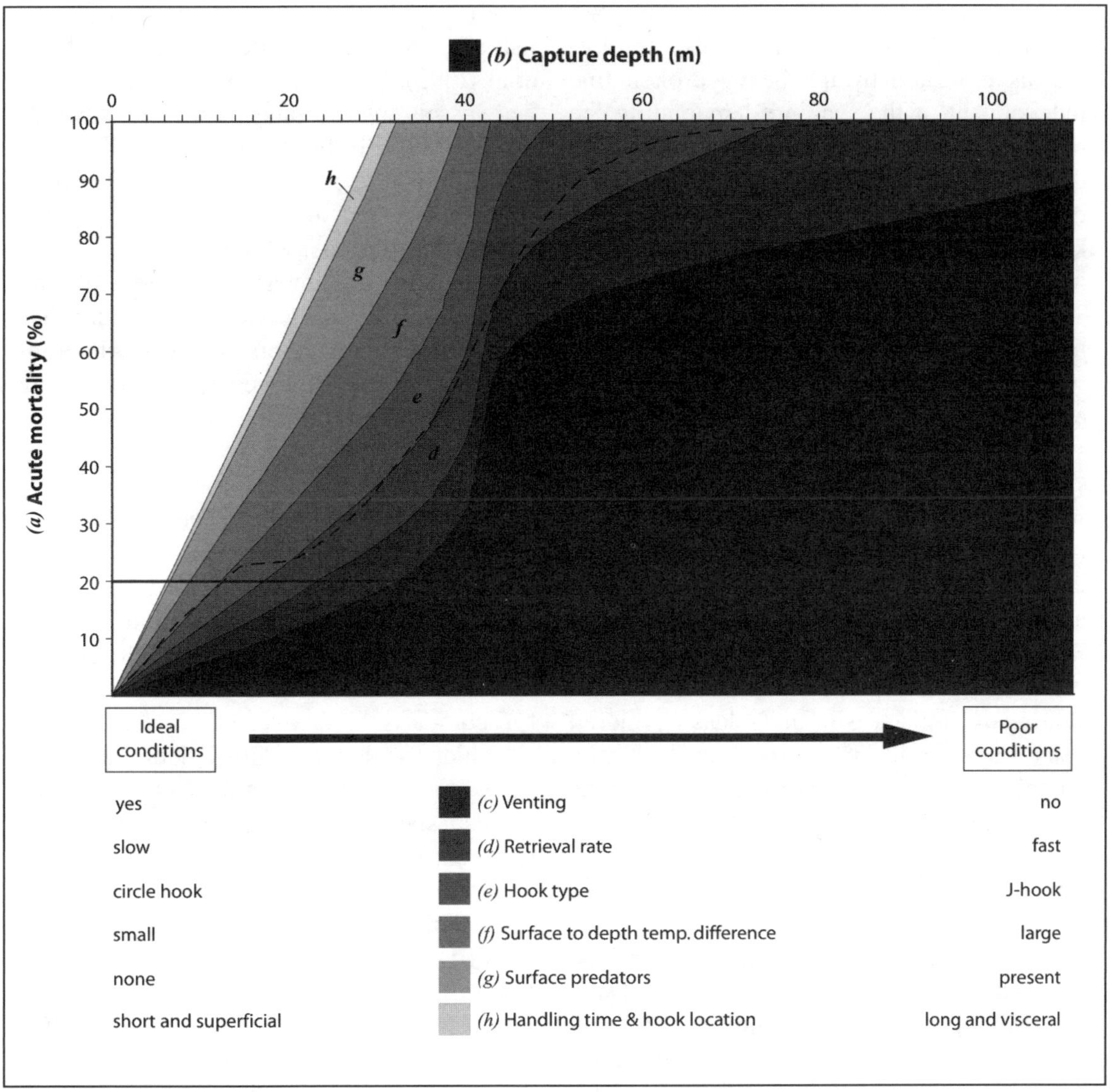

Figure 5. Graphical display of mortality (%) represented by (*a*) as a function of capture depth (m) represented by (*b*) in red snapper *Lutjanus campechanus* (Rummer and Bennett 2005) as well as six other factors represented as (*c–h*), when information from the logistical regression analysis performed for this review was incorporated into overall catch and release (CAR) mortality. Each additional factor affecting CAR was considered additive to the baseline mortality associated with capture depth and treated as a binary variable, e.g. poor and ideal conditions; data were compiled from information on CAR in other freshwater and seawater physoclist species as well as red snapper. See text in figure for further details. Note: The letters used to demarcate the dependent variable of percent mortality (*a*) and seven independent variables (*b–h*) were to maintain consistency with the abbreviations for the logistical regression model. The heavy dashed line represents actual field data from research on the Westralian jewfish, *Glaucosoma hebraicum*, used as a reference and for verifying the model with existing field data on a non red snapper species (St. John and Syers 2005). The solid horizontal line parallel to the *x*-axis at 20% represents the generally accepted mortality rate based on CAR mortality.

St. John and Syers (2005) on Westralian jewfish, *Glaucosoma hebraicum*, was also plotted, as indicated by the heavy broken line independent of the shading demarcating the parameters used in the theoretical model. To date, St. John and Syers (2005) have made the closest attempt to investigating all of the key parameters involved in CAR mortality in their study on the Westralian jewfish. The curve plotted with their data fell near the middle of the plots generated from the theoretical model, suggesting that the theoretical estimates generated from the model are reasonable (St. John and Syers 2005).

As red snapper are retrieved from deep water via traditional angling gear, the majority of the injuries that will dictate release mortality due to SB overinflation and CDS have already occurred (Figure 5) (Rummer and Bennett 2005). Numerous studies suggest that modifying gear type, slowing retrieval times, venting overinflated SBs, and releasing injured fish in cages results in improved survival rates, but this is only beneficial to some fish over the short-term and does not address the CD injuries that comprise the bulk of the overall detriment experienced by the fish. Realizing the extent of injury that has occurred when a fish undergoes decompression from depth and the potential repercussions of CDS is the first step to clarifying long-term and thus overall CAR mortality (Nichol and Chilton 2006). Retrieval conditions, species and size relationships, handling, and release conditions play a key role in the extent of injury incurred by the fish, and studies should be designed with this comprehensive CAR mortality framework in mind.

Concluding Thoughts

Management strategies are typically designed assuming CAR mortality is below 20% (Muoneke and Childress 1994). Recent field studies suggest this may be possible for red snapper retrieved from depths ranging 20–40 m (Patterson et al. 2001; Burns et al. 2004) but rates may be greater than 70% if fish are retrieved from deeper depths (Burns et al. 2002). Logbook records from commercial vessels suggest rates may range upwards of 72–78% (Poffenberger and McCarthy 2004). In this review, it is evident that current estimates for red snapper CAR mortality are indeed multifarious, which may be the trend for other fish species as well (Nieland et al. 2007, this volume). CAR mortality depends on a multitude of factors, some of which interact with others; however, the underlying cause of CAR mortality, especially in a deepwater physoclistous fish, is directly related to capture depth.

Clearly, the regulations needed in order to ensure CAR mortality remains low would have to be extremely conservative. In toxicology studies, the lethal concentration of a toxicant at which 50% mortality would be predicted (LC50) is commonly reported and used for comparisons between species. Perhaps a similar threshold, a lethal depth at which CAR mortality is 20%, or the LD_{20}, is desired for fish species where CAR is common or required. Because fisheries models are commonly based on CAR mortality of 20% or below, this seems a reasonable starting point. If an LD_{20} was assigned to red snapper based solely on capture depth related mortality, it would be approximately 30 m (data plotted from Rummer and Bennett 2005), which is realistic and a depth where red snapper are fished in some areas of the Gulf of Mexico (Moran 1988; Workman and Foster 1994; Manooch et al. 1998; Dorf 2003; Burns et al. 2004). However, those data only account for capture depth in fish decompressed in a laboratory decompression chamber. If the remaining factors utilized to build the predictive model were considered as confounding this baseline mortality, the predicted LD_{20} for red snapper could be as shallow as 6 m (Figure 5). Again, this estimate dictates that the factors considered exhibit a linear relationship and an additive effect on one another;

multiplicative interactions could result in an even shallower estimate.

For red snapper or any other high profile fishery species, no single investigation considering all of the parameters described in the CAR mortality framework (Figure 3) has yet been executed. It is promising that results from field studies, however, fall within the range of depth related mortality rates calculated from the predictive model. For example, if St. John and Syers' (2005) data for CAR mortality in the Westralian jewfish, *G. hebraicum*, a species with similar depth profile to red snapper, was re-plotted so depth was the independent variable, the deeper end of the LD_{20} range would be close to 21 m. If CAR mortality data for Pacific cod, *Gadus macrocephalus*, a particularly deep-dwelling species, were re-plotted against capture depth, the LD_{20} would be 10 m (data plotted from Nichol and Chilton 2006). The former of the two field studies highlighted used caged-release protocols, and the latter a traditional mark–recapture protocol, making it difficult to make concrete conclusions regarding long-term effects and direct causes of mortality. This information speaks well for the model but poorly for the fate of the fishery. Limiting the red snapper CAR fishery to depths between 6 and 30 m, the most conservative to the most liberal depths from which we can expect 20% CAR mortality, may not be the best course of action. Red snapper occasionally occupy depths deeper than 200 m, and depth and age-class distribution are closely linked, meaning a depth limitation would alter current population structure (Moran 1988; Workman and Foster 1994; Manooch et al. 1998). Furthermore, as seen in Figure 5, as each fishing factor is accounted for, the LD_{20} decreases representing depths where fishing practices probably cannot be sustained economically. This overview illuminates the necessary approach to understanding the root of CAR mortality, starting with capture depth and SB morphology and physiology. This area of research will, undoubtedly be more heavily investigated as stocks continue to decline necessitating modifications to current management strategies, and fits well within the context of conservation physiology, an emerging discipline where physiological responses of organisms to human influences that may contribute to population declines are directly investigated (Wikelski and Cooke 2006).

The most effective way to manage a fishery that succumbs so seriously to CAR mortality is not via size limits, season closures, and bag limits. Countless combinations have been proposed in an effort to maintain status quo while rebuilding the fishery (SEDAR 7 2005). All combinations, however, seem problematic; either the socioeconomics of the red snapper fishery or the potential for population growth and recovery are negatively impacted. Furthermore, sufficient information is not yet available to, with confidence, impose maximum fishing depths, which would be difficult to monitor and may vary too greatly between species. It would seem logical to impose fishing restrictions to discrete areas such that fish populations can be safeguarded from all fishing and CAR activity, therefore protecting age, size, and sex classes and ratios simultaneously. Aquatic protected areas (APAs) and marine protected areas (MPAs) are generally restricted areas, and no-take reserves (NTRs) are extremely restricted and encompass areas where all fishing and extractive activities are banned and human impact is minimal (Bohnsack 1998). Both APA or MPAs and NTRs are modern management strategies with growing acceptance and have been particularly successful along the North American West Coast in protecting long-lived, slow-growing rockfish (*Sebastes* spp.) (Coleman et al. 2000; Soh et al. 2001; Schroeder and Love 2002; Roberts 2003; Berkeley et al. 2004; Bartholomew and Bohnsack 2005; Smith et al. 2006). Average values for fish density, biomass, organism size, and biodiversity increase by almost four-fold, develop quickly, and persist, compared to areas outside reserves (Mosqueira et al. 2000; Halpern and Warner 2002; Ault et al. 2005). Additionally, this strategy protects genetic diversity, ecosystem structure, function and integrity, increases scientific and public knowledge and understanding of aquatic systems, enhances nonconsumptive opportunities, as well as provides fishery benefits (Bohnsack 1998; Coleman et al. 2000). CAR mortality and serial overfishing can be reduced therefore supporting sustainable fisheries without reducing current catch levels (Soh et al. 2001; Roberts 2003). For fish species like red snapper, where overfishing is widespread and CAR mortality is high, or other species where CAR is unclear or a thorough investigation of depth-related CAR mortality has not been performed, strategies based on space (MPAs and NTRs) rather than time or numbers (i.e., season closures, size limits, bag limits, etc.), have the greatest potential for overall conservation and sustainability and should be seriously considered (Coleman et al. 2000).

Acknowledgments

The author wishes to acknowledge Colin Brauner of the University of British Columbia, Department of Zoology for numerous brainstorming sessions as well as helpful editorial comments throughout the process of this project. Appreciation is also due to David Nieland and colleagues at Louisiana State University, Coastal Fisheries Institute as well as Michael Davis and Steven Parker of Oregon State University, Hatfield Marine Science Center and Oregon Department of Fish and Wildlife for helpful suggestions and inspiring conversation integral to this review. The author also wishes to acknowledge D. Baker, C. Fu, both of the University of British Columbia, and two anonymous reviewers whose comments and suggestions resulted in dramatic improvements to this contribution. A great deal of appreciation is due to Kim Kamo for countless hours designing and formatting graphics as well as continuous encouragement. Finally, the author thanks the American Fisheries Society Southern Division and Florida Chapter for support during the earlier projects incorporated into this piece as well as the University of British Columbia Graduate Fellowship Fund and National Sciences and Engineering Research Council for continued financial support.

References

Aalbers, S. A., G. M. Stutzer, and M. A. Drawbridge. 2004. The effects of catch-and-release angling on the growth and survival of juvenile white seabass captured on offset circle and j-type hooks. North American Journal of Fisheries Management 24:793–800.

Alexander, R. M. 1972. The energetics of vertical migration by fishes, Pages 273–294 *in* M. A. Sleigh and A. G. Macdonald, editors. Symposium 26, Society for Experimental Biology 26:273-294.

Alexander, R. M. 1993. Buoyancy. Pages 75–97 *in* The Physiology of Fishes. D. H. Evans, editor. CRC Press, Boca Raton, Florida.

Ault, J. S., J. A. Bohnsack, S. G. Smith, and J. G. Luo. 2005. Towards sustainable multispecies fisheries in the Florida, USA, coral reef ecosystem. Bulletin of Marine Science. 76:595–622.

Barimo, J. F., and M. L. Fine. 1998. Relationship of swim-bladder shape to the directionality pattern of underwater sound in the oyster toadfish. Canadian Journal of Zoology 76:134–143.

Bartholomew, A., and J. Bohnsack. 2005. A review of catch-and-release angling mortality with implications for no-take reserves. Reviews in Fish Biology and Fisheries 15:129–154.

Beggs, G. L., G. F. Holeton, and E. J. Crossman. 1980. Some physiological consequences of angling stress in muskellunge, *Esox masquinongy* Mitchill. Journal of Fish Biology 17:649–659.

Berkeley, S. A., M. A. Hixon, R. J. Larson, and M. S. Love. 2004. Fisheries sustainability via protection of age structure and spatial distribution of fish populations. Fisheries 29:23–32.

Bettinger, J. M., J. R. Tomasso, and J. J. Isely. 2005. Hooking mortality and physiological responses of striped bass angled in freshwater and held in live-release tubes. North American Journal of Fisheries Management 25:1273–1280.

Bettoli, P. W., and R. S. Osborne. 1998. Hooking mortality and behavior of striped bass following catch and release angling. North American Journal of Fisheries Management 18:609–615.

Bettoli, P. W., C. S. Vandergoot, and P. T. Horner. 2000. Hooking mortality of saugers in the Tennessee River. North American Journal of Fisheries Management 20:833–837.

Bohnsack, J. A. 1998. Application of marine reserves to reef fisheries management. Australian Journal of Ecology 23:298–304.

Bohnsack, J. A., and D. E. Harper. 1988. Length-weight relationships of selected marine reef fishes from the southeastern United States and the Caribbean. National Oceanic and Atmospheric Administration (NOAA), Technical Memorandum NMFS-SEFC–215:31.

Bruesewitz, R. E., D. W. Coble, and F. Copes. 1993. Effects of deflating the expanded swimbladder on survival of burbot. North American Journal of Fisheries Management 13:346–348.

Bugley, K., and G. Shepherd. 1991. Effect of catch-and-release angling on the survival of black sea bass. North American Journal of Fisheries Management 111:468–471.

Burns, K. M., N. F. Parnell, and R. R. Wilson Jr. 2004. Partitioning release mortality in the undersized red snapper bycatch: comparison of depth vs. hooking effects. Mote Marine Laboratory Technical Report 932. MARFIN Grant NA97FF0349.

Burns, K. M., C. C. Koenig, and F. C. Coleman. 2002. Evaluation of multiple factors involved in release mortality of undersized red grouper, gag, red snapper, and vermilion snapper. Mote Marine Laboratory Technical Report No. 814. MARFIN Grant NA87FF0421.

Busk, M., J. Overgaard, J. W. Hicks, A. F. Bennett, and T. Wang. Effects of feeding on arterial blood gases in the American alligator *Alligator mississippiensis*. Journal of Experimental Biology 203(20):3117–3124.

Carpenter, K. E., T. M. Berra and J. M. Humphries Jr. 2004. Swim bladder and posterior lateral line nerve of the nurseryfish, *Kurtus gulliveri* (Perciformes: Kurtidae). Journal of Morphology. 260:193–200.

Casselman, S. J. 2005. Catch-and-release angling: a review with guidelines for proper fish handling practices, p. 26. Fish and Wildlife Branch, Ontario Ministry of Natural Resources. Peterborough, Ontario, Canada.

Coleman, F. C., C. C. Koenig, G. R. Huntsman, J. A. Musick, A. M. Eklund, J. C. McGovern, R. W. Chapman, G. R. Sedberry, and C. B. Grimes. 2000. Long-lived reef fishes: The grouper-snapper complex. Fisheries 2000:14–21.

Collins, M. R., J. C. McGovern, G. R. Sedberry, H. S. Meister, and R. Pardieck. 1999. Swim bladder deflation in black sea bass and vermillion snaper: Potential for increasing post-release survival. North American Journal of Fisheries Management 19:828–832.

Cooke, S. J., B. L. Barthel, C. D. Suski, M. J. Siepker, and D. P. Philipp. 2005. Influence of circle hook size on hooking efficiency, injury, and size selectvity of bluegill with comments on circle hook conservation benefits in recreational fisheries. North American Journal of Fisheries Management 25:211–219.

Cooke, S. J., K. G. Ostrand, C. M. Bunt, J. F. Schreer, D. H. Wahl, and D. P. Philipp. 2003a. Cardiovascular responses of largemouth bass to exhaustive excersise and brief air exposure over a range of water temperatures. Transactions of the American Fisheries Society 132:1154–1165.

Cooke, S. J., D. P. Philipp, K. M. Dunmall, and J. F. Schreer. 2001. The influence of terminal tackle on injury, handling time, and cardiac disturbance of rock bass. North American Journal of Fisheries Management 21:333–342.

Cooke, S. J., J. F. Schreer, K. M. Dunmall, and D. P. Philipp. 2002a. Strategies for quantifying sublethal effects of marine catch-and-release angling: Insight from novel freshwater applications. Pages 121–134 *in* J. Lucy and A. Studholme, editors. Catch and release in marine recreational fisheries. American Fisheries Society, Symposium 30, Bethesda, Maryland.

Cooke, S. J., J. F. Schreer, D. H. Wahl, and D. P. Philipp. 2002b. Physiological impacts of catch-and-release angling practices on largemouth bass and smallmouth bass. Pages 489–512 *in* D. P. Philipp and M. S. Ridgway, editors. Black bass: ecology, conservation, and management. American Fisheries Society, Symposium 31, Bethesda, Maryland.

Cooke, S. J., and C. D. Suski. 2005. Do we need species-specific guidelines for catch-and-release recreational angling to effectively conserve diverse fishery resources? Biodiversity and Conservation 14:1195–1209.

Cooke, S. J., C. D. Suski, B. L. Barthel, K. G. Ostrand, B. L. Tufts, and D. P. Philipp. 2003b. Injury and mortality induced by four hook types on bluegill and pumkinseed. North American Journal of Fisheries Management 23:883–893.

Davenport, J. 1999. Swimbladder volume and body density in an armoured benthic fish, the streaked gurnard. Journal of Fish Biology 55:527–534.

Davis, M. W. 2002. Key principles for understanding fish bycatch discard mortality. Canadian Journal of Fisheries and Aquatic Sciences 59:1834–1843.

Dorf, B. A. 2003. Red snapper discards in Texas coastal waters—a fishery dependent onboard survey of recreational headboat discards and landings. Pages 155–166 *in* D. R. Stanley and A. Scarborough-Bull, editors. Fisheries, reefs, and offshore development. American Fisheries Society, Symposium 36, Bethesda, Maryland.

Fable, W. A. J. 1980. Tagging studies of red snapper (*Lutjanus campechanus*) and vermilion snapper (*Rhomboplites aurorubens*) off the South Texas coast. Contributions in Marine Science 23:115–121.

Ferguson, R. A., and B. L. Tufts. 1992. Physiological effects of brief air exposure in exhaustively exercised rainbow trout (*Oncorhynchus mykiss*): implications for "catch-and-release" fisheries. Canadian Journal of Fisheries and Aquatic Sciences 49:1157–1162.

Gitschlag, G. R., and M. L. Renaud. 1994. Field experiments on survival rates of caged and released red snapper. North American Journal of Fisheries Management 14:131–136.

GMFMC. 2000. Amendment to the reef fish fishery management plan to set total allowable catch and management measures for red snapper for the 2000 and 2001 seasons, Pages 1–59 *in* Gulf of Mexico Fishery Management Council of the National Oceanic and Atmospheric Administration (NOAA), Tampa, Florida.

Gotshall, D. W. 1964. Increasing tagged rockfish (Genus Sebastodes) survival by deflating the swimbladder. California Fish and Game 50:253–260.

Gustaveson, A. W., R. S. Wydoski, and G. A. Wedemeyer. 1991. Physiological response of largemouth bass to angling stress. Transactions of the American Fisheries Society 120:629–636.

Haight, W. R., F. A. Parrish, and T. A. Hayes. 1993. Feeding ecology of deepwater lutjanid snappers at Penguin Bank, Hawaii. Transactions of the American Fisheries Society 122:328–347.

Halpern, B. S., and R. R. Warner. 2002. Marine reserves have rapid and lasting effects. Ecology Letters 5:361–366.

Harden-Jones, F. R. 1952. The swimbladder and the vertical movements of teleostean fishes. I. Physical Factors. Journal of Experimental Biology 29:553–556.

Harden-Jones, F. R., and P. Scholes. 1985. Gas secretion and resorption in the swimbladder of the cod *Gadus morhua*. Journal of Comparative Physiology B 155:319–331.

Helm, G., and S. Smullen. 1997. Report finds Gulf of Mexico red snapper severely overfished, NOAA (National Oceanic and Atmospheric Administration), Washington, D.C.

Hicks, J. W., and A. F. Bennett. 2004. Eat and run: prioritization of oxygen delivery during elevated metabolic states. Respiratory Physiology and Neurobiology 144(2–3):215–224.

Keniry, M. J., W. A. Brofka, W. H. Horns, and J. E. Marsden. 1996. Effects of Decompression and puncturing the gas bladder on survival of tagged yellow perch. North American Journal of Fisheries Management 16:201–206.

Lee, D. P. 1992. Gas bladder deflation of depressurized largemouth bass. North American Journal of Fisheries Management 12:662–664.

Lindsay, R. B., R. K. Schroeder, K. R. Kenaston, R. N. Toman, and M. A. Buckman. 2004. Hooking mortality by anatomical location and its use in estimating mortality of spring chinook salmon caught and released in a river sport fishery. North American Journal of Fisheries Management 24:367–378.

Love, R. H. 1969. Maximum side-aspect target strength of an individual fish. Journal of the Acoustical Society of America. 46:746–752.

Love, R. H. 1971. Dorsal-aspect target strength of an individual fish. Journal of the Acoustical Society of America. 49:816–823.

Love, R. H. 1977. Target strength of an individual fish at any aspect. Journal of the Acoustical Society of America 62:1397–1403.

Low, R. A., Jr. 1981. Mortality rates and management strategies for black sea bass off the southeast coast of the United States. North American Journal of Fisheries Management 1:93–108.

Lupes, S. C., M. W. Davis, B. L. Olla, and C. B. Schreck. 2006. Capture-related stressors impair immune system function in sablefish. Transactions of the American Fisheries Society 135:129–138.

Machias, A., and N. Tsimenides. 1996. Anatomical and physiological factors affecting the swimbladder cross-section of sardine *Sardina pilchardus*. Canadian Journal of Fisheries and Aquatic Sciences 53:280–287.

Manire, C., R. Hueter, E. Hull, and R. Spieler. 2001. Serological changes associated with gill-net capture and restraint in three species of sharks. Transactions of the American Fisheries Society 130:1038–1048.

Manooch, I., C. S., J. C. Potts, D. S. Vaughan, and M. L. Burton. 1998. Population assessment of the red snapper from the southeastern United States. Fisheries Research 38:19–32.

Millard, M. J., and J. W. Mohler. 2005. Mortality associated with catch-and-release angling of striped bass in the Hudson River. North American Journal of Fisheries Management 25:1533–1541.

Moran, D. 1988. Species profiles: Life histories and environmental requirements of coastal fishes and invertebrates (Gulf of Mexico)—red snapper. U.S. Fish and Wildlife Service, Washington, D.C.

Morrissey, M. B., C. D. Suski, K. R. Esseltine, and B. L. Tufts. 2005. Incidence and physiological consequences of decompression in smallmouth bass after live-release angling tournaments. Transactions of the American Fisheries Society 134:1038–1047.

Mosqueira, I., I. M. Cote, S. Jennings, and J. D. Reynolds. 2000. Conservation benefits of marine reserves for fish populations. Animal Conservation 3:321–332.

Muoneke, M. I., and W. M. Childress. 1994. Hooking mortality: A review for recreational fisheries. Reviews in Fisheries Science 2:123–156.

Murphy, M. D., R. F. Heagey, V. H. Neugebauer, M. D. Gordon, and J. L. Hintz. 1995. Mortality of spotted seatrout released from gill-net or hook-and-line gear in Florida. North American Journal of Fisheries Management 15:748–753.

Nelson, K. L. 1998. Catch-and-release mortality of striped bass in the Roanoke River, North Carolina. North American Journal of Fisheries Management 18:25–30.

Neufeld, M. D., and C. R. Spence. 2004. Evaluation of a simple decompression procedure to reduce decompression trauma in trap-caught burbot. Transactions of the American Fisheries Society 133:1260–1263.

Nieland, D. L, A. J. Fischer, M. S. Baker, Jr., and C. A. Wilson. 2007. Red snapper in the northern Gulf of Mexico: age and size composition of the commercial harvest and mortality of regulatory discards. Pages 301–310 *in* W. F. Patterson, III, J. H. Cowan, Jr., G. R. Fitzhugh, and D. L. Nieland, editors. Red snapper ecology and fisheries in the U.S. Gulf of Mexico. American Fisheries Society, Symposium 60, Bethesda, Maryland.

Nichol, D. G., and E. A. Chilton. 2006. Recuperation and behaviour of Pacific cod after barotrauma. ICES Journal of Marine Science 63:83–94.

Pankhurst, N. W., and M. Dedual. 1994. Effects of capture and recovery on plasma levels of cortisol, lactate and gonadal steroids in a natural population of rainbow trout. Journal of Fish Biology 45(6): 1013–1025.

Parker, R. O. 1985. Survival of released red snapper. South Atlantic and Gulf of Mexico Fishery Management Councils, Charleston, South Carolina and Tampa, Florida.

Parrish, F. A., and R. B. Moffitt. 1993. Subsurface fish handling to limit decompression effects on deepwater species. Marine Fisheries Review 54:29–32.

Patterson, W. F., III, J. C. Watterson, R. L. Shipp, and J. H. Cowan, Jr. 2001. Movement of tagged red snapper in the Northern Gulf of Mexico. Transactions of the American Fisheries Society. 130:533–545.

Pauly, D., V. Christensen, S. Guenette, T. J. Pitcher, U. R. Sumaila, C. J. Walters, R. Watson, and D. Zeller. 2002. Towards sustainability in world fisheries. Nature 418:689–695.

Pelzman, R. J. 1978. Hooking mortality of juvenile largemouth bass *Micropterus salmoides*. California Fish and Game 64:185–188.

Poffenberger, J. and K. J. McCarthy. 2004. Estimates of red snapper discards by vessels with federal permits in the Gulf of Mexico. Southeast Fishery Science Center, Miami. Sustainable Fisheries Division Contribution SFD 2004–023.

Porch, C. E. 2005. Projected effects of changes in minimum size regulations on the future status of the red snapper (*Lutjanus campechanus*) fishery in the U.S. Gulf of Mexico. Southeast Fisheries Science Center, Miami. Sustainable Fisheries Division Contribution SFD-2005–009.

Pritchard, E. S. 2005. Fisheries of the United States 2004. National Marine Fisheries Service (NMFS), Office of Science and Technology, Silver Spring, Maryland.

Render, J. H., and C. A. Wilson. 1994. Hook-and-line mortality of caught and released red snapper around oil and gas platform structural habitat. Bulletin of Marine Science. 55:1106–1111.

Render, J. H., and C. A. Wilson. 1996. Effects of gas bladder deflation on mortality of hook-and-line caught and released red snappers: Implications for management. *In* F. Arreguin-Sanchez, J. L. Munro, M. C. Balgos, and D. Pauly, editors. International Workshop on Tropical Snappers and Groupers, Volume 48. ICLARM, Campeche, Mexico.

Ribbink, A. J., and B. J. Hill. 1979. Depth equilibration by two predatory cichlid fish from Lake Malawi. Journal of Fish Biology 14:507–510.

Roberts, C. M. 2003. Our shifting perspectives on the oceans. Oryx 37:166–177.

Rogers, J. H., H. T. Langston, and T. E. Targett. 1986. Anatomical trauma to sponge-coral reef fishes captured by trawling and angling. Fisheries Bulletin. 84:697–703.

Rummer, J. L., and W. A. Bennett. 2005. Physiological effects of swim bladder overexpansion and catastrophic decompression on red snapper. Transactions of the American Fisheries Society 134:1457–1470.

Schirripa, M. J., and C. M. Legault. 1999. Status of red snapper in US waters of the Gulf of Mexico updated through 1998. National Marine Fisheries Service, Southeast Fisheries Science Center, Sustainable Fisheries Division, Contribution SFD-99/00–75. Miami, Florida.

Schmidt-Nielsen, K. 1997. Animal Physiology. Cambridge University Press, New York.

Schroeder, D. M., and M. S. Love. 2002. Recreational fishing and marine fish populations in California. California Cooperative Oceanic Fisheries Investigations Reports 43:182–190.

SEDAR 7. 2005. Stock assessment report of SEDAR 7 Gulf of Mexico Red Snapper. SEDAR (http://www.sefsc.noaa.gov/sedar/), Charleston, South Carolina.

Shasteen, S. P., and R. J. Sheehan. 1997. Laboratory evaluation of artificial swim bladder deflation in largemouth bass: Potential benefits for catch-and-release fisheries. North American Journal of Fisheries Management 17:32–37.

Smith, M. D., J. J. Zhang, and F. C. Coleman. 2006. Effectiveness of marine reserves for large-scale fisheries management. Canadian Journal of Fisheries and Aquatic Sciences 63:153–164.

Soh, S., D. R. Gunderson, and D. H. Ito. 2001. The potential role of marine reserves in the management of shortraker rockfish (*Sebastes borealis*) and rougheye rockfish (*S. aleutianus*) in the Gulf of Alaska. Fishery Bulletin. 99:168–179.

St. John, J., and C. J. Syers. 2005. Mortality of the demersal West Australian dhufish, *Glaucosoma hebraicum* (Richardson 1845) following catch and release: The influence of capture depth, venting and hook type. Fisheries Research 76:106–116.

Stephens, F. J., J. J. Cleary, G. Jenkins, B. Jones, S. R. Raidal, and J. B. Thomas. 2002. Haemoglobin and oxygen transport of the West Australian dhufish, *Glaucosoma hebraicum* Richardson, and other species. Journal of Fish Diseases 25:409–414.

Stevens, M. M. 2004. Commercially important Gulf of Mexico/South Atlantic snappers, p. 1–29. Monterey Bay Aquarium, Monterey, California.

Strand, E., C. Jorgensen, and G. Huse. 2005. Modelling buoyancy regulation in fishes with swimbladders: bioenergetics and behaviour. Ecological Modelling 185:309–327.

Suski, C. D., S. S. Killen, M. B. Morrissey, S. G. Lund, and B. L. Tufts. 2003. Physiological changes in largemouth bass caused by live-release angling tournaments in Southeastern Ontario. North American Journal of Fisheries Management 23:760–769.

Tufts, B. L., Y. Lang, K. Tufts, and R. G. Boutilier. 1991. Exhaustive exercise in "wild" Atlantic salmon (*Salmo salar*): acid-base regulation and blood gas transport. Canadian Journal of Fisheries and Aquatic Sciences 48:868–874.

Watterson, J. C., I. Patterson, W. F., III, R. L. Shipp, and J. H. Cowan, Jr. 1998. Movement of red snapper, *Lutjanus campechanus*, in the north central Gulf of Mexico: potential effects of hurricanes. Gulf of Mexico Science. 16:92–104.

Wedemeyer, G. A., B. A. Barton, and D. J. McLeay. 1990. Stress and acclimation. Pages 213–272 *in* C. B. Schreck and P. B. Moyle. Methods for Fish Biology. American Fisheries Society, Bethesda, Maryland.

Wells, R. M. G. and R. E. Weber. 1989. The measurement of oxygen affinity in blood and hemoglobin solutions. Cambridge University Press, New York.

Wikelski, M., and S. J. Cooke. 2006. Conservation physiology. Trends in Ecology & Evolution 21:38–46.

Wilde, G. R. 1998. Tournament-associated mortality in black bass. Fisheries 23:12–22.

Wilde, G. R., M. I. Muoneke, P. W. Bettoli, K. L. Nelson, and B. T. Hysmith. 2000. Bait and temperature effects on striped bass hooking mortality in freshwater. North American Journal of Fisheries Management 20:810–815.

Williamson, D. H., G. R. Russ, and A. M. Ayling. 2004. No-take marine reserves increase abundance and biomass of reef fish on inshore fringing reefs of the Great Barrier Reef. Environmental Conservation 31:149–159.

Wilson Jr., R. R., and K. L. Smith, Jr. 1985. Live capture, maintenance, and partial decompression of a deep-sea grenadier fish (*Coryphaenoides acrolepsis*) in a hyperbaric trap aquarium. Deep Sea Research 32:1571–1582.

Wilson, R. W., and K. M. Burns. 1996. Potential survival of released groupers caught deeper than 40 m based on shipboard and in-situ observations, and tag-recapture data. Bulletin of Marine Science 58:234–247.

Wittenberg, J. B., and B. A. Wittenberg. 1974. The choroid rete mirabile of the fish eye. I. Oxygen secretion and structure: comparison with the swimbladder rete mirabile. Biological Bulletin 146:116–136.

Wood, C. M., J. D. Turner, and M. S. Graham. 1983. Why do fish die after severe exercise? Journal of Fish Biology 22:189–201.

Workman, I. K., and D. G. Foster. 1994. Occurrence and behavior of juvenile red snapper, *Lutjanus campechanus*, on commercial shrimp fishing grounds in the northeastern Gulf of Mexico. Marine Fisheries Review 56(2):9–11.

Population Dynamics and Structure Overview

David L. Nieland

Over the past two decades we have learned much about red snapper *Lutjanus campechanus* biology, life history, and ecology, all of which has improved our understanding of the red snapper in the Gulf of Mexico (GOM) both as a species and as a resource. We have also applied new technologies and new techniques to our research efforts that have allowed us to look at red snapper populations at much finer scales of differentiation. Section II of this book is composed of seven papers addressing population dynamics and population structure of red snapper in the U.S. GOM. Together they illustrate the changing approaches to our study of red snapper populations and the wide variety of techniques, from traditional to molecular to mathematical, employed in our collective research on the species; individually they provide information vital to our understanding both of red snapper populations and of the strategies that might be appropriately applied to red snapper management.

The first two papers by Andrew Strelcheck et al. and Sandra Diamond et al. examine the site fidelity and movements of tagged/recaptured red snapper, the former at low relief artificial reefs off the coast of Alabama and the latter at both natural reefs and artificial reefs (mostly petroleum platforms) off the Texas coast. The moderate site fidelity and the comparatively low vagility reported in both studies suggest that local populations of red snapper in the GOM may be relatively isolated from one another and may warrant management on a more regional, rather than Gulf-wide, basis. Strelcheck et al. further state that the artificial reefs off Alabama are suitable habitat for adult red snapper, yet they support a net loss of biomass due to high fishing mortality.

The third paper is Andrew Fischer's synopsis of the evolution of fisheries ageing and validation techniques as they have been applied to red snapper in the GOM. He begins with early efforts to age red snapper from scale annuli and continues through to the current practice of counting annuli in sectioned otoliths. The increase in potential longevity of red snapper, from less than 10 years to almost 60 years, realized over the span of the studies discussed may have impacted red snapper management as much, or more, than any other life history datum. Fischer also includes sections addressing ageing precision and accuracy and the application of (and pitfalls of) the von Bertalanffy growth model to red snapper length at age data.

John Gold and Eric Saillant contribute the fourth paper of this section, a study of the population structure of red snapper based on analyses of allelic variation in nuclear- encoded microsatellites and of haplotype variation in mtDNA. They conclude that red snapper in the GOM maintain a complex of semi-isolated populations in which relatedness is maintained over geologic time by gene flow, yet the populations are demographically independent over the short term. This would appear not only to support the limited movement of red snapper reported by Strelcheck et al. and Diamond et al., but also to argue for regional management of red snapper populations in the GOM.

The fifth paper in this section, authored by Melissa Woods Jackson et al., provides an example of how two populations of red snapper in the northern GOM are behaving in response to their respective environments and selective regimes. The maturation of female red snappers in Alabama waters both at lesser length and at younger age than females in Louisiana waters is reported to result from reductions in population size and juvenescence of the population, presumably due to greater fishing mortality in Alabama waters. Demonstrable demographic differences such as these between adjacent populations provides further evidence that management of red snapper in the GOM as a single stock may not be the best approach.

An examination of the interactive effect of age and length on red snapper batch fecundity is contributed by Clay Porch et al. Among those data necessary for constructive assessment of fish populations, batch fecundity (estimated number of ova spawned during successive spawning events) is perhaps the most elusive. Even with a good strategy for sampling a fish population in place, the transitory nature of the hydrated stage of oocyte maturation generally limits the number of females available for batch fecundity estimations. Batch fecundity is often estimated, for assessment purposes and others, from fish length, the two variables being highly correlated in most cases. The relationship between batch fecundity and fish age is usually not as good, but if considered in concert with fish length, the ability to improve estimates of batch fecundity from both length and age data would be of great utility for applications such as age-based population assessments. Through analyses of various models for the estimation of batch fecundity, the authors ultimately establish that red snapper length alone is as good an estimator of batch fecundity as would be length and age together.

Finally, Will Patterson takes the mosaic of contemporary red snapper literature, extracts the relevant information, and integrates it all to support a hypothesis of population structuring in the northern GOM. Patterson's cogent synthesis of larval and juvenile dispersal, ontogenetic shifts in habitat usage, site fidelity and movements, and genetics of red snapper ultimately leads to the inescapable conclusion that the species is likely composed of distinct subpopulations among which dispersal mechanisms exist, yet these subpopulations exhibit asynchrony in population demographics such as growth rates, size/age at maturity, and genetic composition: a metapopulation.

As the seven contributions in the following section certainly illustrate, the management of a widely distributed, long-lived, highly fecund, demographically disparate species such as red snapper is fraught with complexities. The single stock approach that has always been applied to red snapper in the GOM has the advantage of simplicity in both application and enforcement of regulations. Yet it is becoming increasingly evident that differential management over limited geographic areas may be more appropriate for the species, if not more complicated for both management and enforcement personnel. The collective efforts of red snapper researchers from all along the Gulf Coast will continue both to improve our knowledge base and to provide resource managers with the best possible information and advice for the conservation and wise use of the red snapper in the GOM.

American Fisheries Society Symposium 60:147–162

Site Fidelity, Movement, and Growth of Red Snapper: Implications for Artificial Reef Management

ANDREW J. STRELCHECK[1]
NOAA Fisheries Service
Southeast Regional Office
263 13th Avenue South, St. Petersburg, Florida 33701 USA

JAMES H. COWAN, JR.
Coastal Fisheries Institute and Department of Oceanography and Coastal Sciences
School of the Coast and Environment
Louisiana State University
Baton Rouge, Louisiana 70803 USA

WILLIAM F. PATTERSON, III
Department of Biology
University of West Florida
11000 University Parkway, Pensacola, Florida 32514 USA

Abstract.—Red snapper, *Lutjanus campechanus*, ($n = 4{,}317$) were captured and tagged at 14 experimental artificial reefs of two designs during quarterly research cruises ($n = 17$) off coastal Alabama between January 1999 and October 2002. Six-hundred and twenty nine recaptures were reported, representing 578 tagged red snapper. Sixty-five percent of recaptures ($n = 412$) were made at the site of release on subsequent research cruises, while 217 recaptures were reported by fishers. Eighty-six percent of individuals with known recapture locations moved 2 km or less from the site of release; mean and maximum distances moved were 2.1 km and 201 km, respectively. Nine red snapper moved greater than 80 km. Mean dispersion rate from release sites was 8.6 m d^{-1}. Annual site fidelity of tagged fish was estimated using nonlinear decay models. Estimated annual site fidelity ranged from 48% to 52% $year^{-1}$ and was not significantly affected by artificial reef design, reef fish biomass at the site of release, or artificial reef densities surrounding each tagging site. Growth rates were estimated by regressing the change in red snapper total length versus the days a fish was at liberty. Mean growth rate for all recaptured fish was 0.206 mm d^{-1}. Growth rates were significantly affected by reef size (faster at larger experimental reefs) and reef fish biomass (slower at tagging sites supporting low reef fish biomass), but were not affected by artificial reef density. Moderate site fidelity and low dispersion rates during our study provide support for the hypothesis that artificial reefs off Alabama are suitable habitat for adult red snapper. However, characteristics of artificial reefs, such as reef size and standing stock biomass, may affect red snapper growth. Furthermore, ratios of instantaneous growth in weight to total mortality (G/Z) suggest artificial reefs off Alabama serve as net sinks (i.e., $G/Z < 1$) of red snapper biomass under current fishing mortality rates.

[1] Corresponding author: andy.strelcheck@noaa.gov.

Introduction

Tagging studies are used both to assess fish migration and movement and to estimate fish growth, mortality, and abundance (Hilborn et al. 1990). In artificial reef research, tagging studies often are used to assess experimental design assumptions (e.g., independence), homing, and movement of reef fishes (Hixon and Beets 1989, 1993; Beets and Hixon 1994; Eggleston et al. 1997; Watterson et al. 1998; Patterson and Cowan 2003). Tag-recapture studies also are used to estimate site fidelity of reef fishes at artificial and natural reefs (Lindberg and Loftin 1998; Szedlmayer 1997; Patterson and Cowan 2003; Szedlmayer and Schroepfer 2005; Schroepfer and Szedlmayer 2006). Site fidelity provides an annual estimate of reef fish immigration or emigration from an artificial reef. Estimates of site fidelity, distance moved, and reef fish growth rates obtained from tagging studies all can be used to make inferences about the resource value of a particular habitat (in this case an artificial reef) or complex of habitats (Lindberg et al. 1990).

It has been hypothesized that site fidelity of reef-associated organisms is dependent on both prey availability and the availability of suitable refuge (resource mosaic hypothesis: Lindberg et al. 1990; Frazer and Lindberg 1994; density-dependent habitat selection: see Lindberg et al. 2006). Reef-associated fish species that rely on benthic prey as a primary component of their diet, such as young-adult red snapper *Lutjanus campechanus*, create a gradient of prey depletion (i.e., feeding halo) around artificial reef structures (Frazer and Lindberg 1994; Lindberg 1996; Bortone et al. 1998) resulting in negative feedbacks to reef fish energetics, residence times, and local abundance, especially if the feeding halos of closely spaced reefs overlap (Lindberg et al. 2006). As a result, the degree of prey depletion and associated negative feedbacks alters the potential for sustained productivity of an artificial reef and artificial reef complexes. It is theorized that bioenergetic demands increase as foraging area increases resulting in increased reef fish emigration from resource-depleted habitats to habitats containing a greater abundance of prey (optimal foraging theory, Charnov 1976).

Artificial reefs or artificial reef complexes that sustain prey resources over time may potentially benefit reef fishes and fishery productivity more by reducing the energetic costs of foraging, increasing growth rates, and increasing site fidelity. While past research has demonstrated reef fish abundance increases both with increasing reef size (see review by Pickering and Whitmarsh 1997) and with spacing (Schroeder 1987; Frazer and Lindberg 1994; Lindberg et al. 2006), the size and spacing of artificial reefs can alter growth rates, site fidelity, and population dynamics of reef fishes (Lindberg 1996; Lindberg and Loftin 1998; Lindberg et al. 2006). Although larger, more widely dispersed reefs may hold greater benefit to fishers (increased catch rates), smaller, more isolated reefs may serve to better benefit marine resources through increased growth rates. In theory, this occurs through reductions in competition and bioenergetic demands at more widely spaced reefs provided that mortality rates do not change as a function of spacing.

In the current study, information obtained from a mark–recapture study was used to estimate site fidelity, movement, growth, and productivity of red snapper at artificial reefs off coastal Alabama. Movement and growth parameters were evaluated in relation to the distribution, abundance, and demographic characteristics of artificial reefs. We first evaluated site fidelity, movement, and growth of all tagged fish captured during our study. We then examined whether habitat characteristics (e.g., density of artificial reefs, reef design/size, and biomass of reef fish residing at tagging sites) affected site fidelity and growth rates of red snapper. Finally, we compared instantaneous rates of growth in weight to total mortality estimates for red snapper from the eastern Gulf of Mexico (SEDAR 2005). We hypothesized red snapper residing at smaller reefs, surrounded by lower densities of artificial reefs, would have higher site fidelity and growth rates than red snapper residing at larger reefs, surrounded by greater densities of artificial reefs.

Methods and Materials

We tagged red snapper at 14 experimental artificial reef sites of two different designs (Tetrahedrons and Reefballs; Table 1) during quarterly research cruises conducted from February 1999 to October 2003. Tagging sites were located 25 to 35 km south-southeast of Dauphin Island, Alabama, in the Hugh Swingle General Permit Area (HSGPA) at depths ranging from 22 to 31 m (Figure 1A). Locations of tagging sites were unpublished and assumed to be unknown to recreational and commercial fishers during the course of our study.

Red snapper were captured from a charter fishing vessel with rod-and-reel and placed in 185-l holding tanks. All red snapper were measured to the nearest mm fork length (FL) and total length (TL) and tagged with a green Floy stainless steel-core internal anchor tag. Tags were inserted through a small incision in the left wall of the abdominal cavity. Each tag included a unique identification number, the word "reward", and a toll-free telephone number to report the fish upon capture. After tagging, fish were released alive at the site of capture. A $5 reward was offered for each tag return.

Tagged fish were captured during subsequent research cruises and by recreational and commercial fishers. Fish recaptured during research cruises were measured and released; internal anchor tags were replaced if damaged. When available, information on recapture location (e.g., private or public reef; Loran C or GPS coordinates), recapture date, and recapture length were obtained from commercial and recreational fisher recoveries.

Distance Moved and Dispersion Rate

Distance moved and dispersion rate were computed for all recaptures with known recapture locations. Distance moved was determined by measuring the linear distance between the tagging site and the recapture site. Dispersion rate was estimated by dividing the linear distance moved by the number of days at liberty. The delta method (Aitchison 1955; Pennington 1983) was used to obtain unbiased estimates of mean distance moved and rate of dispersion because many fish were recaptured at their site of release, resulting in many zeros for movement and dispersion data.

Negative binomial regressions were computed with PROC GENMOD in SAS (Hilbe 1994) to test the effects of time at liberty, length at tagging, artificial reef density, and season of tagging on red snapper movement and dispersion. Models first were fitted for each independent variable and the single-variable model with the lowest significant *P*-value ($\alpha = 0.05$) was chosen as the base model. Variables were added to the base model in order of significance to assess whether or not the deviance of the model was significantly reduced (Agresti 1990). Variables continued to be added to the model until the fit of the model was not significantly improved.

Table 1. Description and specifications of experimental artificial reefs used as tagging sites.

	Reefball	Tetrahedron
Description	Concrete, dome-shaped reef with a hollow interior cavity and numerous exterior holes	Reinforced, hollow, floorless concrete tetrahedron with four sides containing holes
No. of reef complexes	7	7
No. of reefs per complex	3	3
Height	1.22	1.52
Base diameter	1.52	2.43
Interior cavity diameter (cm)	60	N/A
Exterior hole diameter (cm)	22	30
Number of exterior holes	35–41 per reef	5 per side (20 per reef)
Weight (kg)	1050	1800
Footprint (m^2)	1.82	5.9
Surface area (m^2)	5.25	8.75
Volume (m^3)	1.8	3.5

Site Fidelity

Site fidelity was estimated by modeling the decay in recaptures made at tagging sites over time (Patterson and Cowan 2003). Nonlinear decay models were initially computed for all recaptures at tagging sites and for all recaptures at tagging sites by reef design (Tetrahedrons and Reefballs). For each model, recaptures were grouped into intervals of days at liberty based on the average time (102 d) between visits to tagging sites. Because fewer red snapper were available for recapture during subsequent time intervals, a correction factor was used to adjust for the reduced number of fish at liberty. The correction factor for each time interval was calculated as the ratio of the number of red snapper at liberty during that particular time interval divided by the total number of fish tagged. The total number of recaptures made was then adjusted by dividing the total number of recaptures made during each sampling interval by that interval's correction factor.

To estimate site fidelity, a nonlinear decay model was fit to the adjusted number of recaptures with PROC NLIN in SAS (SAS Institute, Inc. 1999):

$$N_t = N_0 e^{-Dt} \quad (1)$$

where N_t is the number of fish recaptured in time interval t, N_0 is the number of fish recaptured in the first time interval, D is the instantaneous rate in decline of recaptures, and t is time in days.

The instantaneous rate of decline (D) equals the instantaneous rate of mortality (Z) plus the instantaneous rate of emigration (E), while instantaneous total mortality equals the sum of natural (M) and instantaneous fishing (F) mortalities. Since no recaptures by fishers were reported from tagging sites, fishing mortality was assumed to be zero. Estimates of D were not affected by tag loss because damaged stainless steel-core tags were readily observed during research cruises. Therefore, D equaled $M + E$. An M of 0.1 was assumed for all analyses (SEDAR 2005), allowing E to be computed by subtraction. Site fidelity (SF), as an annual percentage then was estimated with the following equation:

$$SF = e^{-E} \quad (2)$$

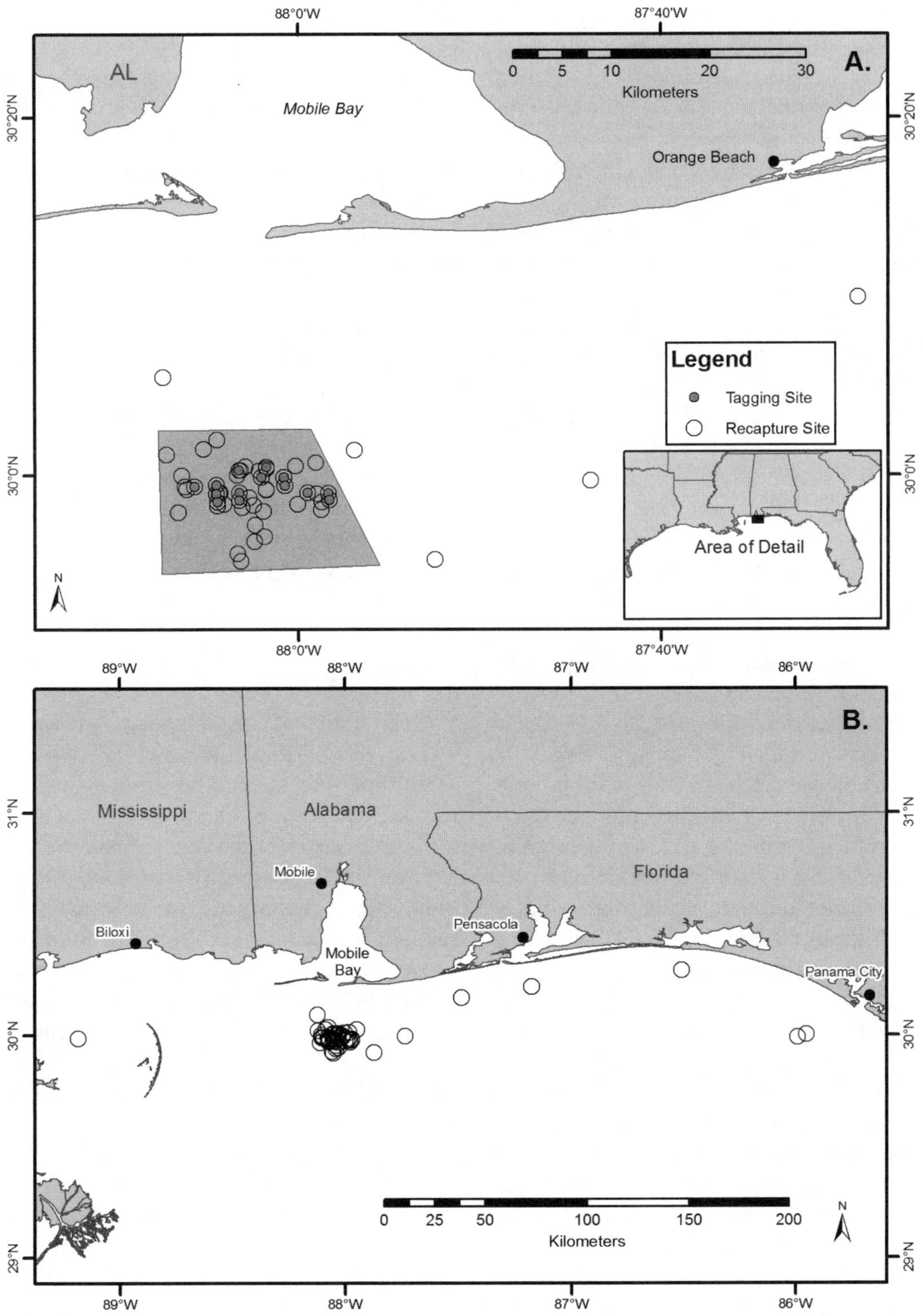

Figure 1. Geographical distribution of red snapper *Lutjanus campechanus* recaptures off coastal Alabama (A) and all red snapper recaptures Gulfwide (B). Open circles may represent multiple recaptures at the same location. Tagging sites are indicated by closed circles. The Hugh-Swingle General Artificial Reef Permit Area is shaded as a gray polygon.

Two additional nonlinear decay models were computed to examine both site fidelity at tagging sites surrounded by different densities of artificial reefs (within 0.1 km^2) and site fidelity at tagging sites supporting different biomasses/abundances of reef fish. Because there were few tag returns ($n < 15$) at some artificial reef sites, data were pooled to increase sample sizes. Tagging sites supporting various biomasses of reef fish were pooled as follows: low reef fish biomass = <14 kg $reef^{-1}$, intermediate reef fish biomass = 15–26 kg $reef^{-1}$, and high reef fish biomass = >26 kg $reef^{-1}$. Tagging sites surrounded by varying densities of artificial reefs were pooled as follows: low artificial reef density = <5 reefs/10,000 m^2, intermediate artificial reef density = 6–10 reefs/10,000 m^2, and high artificial reef density = >10 reefs per 10,000 m^2. Reef fish biomasses and artificial reef densities were determined from diver visual surveys and side-scan sonar imagery as reported in Strelcheck et al. (2005).

The effect of tagging site characteristics (reef design, artificial reef density, or reef fish biomass) on site fidelity was tested by transforming (*ln* + 1) the adjusted number of recaptures for each model and then computing an analysis of covariance (ANCOVA) to test for significant differences in model slopes (rates of decay). A significance level of 0.05 was used for all statistical tests.

Growth Rate

Growth rates for red snapper recaptures were estimated with simple linear regression because most red snapper tagged were small and growth is relatively linear for the first few years of life (Patterson et al. 2001a). Change in TL (length at tagging – length at recapture) was regressed against days at liberty to estimate average growth per day. Growth rates were calculated for all recaptures with known recapture lengths, for recaptures at various tagging sites (Tetrahedrons, Reefballs), for recaptures at tagging sites surrounded by differing densities of artificial reefs, for recaptures at tagging sites supporting various biomasses of reef fish, and for recaptures that did or did not move. Analyses of covariance were applied to test for differences in slopes (growth rates) among reef designs, artificial reef densities (low, intermediate, and high), reef fish biomasses (low, intermediate, and high), and movement (recaptured at site of release versus away from the site of release). If the assumption of homogeneity of slopes was rejected when comparing slopes among artificial reef densities or reef fish biomasses, additional analyses of covariance ($\alpha = 0.016$) were applied to test for differences in slopes between treatments (i.e., low versus intermediate, low versus high, and intermediate versus high).

G/Z ratio

The ratio of instantaneous growth in weight (G) to total mortality (Z) over the same time interval can be used as an index for assessing fish production (Houde 1996). G/Z ratios greater than one indicate net production, or increasing fish biomass. To estimate G, mean growth rate for tagged red snapper with known recapture locations was converted to annual instantaneous growth in weight with the length-weight equation in Patterson et al. (2001a). A 406 mm TL red snapper was assumed for the starting length, because this was the legal minimum size limit for recreational harvest during the course of our study. Because growth is basically linear to about 500–600 mm TL, the starting fish length was increased by the mean daily growth rate for recaptured red snapper for one year (481.2 mm TL). The starting and final length were then converted to weight in gm and the following formula was used to calculate G:

$$G = lnW_0 - lnW_t \ (3)$$

where W_0 is the predicted weight at 406 mm TL and W_t is the predicted weight of that fish at the end of one year (1.49 kg). Average *F* for 2001–2003 were obtained from SEDAR (2005) for the most vulnerable age-class to the entire eastern Gulf of Mexico red snapper fishery ($F = 0.80$) and for the most vulnerable age-class to the recreational fishery in the eastern Gulf ($F = 0.63$). Alabama and Northwest Florida account for nearly all eastern Gulf recreational landings and therefore are likely representative of *F* occurring off Alabama, although spatial differences in *F* may exist in the eastern Gulf. Total mortality was calculated by summing fishing mortality rates with the estimate of natural mortality for red snapper ($M = 0.1$ for age 2+; SEDAR 2005).

Results

Four thousand three hundred and seventeen red snapper were tagged; 2,650 red snapper were tagged at Tetrahedron reefs and 1,667 red snapper were tagged at Reefballs (Table 2). A total of 629 red snapper recaptures were recorded, representing 578 individuals. A majority of these fish were recaptured once ($n = 530$), 45 fish were recaptured twice, and 3 fish were recaptured three times. Recaptures of 363 red snapper were made at release sites, 199 were reported as recoveries by fishers, and 16 were initially recaptured at release sites and subsequently reported by fishers as being captured away from the site of release.

Mean TL at tagging for recaptures was 335 mm (±63.3 mm SD) and mean TL at recapture for fish with known recapture length was 400 mm (±69.8 mm SD). Fish were at liberty on average 401 d (±280 d SD); minimum and maximum days at liberty were 1 and 1,587 d, respectively. A majority of recaptures were at liberty 1 year or less (51.8%); however, 34.5% of red snapper recaptured were at liberty for 1–2 years, and 13.7% of red snapper recaptured were at liberty for greater than 2 years.

Table 2. Number of red snapper *Lutjanus campechanus* tagged at each tagging site over time. T = Tetrahedron; RB = Reefball.

Date	Tagging Site														
Tagged	T2	T3	T4	T5	T6	T7	T8	RB1	RB2	RB4	RB5	RB6	RB7	RB8	Total
Feb-99	49	42	24	50	28	40	47	18	14	13	20		32	8	385
May-99	12	23	9	55	11	21	26	25	18	18	7		35	4	264
Sep-99	41	25	20	29	1	19	15	15	11	19	2	4	24	8	233
Dec-99	56	31	43	50	13	13	17	20	17	28	7	21	11	6	333
Feb-00	15	19	21	37	6	21	32	9	22	11		2	11	25	231
May-00	19	21	15	22	1	4	19	19	18	8	2	11	33	14	206
Sep-00	6	52	26	15	50	23	40	13	9	18	7	11	30	24	324
Dec-00	37	37	69	64	43	53	62	49	30	38	22	33	59	36	632
Apr-01	20	12	28	44	7	14	29	21	29	13	21	10	41	11	300
May-01		2	13				10	8	6				26		65
Jul-01	9			6						1		1		4	21
Sep-01	19	24	15	33	12	30	26	37	31	10	22	16	36	23	334
Dec-01	28	33	27	42	32	46	45	24	21	6		10	17	15	346
Mar-02	10				5					1		6			22
Jun-02	16	57	22	5	3	20	7	10	22	2		4	50	10	228
Sep-02		57	26	32		38			23				25	46	247
Oct-02	34				35		38	3		14		22			146
Total	371	435	358	484	247	342	413	271	271	200	110	151	430	234	4317

Distance moved and rate of dispersion

Specific recapture location was reported by fishers for 116 recaptures and was known for all recaptures made by us at tagging sites (total n = 479). Mean (±SD) distance moved for all recaptures was 2.08 km (±0.46). Mean (±SD) distance moved for recaptures reported by fishers was 2.86 km (±0.13). Maximum distance moved was 201 km due east of the release site. Nine fish moved greater than 80 km away from the release site; eight were recaptured between Pensacola and Panama City, Florida, and one was recaptured west of the Chandelier Islands, Louisiana (Figure 1B). Two red snapper tagged at Tetrahedron-7 were recaptured at the same location off Destin, Florida (113 km from release site) and two other red snapper tagged at Reefball-4 were recaptured at the same location off Panama City, Florida (195 km from release site). Eighty-six percent (n = 412) of recaptures with known recapture locations were recaptured within 2 km of their release site, 94.6% within 5 km of the release site, and 96.9% within 10 km of the release site. Nearly 97% of recaptures with known recapture locations were from the HSGPA (Figure 1A).

In the single-variable, negative binomial regression models, distance moved was significant both for days at liberty (X^2 = 16.47, p < 0.0001) and for artificial reef density (X^2 = 7.76, p < 0.005). Length at tagging and season did not significantly affect the distance red snapper moved. When artificial reef density was added to the days at liberty model, model deviance increased. The final model included only days at liberty, which had a positive effect on distance moved.

Mean rate of dispersion for all recaptures was 8.6 m (±2.53) per day. Rate of dispersion for recaptures made away from the site of release ranged from 0.22 to 8,080 m per day. The maximum rate of dispersion was by a red snapper that was tagged at Tetrahedron-4 and recaptured the next day at Tetrahedron-8. In the single-variable, negative binomial regression models, rate of dispersion was significant both for days at liberty (X^2 = 26.06, p < 0.0001) and for artificial reef density surrounding tagging sites (X^2 = 6.43, p < 0.011). When artificial reef density was added to the days-at-liberty model, model deviance increased. Therefore, the final model included only days at liberty, which had a positive effect on rate of dispersion.

Site Fidelity

The decline in recaptures over time for all nonlinear decay models was highly significant (p < 0.001, r^2 > 0.94) (Figure 2). Annual site fidelity was 51.5% year^{-1} for all recaptures. Site fidelity ranged from 48.3% year^{-1} to 54.6% year^{-1} for the various levels of reef type, reef fish biomass, and reef density factors. There were no significant differences in annual site fidelity among levels within any single factor.

Growth Rate

Mean growth rate for all recaptures was 0.206 mm d^{-1}. Growth rates were significantly greater for red snapper recaptured at Tetrahedrons (0.215 mm d^{-1}) than for red snapper recaptured at Reefballs (0.194 mm d^{-1}; Figure 3A) (p = 0.029). Fish movement did not affect growth rate; mean growth rate of red snapper that moved (0.213 mm d^{-1}) was similar to mean growth rate of red snapper recaptured at the site of release (0.206 mm d^{-1}; Figure 3B) (p = 0.40). Growth rates were significantly different for red snapper recaptured at tagging sites supporting different reef fish biomasses (p = 0.026). Red snapper recaptured at tagging sites support-

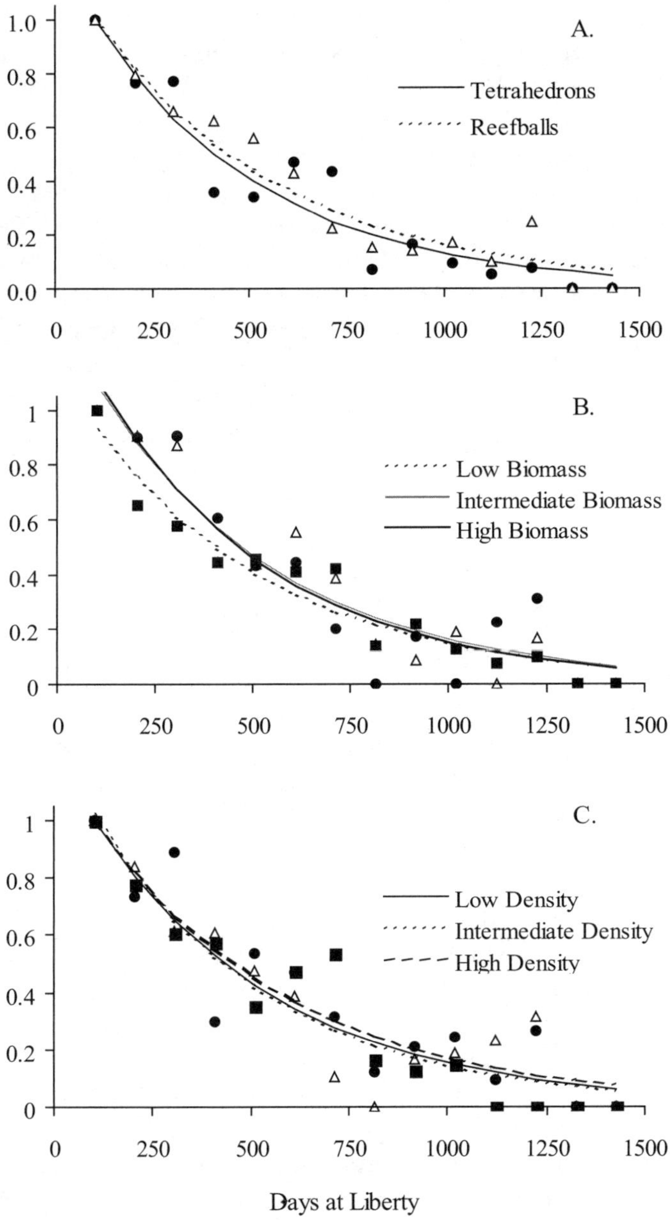

Figure 2. Relative number of red snapper *Lutjanus campechanus* recaptures made on tagging trips over time fitted with non-linear decay models for (A) two artificial reef designs, (B) different levels of reef fish biomass, and (C) different densities of artificial reefs surrounding tagging sites.

ing low biomasses of reef fish grew significantly slower than red snapper recaptured at tagging sites supporting intermediate and high biomasses of reef fish (low versus intermediate, $p < 0.001$; low versus high, $p = 0.006$). Growth rates did not significantly differ between tagging sites supporting low and intermediate reef fish biomasses ($p = 0.58$) or among tagging sites surrounded by different densities of artificial reefs ($p = 0.13$; Figure 3C).

G/Z ratios

The instantaneous growth coefficient estimated over the range of lengths considered (406–481 mm TL) was 0.54 year^{-1}. Ratios of *G/Z* were less than 1 for both total mortality rates considered ($Z = 0.73$ or 0.90 year^{-1}), ranging from 0.60 to 0.74.

Discussion

Several tagging studies have been conducted off coastal Alabama since 1990 to evaluate growth, movement, and site fidelity of red snapper (Szedlmayer and Shipp 1994; Patterson and Cowan 2003; Szedlmayer and Schroepfer 2005). Each of these studies tagged fish residing on artificial reefs to assess red snapper population dynamics, and in some instances, artificial reef ecology. In the current study, our objective was to examine differences in red snapper population dynamics in relation to various artificial reef characteristics (reef design, densities, carrying capacities). By examining habitat related differences in site fidelity, movement, and growth, this study sought to make inferences about the habitat quality of particular reef designs and locations.

Distances moved and mean dispersion rate were comparable to two previous studies (Szedlmayer and Shipp 1994; Watterson et al. 1998), but were much lower than movement and dispersion rates reported by Patterson and Cowan (2003). Szedlmayer and Shipp (1994) reported recapturing 76% of tagged red snapper within 2 km of release sites and maximum distance moved was 32 km. Watterson et al. (1998) reported recapturing 80% of red snapper not at liberty during Hurricane Opal at their release site; mean distance moved for fish not at liberty during Hurricane Opal ranged from 1.7 to 2.5 km. In this study, mean distance moved by red snapper was 2.1 km and 86% of red snapper were recaptured within 2 km of their release site. Our results are an order of magnitude less than reported for tagged red snapper at liberty during hurricanes (Watterson et al. 1998; Patterson and Cowan 2003). Fish at liberty during Hurricane Opal moved a mean distance of 32.6 km (Watterson et al. 1998). Mean distance moved by tagged red snapper at liberty during Hurricanes Opal and Georges was 29.6 km (Patterson and Cowan 2003). Despite these differences, our results indicate red snapper do move long distances in the absence of hurricane activity. Maximum distance moved in this study was 201 km, compared to 252 km and 352 km reported by Watterson et al. (1998) and Patterson and Cowan (2003), respectively.

Overall, site fidelity for red snapper in this study was similar to the results of Szedlmayer and Shipp (1994) and Watterson et al. (1998) who qualitatively concluded red snapper exhibit strong site fidelity during nonhurricane periods. In contrast, site fidelity estimates presented in this study were nearly twice as high as those presented in Patterson and Cowan (2003), which is the only other study of red snapper that has quantitatively estimated site fidelity. Site fidelity in the Patterson and Cowan (2003) study was approximately 25% year^{-1}, regardless of whether or not fish were at liberty during hurricanes. Although the same methods and similar sized fish were used to quantify site fidelity between this latter study and the current study, several key differences exist between these two studies,

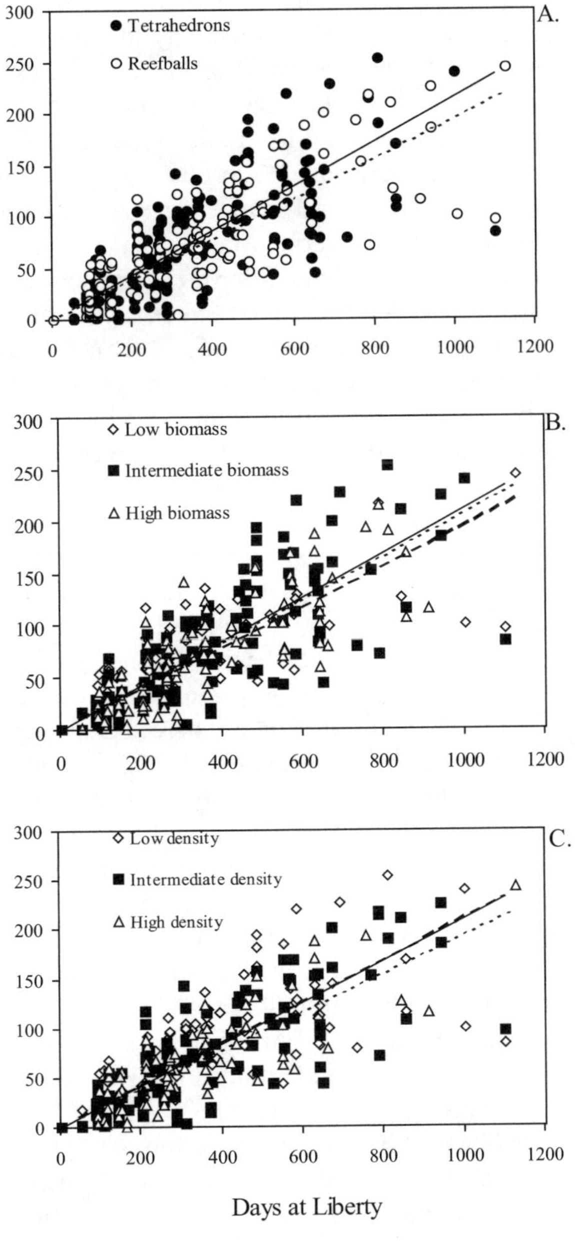

Figure 3. Linear regressions of change in red snapper *Lutjanus campechanus* total length (mm) versus days-at-liberty for: (A) two artificial reef designs, (B) different levels of reef fish biomass, and (C) different densities of artificial reefs surrounding tagging sites.

including the absence of hurricanes and more frequent sampling during our study. Although Patterson and Cowan (2003) did estimate site fidelity for red snapper not at large during hurricanes, sample size was small (n = 121) relative to this study (n = 340). In addition, mean time between sampling intervals was much less for the current study (102 d versus 150 d). Increased sampling effort potentially allowed a greater number of red snapper to be recaptured, as well as a greater number of red snapper to be tagged per sampling period, thereby increasing estimates of site fidelity.

Several telemetry studies off coastal Alabama have also examined residence time and site fidelity of red snapper (Szedlmayer 1997; Szedlmayer and Schroepfer 2005; Schroepfer and Szedlmayer 2006). Szedlmayer (1997) reported red snapper residence times ranging from 17 to 597 d, while Schroepfer and Szedlmayer (2006) reported red snapper residence times ranging from 1 to 595 d, with a mean residence time of 218 d. Both of these studies concluded red snapper exhibit high site fidelity. Szedlmayer and Schroepfer (2005) used 'event analysis' to reanalyze residency data from both of the previously published telemetry studies, as well as unpublished tagging data. Using event analysis, mean residence time was estimated to be 373 d, which is similar to the results of this study. The authors concluded red snapper exhibit high site fidelity and stated their estimate of mean residence time was likely underestimated. However, their results indicate only a 50% probability of detecting ultrasonically tagged red snapper one year after release. Additionally, because fish lost immediately after tagging were not used for analysis, it appears site fidelity was likely overestimated by Szedlmayer and Schroepfer (2005) and may have been both less than the results of this study and more similar to the results of Patterson and Cowan (2003) if all tagged fish were used for analysis.

Despite variable estimates of site fidelity observed in this study and other studies at small artificial reefs, we conclude site fidelity is relatively low for red snapper captured at small artificial reefs, especially given the fact that red snapper live in excess of 50 years (Wilson and Nieland 2001). There was only a 50% probability of recapturing a tagged red snapper one year after release. Our study largely tagged small, sub-legal red snapper, which have previously been shown to move less than larger, legal-sized red snapper (Patterson et al. 2001b). As red snapper size increases, movement is expected to increase, resulting in even lower estimates of site fidelity. Even if movement away from the site of capture does not increase with increasing size, the cumulative effect of a 50% decrease in site fidelity per year is substantial (e.g., 3% probability after 5 years, 0.1% probability after 10 years). Therefore, regional rather than site-specific site fidelity may be more significant, especially if movement occurs in a limited area where red snapper are subjected to high rates of fishing mortality. As indicated by our study, 97% of tagged fish caught were recaptured within the HSGPA, which is a designated artificial reef area heavily fished primarily by recreational anglers. Limited movement within this area might actually be detrimental to red snapper given high fishing mortality rates.

Artificial reef design, reef fish biomass, and artificial reef densities did not significantly affect site fidelity of red snapper. Based on the resource mosaic hypothesis (Lindberg et al. 1990), we expected smaller, more isolated reefs supporting lower biomasses of reef fish to have the highest site fidelity. This is predicted to occur because reef fishes are potentially less likely to move between reefs during foraging due to increased risks of predation and reduced proximity to shelter. When reef densities are high, distances between reefs are shorter and reef fish may move among reefs more readily, resulting in increased movement and an expanded 'home' range.

Our intent was to evaluate site fidelity of red snapper at individual tagging sites. However, data were pooled among sites because of low sample sizes. Pooling data may have limited our ability to determine whether artificial reef characteristics, such as size and spatial proximity to other artificial reefs, affected red snapper site fidelity. Additionally, there was only a two-fold difference in artificial reef sizes used in this study (Table 1). Lindberg et al. (2006) found gag, *Mycteroperca microlepis*, residence times were greater on large, more widely spaced artificial patch reefs when compared to smaller, more closely spaced artificial patch reefs that were 4X smaller in size (4 m^3). The small patch reefs used by Lindberg and Loftin (1998) were 2X larger than the largest artificial reefs (Tetrahedrons) used in this study. Because our artificial reefs were small, bioenergetic demands might have not have been significantly different to reduce red snapper residency. Also, McCawley (2003) found a major portion of red snapper diet (41% by weight) may be derived from food/prey residing in the water column. Although a majority of a red snapper's diet is still from benthic prey (55% by weight), the dependence on water column prey for food could reduce the negative energetic feedbacks hypothesized by the resource mosaic hypothesis (Lindberg et al. 1990). If this were the case, then the abundance and distribution of pelagic prey may have played an important role in structuring our artificial reef fish assemblages. Nevertheless, recent work done by Shipley and Cowan (NOAA Fisheries, unpublished data) both demonstrates the likelihood of foraging halos around reefs of the size studied here and shows that prey concentrations diminish when the halos overlap. These results are based upon a synthesis of results from multiple studies in this region and suggest that reefs should be spaced about 600 m apart to reduce competitive interference.

Growth rates of red snapper recaptured in our study were comparable to previous studies (Szedlmayer and Shipp 1994; Watterson et al. 1998; Patterson et al. 2001a), which indicated growth rates of tagged red snapper range from 0.22 to 0.25 mm d^{-1}. Our study estimated the mean growth rate of all recaptures to be 0.21 mm d^{-1}. Growth rates were significantly different between artificial reef designs and among reefs supporting different biomasses of reef fish. The significantly higher growth rate observed at Tetrahedrons, the larger of the two artificial reefs in our study, was an unexpected result; the resource mosaic hypothesis predicts slower growth rates at larger artificial reefs (Lindberg et al. 1990). Similarly, slower growth rates were observed at tagging sites supporting lower biomasses of reef fish; however, ecological theory predicts tagging sites supporting higher biomasses of reef fish would have slower growth rates. As discussed above, differences in our results relative to those predicted by ecological theory include the size of artificial reefs used, data pooling, and the relative importance of pelagic prey. With respect to pelagic prey, Tetrahedrans could simply attract more pelagic prey than Reefballs, resulting in little or no expected negative feedbacks from benthic foraging. Additionally, artificial reef sites with lower reef fish biomasses may have lower available prey bases for reasons unrelated to the size and design of our experimental artificial reefs. Furthermore, growth rates of red snapper residing at reefs surrounded by relatively low artificial reef densities could simply reflect differences in prey availability or quality and not represent negative density dependent processes.

Our results are contrary to those of Lindberg et al. (2006), who found gag growth to be greater at smaller rather than larger artificial reefs. They concluded gag selected shelter at the expense of maximizing growth. Artificial reef size and spacing, reef fish biomass, and the ability of anglers to easily locate artificial reefs all likely affect the overall level of red snapper productivity achieved at an artificial reef or complex of artificial reefs. Unfortunately, because of the design and scale of our study, such differences were not observed. Further research is needed to elucidate whether artificial reef spacing and reef fish biomass play an important role in structuring artificial reef fish assemblages in the northern Gulf of Mexico.

In summary, site fidelity was higher and both distances moved and rates of dispersion were lower for red snapper when compared to previous studies. The low rate of dispersion of red snapper, at least during nonhurricane years, within the complex of artificial reefs (n = 20,000) off coastal Alabama supports the potential for localized production of *adult* red snapper, assuming mortality does not exceed growth. Although there is no evidence that natural habitat limits the current population size of red snapper, artificial reefs off coastal Alabama have allowed reef fish to inhabit areas that previously attracted few adult fishes of recreational or commercial value. If red snapper are retained along the Alabama shelf and artificial reefs provide suitable habitat for red snapper growth, reproduction, and survival, it is plausible for enhancement of production to occur. However, characteristics of artificial reefs, such as reef size and standing stock biomass, may affect the overall rate of red snapper growth. In addition, if fishing mortality exceeds either productivity or recruitment and if red snapper production is not limited by the availability of habitat (Cowan et al. 1999), then high fishing mortality rates may offset or diminish any net gains in productivity resulting from artificial reef construction. Based on a recent stock assessment of red snapper (SEDAR 2005), fishing mortality rates in the eastern Gulf of Mexico were estimated to be well above levels producing maximum sustainable yield, thus the likelihood artificial reefs off Alabama have a net positive effect on red snapper population productivity appears to be remote at this time. Instead, Alabama's artificial reef program may be serving as a net sink for red snapper production. The lower rates of dispersion and higher site fidelity of red snapper observed in this study may further diminish productivity during nonhurricane periods when red snapper dispersion rates are lower, thus making them even more vulnerable to fishing mortality along the Alabama shelf. However, movement during both hurricane and nonhurricane periods likely diminishes the extent of spatial differences in F occurring in the eastern Gulf red snapper population.

As a final thought, our study results imply that increases in fish mortality attributable to artificial reefs could be offset to result in a zero sum game, or that reefs could be used to produce new fish biomass if some fraction of reefs were deployed in areas of limited or no fishing, such as no take reserves. As our understanding continues to improve, we think we will be well positioned in the future to use artificial reefs as conservation management tools designed to do more than simply create new fishing opportunities.

Acknowledgments

We thank the Alabama Department of Conservation and Natural Resources Marine Resources Division for providing funding for this project. We thank Capt. M. Thierry, A. Thierry, and S. Thierry for allowing us to tag red snapper onboard their charter fishing vessel *Lady Ann*. We thank R. Shipp, J. McCawley, W. Ingram, J. Jackson, M. Woods, J. O'Neal, B. Sauls, and the many volunteer anglers who assisted in capturing and tagging fish. We thank the many recreational and commercial fishers who reported tag-recaptures. Finally, we thank Amanda Frick for preparing GIS figures for this manuscript.

Notice: NOAA Fisheries Service does not approve, recommend, or endorse any proprietary product or material mentioned in this publication. The views expressed by the principal author of this manuscript do not necessarily reflect those of NOAA Fisheries Service.

References

Agresti, A. 1990. Categorical data analysis. Wiley, New York.

Aitchison, J. 1955. On the distribution of a positive random variable having a discrete probability mass at the origin. Journal of the American Statistical Association 50:901–908.

Beets, J. and M. A. Hixon. 1994. Distribution, persistence, and growth of groupers (Pisces: Serranidae) on artificial and natural patch reefs in the Virgin Islands. Bulletin of Marine Science 55(2–3):470–483.

Bortone, S. A., R. C. Cody, R. K. Turpin, and C. M. Bundrick. 1998. The impact of artificial-reef fish assemblages on their potential forage area. Italian Journal of Zoology 65: 265–267.

Charnov, E. L. 1976. Optimal foraging: the marginal value theorem. Theoretical Population Biology 9:126–136.

Cowan, J. H., Jr., W. Ingram, J. McCawley, B. Sauls, A. Strelcheck, and M. Woods. 1999. The attraction vs. production debate: does it really matter from the management perspective? A response to the commentary by Shipp, R.L. 1999. Gulf of Mexico Science 17:51–55.

Eggleston, D. B., R. N. Lipcius, and J. J. Grover. 1997. Predator and shelter-size effects on coral reef fish and spiny lobster prey. Marine Ecology Progressive Series 149:43–59.

Frazer, T. K. and W. J. Lindberg. 1994. Refuge spacing similarly affects reef-associated species from three phyla. Bulletin of Marine Science 55(2–3): 388–400.

Hilbe, J. M. 1994. Log negative binomial regression using the GENMOD procedure in SAS/STAT software. Pages 1199–1222 *in* Proceedings of Nineteenth Annual SAS Users Group International Conference. SAS Institute, Inc., Cary, North Carolina.

Hilborn, R., C. J. Walters, and D. B. Jester, Jr. 1990. Value of fish marking in fisheries management. Pages 5–7 *in* N. C. Parker, A. E. Giorgi, R. C. Heidinger, D. B. Jester, Jr., E. D. Prince, and G. A. Winans, editors. Fish-marking techniques. American Fisheries Society, Symposium 7, Bethesda, Maryland.

Hixon, M. A. and J. Beets. 1989. Shelter characteristics and Caribbean fish assemblages: experiments with artificial reefs. Bulletin of Marine Science 44: 666–680.

Hixon, M. A., and J. Beets. 1993. Predation, prey refuges, and the structure of coral-reef fish assemblages. Ecological Monographs 63:77–101.

Houde, E. D. 1996. Evaluating stage-specific survival during the early life of fish. Pages 51–66 *in* Y. Watanabe, Y. Yamashita, and Y Oozeki, editors. Survival stages in early life stages of marine resources. AA Balkema, Rotterdam, Netherlands.

Lindberg, W. J. 1996. Fundamental design parameters for artificial reefs: interaction of patch reef spacing and size. Final Project Report, March 1996, to the Florida Department of Environmental Protection, Division of Marine Resources. MARFIN Grant C-6729.

Lindberg, W. J., T. K. Frazer, and G. R. Stanton. 1990. Population effects of refuge dispersion for adult stone crabs (Xanthidae, Menippe). Marine Ecological Progressive Series 66:239–249.

Lindberg, W. J., T. K. Frazer, K.M. Portier, F. Vose, J. Loftin, D.J. Murie, D.M. Mason, B. Nagy, and M. K. Hart. 2006. Density-dependent habitat selection and performance by a large mobile reef fish. Ecological Applications 16:731–746.

Lindberg, W. J. and J. L. Loftin. 1998. Effects of artificial reef characteristics and fishing mortality on gag (*Mycteroperca microlepis*) productivity and rccf fish community structure. Final Project Report, June 1998, to the Florida Department of Environmental Protection, Division of Marine Resources. MARFIN Grant Number MR-073.

McCawley, J. R. 2003. Diet and prey demand of red snapper, *Lutjanus campechanus*, on Alabama artificial reefs. Master's thesis. University of South Alabama, Mobile, Alabama.

Patterson, W. F., III, J. H. Cowan, Jr., C. A. Wilson, and R. L. Shipp. 2001a. Age and growth of red snapper, *Lutjanus campechanus*, from an artificial reef area off Alabama in the northern Gulf of Mexico. Fishery Bulletin 99:617–627.

Patterson, W. F., III, J. C. Watterson, R. L. Shipp, and J. H. Cowan, Jr. 2001b. Movement of tagged red snapper in the northern Gulf of Mexico. Transactions of the American Fisheries Society 130:533–545.

Patterson, W. F., III, and J. H. Cowan, Jr. 2003. Site fidelity and dispersion of red snapper associated with artificial reefs in the northern Gulf of Mexico. Transactions of the American Fisheries Society 36:181–193.

Pennington, M. 1983. Efficient estimators of abundance for fish and plankton surveys. Biometrics 39:281–286.

Pickering, H., and D. Whitmarsh. 1997. Artificial reefs and fisheries exploitation: a review of the 'attraction versus production' debate, the influence of design and its significance for policy. Fisheries Research 31:39–59.

SAS Institute, Inc. 1999. Statistics version 8.0. SAS Institute Inc., Cary, North Carolina.

Schroeder, R. E. 1987. Effects of patch size and isolation on coral reef fish recruitment. Bulletin of Marine Science 41(2):441–451.

Schroepfer, R. L. and S. T. Szedlmayer. 2006. Estimates of residence and site fidelity for red snapper *Lutjanus campechanus* on artificial reefs in the northeastern Gulf of Mexico. Bulletin of Marine Science 78(1): 93–101.

SEDAR (Southeast Data, Assessment, and Review). 2005. SEDAR 7: Gulf of Mexico red snapper complete stock assessment report. SEDAR, Charleston, South Carolina.

Strelcheck, A. J., J. H. Cowan, Jr., and A. Shah. 2005. The influence of reef location on artificial-reef fish assemblages in the north-central Gulf of Mexico. Bulletin of Marine Science 77: 425–440.

Szedlmayer, S. T. 1997. Ultrasonic telemetry of red snapper, *Lutjanus campechanus*, at artificial reef sites in the northeast Gulf of Mexico. Copeia 4:846–850.

Szedlmayer, S. T., and R. L. Schroepfer. 2005. Long-term residence of red snapper on artificial reefs in the northeastern Gulf of Mexico. Transactions of the American Fisheries Society 134:315–325.

Szedlmayer, S. T. and R. L. Shipp. 1994. Movement and growth of red snapper, *Lutjanus campechanus*, from an artificial reef area in the northeastern Gulf of Mexico. Bulletin of Marine Science 55:887–896.

Watterson, J. C., W. F. Patterson, III, R. L. Shipp, and J. H. Cowan, Jr. 1998. Movement of red snapper, *Lutjanus campechanus*, in the north central Gulf of Mexico: potential influence of hurricanes. Gulf of Mexico Science 16:92–104.

Wilson, C. A., and D. L. Nieland. 2001. Age and growth of red snapper, *Lutjanus campechanus*, from the northern Gulf of Mexico off Louisiana. Fishery Bulletin. 99:653–664.

American Fisheries Society Symposium 60:163–187, 2007

Movers and Stayers: Individual Variability in Site Fidelity and Movements of Red Snapper off Texas

SANDRA L. DIAMOND[1, 2], MATTHEW D. CAMPBELL, DEREK OLSON, AND YUNING WANG
Department of Biological Sciences
Texas Tech University, Lubbock, Texas 79409 USA

JASON ZEPLIN
Texas Parks and Wildlife
95 Fish Hatchery Road, Brownsville, Texas 78412 USA

STEVEN QUALIA
Fish Trackers, Inc.
6710 Holly Road, Corpus Christi, Texas 78412 USA

Abstract.—Site fidelity and movements of red snapper *Lutjanus campechanus* off the coast of Texas were estimated from two tagging programs conducted aboard recreational boats. In one program, we tagged 5,614 red snapper at over 200 fishing sites between July 2002 and August 2005 using hook-and-line gear; fishermen and others reported 130 recaptured fish (a 2.3% return rate), 82 of which included location data. About 54% of the recaptured fish had moved, with an average movement of 20.4 km. In the other program, over 9,000 fish were tagged between 1983 and 2006, but only 68 records of fish tagged between 1986 and 2000 could be analyzed due to computer problems. Of the recaptured fish, 60 could be analyzed for movement, and 17 fish (28%) had moved an average of 19.1 km. Important predictors of movement in at least one program or analysis were depth, habitat type, isolation of the initial capture location, time at liberty, and fish size. Red snapper showed a higher probability of movement from capture sites that were in deep water (>40 m), natural rather than artificial structures, and isolated from other sites rather than clustered. Differences in movement from natural versus artificial sites may have been related to red snapper 'sub-cohort' behavior, (the tendency of small groups of red snapper to associate together), as the majority of fish that moved were members of sub-cohorts. Larger fish had a higher probability of movement than smaller fish. In one program, fish movements increased with increasing time at liberty, which is similar to past studies on red snapper site fidelity. In the other program, movements decreased with increasing time at liberty, but this was probably a bias caused by the spatial and temporal scale of the recapture effort. Red snapper that moved showed a higher average daily growth rate between capture and recapture than fish that were stationary,

[1] Corresponding author: s.diamond@uws.edu.au.
[2] Current address: University of Western Sydney,
Locked Bag 1797, Penrith South DC, NSW 1797, Australia

suggesting that movements may be beneficial for red snapper. The spatial scale of fish movements in this study, although larger than in other studies without hurricanes or translocation of fish, was still small enough to support the idea that red snapper populations in the northern Gulf of Mexico are relatively isolated, and that there may be a separate demographic stock off Texas.

Introduction

Variability among individuals within a population is one of the conditions necessary for natural selection to occur. One way in which variability among individuals can manifest itself is in differences in mobility and movements. For example, in many species some individuals within a population show a high degree of site fidelity, while others disperse, sometimes over large temporal or spatial scales. The decision to stay in one location or to move is influenced by the relative costs and benefits of these choices (Solomon 2003) and has consequences for survival and fitness at both the individual and the population levels (Steingrimsson and Grant 2003).

Most studies of movements in marine fishes have simply reported the proportion of fish within a population exhibiting high or low mobility, rather than explored factors affecting this individual variability. In stream fishes, factors that have been found to influence movements and site fidelity are: fish size and growth rate (Skalski and Gilliam 2000), fish condition (Hilderbrand and Kershner 2004), population density (Steingrimsson and Grant 2003), resource availability (Bujold et al. 2004), habitat type (Belanger and Rodriguez 2002), and inherent 'boldness' (Fraser et al. 2001). While these factors may also influence movements and site fidelity in marine fishes, large-scale weather and ocean phenomena such as storms, currents, and hypoxic conditions may also play an important role (Watterson et al. 1998; Patterson et al. 2001, Lenihan et al. 2001).

Red snapper *Lutjanus campechanus* is a long-lived reef fish that exhibits considerable individual variability in site fidelity and movement. Since the 1960s, red snapper has been the focus of many studies on habitat preferences and movement, and estimates of site fidelity within a study have ranged from about 25% to more than 90%. Factors suggested to be important in the percentage of fish that move compared to fish that remain at a site have included either size or age of fish (Moseley 1965), depth of capture (Beaumariage 1969; Watterson et al. 1998), seasonal patterns due to water temperature or reproductive condition (Topp 1963; Beaumariage and Bullock 1976), hurricanes (Watterson et al. 1998; Patterson et al. 2001), and translocation, or releasing fish at a different location than the original tagging site (Watterson et al. 1998; Patterson et al. 2001, Peabody 2004).

Site fidelity and movements have important implications for red snapper management and conservation. Red snapper has been overfished since the mid-1980s, and is thought to be at less than 10% of sustainable levels (Stock Assessment Workshop Report 2005). Currently red snapper in the Gulf of Mexico (the Gulf) is managed as a unit stock, although recent studies have found regional differences in size-at-age and growth rate (Fischer et al. 2004), size and age of females at maturity (Woods et al. 2003), and effective population size (Saillant and Gold 2006). The frequency and scale of movements are important indicators of stock structure and the potential for genetic mixing. Red snapper populations have also been affected by an increase in the amount of hard structure in the Gulf, including artificial reefs and oil platforms (Peabody 2004; Szedlmayer and Schroepfer 2005). High site fidelity over long time periods could indicate whether artificial habitats contribute to increases in production or simply attract fish from other locations (Schroepfer and Szedlmayer 2006). In addition, marine protected areas around preferred habitats are potential management measures that may be used in the future to rebuild red snapper populations. More information on the proportion of red snapper that exhibit site fidelity and factors influencing movements is critical for the proper design and evaluation of marine protected areas for red snapper.

The purpose of our study was to investigate individual variability in site fidelity and dispersal of red snapper living in the Gulf off the coast of Texas. Based on previous studies, we expected that the majority of tagged red snapper would stay in the vicinity of tagging, while a smaller proportion would move. Our goal was to examine potential factors that influence movement. We focused on internal factors such as fish size at capture and growth rate between capture and recapture, and features of the capture habitat such as depth, isolation, type of habitat and vertical relief. Our hypotheses were: 1) larger fish will move more often than smaller fish since they require more resources and are less vulnerable to predation in the open areas between habitats; 2) mobile fish will grow faster than sedentary fish because they move to avoid competition; 3) habitat isolation will increase movements, as potential habitats cannot be explored easily, and fish will have to leave the area to explore; 4) fish will move less frequently from natural than artificial habitats; and 5) red snapper will move less often from habitats with high vertical relief than low relief because of their attraction to structure.

Methods

Tagging and Recapture

Tagging and recapture of red snapper were carried out by two different tagging programs using similar methods. Both programs involved tagging fish at various locations throughout Texas state and federal waters of the Gulf and incidental recaptures reported primarily by anglers, commercial fishermen, and fish dealers.

Texas Tech University (TTU)

Red snapper tagging took place aboard volunteer charterboats and headboats from 1 July 2002 to 31 August 2005. In 2003–2005, summer tagging (May through August) took place on fishing vessels operating out of Port Isabel, Port Aransas, Freeport, and Galveston. During these summer tagging periods, each port had coverage at least one week per month except for a few occasions when inclement weather kept fishing vessels at the dock. Winter tagging (September through April) and minimal tagging during summer 2002 took place on vessels fishing out of Port Aransas exclusively. Winter tagging was less consistent due to recreational fishing closures, lack of bottom fishing trips, and frequent inclement weather cancellations.

Red snapper were captured with hook and line and tagged by TTU personnel. Fishing locations were chosen by the captain of the participating vessel and ranged from 15 to 100 m in depth (Figure 1). Information on the location of tagging operations was recorded at the discretion of the fishing captain (i.e., captains were encouraged to give exact positions, but many chose not to disclose this information). Manual reels were most frequently used aboard the vessels; however, one vessel out of Port Aransas exclusively used electric reels. Circle or j-hooks were used depending on the preference of the captain or customer. All tagged fish were measured in total length (TL) or fork length (FL), tagged with bright yellow T-bar anchor tags (Floy tags), vented (expanded air bladder punctured) with syringes or the tagging gun, and released. Tags were anchored in the dorsal fin musculature to prevent tag loss. Some tagged

tagged fish had <0.5 cc of blood drawn from the caudal vein for a separate study.

Fish were recaptured by recreational and commercial anglers who reported tags to a phone number written on the tag. Participating fishermen were entered into a lottery for $50, but there were no individual rewards for tagged fish. The tagging program was publicized by flyers sent to most of the recreational fishing companies in Texas, by speaking to individual fishermen at docks, and by news articles in local fishing newspapers.

Fish Trackers (FT)

The major differences between the Fish Trackers (*FT*) tagging program and the TTU program were: 1) in the *FT* program, fish were tagged by both research personnel (around 25% of fish) and by volunteer anglers aboard charterboats, headboats, and private boats, 2) yellow plastic dart tags were used instead of T-bar tags, and 3) no blood was taken from any fish. Tagging began in 1983 and fish have been tagged sporadically until the present time. Rewards such as hats were offered for recaptures early in the program, but rewards were stopped by 1987. The program was advertised with flyers, newspaper articles, and by word of mouth. Due to problems with computerized data files, only a random subset of the recaptured fish was used in the analysis, representing records from 1986 through 2000. Fish were tagged in both Texas and Florida, but only the Texas data were used in this analysis. Fish may or may not have been vented before release.

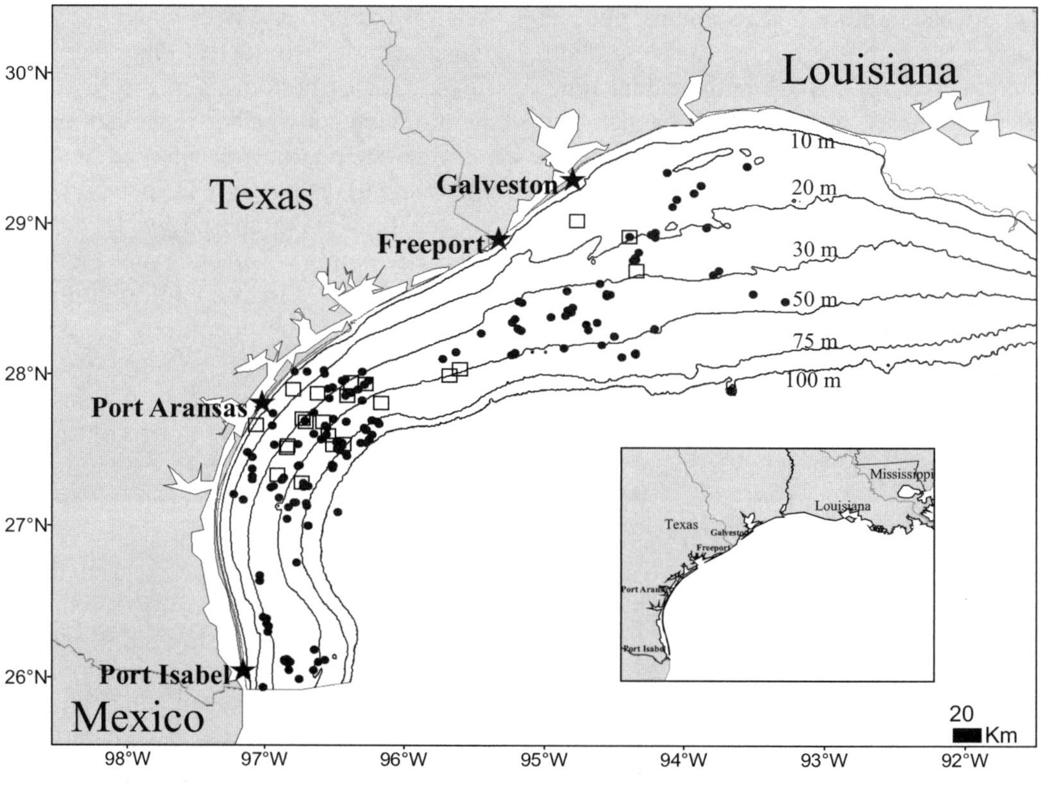

Figure 1. Red snapper *Lutjanus campechanus* tagging sites for TTU (solid circles) and Fish Trackers (open squares). Depth contours are in meters.

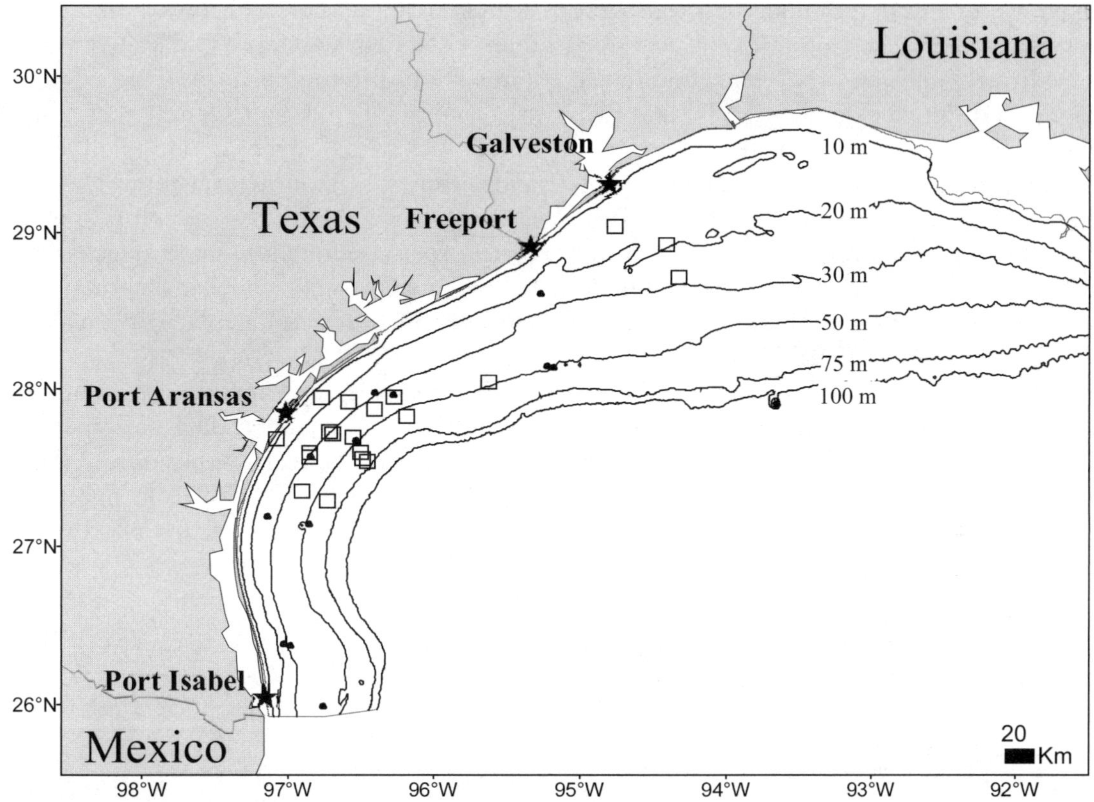

Figure 2. Locations of red snapper *Lutjanus campechanus* tagged by TTU (solid circles) and Fish Trackers (open squares) that were recaptured at the intitial tagging location ("stayers"). Depth contours are in meters.

Data Analysis

Fish from both tagging programs were classified as 'stayers' or 'movers' based on the locations of initial tagging (Figure 1) and first recapture (Figures 2 and 3). The criteria for moving or staying differed among fish, based on the degree of certainty we had in the location information, as described below. Locations that were at a recognizable landmark such as a labeled oil platform and locations where exact coordinates were disclosed by the fishing captain were considered 'certain.' Fish with certain locations for both tagging and recapture were considered to have moved if their coordinates differed. Locations such as verbal descriptions of reefs or banks with a large spatial extent or complex structure, GPS coordinates that did not include seconds or fractions of minutes, and positions based on distance and direction (in degrees) were considered 'probable', and coordinates as close as possible to the given descriptions were chosen for these locations. Fish with probable locations were considered to have moved if their verbal descriptions differed. If we had only verbal descriptions of capture and recapture locations for a particular fish (such as 'East Bank' off Port Isabel, which has a large spatial area), it was not possible to tell whether the fish moved small distances within the same reef complex, and we therefore considered these fish to have stayed in the same location. Locations with less information than described above were considered 'uncertain' and were not used.

Movers and stayers were contrasted based on differences in length (TL and FL), days at liberty, average growth rate between tagging and recapture, and habitat at tagging. Depth of capture was obtained from the captain or determined from the Hilton's Offshore Atlas for recognized landmarks such as oil platforms. Habitat differences included measures of habitat isolation and qualitative measures. Measures of isolation were the distance to the nearest potential habitat (nearest "neighbor") and the number of neighbors within 1, 2, 5, 10, and 25 km; isolated habitats were farther from the nearest habitat and had fewer neighbors than clustered habitats. Qualitative measures were differences in the type of habitat (natural—banks and natural reefs; artificial—platforms, wrecks, and wellheads) and differences in structural relief (high or above water—platforms and buoys; moderate—reefs, seamounts, large wrecks, and banks; small—small wrecks, obstructions). Distance moved was calculated by ArcGIS 9.0 (ESRI, Redlands, CA).

To provide information on potential habitats for red snapper, data on structures such as platforms, reefs, banks, wrecks, and obstructions were digitized from published charts (Table 1). Because exact coordinates

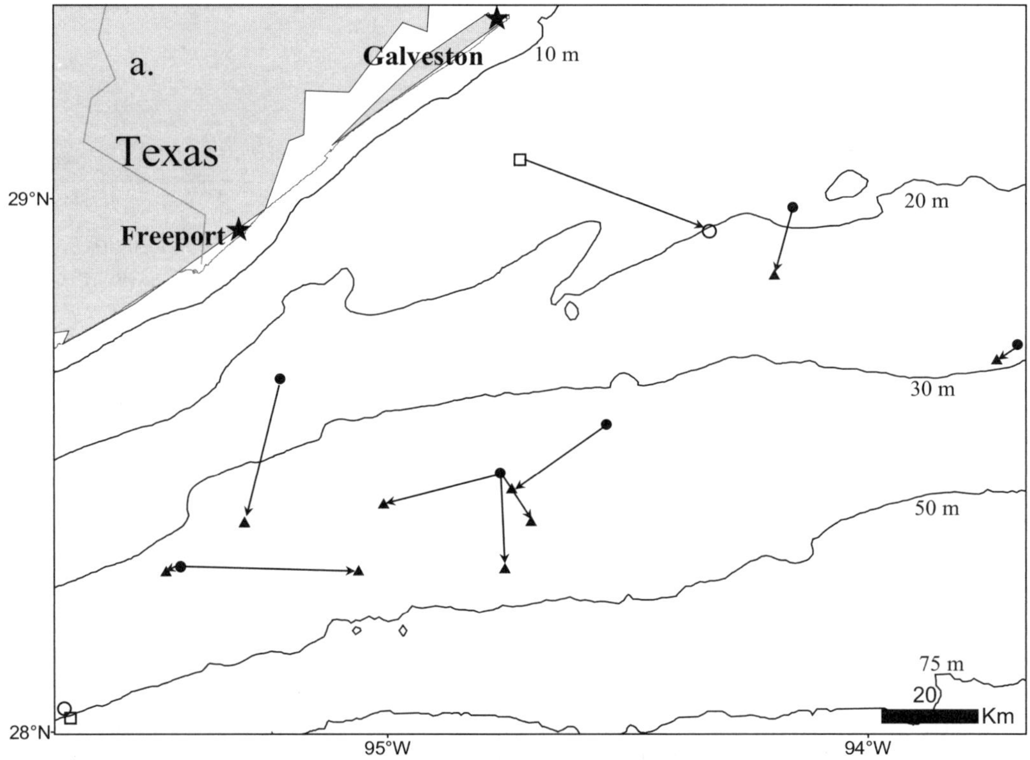

Figure 3. Movements of red snapper *Lutjanus campechanus* tagged by TTU (solid circles) and Fish Trackers (open squares) and recaptured by fishermen (solid triangles, open circles, respectively) in state and federal waters of the Gulf of Mexico out of four Texas ports: (a) Freeport and Galveston, (b) Port Aransas, and (c) Port Isabel. Some movements cannot be shown on maps because the moving distances are too short. The arrows indicate the direction of movement. The length of black arrows does not necessarily represent the distance moved. Depth contours are in meters.

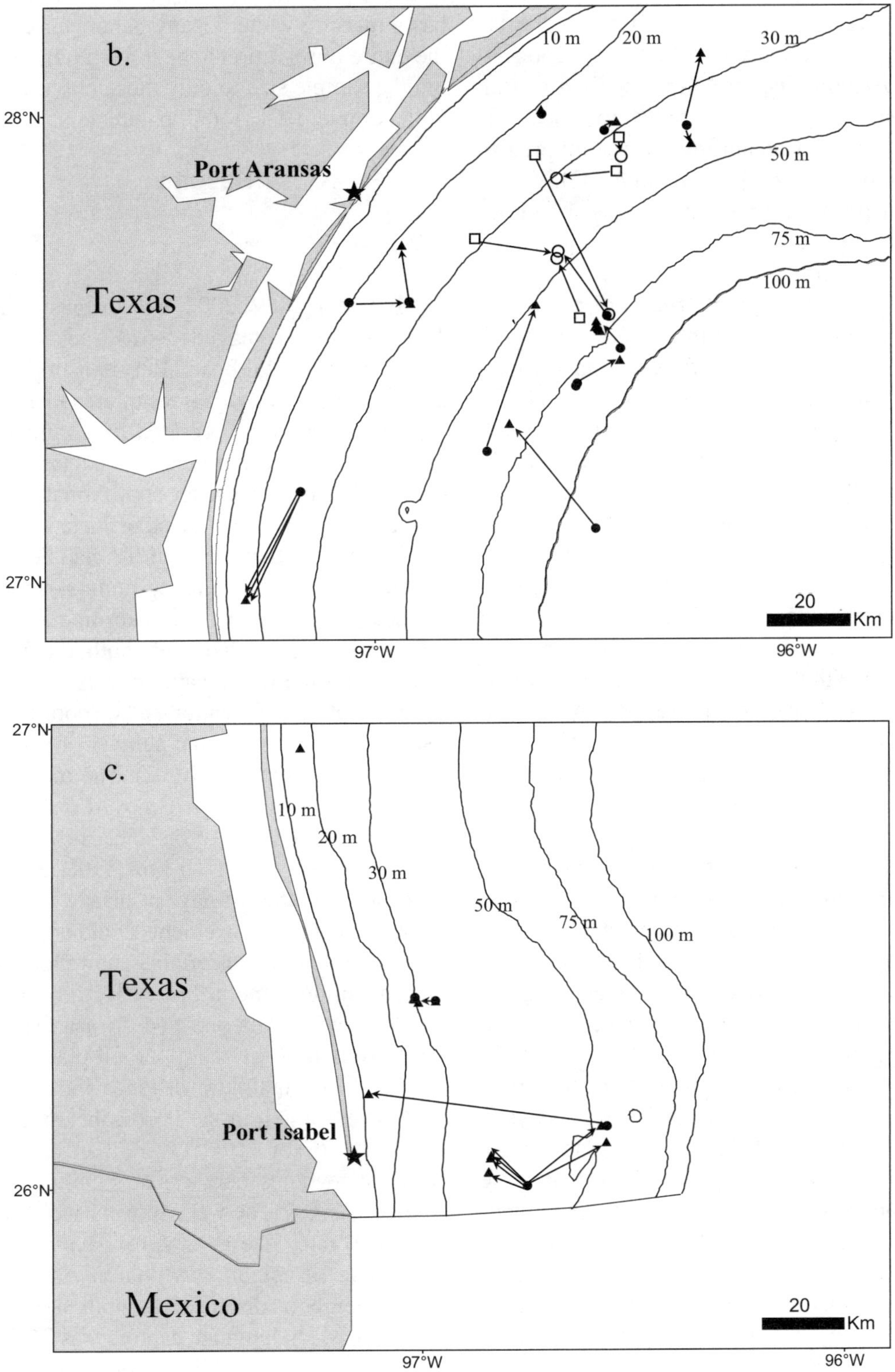

Figure 3. (Continued)

for the same landmarks differ on published charts, we used the Hilton's Offshore Atlas as our primary source, followed by Hook-N-Line Offshore Fishing Maps and Fish-n-Map Company maps. Coordinates for locations of tagging and recapture were compared to the file of potential habitats in ArcGIS 9.0 to obtain the distance to the nearest neighbor and the number of neighbors within 1, 2, 5, 10 and 25 km.

We used two-sample *t*-tests of means using SAS 8.0 (SAS Institute, Inc. Cary, North Carolina) to compare differences in parameters between movers and stayers separately for each tagging program. Metrics tested for both programs were: number of neighbors within 1, 2, 5, 10, and 25 km, nearest neighbor distance, depth of capture, days at liberty, size at capture, and average daily growth rate. *T*-test probabilities were corrected with the Bonferroni correction for experimentwise error. Full and stepwise logistic regressions (PROC LOGISTIC) with the probit link were used to test the effects of size at capture, average growth rate per day, days at liberty, depth at capture, and the number of neighbors within 2 km. When sample sizes permitted, the data in the TTU study were analyzed separately by port, since fishery characteristics and habitat differ by port. Chi-square tests and Fisher's exact tests were used to determine whether there were significant differences in habitat type and structural relief between movers and stayers, and in sub-cohort grouping behavior between artificial and natural habitats. Because of the small sample sizes, we also used categorical and regression tree analysis (CART analysis) in CART 5.0 (Salford Systems, San Diego, California) to look for variables that were good predictors of fish movement. One major benefit of using a CART model over logistic regression is that if a value is missing for a parameter, then CART uses a surrogate, a substitute parameter that splits the data in the same way as the primary parameter. Thus, all observations are used in the analysis. In contrast, in the logistic regression, an observation that has a missing value for any parameter causes the entire observation to be thrown out, reducing the sample size to only those observations with complete data for all parameters.

Results

TTU

Between July 2002 and August 2005, TTU personnel tagged 5,614 red snapper aboard recreational boats. Between July 2002 and December 2005, 130 recaptures identified by tag number were reported by commercial and recreational fishermen, boat captains, and fish market personnel, an approximate 2.3% return rate. In at least 22 other instances, we received notice that a tagged fish had been recaptured, but the specific tag number was not obtainable. Eighty-two fish had some location information associated with both the initial capture and the first recapture. Based on verbal descriptions of location and on coordinates (when available), an equal number of fish was mobile or sedentary (Table 2). The mean distance traveled by fish that moved was about 20 km, and the average distance moved over all 82 fish was about 10 km (Table 3). The maximum number of days at liberty was 630 d for a fish whose movements could not be determined due to an uncertain capture location. The maximum time at liberty for a fish with usable location data was 564 d for a fish that had moved 14.8 km from one oil platform to another. The maximum distance moved was 58.3 km for a fish in the Port Isabel area that was at liberty for 119 d. Over all areas within Texas (statewide), there was no significant difference ($P > 0.05$) between movers and stayers in growth rate, size at capture, days at liberty, number of neighbors within any distance, nearest neighbor distance, or depth at capture (Table 3). Mean length at capture was 36.3 cm TL for movers and 35.3 cm for stayers, both under the minimum recreational size limit of

Table 1. Potential habitat for red snapper *Lutjanus campechanus* off the Texas coast. Potential habitats were digitized from nautical charts, mapped in ArcGIS (ESRI, version 9.0), and analyzed for isolation or clustering. Type: A = Artificial structure, N = Natural structure. Obstructions could not be classified as to habitat type because no details concerning obstructions were provided on the nautical charts. Relief: L = Low (near the bottom), M = Moderate (rising some height above the bottom), H = High (extending above the surface). CV = Coefficient of Variation (standard deviation/mean). Nearest Neighbor (km) = the distance to the nearest neighbor; # Neighbors = the number of neighbors within the specified distance.

Potential Habitat	Count	Type	Relief	Nearest Neighbor (km)	CV	# Neighbors (1 km)	CV	# Neighbors (2 km)	CV	# Neighbors (5 km)	CV	# Neighbors (10 km)	CV	# Neighbors (25 km)	CV
Artificial reef	123	A	L	1.10	1.43	2.72	1.16	3.49	0.92	6.58	0.70	14.37	0.52	50.82	0.37
Banks, reefs, rocks, structures	323	N	M	3.79	1.90	0.44	1.95	0.93	1.33	2.83	0.93	6.76	0.84	29.27	0.72
Buoy	24	A	H	2.38	1.20	1.00	1.41	2.54	1.05	4.46	0.69	9.54	0.52	41.67	0.46
Obstruction	120	--	L	1.81	1.16	1.10	1.87	2.59	1.60	7.53	0.83	19.37	0.46	80.28	0.31
Platform	965	A	H	3.27	0.91	0.14	3.61	0.56	2.27	2.82	1.00	9.89	0.68	50.55	0.45
Underwater feature	14	N	L	3.65	1.05	0.43	1.20	1.00	1.36	1.93	0.90	6.07	0.67	26.43	0.48
Wellhead	13	A	L	2.32	0.70	0.15	2.44	0.77	1.20	2.31	0.51	11.46	0.64	56.92	0.33
Wreck	187	A	M	3.89	6.20	2.48	2.07	2.96	1.74	4.95	1.03	12.39	0.60	49.40	0.50

Table 2. Numbers of red snapper *Lutjanus campechanus* used in the two tagging programs. Movement was based on either the verbal description of location or on coordinates. Distance was only analyzed for 'certain' or 'probable' coordinates (see text).

	Number of Fish	
	TTU	Fish Trackers
Tagged	5,614	~9,000 [b]
Returned	130 [a]	68 [b]
Analyzed for movement	82	60
Analyzed for distance	74	51
Movers	43	17
Stayers	39	43

[a] There were at least 22 unconfirmed reports of additional recaptures (no tag number, no return phone number for fishermen, etc.)

[b] Only a random subset of the data was analyzed due to computer malfunctions.

Table 3. Differences in movers versus stayers, or mobile versus non-mobile red snapper *Lutjanus campechanus*. The movers and stayers categories are based either on verbal descriptions or GPS coordinates. The distance moved is only for fish with 'certain' or 'probable' location data. 'Stayers' were given moving distances of 0 if their verbal descriptions were the same, but fish may have moved between different locations on the same landmark, such as different spots on a bank or different legs on an oil platform. Depth at recognized landmarks was taken from nautical charts. CV = Coefficient of Variation (standard deviation/mean); n = number of fish.

		TTU		*FT*	
Parameter		**Movers**	**Stayers**	**Movers**	**Stayers**
Distance moved (km)	mean	20.4	0	19.1	0
	cv	1.25	0	0.77	0
	n	35	38	9	42
Nearest neighbor (m)	mean	3.32	2.34	2.09	2.40
	cv	1.01	0.70	0.38	0.46
	n	36	34	9	42
Neighbors (1 km)	mean	1.50	2.65	0.22	0.26
	cv	2.65	2.11	3.00	3.38
	n	36	34	9	42
Neighbors (2 km)	mean	1.67	2.76	0.67	0.45
	cv	2.43	2.02	1.30	2.02
	n	36	34	9	42
Neighbors (5 km)	mean	4.86	6.94	4.78	3.67
	cv	1.38	1.05	0.45	0.69
	n	36	34	9	42
Neighbors (10 km)	mean	8.92	11.29	12.89	11.10
	cv	0.90	0.78	0.55	0.56
	n	36	34	9	42
Neighbors (25 km)	mean	36.31	32.61	60.33	59.26
	cv	0.71	0.73	0.39	0.36
	n	36	33	9	42
Avg. Growth/Day (cm)	mean	0.046	0.049	0.049	0.039
	cv	0.63	1.37	0.55	0.66
	n	38	33	16	32
Days at liberty	mean	147.17	175.79	173.94	87.98
	cv	0.95	0.83	0.57	1.00
	n	42	38	17	42
TL at capture	mean	36.33	35.31	28.99	31.28
	cv	0.09	0.09	0.12	0.16
	n	40	39	17	43
Depth at capture (m)	mean	38.48	34.79	40.91	34.62
	cv	0.14	0.08	0.38	0.43
	n	42	39	9	30

40.6 cm. There was no detectable pattern in the direction of movement in any of the regions of Texas, but most movements were across depth contours rather than parallel to depth contours (Figure 3).

In the logistic regression using the statewide TTU dataset, the only parameter close to significance was days at liberty ($P = 0.0504$). The probability of movement decreased slightly with increasing days at liberty. Only Port Aransas and Port Isabel had sufficiently large sample sizes for logistic regression analyses. For Port Aransas, habitat type was the only significant parameter ($P = 0.0451$) when it was included as a class variable in a stepwise logistic regression. Fish on natural habitats had a higher probability of movement than fish on artificial habitats.

The CART analysis on the statewide TTU data showed that habitat isolation (neighbors within 5 km), depth, and fish size were good predictors of movement (Figure 4). Seventy-six percent of the fish that were initially captured at isolated habitats (<2.5 neighbors within 5 km) were classified as "movers." Fish in more clustered habitats (more neighbors) showed a large effect of depth and size, with 96% of tagged fish that were small and in shallow water (TL < 37.9 cm and depth <40 m) classified as "stayers." Larger fish (TL > 37.9 cm) tagged in <40 m of water were slightly more likely to move than remain at a site (58.3%), while 74% of the fish tagged in water deeper than 40 m were classified as movers rather than stayers (Figure 4). Habitat type was a good predictor of movement in the Port Aransas area. On artificial habitats, fish were almost equally likely to move or stay, but on natural habitats, 93% of fish moved. When habitat type was removed as a possible predictor, the days at liberty, number of neighbors within 10 km, and depth were the best predictor variables. Fish that were recaptured within the first 29.5 d were more likely to have stayed at the tagging location, while movements of fish that were at liberty longer were best predicted by habitat isolation (fewer than 9.5 neighbors within 10 km) or depth (more than 51 m). The CART model for Port Isabel showed that fish size and depth were good predictors of movement. Small fish (<37.9 cm TL), particularly in shallow water (depth <38.5 m), were very likely to stay at the capture site, while small fish in deep water and large fish were more likely to move.

Fish Trackers

Although *FT* tagged almost 9,000 red snapper between 1983 and 2005, only 68 records of red snapper tagged between 1986 and 2000 were available for analysis. The red snapper records were part of a larger dataset of almost 90,000 fish of over 100 species tagged since 1983. Corruption of computerized data and inaccessibility of data entered into older, nontransferable computer platforms using now-obsolete software made most of the records inaccessible. The records analyzed here are those that could be reentered from the original data sheets, and most of the available data were for species other than red snapper. Of the 68 red snapper records available for analysis, only 60 had enough location data to determine whether the fish had moved or stayed; 17 fish out of the 60 (28%) had moved (Table 2). Fish that moved traveled an average of about 19 km. There were no detectable patterns in the direction of movement but, as with the TTU data, movements were across rather than parallel to depth contours (Figure 3). The maximum time at liberty was 730 d for a fish whose movements could not be determined due to an uncertain recapture location. The second longest time at liberty was 349 d for a fish that remained at the same oil platform. The greatest movement in this dataset was a fish that was captured at an oil platform off Port Aransas and moved offshore 42.6 km to a natural bank, where it was recaptured 281 d later. There were no significant differences between movers and stayers in growth rate, number of neighbors at any distance, distance to nearest neighbor, depth at capture, or size at capture, although there was a significant difference in days at liberty, even when corrected for experimentwise error. Movers were at liberty almost twice as long before recapture as stayers ($P = 0.0039$). Fish tagged by *FT* were generally smaller than fish tagged by TTU personnel, with mean sizes about 5–6 cm shorter for *FT* fish (Table 3). Since tagging began in 1983 (one year before size limits were instituted), size limits for recreational red snapper fishing have varied between 33.0 and 45.7 cm TL, but have been set at 40.6 cm TL since 1999.

Thirty-nine *FT* records had complete data for the logistic regression: nine movers and 30 stayers. Results of the stepwise logistic regression showed that the time at liberty was highly significant in predicting movement ($P = 0.0137$), and that depth may also be important, although with such a small number of movers, this result is preliminary ($P = 0.0426$). The probability of movement increased with increasing time at liberty and increasing depth (Table 4). CART analysis showed that the best predictor variables of red snapper movement were days at liberty and average growth rate. Ninety percent of the fish that were at liberty less than 80 d were

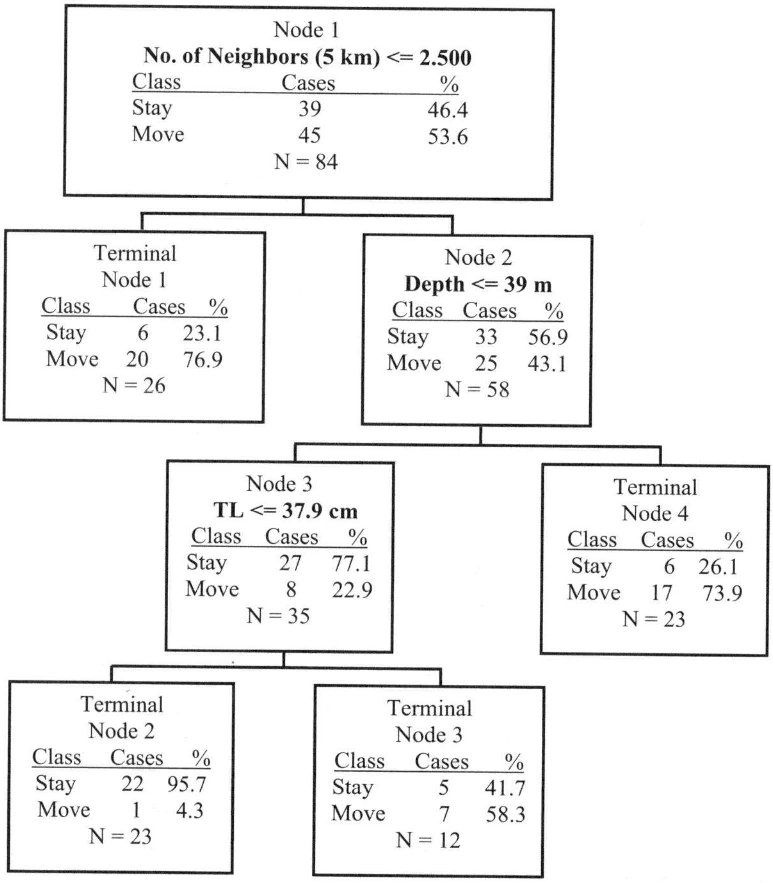

Figure 4. An example of CART results: Analysis for TTU statewide data. If the condition is true, then proceed to the left split. If the condition is not true, proceed to the right split. The tree is based on finding the most homogeneous child nodes while reducing the misclassification error rate.

stayers. Of the fish that were at liberty more than 80 d, those with an average growth rate of less than 0.03 cm/d were all stayers, while fish with an average growth rate of more than 0.03 cm/d were more likely to have moved.

Discussion

Individual Variability

Factors that were important in explaining the variability among individuals in the probability of movement in at least one dataset or analysis were time at liberty, depth, habitat type, habitat isolation, fish size, and growth rate. Because the numbers of recaptured red snapper in both the TTU and *FT* programs were small, the CART results provide support for the results of the logistic regressions. Time at liberty was a significant factor in movement in both the TTU program and the *FT* program, but in different ways. The results of the *FT* study and the Port Aransas CART model indicate that while very short-term site fidelity was high (29.5 d for Port Aransas and 80 d for *FT* data), most fish would move eventually, given sufficient time after capture. An increased probability of movement with time at liberty was also found by other researchers using both passive and acoustic tags (Patterson et al. 2001; Peabody 2004; Schroepfer and Szedlmayer 2006). Consequently, while relatively brief studies may accurately describe site fidelity in the short term, these descriptions do not adequately represent long-term residency patterns. Increasing time at liberty also increases the chances that other events influencing movements, such as hurricanes (Watterson et al. 1998; Patterson et al. 2001), will occur between the capture and recapture events, changing the estimates of site fidelity. The temporal scale of site fidelity studies, which is usually about one year (Table 5), is much shorter than the lifespan of red snapper, which can live over 50 years (Wilson and Nieland 2001). Even with heavy fishing pressure, about 20% of red snapper live longer than 5 years (Wilson and Nieland 2001), three years after recruiting to structure, so most site fidelity studies do not encompass a significant fraction of the adult life of an average fish. Thus, it is likely that even in areas where very high site fidelity has been found, almost all red snapper will relocate at some time during their lives if they survive long enough.

In contrast to the results discussed above, the statewide TTU logistic model showed the opposite trend—that the likelihood of movement decreased as the time at liberty increased. This trend was also illustrated by the higher average time at liberty of stayers compared to movers (Table 3). However, this result should be considered exploratory because of the relatively small number of recaptured fish that could be analyzed in the logistic regression. The most likely reason for a decreasing probability of movement with increased time at liberty is the sampling design of our study combined with the large number of potential habitats for red snapper in Texas. The majority of the TTU tagging study was conducted from headboats and charterboats during the summer, when fishing effort is highest. Many of our recaptures were reported from these same boats. Captains on these recreational for-hire boats do not want to deplete particular fishing spots, so they do not return to the same location immediately. This spatial variability in recapture effort makes it more likely that movers will be recaptured faster than fish that remained at the tagging site, biasing the results in favor of movers over short time scales. A higher recapture rate and a larger number of participating vessels would be needed to better test the effects of time at liberty on fish movement.

Depth was a consistently important habitat parameter, with increased movement related to increased depth of capture in the logistic model for the *FT* data and serving as a splitter in the CART models for the TTU statewide, Port Aransas, and Port Isabel data (Table 4). Depth has been found to be a significant factor in past studies, with fish taken from shallower water (≲30 m) moving less than fish from deep water (see Beaumariage 1969 for early studies; Watterson et al. 1998). This depth threshold is shallower than, but consistent with, the results of our study, as we found a depth threshold for movement of about 40 m. Shallow habitats generally have higher productivity than deeper habitats due to the relationship between primary productivity and light attenuation (Bierman et al. 1994). Resources may, therefore, be less limiting in shallow water, inducing fish to stay in shallow habitats. Another potential reason why fish captured in deeper water have a tendency to move rather than stay is horizontal displacement of fish brought up from deep water. Fish from deep water suffer from air bladder expansion caused by the rapid ascent during capture, causing them to struggle or float at the surface before returning to depth, especially if the air is not vented from the air bladder before the fish is released (Diamond, unpublished data; Rummer and Bennett 2005). Floating on the surface before returning to depth may move fish away from the original capture site, making it more likely that these fish will be movers rather than stayers.

Other important habitat parameters were habitat isolation and the type of habitat (Table 4). Red snapper were more likely to move from isolated habitats (fewer neighbors) than clustered habitats, supporting our hypothesis. Fish were also more likely to move from natural habitats than from artificial habitats, although structural relief was not a significant factor in any of the analyses. The increase in movements with habitat isolation may be a result of exploratory behaviors used by individuals to assess the costs and benefits of staying compared to moving (Doerr and Doerr 2005). Red snapper in clustered habitats may be able to explore nearby alternative habitats with very little cost. For red snapper in isolated habitats, the cost of exploration is the same as the cost of moving, so fish that risk exploring alternative habitats most likely remain at the new location.

The tendency for fish to move from natural habitats was unexpected, but significant in the Port Aransas area. In fact, the probability of movement away from natural habitats was actually underestimated in this study. We often received verbal descriptions rather than GPS positions of capture and recapture locations, particularly for natural habitats such as banks or reefs. Without exact coordinates, we considered these fish to have stayed at the capture location if the verbal description was the same, although the fish may have moved hundreds of meters on the same bank or reef. Thus, we have potentially biased the classifications toward stayers in natural habitats. Although there may have been similar instances for artificial habitats (such as the Liberty Ship artificial reefs, which have several large wrecks together), most artificial habitats encompass a much smaller spatial area than natural banks and reefs, so fish on artificial habitats were more likely to be classified as movers when they moved small distances than fish on natural structures. Interestingly, fish that moved from natural habitats also tended to be recaptured on other natural habitats, while fish that moved away from artificial habitats tended to be recaptured on other artificial habitats. The most common artificial habitats off Port Aransas are oil platforms (Table 1). While oil platforms are large, complex structures that have much in common with natural reefs in terms of structural complexity and biodiversity (Stanley and Wilson 1990), there may be some differences that are discernable to red snapper between these artificial habitats and natural reefs and banks. Natural habitats are also more isolated on average than artificial habitats such as oil platforms, artificial reefs, and wellheads (Table 1), so higher movement away from natural habitats may partially be a function of their isolation.

The apparent difference in moving probability based on habitat type may also reflect an underlying difference in red snapper behavior, rather than a difference in habitat type alone. Red snapper have a tendency to associate together in small groups, called 'sub-cohorts.' Members of sub-cohorts are tagged together at the same time and location and are often recaptured together, at either the same or different sites (Figure 3). The majority of movers off Port Aransas were members of sub-cohorts (67%), and fish in sub-cohorts were found significantly more often on natural habitats than artificial habitats ($P = 0.023$). Therefore, movement may have been related to behavior as well as habitat type. The prevalence of sub-cohorts in the recapture data are important for several reasons. First, the probability of movement for an individual fish in a sub-cohort may not be an independent event, although we analyzed movements as individual decisions. This would lead to biases in the probability of movement and the proportions of movers versus stayers if decisions regarding moving or staying among members of a sub-cohort are made differently than decisions by individual fish. However, any biases are difficult to quantify because it is uncertain whether all snapper form sub-cohorts, what factors affect this behavior, and whether fish make choices regarding movement as individuals or as members of the group. Second, it is obvious from the recapture data that individuals from some sub-cohorts are recaptured more than others. This may reflect higher survival of some sub-cohorts due to variability in tagging conditions (water temperature, presence or absence of thermocline, handling time, depth of tagging, etc.), unequal recapture effort by location or habitat type, or unequal reporting rates by fishermen (i.e., some sub-cohorts may be reported when recaptured while others were not reported). Third, red snapper associations in sub-cohorts may introduce a size or age bias in the recapture data if sub-cohorts consist only of individuals of the same age or size. Red snapper sub-cohort behavior has not been well

Table 4. Summary of important factors predicting red snapper *Lutjanus campechanus* movements for Texas Tech University (TTU) and Fish trackers (FT) data. TTU data were analyzed in its entirety (TX) and by port, however only Port Aransas and Port Isabel had sufficient sample sizes for analysis. Due to small sample sizes on all data sets, the results of the logistic regressions may not replicate well with larger data sets. The types of analysis were: L = logistic model, C = CART model. The probability of movement is higher for the term on the left and lower for the term on the right. CART analysis shows the results of the optimal tree, which is the tree that maximizes homogeneity in the child nodes while minimizing the misclassification rate.

Parameter	Data	Analysis	Probability of Movement	Level of Significance	Comments
Time at Liberty	TTU – TX	L	Short time > Long time	0.0504	
	TTU – Pt Aransas	C	Long time > Short time		
	FT	L	Long time > Short time	0.0137	
	FT	C	Long time > Short time		
Depth	TTU – TX	C	Deep water > Shallow water		
	TTU – Pt Aransas	C	Deep water > Shallow water		
	TTU – Pt Isabel	C	Deep water > Shallow water		
	FT	L	Deep water > Shallow water	0.0426	
Habitat Type	TTU – Pt Aransas	L	Natural > Artificial	0.0451	
	TTU – Pt Aransas	C	Natural > Artificial		
Habitat Isolation	TTU – TX	C	More isolated > Less isolated		Isolated = few neighbors within 5 km
	TTU – Pt Aransas	C	More isolated > Less isolated		Isolated = few neighbors within 10 km
Fish Length[a]	TTU – TX	C	Larger fish > Smaller fish		
	TTU – Pt Isabel	C	Larger fish > Smaller fish		
Growth Rate/Day	FT	C	Faster growth > Slower growth		

[a] Almost all fish tagged were less than the legal size limit of 40.6 cm TL.

studied, but has potential implications for management, particularly if fish in a particular sub-cohort are genetically related. If so, then localized depletion of sub-cohorts could lead to variance in the number of surviving offspring by different parents, reducing the effective population size and affecting stock structure (Saillant and Gold 2006).

Fish characteristics such as size and average growth rate were also significant parameters in some, but not all, of the models. Our hypothesis that larger fish would move more than smaller fish was supported by the CART model results of the TTU statewide study and the Port Isabel study. Although the difference in the average length between movers and stayers in the TTU study was very small and not statistically significant (Table 3), we observed a distinct difference in the robustness of smaller undersized fish (≲35 cm) compared to larger undersized fish (close to the size limit). Smaller fish were more likely to die during the initial capture procedure, possibly due to heat shock caused by thermally stratified water in the summer (Diamond, unpublished data), so the size threshold for movement of 37.9 cm TL found in the CART models may indicate some type of physiological breakpoint in fish survival. The stress of moving to a new location when added to capture stress may have produced differential mortality rates that caused small fish to be under-represented in the movers category. Conversely, the increased likelihood of movement in larger fish may be due to ontogenetic shifts in habitat preference, as larger fish are known to actively defend structure from smaller fish (Bailey et al. 2001). Thus, smaller fish may be forced out into marginal habitat and only move when they grow large enough to defend a more preferred habitat.

Growth rate was associated with movement in the *FT* study (CART model), but since growth rate was measured between capture and recapture, a higher growth rate is a result rather than a cause of movement. Dispersal theory predicts that fitness should be higher for movers than stayers when safe habitats are patchily distributed within unsafe habitats, otherwise the costs would outweigh the benefits and animals would not disperse (Fraser et al. 2001). Our data support this prediction. Fish with average growth rates higher than 0.03 cm per day were more likely to have moved and those with lower average growth rates more likely to have remained at the same location. For fish that moved, the more isolated the location of capture, the faster the growth rate between capture and recapture, indicating that these fish may have benefited greatly from moving. Hilderbrand and Kershner (2004) found a similar result with cutthroat trout *Oncorhynchus clarkii*. Fish that moved were larger than the average fish, and the largest of the mobile fish were in poorer condition on initial capture than sedentary fish of the same size. After one year, the condition of movers was equal to or better than the condition of similar sized fish that remained, indicating that movement was beneficial particularly for the largest fish.

Although we did not assess environmental factors in our study, other authors have found that environmental conditions can play a role in the proportion of movers and stayers. Large-scale climate events such as hurricanes have been shown to increase the proportion of movers (Watterson et al. 1998; Patterson et al. 2001), although acoustic studies conducted in Louisiana and Alabama showed no effects of tropical storms on the probability of movement (Peabody 2004; Szedlmayer and Schroepfer 2005). Abundance and average size of red snapper at inshore artificial reefs have been shown to increase after hurricanes (Turpin and Bortone 2002), which reflects the tendency of larger fish to move and fish from deeper water to move across depth contours to shallow water. Smaller scale or transient environmental parameters such as cold fronts may also play a role in movements (Moseley 1965). Juvenile red snapper show recogniz-

Table 5. Site fidelity and movements in red snapper *Lutjanus campechanus* from published tagging studies. Studies are divided by significant moving factors whenever possible. For some studies depending on tag reporting by fishermen (Tag = F), not all recapture reports included location information, so the # or % Recaptured or Relocated = number or percent reported with location information (total number or percent reported); otherwise the locations were assumed to be known for all recaptured or relocated fish. % Movers and % Stayers were calculated only from fish with known location information. Average distance moved includes movers only, which overestimates the actual average distance moved over all tagged fish. Tags: A = Acoustic tags, P = Passive tags, F = Fishermen recaptures, R = Researcher recaptures or relocation at or near the tagging site. Studies using both passive tags and researcher captures at the tagging site (Tag = PR) measure minimum site fidelity since fish that are not recaptured are assumed to have moved. Some studies list two estimates of site fidelity, one at the tagging site and one within a specified distance; both are included here. '—' indicates that information was not included in the published study.

Author[a]	Year Tagged	# Tagged	Location	# Recaptured or Relocated [w/location (total)]	% Recaptured or Relocated [w/location (total)]	Avg. Time at Liberty (days)	% Movers	% Stayers	Avg. Distance Moved	Tags	Comments
Beaumariage (1969)	1962-1965	1,372	FL	384	28.0	131	--	--	--	PF	Total program
Beaumariage (1964)	1963	336	FL	111	33.0	--	0	100	0	PF	Subset of Beaumariage (1969), all recaptures in 'same approximate area of release,' cited in Fable (1980)
Beaumariage (1969)	1965	312	FL	82	26.3	113	9.8	90.2	107.6 km	PF	Subset of Beaumariage (1969)
Fable (1980)	1977	299	TX	17	5.6	125	5.9	94.1	5 km	PF	Electric reels
Szedlmayer and Shipp (1994)	1990-1991	1,155	AL	37 (146)	3.2 (12.6)	--	43	57	--	PF	Recaptures at tagging site
"	"	"	"	"	"	--	24	76	~15 km	"	Recaptures within 2 km of tagging site
Szedlmayer (1997)	1992-1994	23	AL	19	82.6	150	37	63	--	AR	Relocated at tagging site
"	"	"	"	"	"	"	0	100	396 m	"	Relocated within 1 km of tagging site
Culbertson and Peter (1998)	1996-1997	683	TX	103	15.1	61.6	NA	$\geq$15.1[b]	NA	PR	Researcher recaptures at tagging site within 233 days[b]
Watterson et al. (1998)	1995-1996	1,604	AL	167	10.4	207	45	55	--	PRF	Total program, subset of Patterson et al. (2001), recaptures within 2 km of tagging site
"	"	790	AL	100	12.7	--	21	79	~ 2 km	"	Not at liberty during hurricanes
"	"	814	AL	74	9.1	--	77	23	32.6 km	"	At liberty during hurricane
"	"	1017	AL	139	13.7	--	39	61	0.9 km	"	Not translocated
"	"	587	AL	35	6.0	--	77	23	5.1 km	"	Translocated

Table 5. (Continued)

Author[a]	Year Tagged	# Tagged	Location	# Recaptured or Relocated [w/location (total)]	% Recaptured or Relocated [w/location (total)]	Avg. Time at Liberty (days)	% Movers	% Stayers	Avg. Distance Moved	Tags	Comments
Patterson et al. (2001)	1995-1998	2,932	AL	519	17.7	404	64	36	29.4 km	PRF	Total program, extension of Watterson et al. (1998), recaptures within 2 km of tagging site
"	"	--	AL	264[c]	--	247[c]	--	--	7.4 km	"	Not at liberty during hurricanes
"	"	--	AL	--	--	562[c]	--	--	42.4 km	"	At liberty during hurricanes
"	"	2,053	AL	441[c]	21.5	--	--	--	--	"	Not translocated
"	"	879	AL	114[c]	13.0	--	--	--	--	"	Translocated
Peabody (2004)	2003	125	LA	97	77.6	--	7.2	92.8	NA	ARF	Fidelity decreased by 29% for every 10 day period
Szedlmayer and Schroepfer (2005)	2000-2004	54	AL	31	57.4	218	33	67	--	AR	Relocated at tagging site but 'most' fish relocated within 1.6 km of tagging site
Schroepfer and Szedlmayer (2006)	2006	15	AL	14	93.3	373[d]	13	87	--	AR	Relocated within 200 m of receiver, estimated by event analysis
This study – TTU	2002-2005	5,614	TX	82 (130)	1.5 (2.3)	165.6	52.4	47.6	20.4 km	PF	Tagged by researchers aboard charter and headboats
This study – FT	1986-2000	~9000[e]	TX	60 (68)[e]	e	112.7	28.3	71.7	19.1 km	PF	Tagged by fishermen and researchers

[a] Some studies are subsets or extensions of other studies and are listed separately.
[b] Data exclude fishermen recaptures.
[c] Updated by Patterson and Cowan (2003).
[d] Time = median time at liberty
[e] Only a subset of the data could be analyzed due to computer problems (see text for explanation).

able habitat preferences for water temperatures of 26–27°C, dissolved oxygen levels of 5 mg/L or higher, and salinity levels of 35 psu (Gallaway et al. 1999), so it is likely that adult snapper have distinct habitat preferences as well. Local changes in these values due to changes in weather, current flow, or hypoxia could motivate red snapper to move to areas of more optimal habitat conditions.

One other fish characteristic that has been postulated to cause heterogeneity in movements within a population is the genetic tendency for exploratory behavior (Fraser et al. 2001). While 'boldness' may explain individual differences in movements, this is not a quality that can be tested in a field study such as ours. Future studies using selective breeding or common garden experiments in the laboratory may indicate whether boldness is a factor in red snapper movements.

Variability Among Site Fidelity Studies

Early studies of red snapper movement based on fishermen recaptures showed very low frequencies of movement, with movers often making up 10% or less of the recaptured fish (Beaumariage 1969; Fable 1980). More recent estimates of dispersal are much higher and more variable, ranging from less than 25% movers to over 70% movers (Table 5). Our study also reflects this variability, with the TTU data showing a higher frequency of movement (52.4%) than the *FT* data (28.3%). TTU data also showed a higher frequency of movement than most other studies where hurricanes or translocation is not a factor, particularly since many of the stayers in our study would likely have been considered movers in studies with more focused recapture effort at specific tagging sites (Table 5). Both programs also showed a greater average distance of movement than studies without hurricanes or translocation of fish.

The interplay between habitat isolation, depth, and habitat type may offer an explanation for the high degree of variability seen in previous tagging studies. At shallow locations characterized by relatively clustered habitats, particularly artificial reefs, red snapper show strong patterns of site fidelity, especially when translocation and hurricanes are not factors (Szedlmayer and Shipp 1994; Szedlmayer 1997, Watterson et al. 1998; Szedlmayer and Schroepfer 2005; Schroepfer and Szedlmayer 2006). The Hugh Swingle General Permit Area off Alabama, where many of the studies in the northeastern Gulf have been conducted, is an example of a shallow-water cluster of artificial reefs. Studies conducted there have shown high levels of site fidelity. However, at deeper locations with low habitat density, such as much of the northwestern Gulf, these site fidelity patterns are weaker, as shown in the higher proportion of movers in Peabody (2004) and this study.

It is difficult to reconcile the contrasting results of high site fidelity shown by some acoustic tag studies (Szedlmayer 1997; Szedlmayer and Schroepfer 2005; Schroepfer and Szedlmayer 2006) with studies conducted in the same areas using passive tags that show much weaker site fidelity (Patterson et al. 2001; Patterson and Cowan 2003). These differences could be due to the size or age of fish in these studies or due to differences in methodology that bias the estimates of site fidelity. For example, in studies using passive tags where recaptures are made only by researchers and only at specific tagging sites (Culbertson and Peter 1998; Patterson and Cowan 2003), site fidelity can be underestimated due to the assumption that fish that are not recaptured have moved. With passive tags, it is not possible to state whether tagged fish are not present or are present but not caught, and some correction for this factor must be made in the calculations of fish present and site fidelity. However, this caveat does not apply to acoustic studies (Peabody 2004) or studies with recapture efforts at different locations (Watterson et al. 1998;

Patterson et al. 2001, this study) which use counts of fish that moved as well as fish that stayed and still showed high rates of movement.

Other differences in methods, including the temporal and spatial scales of the study design and the tagging and recapture methods, may contribute to the differences in estimated movements among different studies. For example, in many studies, red snapper were tagged by research personnel at a few specific locations, usually from a charter or research vessel (Beaumariage 1969; Fable 1980, Szedlmayer and Shipp 1994; Culbertson and Peter 1998; Watterson et al. 1998; Patterson et al. 2001). Tagging at a few specific locations allows the researcher to measure dispersal distance and direction from a point location and may increase the number of fish recaptured, but it may also bias the sample towards recapturing fish that stay rather than movers due to unequal recapture effort at locations away from the dispersal site. Depending on recapture methods, the spatial scale of staying or moving may also be very small and fish movements may be measured in meters rather than kilometers. This is particularly true in acoustic studies, and increases the proportion of movers relative to stayers compared to studies with larger spatial scale (Szedlmayer and Shipp 1994; Peabody 2004, Szedlmayer and Schroepfer 2005; Schroepfer and Szedlmayer 2006).

In our study, red snapper were tagged during normal recreational fishing operations, sometimes by volunteer recreational fishermen, at any location that recreational fishermen chose to fish. The advantage of this method is that it covers a much larger spatial scale than point-dispersal studies, but the disadvantages are that the coverage of any particular site is sparse, and that location information is often vague. Classification as a mover required an individual fish to have moved far enough from the capture site to either be recaptured at a recognizably different location, or at a different landmark. The definition of movement may therefore be a fish that moves kilometers rather than meters. This difference in method and spatial scale causes an underestimate of the proportions of movers in studies relying on recaptures made by fisherman (i.e., fish are counted as stayers when they actually moved) as compared to acoustic studies or studies where researchers recaptured most of the fish. However, this bias may be balanced by the potentially low recapture effort at the tagging site, which would cause an underestimate in the proportion of stayers.

Tagging during research operations has several advantages over fishery-dependent tagging, including both the ability to reduce handling time and stress and the use of less stressful capture methods. For example, in our study, one of the headboats used electric reels, which bring fish to the surface more quickly and may increase stress and mortality from barotrauma. Also, on headboats, an individual fish may have been handled by several people before being tagged by the researcher. In contrast, one fishery-independent study used collapsible traps to capture and tag fish underwater, which reduced stress and increased the recapture rate from 8.4% to 29.1% compared to hook and line, although the traps caught 50% fewer fish than hook and line fishing during the initial capture phase (Culbertson and Peter 1998). Increased handling stress and barotrauma affect the rate of recapture because they increase the mortality rate of tagged fish, and they may also increase the proportion of movers in a population, as illustrated by translocated fish. Translocated fish suffer from increased handling stress compared to nontranslocated fish since they are on the surface longer, and on average they are in poorer condition upon release even when captured at the same depth (Watterson et al. 1998), meaning that fewer fish are able to swim down immediately. They also have a higher probability of movement than nontranslocated fish (Watterson et al. 1998; Patterson et al. 2001). The correlation between movement and stress may be due to horizontal displacement of fish that float at the surface before being able to return to depth, but physiological stress may also be a trigger for movement.

Tag Returns

Although the tag return rate of 2.3% in the TTU study was low compared to past studies, some of which reported tag return rates of over 20% (Watterson et al. 1998), there was probably little bias in return rate between fish that moved and fish that stayed. There are several possible explanations for our low return rate. First, in most studies researchers tagged fish at one or a few specific sites and much of the recapture effort was near that site. In contrast, we tagged fish at almost 200 discrete sites and recapture effort was diluted across a much larger matrix of potential habitats (Figure 1). Diluted fishing effort means that the chance of recapturing any individual fish, whether it moved or stayed, is very small. Second, we tagged fish in waters up to about 100 m in depth, and our average depth of 42 m was much greater than the tagging studies in the northeastern Gulf, which usually took place in 20–30 m. Fish captured from deep water can experience more severe barotrauma, with a higher proportion of fish exhibiting everted stomachs, intestines bulging out of the anus, and bulging eyes (Rummer and Bennett 2005). Fish with these conditions have a much poorer chance of survival than do fish without these symptoms, particularly the trauma to the intestines and eyes (Diamond, unpublished data). Due to the water depth in the TTU study, almost 25% of tagged fish showed trauma to intestines or eyes and an additional 25% had everted stomachs, so many of these fish may not have survived the initial catch procedure, lowering the potential recapture rate. Sublethal effects of barotrauma may have also increased postrelease predation. Tag recognition and reporting in our study was also low. Although our tags were bright yellow, an encrusting red organism often grew on the tags making them hard to see if the fish was at liberty for several months. In fact, several tags were reported by grocery store employees, after fish had gone through the fishermen and one or more wholesalers. In addition, we did not offer a reward, so reporting was probably lower than other tagging studies (Pollock et al. 2001). Finally, although our fishermen volunteers were extremely cooperative and we advertised heavily regarding the tagging program, we did not get complete tag returns. We were often told by captains and deckhands on recreational for-hire boats that they had collected tags from their passengers and then lost them before reporting them to us.

Assumptions

One of the largest assumptions inherent in a study such as this one is that fish that were recaptured at the initial capture location actually stayed in that location, rather than moving and returning to the same location at a later date. Although it is possible that fish are moving continuously, evidence from acoustic studies of red snapper movements supports the idea that stayers do in fact remain at a site. One study using manual acoustic tracking or automatic data recorders to continuously log movements showed that most red snapper stay at one location from day to day, and may stay within a few hundred meters of the same structure most of the time (Szedlmayer and Schroepfer 2005). However, some red snapper may move off structure at night, presumably to forage (Peabody 2004; Szedlmayer and Schroepfer 2005), and rare individuals may be away for up to 24 h before returning to the same structure (Szedlmayer and Schroepfer 2005). Once red snapper move to a new location, they are rarely seen at the initial tagging location again. In addition, translocated fish do not show homing behavior (Watterson et al. 1998; Patterson et al. 2001, Peabody 2004), further indicating that red snapper are not moving regularly among locations.

A second assumption in our study is that fishermen are accurately reporting capture and recapture locations. This assumption is difficult to validate since we have no way to pinpoint locations other than with the information given to us by fishermen. It is possible that some fishermen give incorrect locations either by accident or by design. However, because we have tagged aboard the same charterboats and headboats for several years, we have developed close relationships with many boat captains who have provided most of our recapture information and we think that the data they provide are correct.

Differences between TTU and *FT* Data

Although the methods were similar, data from the TTU and *FT* tagging programs showed interesting contrasts. The average distance moved was very similar between the two programs, but the habitats where fish were tagged were different. In the TTU program there was greater distance to the nearest neighbor than in the *FT* program, but the number of neighbors within 1, 2, and 5 km was greater in the TTU program. These differences are partly due to changes that have occurred in the location of platforms in the Gulf. Since *FT* tagging began, many platforms have been added and some have been removed, so our use of modern charts to plot the location of neighboring habitats has biased these measurements somewhat; however it is difficult to say in which direction the bias runs. Removal of platforms also results in large-scale mortalities of fish (Nieland and Wilson 2003), which may have lowered the probability of recapture for *FT* fish that stayed at those platforms.

The size differences between TTU and *FT* fish could have been due either to differences in the time span of the studies or to the differences between recreational for-hire and private recreational fishermen. TTU data were collected between 2002 and 2005 when the minimum recreational size limit was 40.6 cm, so recreational-for-hire companies used larger hooks to reduce the number of discarded undersized fish. The *FT* data were collected from 1986 to 2000; size limits during this time varied from none to 40.6 cm, which may have influenced the sizes of fish caught. With no minimum size limits or smaller size limits, fishermen tag and release only the smallest fish. The fact that we can retrieve so little of the *FT* data is unfortunate, but we do not think that the data that we did retrieve were biased in terms of the proportion of movers versus stayers. The data that were retrieved were randomly stored by time and tagging location and they were in no way divided into categories based on movements. The differences in the proportions of movers versus stayers between the two tagging programs could have been due to the differences in isolation, time at liberty, or fish size, all of which were significant factors in movement probability.

Movements and Red Snapper Management

Red snapper commercial and recreational fisheries are managed by a host of regulations including closed seasons and size limits, which require fish of certain sizes or during certain times of year to be discarded overboard. Juvenile red snapper are also discarded from shrimp trawls. Discarding due to management regulations not only causes an added degree of mortality, but it also relocates fish vertically, and may be relocating fish horizontally due both to barotraumas and boat movements between catch and discard locations. Since translocation is a factor in red snapper movements (Watterson et al. 1998; Patterson et al. 2001), regulations in recent years may be changing the frequency of movements and mixing, but whether this added movement is detrimental or beneficial to the population is difficult to assess.

The spatial scale of movements in both the TTU and *FT* programs was relatively large compared to other studies where hurricanes or translocation of fish were not factors (Table 5, Watterson et al. 1998; Patterson et al. 2001), but still small relative to the extensive coastline of Texas. In addition, no fish moved far enough to be caught by fishermen from another Texas port and there was no particular direction to movements. Although we did receive notice of recaptured fish from employees of fish wholesalers in western Louisiana, it is likely that the commercial fishermen who caught these fish were fishing in federal waters off Texas, according to the people who reported the tags. The scale of movements in our study therefore supports the hypothesis that red snapper stocks in the northern Gulf are relatively isolated, with periodic long-range dispersal (Pruett et al. 2005), caused either by hurricanes (Patterson et al. 2001) or by some other factor that triggers long-range movements. This stock isolation would explain the smaller sizes-at-age, smaller maximum sizes, and higher abundances of young fish seen in Texas compared to other locations off Louisiana and Alabama (Fischer et al. 2004). Our data, therefore, support the idea of a separate demographic stock off Texas (Saillant and Gold 2006).

Acknowledgments

This study would not have been possible without the assistance of a large number of fishermen, boat captains, fish wholesalers, and students who helped with the tagging and recapture efforts. We particularly thank Stephen Murphy (Murphy's Law), Marvin Horner (Scat Cat), Terry Trammel (Captain Elliot's), Keith Schoolcraft (Muy Caliente), Kirk Vossler (Susan Nell), Brad Barwise (Water World), and the other recreational captains, owners, and companies that either allowed us to tag fish aboard their boats, or tagged fish for us. We also thank Chris Rigaud, Todd Neahr, Nicole Ekstrom, Jesse Schuck, Saket Waghmode, Andrea Huntoon, Hedrick Han, Serena Ahrens, Preston Bean, Corley Hodges, Sean Connell and Kristin Reagan for participating in the tagging studies and Matt White, Angela Wright, Lec Wells, Ryan Riley, Michelle Ribera, Brian Ducas, Denice Frausto, and Melissa Muenzler for data entry and validation. We particularly thank Laura Panto for her work on finding the coordinates of verbal locations and data validation. The manuscript was improved by thoughtful comments from two anonymous reviewers, and we also thank David Nieland for his editorial assistance. This work was funded through MARFIN Grant NA17FF2012 from the National Marine Fisheries Service.

References

Bailey, H. K., IV., J. H. Cowan, Jr., and R. L. Shipp. 2001. Experimental evaluation of potential effects of habitat size and presence of conspecifics on habitat association by young-of-the-year red snapper. Gulf of Mexico Science 19:119–131.

Beaumariage, D. S. 1969. Returns from the 1965 Schlitz tagging program including a cumulative analysis of previous results. Florida Department of Natural Resources. Technical Series 59:1–38.

Beaumariage, D. S., and L. H. Bullock. 1976. Biological research on snappers and groupers as related to fishery management requirements. Pages 86–94 *in* H. R. Bullis, Jr., and A. C. Jones, editors. Proceedings: colloquium on snapper-grouper fishery resources of the western central Atlantic Ocean. Florida Sea Grant College Program Report 17.

Belanger, G., and M. A. Rodriguez. 2002. Local movement as a measure of habitat quality in stream salmonids. Environmental Biology of Fishes 64: 155:164.

Bierman, V. J., Jr., S. C. Hinz, D. W. Zhu, W. J. Wiseman, Jr., N. N. Rabalais, and R. E. Turner. 1994. A preliminary mass balance model of primary productivity and dissolved oxygen in the Mississippi River plume/inner Gulf shelf region. Estuaries 17:886–899.

Bujold, V., R. A. Cunjak, J. P. Dietrich, and D. A. Courtemanche. 2004. Drifters versus residents: Assessing size and age differences in Atlantic salmon (*Salmo salar*) fry. Canadian Journal of Fisheries and Aquatic Sciences 61:273–282.

Culbertson, J. C., and D. D. Peter. 1998. Development of tagging techniques for monitoring fish populations at Texas artificial reefs. Gulf of Mexico Science 16:46–53.

Doerr, E. D., and V. A. J. Doerr. 2005. Dispersal range analysis: Quantifying individual variation in dispersal behaviour. Oecologia 142:1–10.

Fable, W. A., Jr. 1980. Tagging studies of red snapper (*Lutjanus campechanus*) and vermilion snapper (*Rhomboplites aurorubens*) off the South Texas coast. Contributions in Marine Science 23:115–121.

Fischer, A. J., M. S. Baker, Jr., and C. A. Wilson. 2004. Red snapper (*Lutjanus campechanus*) demographic structure in the northern Gulf of Mexico based on spatial patterns in growth rates and morphometrics. Fishery Bulletin 102:593–603.

Fraser, D. F., J. F. Gilliam, M. J. Daley, A. N. Le, and G. T. Skalski. 2001. Explaining leptokurtic movement distributions: intrapopulation variation in boldness and exploration. American Naturalist 158:124–135.

Gallaway, B. J., J. G. Cole, R. Meyer, and P. Roscigno. 1999. Delineation of essential habitat for juvenile red snapper in the northwestern Gulf of Mexico. Transactions of the American Fisheries Society 128:713–726.

Hilderbrand, R. H., and J. L. Kershner. 2004. Are there differences in growth and condition between mobile and resident cutthroat trout? Transactions of the American Fisheries Society 133:1042–1046.

Lenihan, H. S., C. H. Peterson, J. E. Byers, J. H. Grabowski, G. W. Thayer, and D. R. Colby. 2001. Cascading of habitat degradation: oyster reefs invaded by refugee fishes escaping stress. Ecological Applications 11:764–782.

Moseley, F. N. 1965. Biology of the red snapper, *Lutjanus aya* Bloch, of the northwestern Gulf of Mexico. Master's thesis, University of Texas, Austin, Texas.

Nieland, D. L., and C. A. Wilson. 2003. Red snapper recruitment to and disappearance from oil and gas platforms in the northern Gulf of Mexico. Pages 73–81 *in* D. R. Stanley and A. Scarborough-Bull, editors. Fisheries, reefs, and offshore development. American Fisheries Society, Symposium 36, Bethesda, Maryland.

Patterson, W. F., III, J. C. Watterson, R. L. Shipp, and J. H. Cowan, Jr. 2001. Movement of tagged red snapper in the northern Gulf of Mexico. Transactions of the American Fisheries Society 130:533–545.

Patterson, W. F., III, and J. H. Cowan, Jr. 2003. Site fidelity and dispersion of red snapper associated with artificial reefs in the northern Gulf of Mexico. Pages 181–194 *in* D. R. Stanley and A. Scarborough-Bull, editors. Fisheries, reefs, and offshore development. American Fisheries Society, Symposium 36, Bethesda, Maryland.

Peabody, M. B. 2004. The fidelity of red snapper (*Lutjanus campechanus*) to petroleum platforms and artificial reefs in the northern Gulf of Mexico. Master's thesis, Louisiana State University, Baton Rouge, Louisiana.

Pollock, K. H., J. M. Hoenig, W. S. Hearn, and B. Calingaert. 2001. Tag reporting rate estimation: 1. An evaluation of the high-reward tagging method. North American Journal of Fisheries Management 21:521–532.

Pruett, C. L., E. Saillant, and J. R. Gold. 2005. Historical population demography of red snapper (*Lutjanus campechanus*) from the northern Gulf of Mexico based on analysis of sequences of mitochondrial DNA. Marine Biology 147:593–602.

Rummer, J. L., and W. A. Bennett. 2005. Physiological effects of swim bladder overexpansion and catastrophic decompression on red snapper. Transactions of the American Fisheries Society 134:1457–1470.

Saillant, E., and J. R. Gold. 2006. Population structure and variance effective size of red snapper (*Lutjanus campechanus*) in the northern Gulf of Mexico. Fishery Bulletin 104:136–148.

Schroepfer, R. L., and S. T. Szedlmayer. 2006. Estimates of residence and site fidelity for red snapper *Lutjanus campechanus* on artificial reefs in the northeastern Gulf of Mexico. Bulletin of Marine Science 78:93–101.

Skalski, G. T., and J. F. Gilliam. 2000. Modeling diffusive spread in a heterogeneous population: a movement study with stream fish. Ecology 81:1685–1700.

Solomon, N. G. 2003. A reexamination of factors influencing philopatry in rodents. Journal of Mammalogy 84:1182–1197.

Stanley, D. R., and C. A. Wilson. 1990. A fishery-dependent based study of fish species and composition and associated catch rates around oil and gas structures off Louisiana. Fishery Bulletin 88:719–730.

Steingrimsson, S. O., and J. W. A. Grant. 2003. Patterns and correlates of movement and site fidelity in individually tagged young-of-the-year Atlantic salmon (*Salmo salar*). Canadian Journal of Fisheries and Aquatic Sciences 60:193–202.

Stock Assessment Workshop Report. 2005. Stock Assessment Report of SEDAR 7, Section III: Stock Assessment Workshop Report. Available: http://www.gulfcouncil.org/Beta/GMFMCWeb/downloads/SEDAR7%20AW%20report%20main.pdf (May 2006).

Szedlmayer, S. T. 1997. Ultrasonic telemetry of red snapper, *Lutjanus campechanus*, at artificial reef sites in the northeast Gulf of Mexico. Copeia 1997:846–850.

Szedlmayer, S. T., and R. L. Shipp. 1994. Movement and growth of red snapper, *Lutjanus campechanus*, from an artificial reef area in the northeastern Gulf of Mexico. Bulletin of Marine Science 55:887–896.

Szedlmayer, S. T., and R. L. Schroepfer. 2005. Long-term residence of red snapper on artificial reefs in the northeastern Gulf of Mexico. Transactions of the American Fisheries Society 134:315–325.

Topp, R. W. 1963. The tagging of fishes in Florida, 1962 program. Professional Papers Series 5. Florida State Board of Conservation, Marine Laboratory, Maritime Base, St. Petersburg, Florida.

Turpin, R. K., and S. A. Bortone. 2002. Pre- and post-hurricane assessment of artificial reefs: Evidence for potential use as refugia in a fishery management strategy. ICES Journal of Marine Science 59:S74–S82.

Watterson, J. C., W. F. Patterson, III, R. L. Shipp, and J. H. Cowan, Jr. 1998. Movement of red snapper, *Lutjanus campechanus*, in the north central Gulf of Mexico: potential influence of hurricanes. Gulf of Mexico Science 16:92–104.

Wilson, C. A., and D. L. Nieland. 2001. Age and growth of red snapper, *Lutjanus campechanus,* from the northern Gulf of Mexico off Louisiana. Fishery Bulletin 99:653–664.

Woods, M. K., A. J. Fischer, J. H. Cowan, Jr., and D. L. Nieland. 2003. Size and age of female red snapper *Lutjanus campechanus* in the northern Gulf of Mexico. Proceedings of the Gulf and Caribbean Fisheries Institute 54:526–537.

American Fisheries Society Symposium 60:189–200, 2007

An Overview of Age and Growth of Red Snapper in the Gulf of Mexico

Andrew J. Fischer[1]
Department of Oceanography and Coastal Sciences
School of the Coast and Environment
Louisiana State University
Baton Rouge, Louisiana 70803 USA

Abstract.—The red snapper, *Lutjanus campechanus*, is one of the most economically important fish species in the Gulf of Mexico (GOM). Concerns over the declines in red snapper landings during the 1980s in the GOM exposed the paucity of information regarding the species' age, growth, and population dynamics, all fundamental in fisheries management. This paper reviews the history of red snapper age and growth research in the GOM demonstrating an evolution of fisheries aging and validation techniques. These refinements in aging over time have also impacted management of the red snapper stock in the GOM. Also discussed are efforts to standardize aging techniques throughout the GOM in an effort to improve data accuracy. A number of studies have used the von Bertalanffy growth model to describe a pattern of rapid growth followed by slower growth after the age of ten years for red snapper in the GOM. Additional applications of the von Bertalanffy growth model have also been applied to corroborate red snapper age estimates derived from sectioned otoliths and to discern demographic differences in red snapper growth throughout the GOM.

Introduction

The red snapper, *Lutjanus campechanus*, is one of the most economically important fish species in the Gulf of Mexico (GOM). The commercial fishery for red snapper began in Pensacola, Florida in 1870 (Jarvis 1935; Moseley 1966) and has since grown to become one of the most lucrative in the GOM. Increasing interest in red snapper among recreational anglers in the 1960s (Moseley 1966) has resulted in red snapper becoming one of the most prized sport fishes in the GOM today. Federal management of the red snapper fishery began in 1976 with the implementation of the Magnuson-Stevens Fishery Management and Conservation Act (Hood et al. 2007, this volume). In 1981 the Gulf of Mexico Fishery Management Council implemented the fishery management plan (FMP) for reef fish, of which red snapper is in the management unit, noting that commercial and recreational landings for red snapper were in decline (Hood et al. 2007, this volume). The first stock assessment on red snapper in the GOM followed a few years later (Goodyear 1988) with management of the stock intensifying thereafter.

Heightened concern for the health of the red snapper stock in the GOM exposed the paucity of, and need for, accurate age information for the species. Age is ranked

[1] Corresponding author: afische@lsu.edu.

as among the most influential of biological variables as it forms the basis for calculations of growth rate, mortality rate and productivity (Campana 2001). These variables derived from, and including age, are imperative to monitor year-class strength, conduct stock assessments, and document population recovery (Wilson and Nieland 2001). Therefore, it is essential that any method of aging be validated to ensure data accuracy (Beamish and McFarlane 1983).

Errors in age estimation could result in inflated estimates of total allowable catch and production resulting in over-exploitation of a stock (Lai and Gunderson 1987; Beamish and McFarlane 1995; Campana 2001). Such was the case with orange roughy *Hoplostethus atlanticus* off the coast of New Zealand. Initial otolith based ages of orange roughy estimated a longevity of 20–30 years (Kotlyar 1980; Van den Broek 1983; Gauldie et al. 1989). These longevity estimates resulted in the calculation of a considerably high growth rate thus intensifying fishing pressure on the stock. Refinements in otolith processing and analysis in later studies provided much greater longevity estimates exceeding 125 years resulting in a much slower growth rate (Fenton et al. 1991; Smith et al. 1994). However, the stock had already been fished to the point of near collapse (Campana 2001). Aging errors have had similar impacts on other fish stocks including walleye pollock *Theragra chalcogramma* (Lai and Gunderson 1987; Beamish and Mcfarlane 1995), Pacific ocean perch *Sebastes alutus* (Beamish 1979), and a number of *Sebastes* species off the western coast of Canada (Chilton and Beamish 1982; Campana 2001).

The history of red snapper age and growth research in the GOM demonstrates an evolution of fisheries aging and validation techniques in an effort to improve data accuracy. It also demonstrates how refinements in aging have impacted management of the red snapper stock in the GOM. Although studies of red snapper age and growth in the GOM date back to the 1960s, the majority of studies have occurred after the advent of a red snapper FMP in 1981. This overview is presented as a synopsis of those studies and how they have impacted our understanding of red snapper in the GOM.

Aging and Validation

Several studies have been conducted on red snapper age and growth in the GOM utilizing a variety of approaches and techniques. Moseley (1966) provided the first attempt to age red snapper with the use of scales from 343 adult and juvenile red snapper from the western GOM. He counted "growth rings" on impressions of scales to obtain ages from zero to four years suggesting an accelerated growth rate during the first year. In an early attempt at validation, Moseley (1966) examined the distance of growth rings to the periphery of the scale to determine timing of ring formation. He reported that scale impressions of June sampled fish displayed new checks indicating that growth ring formation may coincide with the spawning season in late spring or early summer. However, a lack of scales throughout the year prevented Moseley (1966) from determining the exact timing of growth ring formation.

A 35% decrease in production in the red snapper commercial fishery from 1965 to 1975 prompted Futch and Bruger (1976) to further examine the age and growth of red snapper in the GOM. The authors used whole sagittal otoliths of 240 red snappers collected off the west coast of Florida to examine age and growth. Only 200 of these (83.3%) were considered legible enough to count "annuli" and obtain ages ranging from one to five years. Additional considerably larger, and presumably older, fish than those aged were also collected. Although the otoliths from these larger specimens were illegible, the authors suggested that red snapper may well live beyond 20 years. In an effort to validate their age estimates, marginal increments of 187 whole otoliths were measured and mean monthly marginal increments plotted to demonstrate timing of annulus formation. The authors reported annulus formation to occur from June to October.

Bortone and Hollingsworth (1980) noted a decline in the red snapper catch during the 1970s that was "substantiated by the, as yet unpublished, Fisheries Management Plan on Reef Fish Resources submitted to the Gulf of Mexico Fishery Management Council." That FMP would be released the following year. In an attempt to "further clarify and describe the life history of red snapper for management purposes" the authors evaluated the comparability of age determination utilizing different hard parts including whole otoliths, scales, and vertebrae. The authors felt it essential to have an accurate database on red snapper biological parameters. "Annuli" or "age-group marks" were counted on all three hard part types with all materials aged at least twice to ensure reliability. Though not all fish were placed into the same age-class with all three aging methods, they found that all three body parts had a statistically similar level of readability (between 75% and 80%). However, Bortone and Hollingsworth (1980) had a sample size of only 46 red snapper, not all of which could be used for aging in all methods. Additionally, only age one and two year old red snapper were represented in their sample population.

Nelson and Manooch (1982) were the first to use sectioned otoliths to examine the age structure of red snapper in the northern GOM (as well as the west-central Atlantic Ocean). Following methods previously used to age bluefin tuna (Berry et al. 1977), the authors sectioned the otoliths laterally through the focus using a Buehler Isomet low-speed saw. Fully 3,323 red snapper sampled from Louisiana and west Florida were sectioned to obtain ages ranging from 1 to 13 years based on counts of "opaque bands" or "annuli." This estimate of longevity was more than double the previously reported maximum age for red snapper in the GOM. The authors compared age estimates from scale and otolith annuli counts for 43 individuals and found 77% agreement. Nelson and Manooch (1982) attempted to validate their ages using monthly mean marginal growth performed on scales. This validation technique indicated that red snapper in the GOM form annuli during June and July. The authors were unable to perform marginal increment analysis with otoliths due to a lack of samples collected throughout all 12 months of the year. However, 18 one-year-old red snapper sampled in the Carolinas during the month of June all exhibited the beginning of new growth (translucence) at the otolith section edge leading the authors to suggest that scale and otolith opaque annuli formation occur simultaneously.

The first stock assessment of red snapper in the GOM was conducted in 1988 (Goodyear 1988) and indicated that the red snapper stock was overfished and undergoing overfishing. A second stock assessment in 1992 (Goodyear 1992) established (in Amendment 3) that the allowable duration of the recovery period for the stock was to be no greater than 1.5 times the unfished generation time. Generation time is a variable that is controlled in part by longevity estimates (Schirripa and Legault 1999) using both age and fecundity data to provide an index of the turnover rate of a population. Based upon available age information at the time, Goodyear (1992) estimated a generation time of 13 years for red snapper resulting in a maximum rebuilding period of 19.5 years (Hood et al. 2007, this volume). Goodyear (1992) stressed that additional collection and analysis of red snapper hard parts for age determination were essential in order to characterize reproductive contribution of females by age.

Analysis of red snapper did continue with

numerous studies building upon known aging techniques to report greater longevity than had been previously observed. Szedlmayer and Shipp (1994) used counts of "annuli" from sectioned otoliths to estimate ages for 409 red snapper sampled with hook and line off Alabama. The authors reported a maximum age of 42 years, a substantial increase over the previously reported maximum age of 13 years (Nelson and Manooch 1982). The authors did not validate their aging method but did exclude any samples in which the two otolith readers were not in agreement after a second reading.

Render (1995) estimated ages from counts of "opaque zones" or "annuli" from sectioned otoliths. Samples were collected from both the Louisiana commercial (n = 339) and recreational (n = 183) fisheries and transverse otolith sections made with a Buehler Isomet low-speed saw. Render observed a maximum longevity estimated at 53 years, greater than any reported estimate in the current literature. Render (1995) validated his age estimates with the use of marginal increment analysis (Beckman et al. 1991). In this validation technique the otolith margin is recorded as opaque or translucent and each is coded for degree of completion. The percent occurrence of otoliths with opaque margins are then plotted by month to determine timing of opaque annulus formation. If one opaque and one translucent zone are formed each year, validation of increments as annual rings is accomplished (Render 1995). Render reported opaque annulus formation occurring from November through May.

The increased longevity estimates reported by Render (1995) and Szedlmayer and Shipp (1994) were available for the 1995 stock assessment (Goodyear 1995). These longevity estimates had direct implications on conservation measures to rebuild the GOM red snapper stock as well as the estimation of the stocks status at the time (Schirripa and Legault 1999). Previous assessments assumed a natural mortality of 0.20 based on longevity estimates in the available literature. Schirripa and Legault (1999) stated that the apparent increased longevity of red snapper argued strongly that natural mortality must be closer to 0.10, lower than estimates used in prior assessments. The increased longevity estimates also resulted in a new estimate of generation time of 19.6 years resulting is a revised rebuilding target date of 2019 (Hood et al. 2007, this volume).

Research into the life history of red snapper in the GOM continued. A new method for processing larger otoliths, including those of red snapper, was first described in Cowan et al. (1995). Using petrographic techniques borrowed from geology, Cowan et al. (1995) described a procedure for sectioning and polishing otoliths with the use of a Hillquist, model 800, thin-sectioning machine. The procedure greatly reduced processing time and cost without sacrificing quality or precision (Cowan et al. 1995). Patterson et al. (2001) used this new processing method to examine red snapper off Alabama. Patterson et al. (2001) obtained a maximum age of 34 years from "opaque zone" counts from otoliths of 1,676 red snapper sampled from the recreational fishery and research cruises. The authors validated their ages using marginal increment analysis (Beckman et al. 1991), finding a clear pattern of opaque zone formation from January through May for most fish.

A number of laboratories in the Gulf region began routine annual aging programs of red snapper in effort to track age structure trends over time for stock assessment purposes (Wilson and Nieland 2001; Allman et al. 2002). Wilson and Nieland (2001) tracked the age structure of red snapper off Louisiana during eight years of variable collection effort from 1989 to 1998. The authors cited speculation of the validity of the reported longevity of over 50 years as an impetus for their research efforts. Otoliths of 3,791 red snapper sampled from the commercial and recreational harvests off Louisiana were sectioned with the use of a Buehler Isomet low speed saw and "opaque annuli" counted for age estimation. Wilson and Nieland (2001) reported a maximum age of 52 years but noted the difficulty in discerning annuli in the otoliths of older individuals. Additionally, the authors stated that the presumptive first annulus posed the most consistent problem for readers as it appeared as a diffuse "smudge" of opaque material variously located relative to the core. The authors validated the periodicity of opaque annulus formation using marginal increment analysis (Beckman et al. 1989) demonstrating annual opaque annulus formation from December through June.

Allman et al. (2002) sampled commercial and recreational fisheries from Texas to the west coast of Florida in 1998 and 1999 collecting a total of 24,626 red snapper. Of those, a sub-sample of 8,169 otoliths was sectioned for age determination following the rapid processing technique of Cowan et al. (1995). The authors counted "annuli" or "opaque zones under reflected light" to observe a maximum age of 47 years. The authors cited the marginal increment analysis of red snapper in Wilson et al. (1988) for validation. Subsequent to this study, Allman et al. (2005) performed marginal increment analysis for 259 red snapper with the use of Photoshop 6.0 equipped with the Andromeda Measurment Filter. The authors defined the marginal increment as "the distance across the translucent zone measured from the last opaque zone to the otolith edge." A plot of edge distance measurements by month indicated that the minimum marginal increment occurred April through July thus indicating opaque zone formation (Allman et al. 2005).

More recently the use of bomb-radiocarbon has been applied to validate otolith section ages of red snapper in the GOM. This method is considered to be the most advanced and accurate method of fish age validation available (Campana 1999). This approach to age validation uses a quantitative measurement of nuclear bomb-produced ^{14}C that is accumulated in carbon-containing hard parts of marine organisms over a known time period of unique atmospheric chemistry (Baker and Wilson 2001). Atmospheric testing of nuclear weapons carried out during the years of 1958–1965 resulted in elevated levels of bomb-produced ^{14}C that could be incorporated in otoliths, coral exoskeletons, bivalves, and calcareous algae accreted during this period. Druffel (1980, 1989) observed a dramatic increase in ^{14}C activity in hermatypic corals that has then been used as a time specific marker. Studies have shown that living organisms accumulate bomb-produced ^{14}C at a similar rate to that of corals from surrounding waters (Kalish 1993; Campana 1997; Frantz et al. 2000). A number of studies have previously used this technique to validate age estimates derived from hard parts of marine fishes (Kalish 1993; Campana and Jones 1998).

Baker and Wilson (2001) applied this technique to validate otolith section based ages of red snapper hatched before, during, and after the nuclear testing period. Ages of red snapper samples ranged from 2 to 55 years with back-calculated years of birth ranging from 1943 to 1996. Accelerator mass spectrometry (AMS) analysis on the red snapper otoliths produced $\Delta^{14}C$ chronologies similar to values recorded in living corals from Bermuda, south Florida, and Belize (Baker and Wilson 2001) as well as to values found in previous radiocarbon-based age validation studies on marine teleosts (Kalish 1993; Campana 1997). Baker and Wilson (2001) stated that their radiocarbon chronologies based on AMS $\Delta^{14}C$ measurements of otoliths provided evidence that otolith section-based age estimates of red snapper from the GOM are valid. The authors noted that this technique is a viable validation alternative for older fish in comparison to marginal increment analysis because it does not require a large sample size. However, costs can be prohibitive and only fishes with birth dates in the 1960s and 1970s are suitable for this age validation technique.

Interpretation of the first annulus in red snapper otoliths remains one of the major sources of disagreement between readers (Wilson and Nieland 2001; Allman 2002) and affects estimates of precision. Wilson and Nieland (2001) hypothesized that variation in the distance from the core to the first annulus is related to the protracted spawning season of red snapper. In an effort to better describe the first annulus, Allman et al. (2005) measured 259 red snapper otolith sections from the core to the distal edge of the presumed first annulus. Although the authors did observe variation in the degree of opacity of the first annulus, they did not see much variation in the distance from the core to the distal edge of the first annulus. Allman et al. (2005) reported a consistent distance of about 1 mm from the core to the distal edge of the first annulus. The authors suggested that a 1 mm distance may therefore be a good guideline for the expected annulus position and aid in interpretation.

Aging Precision and Accuracy

The evaluation of reproducibility of age estimates (precision) is an important component of any aging program (Campana et al. 1995). Precision of age estimates is an important concern for assessing stock condition (Beamish and McFarlane 1995; Campana 2001; Allman et al. 2005). However, Campana (2001) notes that neither precision nor validation of an aging method necessarily result in aging accuracy (Campana 2001). Although a validated aging method may have proved successful for a given study, there is no guarantee that those same structures will be interpreted the same by other readers (Campana 2001). Allman et al. (2005) noted that while validation of aging methods is an important criterion in estimating growth and longevity, age interpretation remains largely subjective. This subjectivity can pose a significant problem in aging red snapper otoliths due to challenges in interpreting the first annulus and in discerning annuli in older individuals.

A measure of precision is a valuable means of assessing the reproducibility of age determinations between individuals or laboratories (Campana 2001). Average percent error (APE) is an index of aging precision that has been widely used in age and growth studies. Average percent error is not independent of age (Beamish and Fournier 1981) and should therefore be a good index to evaluate aging precision for a long lived species like red snapper. A smaller index value indicates increased precision with an APE of 5% or lower considered a good index of precision for a long-lived species (Campana 2001; Allman et al. 2005). A number of the above mentioned red snapper studies have relied on this statistical measure to ensure reproducibility of age estimates reporting APE values of 3.1 (Render 1995), 0.9 (Patterson et al. 2001), 0.09 (Wilson and Nieland 2001), 5.2 (Allman et al. 2002), and 2.5 (Allman et al. 2005). These reported APE values were calculated after readers performed a second reading. In general, APEs decreased after the first reading. Wilson and Nieland (2001) noted that training and experience are critical to achieving accuracy and high between-reader consensus on red snapper annulus counts due to the high level of difficulty in interpretation.

The need for quality control and monitoring of age data are critical as inaccurate age data could lead to serious errors in stock assessment (Lai and Gunderson 1987; Beamish and McFarlane 1995; and Campana et al. 1995). The Gulf States Marine Fisheries Commission (GSMFS) began to conduct yearly aging workshops in 2003 in an effort to bring together all aging laboratories in the GOM to discuss and standardize processing and aging techniques. Evidence from these workshops indicated that differences existed between readers and laboratories in interpretation of red snapper otoliths

Table 1. Age and growth parameters from previous studies of red snapper *Lutjanus campechanus* in the Gulf of Mexico. Sc = scales, SO = sectioned otoliths, $T_0 = 0$ assumes in the modal that length at time 0 = 0.

Study and location	Sample number	Aging structure	Minimum age (year)	Maximum age (year)	L_∞ mm (TL)	K	t_0
Nelson and Manooch (1982); GOM	3,323	Sc and SO	1	15	941	0.17	–0.10
Szedlmayer and Shipp (1994); North central GOM	409	SO	1	42	1,025	0.15	0
Render (1995); Northern GOM	266 females	SO	1	54	771.59 (FL)	0.18	0.0
	256 males				858.64 (FL)	0.09	–2.21
Wilson and Nieland (2001); Northern GOM off Louisiana	3,787	SO	0	53	941	0.18	–0.55
Patterson et al. (2001); Northern GOM off Alabama	1,672	SO	2	34	969	0.19	0.02
Fischer et al. (2004);							
Alabama	1,985	SO	1	35	839 (FL)	0.38	0
Louisiana	1,864		2	37	847.8 (FL)	0.25	
Texas	1,186		1	45	778.2 (FL)	0.49	

(Allman et al. 2005). The National Marine Fisheries Service Panama City Laboratory (in conjunction with Gulf States and the Gulf States Marine Fisheries Commission) compiled a red snapper reference collection as a means to address these aging inconsistencies. Campana et al. (1995) noted that reference collections of otoliths are important elements in maintaining ongoing aging programs to monitor the consistency of aging between individuals and laboratories. The Panama City red snapper reference collection consisted of 300 adult red snapper otolith sections selected to represent most age classes, all seasons, both sexes, different collection years, good to poorly prepared otolith sections, and the entire geographic range (Campana 2001; Allman et al. 2005). The collection was distributed to seven external laboratories for aging. Average percent error was then used to compare age determinations between the seven laboratories and the Panama City laboratory. Initial comparisons of Panama City ages to external laboratory ages indicated that laboratories 1–4 had APEs below the 5% target (2.8, 3.5, 3.7, and 4.5) and laboratories 5 and 6 had APEs slightly above (5.9 and 6.0%) (Allman et al. 2005). Comparison with laboratory 7 generated an APE of 11.6%. This elevated APE was attributed to a new employee at laboratory 7 as well as a slightly different method in otolith preparation (Allman et al. 2005). This effort demonstrated the source of aging errors and highlighted the importance of a reference collection as a training tool (Campana 2001; Allman et al. 2005). Annual GSMFC aging meetings continue to be held and circulation of the red snapper reference set continues each year in an effort to monitor and ensure red snapper aging consistency and data accuracy throughout the GOM.

Growth

A number of the aforementioned studies used the von Bertalanffy model to describe growth of red snapper in the GOM (Table 1). The model fits observed length-at-age data to estimate the parameters L_∞ (maximum theoretical length), k (the growth coefficient describing the rate at which the asymptote is reached), and t_0 (the hypothetical age when total length is zero). Each model described a similar pattern of rapid growth in the first ten years followed by a reduction in growth rate until maximum length is reached. Each study also reported varying estimates of growth parameters. Improvements in data accuracy and increased estimates in longevity resulting from refinements in aging techniques have helped to better define growth of red snapper throughout their life span. When included in the model, older fish (along with a robust sample size) help to pull the asymptote downward better predicting maximum theoretical length (as well as the growth coefficient). Additionally, correct interpretation of the first annulus will result in more accurate estimates of juvenile growth (Wilson and Nieland 2001).

Care must be taken when comparing von Bertalanffy growth models. Parameter estimates may well reflect growth of red snapper in a particular region over a certain time period, but one must assume that the sample population accurately reflects the natural red snapper population that the model represents. This can pose a problem to researchers because both younger red snapper not yet recruited to the fisheries and older red snapper are difficult to collect. This often results in sparse representation of these age classes in sample populations. Under-representation (or over-representation) of a given age-class also will have an affect on parameter estimation. The parameter estimates L_∞ and k are also strongly correlated (Kirkwood and Somers 1984; Wang and Thomas 1995). These parameters are estimated random variables; therefore, the corresponding growth curves are subject to variation (Wang and Milton 2000). Additionally, a number of factors, including year-class strength and within- and between-year differences in growth rates, may also lead to different growth estimates. Francis (1988) suggested that the proper comparisons should be between estimates of *k*, which describes the growth rate, rather than between L_∞,which describes eventual maximum size.

In an effort to corroborate red snapper age estimates derived from sectioned otoliths, Patterson et al. (2001) incorporated von Bertalanffy growth parameters estimated from otolith-aged fish into Fabens' (1965) length increment model to predict total length (TL) at recapture of tagged fish (Labelle et al. 1993; Thompson et al. 1999). Patterson et al. (2001) plotted predicted TL at recapture from Fabens' method against observed TL at recapture to compare growth model predictions to observed values. The authors found that the predicted TL of tagged individuals obtained with Fabens' method and the von Bertalanffy growth parameters estimated from otolith-aged fish corresponded well to the line of 1:1 agreement. The authors felt this comparison between estimated growth of tagged red snapper and otolith-aged fish did corroborate their otolith-based ages as well as provide support for otolith-based estimates of growth.

Growth rates and size-at-age information have also been used to evaluate the stock structure of GOM red snapper. Fischer et al. (2004) modeled growth of red snapper from Alabama, Louisiana, and Texas with weighted mean fork length (FL) at age and mean total weight (TW) at age using the von Bertalanffy growth model. Additionally, the authors compared growth of red snapper aged 1–10 years for each state by comparing linear regressions of mean FL and mean TW at age. The von Bertalanffy models predicted significantly smaller estimates of both maximum theoretical length and maximum theoretical weight for Texas red snapper compared with those from Alabama and Louisiana. Additionally, Texas red snapper were shown to differ significantly in regressions of mean weight at age. This led the authors to conclude that Texas red snapper were significantly smaller in mass (TW) at age and reach smaller maximum sizes than red snapper from Alabama and Louisiana. Fischer et al. (2004) suggested that this demographic variation in growth rates may indicate the existence of separate management units of red snapper in the GOM.

Conclusion

Great strides have been made to characterize the age structure of the red snapper population in the GOM. Previous research efforts have demonstrated that sectioned otoliths are the most effective method for accurately estimating red snapper age. However, the position and timing of the first annulus remains problematic (Wilson and Nieland 2001; Allman 2005)). Further efforts are needed to address first annulus formation to ensure consistent and accurate aging of red snapper. Further, validation of ages across the entire age range would be of great interest as variability in aging can increase with older fish (Campana 2001). Previous efforts examining red snapper growth have been hampered by either small sample sizes or a lack of individuals at both ends of the age spectrum. To accurately describe red snapper growth, more juveniles and older red snappers will need to be included in sample populations. The inclusion of these age classes would greatly improve our knowledge and understanding of red snapper growth in the GOM.

Acknowledgments

I would like to thank Dave Nieland for a critical review of this manuscript. Additional thanks goes to Peter Hood with NOAA Fisheries Southeastern Regional Office for helpful information on red snapper management in the GOM. I would also like to thank the two anonymous reviewers for their helpful comments in improving this paper.

References

Allman, R. J., G. R. Fitzhugh, K. J. Starzinger, and R. A. Farsky. 2005. Precision of age estimation in red snapper (*Lutjanus campechanus*). Fisheries Research 73:123–133.

Allman, R. J., L. A. Limbardi-Carlson, G. R. Fitzhugh, and W.A. Fable. 2002. Age structure of red snapper (*Lutjanus campechanus*) in the Gulf of Mexico by fishing mode and region. Pages 482–495 *in* R. L. Creswell, editor. Proceedings of the Fifty-Third Annual Gulf and Caribbean Fisheries Institute, Fort Pierce, Florida.

Baker, M. S. Jr., and C. A. Wilson. 2001. Use of bomb radiocarbon to validate otolith section ages of red snapper *Lutjanus campechanus* from the northern Gulf of Mexico. Limnology and Oceanography 46:1819–1824.

Beamish, R. J., D. A. Fournier. 1981. A method for comparing the precision of a set of age determinations. Canadian Journal of Fisheries and Aquatic Sciences 38:982–983.

Beamish, R. J., and G. A. McFarlane. 1995. A discussion of the importance of ageing errors, and an application to walleye pollock: the world's largest fishery. Pages 545–565 *in* D. H. Secor, J. M. Dean, and S. E. Campana, editors. Recent developments in fish otolith research. University of South Carolina Press, Columbia, South Carolina.

Beamish, R. J., and G. A. McFarlane. 1983. The forgotten requirement for age validation in fisheries biology. Transactions of the American Fisheries Society 112:735–743.

Beamish. 1979. New information on the longevity of Pacific ocean perch (*Sebastes alutus*). Journal of the Fisheries Research Board of Canada 36:1395–1400.

Beckman, D. W., A. L. Stanley, J. H. Render, and C. A. Wilson. 1991. Age and growth rate estimation of sheepshead *Archosargus probatocephalus* in Louisiana waters using otoliths. Fishery Bulletin 89:1–8.

Beckman, D. W., C. A. Wilson, and A. L. Stanley. 1989. Age and growth of red drum, *Sciaenops ocellatus*, from offshore waters of the northern Gulf of Mexico. Fishery Bulletin 87:17–28.

Berry, F. H., D. W. Lee, and A. R. Bertalino. 1977. Progress in bluefin tuna ageing attempts. International Commission for the Conservation of Atlantic Tunas Scientific Papers 6:305–317.

Bortone, S. A., and C. L. Hollingsworth. 1980. Ageing red snapper, *Lutjanus campechanus*, with otoliths, scales, and vertebrae Northeast Gulf Science 4:60–63.

Campana, S. E., M. C. Annand, J. I. McMillan. 1995. Graphical and statistical methods for determining the consistency of age determinations. Transactions of the American Fisheries Society 124:131–138.

Campana, S. E. 1997. Use of radiocarbon from nuclear fallout as a dated marker in the otoliths of haddock, *Melanogrammus aeglefinus*. Marine Ecology Progress Series 150:49–56.

Campana, S. E., and C. M. Jones. 1998. Radiocarbon from nuclear testing applied to age validation of black drum, *Pogonias cromis*. Fishery Bulletin 96:185–192.

Campana, S. E. 1999. Chemistry and composition of fish otoliths: pathways, mechanisms, and applications. Marine Ecology Progress Series 188:263–297.

Campana, S. E. 2001. Accuracy, precision, and quality control in age determination, including a review of the use and abuse of age validation methods. Journal of Fish Biology 59:197–242.

Chilton, D. E., and R. J. Beamish. 1982. Age determination methods for fishes studied by the Groundfish Program at the Pacific Biological Station. Canadian Special Publication of Fisheries and Aquatic Sciences 60:1–102.

Cowan, J. H., Jr., R. L. Shipp, H. K. Bailey, IV, and D. W. Haywick. 1995. Procedure for rapid processing of large otoliths. Transactions of the American Fisheries Society 124:280–282.

Druffel, E. M.. 1989. Decadal time scale variability of ventilation in the North Atlantic: high-precision measurements of bomb radiocarbon in banded corals. Journal of Geophysical Research 94:3271–3285.

Druffel, E. M.. 1980. Radiocarbon in annual coral rings of Belize and Florida. Radiocarbon 22:363–371.

Fabens, A. J. 1965. Properties and fitting of the von Bertalanffy growth curve. Growth 29:265–289.

Fenton, G. E., S. A. Short, and D. A. Ritz. 1991. Age determination of orange roughy *Hoplostethus atlanticus* (Picses: Trachichthyidae) using ^{210}Pb:^{226}Ra disequilibria. Marine Biology 109:197–202.

Fischer, A. J., M. S. Baker, Jr., and C. A. Wilson. 2004. Red snapper (*Lutjanus campechanus*) demographic structure in the northern Gulf of Mexico based on spatial patterns in growth rates and morphometrics. Fishery Bulletin 102:593–603.

Francis, R. I. C. C. 1988. Are growth parameters estimated from tagging and age-length data comparable? Canadian Journal of Fisheries and Aquatic Sciences 45:936–942.

Frantz, B. R., M. Kashgarian, K. H. Coale, and M. S. Foster. 2000. Growth rate and potential climate record from a rhodolith using ^{14}C accelerator mass spectrometry. Limnology and Oceanography 45:1773–1777.

Futch, R. B., and G. E. Bruger. 1976. Age, growth, and reproduction of red snapper in Florida waters. Pages 165-183 *in* H. R. Bullis Jr. and A. C. Jones, editors. Proceedings: colloquium on snapper-grouper fishery resources of the western central Atlantic Ocean. Florida Sea Grant College Program Report 17.

Gauldie, R. W., L. F. West, and N. M. Davies. 1989. *K*-selection characteristics of orange roughy stocks in New Zealand waters. Journal of Applied Ichthyology 4:140–146.

Goodyear, C. P. 1995. Red snapper in U.S. waters of the Gulf of Mexico, National Marine Fisheries Service, Southeast Fisheries Science Center, Miami Laboratory, MIA-95/96–05, Miami, Florida.

Goodyear, C. P. 1992. Red snapper in U. S. Waters of the Gulf of Mexico. NOAA.NOAA Fisheries Service SEFSC Contribution MIA-91/92–70.

Goodyear, C. P. 1988. Recent trends in the red snapper fishery of the Gulf of Mexico. NOAA Fisheries Service, SEFSC, Miami Laboratory, CRD 87/88–19. Unpublished report.

Hood, P. B., A. J. Strelcheck, and P. Steele. 2007. A history of red snapper management in the Gulf of Mexico. Pages 267–284 *in* W. F. Patterson, III, J. H. Cowan, Jr., G. R. Fitzhugh, and D. L. Nieland, editors. Red snapper ecology and fisheries in the U.S. Gulf of Mexico. American Fisheries Society, Symposium 60, Bethesda, Maryland.

Jarvis, N. D. 1935. Fishery for red snappers and groupers in the Gulf of Mexico. U.S. Department of Commerce, Bureau of Fisheries, Investigational Report 26, Washington, D.C.

Kalish, J. M. 1993. Pre- and post-bomb radiocarbon in fish otoliths. Earth and Planetary Science Letters 114:549.

Kirkwood, G. P., and I. F. Somers. 1984. Growth of the two species of tiger prawn, *Penaeus esculentus* and *Penaeus semisulcatus*, in the western Gulf of Carpentaria. Austrailian Journal of Marine Freshwater Research 35:703–712.

Kotlyar, A. N. 1980. Age and growth speed of the bigheads, *Hoplostethus atlanticus* Collett and *H. mediterraneus* Cuvier (Trachichthydae, Beryciformes). Pages 68-88 *in* P. V. Sherston, editor. Fishes of the open ocean. Moscow Institute of Oceanography, Moscow.

Labelle, M., J. Hampton, K. Bailey, T. Murray, D. A. Fournier, and J. R. Sibert. 1993. Determination of age and growth of Pacific albacore (*Thunnus alalunga*) using three methodologies. Fishery Bulletin 91:649–663.

Lai, H. L., and D. R. Gunderson. 1987. Effects of ageing errors on estimates of growth, mortality, and yield per recruit for walleye pollock (*Theragra chalcogramma*). Fisheries Research 5:287–302.

Moseley, F. N. 1966. Biology of red snapper, *Lutjanus aya* Bloch, of the northwestern Gulf of Mexico. Publications of the Institute of Marine Science 11:90–101.

Nelson, R. S., and C. S. Manooch. 1982. Growth and mortality of red snappers in the west-central Atlantic Ocean and northern Gulf of Mexico. Transactions of the American Fisheries Society 111:465–475.

Patterson, W. F., III, J. H. Cowan, Jr., C. A. Wilson, and R. L. Shipp. 2001. Age and growth of red snapper, *Lutjanus campechanus,* from an artificial reef area off Alabama in the northern Gulf of Mexico. Fishery Bulletin 99:617–627.

Render, J. H. 1995. The life history (age, growth, and reproduction) of red snapper (*Lutjanus campechanus*) and its affinity for oil and gas platforms. Doctoral dissertation. Louisiana State University, Baton Rouge, Louisiana.

Schirripa, M. J. and C. M. Legault. 1999. Status of the red snapper in U.S. waters of the Gulf of Mexico: updated through 1998. National Marine Fisheries Service, Southeast Fisheries Science Center, Miami Laboratory, Miami. Sustainable Fisheries Division contribution SFD-99/00–75.

Smith, D. C., G. E. Fenton, S. G. Robertson, and S. A. Short. 1994. Age determination and growth of orange roughy (*Hoplostethus atlanticus*): a comparison of annulus counts with radiometric ageing. Canadian Journal of Fisheries and Aquatic Sciences 52:391–401.

Szedlmayer, S. T., and R. L. Shipp. 1994. Movement and growth of red snapper, *Lutjanus campechanus*, from an artificial reef area in the northeastern Gulf of Mexico. Bulletin of Marine Science 55(2–3):887–896.

Thompson, B. A., M. Beasley, and C. A. Wilson. 1999. Age distribution and growth of greater amberjack, *Seriola dumerilli*, from the north-central Gulf of Mexico. Fishery Bulletin 97:362–371.

Van Den Broek, W. L. F. 1983. Ageing deepwater fish species: report of a visit to the United Kingdom September-November 1982. Miscellaneous Series Fisheries Research Division, Ministry of Agriculture and Fisheries, unpublished report,

MAF Fisheries, Greta Pt. Library, Wellington, New Zealand.

Wang, Y. G., and D. A. Milton 2000. On comparison of growth curves: How do we test whether growth rates differ? Fishery Bulletin 98:874–880.

Wang, Y. G., and M. R. Thomas. 1995. Accounting for the individual variability in the von Bertalanffy growth model. Canadian Journal of Fisheries and Aquatic Sciences 52:1368–1375.

Wilson, C. A., and D. L. Nieland. 2001. Age and Growth of red snapper, *Lutjanus campechanus*, from the northern Gulf of Mexico off Louisiana. Fishery Bulletin 99:653–664.

Wilson, C. A., D. L. Nieland, A. L. Stanley, and A. J. Fischer. 1988. Age and size distribution of commercially harvested red snapper *Lutjanus campechanus* in the northern Gulf of Mexico. Final report to U.S. Department of Commerce, National Marine Fisheries Service, Marine Fisheries Initiative (MARFIN) Cooperative Agreement NA57FF0287. St. Petersburg, Florida.

American Fisheries Society Symposium 60:201–216, 2007

Population Structure of Red Snapper in the Northern Gulf of Mexico

John R. Gold[1] and Eric Saillant
Center for Biosystematics and Biodiversity, Texas A&M University College Station, Texas 77843 USA

Abstract.—Allelic variation at 19 nuclear-encoded microsatellite loci and haplotype variation in a 590 bp protein-coding fragment of mitochondrial (mt)DNA were assayed among Gulf red snapper sampled from four cohorts at each of three offshore localities (12 samples total) in the northern Gulf of Mexico. Significant heterogeneity in allele and genotype distributions among samples was detected at four microsatellites; six of seven 'significant' pairwise comparisons between samples revealed the heterogeneity to be temporal rather than spatial. Nested-clade analysis of mtDNA variants indicated different temporal episodes of range expansion and isolation by distance. Estimates of variance effective population size (microsatellites) ranged between ~1,000 and >75,000 and differed significantly among localities. The differences in variance effective size likely reflect differences in number of individuals successfully reproducing or differences in patterns and intensity of migration. Collectively, these findings are consistent with the hypothesis that red snapper in the northern Gulf occur as a network (or metapopulation) of semi-isolated assemblages that may be demographically independent over the short term, yet over the long term can influence each other's demographics via gene flow. This type of population structure may be difficult to detect with commonly used, selectively neutral genetic markers.

Introduction

The red snapper *Lutjanus campechanus* is a highly exploited marine fish found primarily on the continental shelf of the Gulf of Mexico (hereafter Gulf) (Smith 1997; Hoese and Moore 1998). The species supports both recreational and commercial fisheries in U.S. waters and has been subjected to intensive management because of precipitous declines in abundance over the last few decades (Goodyear and Phares 1990). As evidenced by this volume, research on red snapper in U.S. waters is now extensive, with the common goal of providing critical information for assessment, allocation, and conservation of red snapper resources. Research in our laboratory has been focused primarily on delineation of stock structure of red snapper in the northern Gulf (Camper et al. 1993; Gold et al. 1997, 2001; Pruett et al. 2005; Saillant and Gold 2006) since management of the fishery, should separate stocks exist, could be subdivided to avoid subregional overexploitation or mortality (Carvalho and Hauser 1995). In addition, different stocks, should they exist, could

[1] Corresponding author: goldfish@tamu.edu.

possess local or subregional adaptations that promote differences in important life history parameters such as growth, fecundity, and disease resistance (Stepien 1995). Failure to recognize occurrence of such stocks potentially could result in localized extinction and loss of unique genetic resources.

Most prior genetic studies of stock structure of red snapper in the northern Gulf involved tests of spatial homogeneity in allele/haplotype distribution at various genetic markers, including nuclear-encoded proteins (allozymes), restriction sites or sequences of mitochondrial (mt)DNA, and nuclear-encoded microsatellites (Johnson 1987; Camper et al. 1993; Gold et al. 1997, 2001; Garber et al. 2004). Almost all of these studies revealed genetic homogeneity across the sampling surface, consistent with the inference that sufficient gene flow to maintain statistically identical allele/haplotype distributions occurs and with the hypothesis of a single, unit stock. However, most of these studies either involved small sample sizes and few loci or included individuals from mixed cohorts. Moreover, the inference regarding gene flow across the northern Gulf was not fully consistent with tag-and-recapture and ultrasonic-tracking experiments (Fable 1980; Szedlmayer and Shipp 1994; Szedlmayer 1997; Patterson et al. 2001) that indicated sedentary behavior and relatively high site fidelity of red snapper adults.

We expanded our genetic studies of red snapper to include estimation of (genetic) effective population size (N_e) and assessment of historical demography (Pruett et al. 2005; Saillant and Gold 2006). Briefly, N_e is defined as the number of individuals in an 'ideal' population that would experience the same magnitude of genetic drift as the actual population (Hartl and Clark 1989). N_e is an important biological parameter because it measures the rate at which a population over time may lose genetic variation and accumulate inbreeding (Turner et al. 2002); populations (or stocks) with small N_e thus may lose genetic resources, become inbred, and suffer from a reduced capacity to respond to changing environmental factors such as intense exploitation or deteriorating habitats. Our interest in historical demography was a consequence of testing the hypothesis proposed by Pruett et al. (2005) that gene flow among red snapper in the northern Gulf was a dynamic process that varied in intensity and duration through both time and space.

In this paper, we synopsize genetic data from a multi-year, interdisciplinary study of red snapper in the northern Gulf. The overall study was focused on stock structure and included data on genetics, age and growth, and reproductive biology. Papers dealing with the latter two areas may be found elsewhere in this volume. Herein, we assess genetic stock structure (based on both nuclear and mitochondrial markers), estimate variance (contemporaneous) genetic effective size, and evaluate historical population demography of red snapper in the northern Gulf. Results of the study support the hypothesis that red snapper in the northern Gulf occur as a network (or metapopulation) of semi-isolated assemblages that may be demographically independent over the short term.

Material and Methods

Adult red snapper belonging to the 1995 and 1997 cohorts were sampled between 1999 and 2001 by angling 40–50 km offshore at each of three localities (Figure 1) in the northern Gulf; young-of-the year (age-0) red snapper belonging to the 1999 and 2000 cohorts were obtained during demersal trawl surveys carried out at the same localities in the fall of each year (1999 and 2000) by the National Marine Fisheries Service (NMFS). Localities were the northwestern Gulf (hereafter Texas), the north-central Gulf (hereafter Louisiana), and the northeastern Gulf

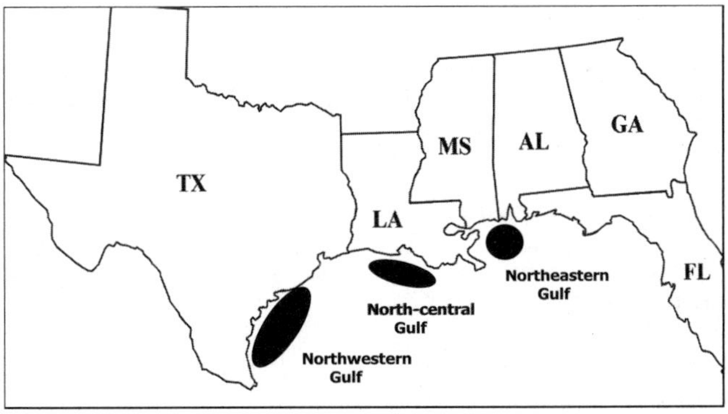

Figure 1. Sample localities for red snapper *Lutjanus campechanus* in the northern Gulf of Mexico.

(hereafter Alabama). Heart and spleen tissues (adults and juveniles) were frozen in liquid nitrogen and stored at –80°C. Adults belonging to the 1995 and 1997 cohorts were identified by otolith-increment analysis (Wilson and Nieland 2001). Sample sizes by cohort and locality are given in Table 1.

Summary statistics for each of 19 microsatellites, including sample sizes, number of alleles, allelic richness, gene diversity, probability of departure from expected Hardy-Weinberg genotypic proportions, and the inbreeding coefficient F_{IS}, were generated as outlined in Saillant and Gold (2006) for each of the 12 samples (four cohorts at each of three localities). Homogeneity of allelic richness and gene diversity among samples was tested with Friedman rank tests. Genotypic disequilibrium between pairs of microsatellites within samples and homogeneity of allele and genotype distributions both at each microsatellite and over all microsatellites were assessed via exact tests; significance of probability values was examined by a Markov-chain method. Statistical programs employed and Markov-chain parameters are outlined fully in Saillant and Gold (2006). Genetic divergence between pairs of samples was evaluated using Weir and Cockerham's (1984) θ. Sequential Bonferroni correction (Rice 1989) was applied to all tests performed simultaneously.

Variance effective population size (N_{eV}) at each locality was estimated via temporal changes (Waples 1989) in allele frequencies between cohorts. The pseudo-maximum-likelihood approach of Wang (2001) was used to obtain estimates of N_{eV} and their 95% confidence intervals. Correction(s) for overlapping generations were generated using the approach developed by Jorde and Ryman (1995, 1996). Specific methods used to correct estimates of N_{eV} for overlapping generations may be found in Saillant and Gold (2006); estimated values for the demographic parameters employed in the correction may be obtained from the authors.

The estimates of N_{eV} generated via the above approach assume that no genetic migration into a locality occurred during the time interval between the cohorts sampled. In order to assess potential effects of migration on the estimates of N_{eV}, the approach of Wang and Whitlock (2003) was employed to simultaneously estimate both N_{eV} and m (the rate of migration). Because the method requires genetic data from all potential sources of migrants into a focal population, estimates of N_{eV} and m in the present data set could only be generated for the locality in the north-central Gulf (see Figure 1). Computation of N_{ev} (Wang 2001) and N_{ev} and m (Wang and Whitlock 2003) employed the software available at http://www.zoo.cam.ac.uk/ioz/software.htm#MLNE. Cor-

Table 1. Samples of red snapper *Lutjanus campechanus* by locality and cohort.

Sample locality	Northwestern Gulf	North-central Gulf	Northeastern Gulf
Adults			
1995 cohort	203	286	377
1997 cohort	211	272	274
Juveniles			
1999 cohort	97	77	63
2000 cohort	65	32	44
Total	576	667	758

rections for overlapping generations were applied as before.

A 590 base-pair (bp) fragment of the mitochondrially encoded NADH dehydrogenase subunit 4 gene (ND-4) was sequenced from each of 30 individuals from each of the four cohorts at each of the three localities (n = 120 per locality, 360 individuals total). Methods used for polymerase-chain-reaction (PCR) amplification and sequencing may be found in Pruett et al. (2005). Summary statistics for the 12 samples, including number of mtDNA haplotypes, haplotype frequencies, and nucleon and nucleotide diversity, were generated as outlined in Pruett et al. (2005). Homogeneity of haplotype distributions among cohorts within regions and among regions (cohorts pooled) was assessed via exact tests and analysis of molecular variance (AMOVA). Statistical programs employed and methods used to estimate fixation indices and probability of significance of exact tests or AMOVA are outlined fully in Pruett et al. (2005).

Nested-clade analysis (Templeton et al. 1995; Templeton 1998) was used to test for geographical association of phylogenetic assemblages (clades) of mtDNA variants. Nested-clade analysis allows one to make inferences regarding historical demographic processes such as contiguous/noncontiguous range expansion, population fragmentation, restricted or recurrent gene flow, and isolation by distance. Details regarding generation of phylogenetic topologies, the nesting of a 95% parsimony network of mtDNA haplotypes, and the permutational contingency analysis used to test the null hypothesis of random geographical distribution of mtDNA clades may be found in Pruett et al. (2005).

Results

Summary statistics, including number of alleles, allelic richness, gene diversity, results of tests of Hardy-Weinberg equilibrium, and F_{IS} values, for the 1995 and 1997 cohorts may be found in Saillant and Gold (2006); summary statistics for the 1999 and 2000 cohorts (not published previously) are given in Appendix Tables 1 and 2. Number of alleles and allelic richness per microsatellite per sample over all four cohorts averaged (±SD) 9.82 ± 4.86 and 7.15 ± 3.04, respectively; gene diversity per microsatellite over all four cohorts averaged (±SD) 0.60 ± 0.22. No significant difference in allelic richness ($X^2_{[11]}$ = 10.90, P = 0.452) or gene

Table 2. Pairwise F_{ST} values (upper diagonal) and probability that $F_{ST} = 0$ (lower diagonal) for pairwise comparisons of 12 samples of red snapper *Lutjanus campechanus* that were significant following Bonferroni correction. Significant probability values are indicated by an asterisk. TX = Texas, LA = Louisiana, AL = Alabama. All comparisons with samples from the 1999 and 2000 cohorts were non-significant (corrected $P > 0.05$).

	TX 95	LA 95	AL 95	TX 97	LA 97	AL 97
TX 95	—	0.001	0.001	0.001	0.001	0.002
LA 95	0.001*	—	0.001	0.001	0.000	0.001
AL 95	0.000*	0.031	—	0.001	0.000	0.001
TX 97	0.000*	0.013	0.000*	—	0.000	0.000
LA 97	0.036	0.756	0.737	0.078	—	0.001
AL 97	0.000*	0.000*	0.000*	0.054	0.073	—

diversity ($X^2_{[11]} = 9.42$, $P = 0.583$) among the 12 samples was detected. Only seven of 248 (2.82%) tests of departure from Hardy-Weinberg equilibrium expectations were significant following Bonferroni correction (Saillant and Gold 2006; Appendix Tables 1 and 2). Of these, two occurred at microsatellite *Prs* 137 (1995 cohort from Alabama; 1999 cohort from Louisiana); the remainder occurred in single samples and involved five different microsatellites. F_{IS} values for the seven tests where departure from Hardy-Weinberg equilibrium expectations were significant ranged from 0.021 to 0.181. None of the pairwise tests of genotypic disequilibrium were significant after Bonferroni correction.

Heterogeneity among all 12 samples in both allele and genotype distributions was found over all microsatellites ($P = 0.000$ for alleles, $P = 0.000$ for genotypes) and, after Bonferroni correction, at four microsatellites: *Lca* 22 ($P = 0.001$ for alleles and $P = 0.000$ for genotypes), *Lca* 91 ($P = 0.000$ for alleles, and $P = 0.001$ for genotypes), *Prs* 240 P = 0.000 for both alleles and genotypes), and *Prs* 303 ($P = 0.001$ for alleles and $P = 0.000$ for genotypes). Pairwise comparisons of allele and genotype distributions among samples (66 comparisons) paralleled one another, with significant heterogeneity following Bonferroni correction found primarily in comparisons involving either the 1995 cohort sampled in Texas waters or the 1997 cohort sampled in Alabama waters (Table 2). These results indicated that the genetic heterogeneity observed over all samples was due primarily to temporal (among cohorts within localities) rather than to spatial (among localities) differences. This also was indicated by the average F_{ST} values among localities (all cohorts separately and summed) of less than 0.001.

Estimates of variance effective size (N_{eV}) and their 95% confidence intervals for each of the three sample localities are given in Table 3. The estimates of N_{eV} are for the time intervals 1995–1997 and 1995–2000. The N_{eV} estimate for the latter was an average over the entire sampling period. In both time intervals, the estimates of N_{eV} for the samples from the northwestern (Texas) and northeastern (Alabama) Gulf fell well within the 95% confidence intervals of one another and were significantly lower than the N_{eV} estimate for the north-central (Louisiana) Gulf. An exact estimate of N_{eV} for the sample from the north-central Gulf during the time interval 1995–1997 could not be obtained as the estimate of N_{eV} with the highest likelihood was over

Table 3. Estimates of variance effective size (N_{eV}) and 95% confidence intervals for red snapper *Lutjanus campechanus* sampled at three geographic localities in the northern Gulf of Mexico. Estimates are given for the time intervals 1995–1997[a] and 1995–2000[b].

Locality	ML N_{eV}	95% low	95% high
Texas[a]	1,098	652	2,706
Louisiana[a]	>75,240	3,275	>75,240
Alabama[a]	1,235	777	2,515
Texas[b]	2,622	1,453	8,792
Louisiana[b]	26,885	3,807	>69,300
Alabama[b]	1,741	1,092	3,576

75,240; the likelihood of high values could not be computed. Regardless, in both time intervals, the estimate of N_{eV} for the sample from the north-central Gulf was an order of magnitude greater than the estimates for the other two sample localities.

Estimates of N_{eV} (incorporating migration) and of m (migration rate) for the sample from the north-central Gulf were estimated using data from the time intervals 1995–1997 and 1995–2000 and the maximum-likelihood approach of Wang and Whitlock (2003). The estimate of N_{eV} for the interval 1995–1997 was 4,887 (95% confidence intervals of 1,543–31,254) and was ~15 times smaller than the estimate generated assuming no migration; m was estimated to be 0.010 (95%confidence intervals of <0.001–0.036). The estimate of N_{eV} for the interval 1995–2000 was 2,835 (95% confidence intervals of 1,486–15,923) and was ~9.5 times smaller than the estimate generated assuming no migration; m was estimated to be 0.021 (95%confidence intervals of <0.001 and 0.042).

A total of 60 unique mtDNA haplotypes was found among the 360 red snapper ND-4 fragments sequenced. Eleven of the haplotypes occurred in all three localities; the number of 'private' haplotypes (those found at only one locality) was 16 (Texas), 10 (Louisiana), and 12 (Alabama). Data on the number and location within codons of synonymous and non-synonymous base substitutions are given in Pruett et al. (2005). Nucleon diversity values (the probability that two haplotypes sampled at random are different) were essentially the same across localities: Texas (0.797 ± 0.028), Louisiana (0.770 ± 0.030), and Alabama (0.793 ± 0.028). Results of exact tests of haplotype-distribution homogeneity among cohorts within localities and among localities (cohorts within localities pooled) were nonsignificant ($P > 0.05$), as were results from AMOVA (among localities $\Phi_{ST} = -0.002$, $P = 0.422$; among year classes, $\Phi_{SC} = 0.003$, $P = 0.278$).

Nesting of the 95% parsimony network (Figure 2) revealed three nesting levels: one- and two-step clades and the entire network. Exact contingency analysis, using geographic distances among sample localities, revealed significant ($P < 0.05$) geographical associations for the entire cladogram, for clades 2–3 and 2–4 at the two-step level, and for clades A and F at the one-step level. Use of the inference key available at <http://darwin.uvigo.es/software/geodis.html> indicated that the sig-

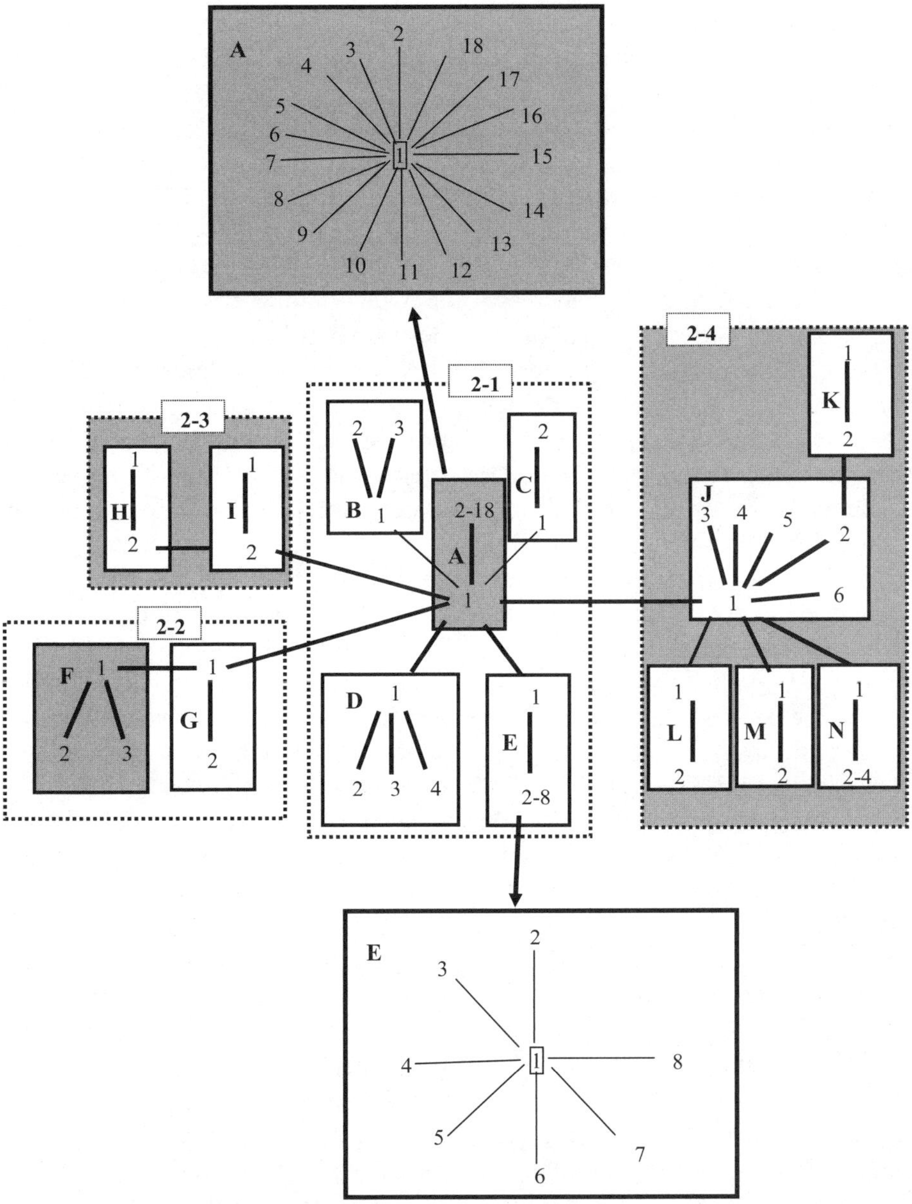

Figure 2. Nested-clade network: numbers correspond to individual haplotypes; letters within boxes surrounded by a solid line represent one-step clades; dashed lines surround two-step clades. Lines between haplotypes correspond to single base-pair substitutions. Shaded boxes represent significant geographical associations within a clade. One-step clades A and E are shown in expanded boxes.

nificant associations for the entire cladogram and for one-step clade F stemmed from restricted gene flow due to isolation by distance, whereas the associations within both two-step clades (2–3 and 2–4) and one-step clade A stemmed from contiguous range expansion or short-distance dispersal across an expanding population front. Details regarding the inference chain and associated clade (D_C) and nested-clade (D_N) distances may be found in Pruett et al. (2005). Closer examination of the spatial distribution of mtDNA haplotypes within each clade further demonstrated the repeated occurrence of these spatial/temporal events. All four two-step clades and several one-step clades (A, J, L, and N) contained haplotypes found at all three sampling localities (indicating range expansion); whereas only one haplotype (found in one individual) from the northwestern Gulf was found in two-step clade 2–2 and a number of one-step clades contained either no or very few haplotypes from one of the three localities. The spatially limited distribution(s) of these haplotypes is consistent with the notion of historically restricted gene flow. Collectively, results from nested-clade analysis indicate a history of recurrent episodes of range expansion and restricted gene flow among red snapper in the northern Gulf.

Discussion

The spatial homogeneity of allele and genotype (microsatellite) and haplotype (mtDNA) distributions observed in this study parallels findings in most prior genetic studies (Johnson 1987; Camper et al. 1993; Gold et al. 1997, 2001; Garber et al. 2004) of red snapper in the northern Gulf. Generally, spatial genetic homogeneity is assumed to indicate occurrence of enough gene flow (migration) to preclude genetic divergence; geographic variation in morphology or life history in these situations is then often inferred to stem from environmental differences between regions. A point largely overlooked, however, is that the genetic markers typically employed in stock-structure studies are presumed to be selectively neutral, which means that they are neither influenced by natural selection nor related to genes impacting an adaptive trait that might impact life history or fitness (McKay and Latta 2002). What this means in theory is that genetic homogeneity observed between or among geographic samples may not necessarily reflect homogeneity in genes affecting life history or fitness traits. In addition, the absence of heterogeneity in selectively neutral genetic markers may not necessarily indicate occurrence of present-day gene flow. Divergence in selectively neutral genetic makers is largely a function of the interaction between gene flow and genetic drift; discrete 'genetic' populations or stocks of a species could thus exist yet be undetectable via 'molecular' markers if there has been insufficient time for isolated lineages to sort into monophyletic assemblages (Arbogast et al. 2002). Finally, gene flow or connectivity over the short term cannot necessarily be estimated accurately based on genetic measures of population differentiation since the latter represent a long-term average rate (Neigel 1997; Kinlan and Gaines 2003).

The significant differences in allele and genotype distributions observed in our studies were largely temporal, reflecting genetic differences among cohorts within localities. These temporal differences account for the significant geographic differences in estimates of genetic effective size (N_{eV}), with red snapper in the north-central Gulf having an effective size that was an order of magnitude larger than red snapper in the northwestern and northeastern Gulf. The spatial differences in N_{eV} indicate the occurrence of different 'demographic' dynamics that potentially reflect spatial differences in the number of adult individuals that successfully produce surviving offspring, differing migration patterns among localities, or a combination of the two (Wang and Whitlock 2003; Fraser et al. 2004). The causes generating these demographic differences are difficult to assess but likely stem in part from differences across the northern Gulf in resource availability and quality or in mortality (Saillant and Gold 2006).

The spatial differences in N_{eV} observed among red snapper at the localities sampled in this study are consistent with reported life history differences. Fischer et al. (2004) found that red snapper sampled at the Texas locality (northwestern Gulf) were significantly smaller at age and reached smaller maximum size than did red snapper sampled at the Louisiana (north-central Gulf) and Alabama (northeastern Gulf) localities, while Woods et al. (2003) found that females sampled at the Alabama locality reached sexual maturity at a younger age and smaller size than did females sampled at the Louisiana locality. The differences in growth rate across localities may reflect differences in nutrient availability (Fischer et al. 2004) but could stem as well from genetic responses to differences in fishing pressure and size-selective mortality (Conover and Munch 2002; Conover et al. 2005). The difference in female age and size at maturity in the northeastern Gulf could signal a stressed population and a compensatory response to overfishing or declining population size (Trippel 1995; Woods et al. 2003).

Results of nested-clade analysis of red snapper mtDNA haplotypes indicated a recurring history of contiguous range expansion and isolation by distance and are consistent with the hypothesis that red snapper across the northern Gulf are not necessarily tied together via continuous gene flow. The timing of the events indicated by nested-clade analysis is problematic in that mutations giving rise to the genetic differences that distinguish individual clades do not necessarily occur at fixed time intervals. However, the two most divergent red snapper mtDNA haplotypes differed by only nine base-pair substitutions (Pruett et al. 2005), suggesting that the events revealed by nested-clade analysis likely occurred within the last million years. This time frame is consistent with notion that multiple factors, including glacial advance or retreat, physical processes such as varying ocean currents and circulation patterns, and differences in habitat all played significant roles in shaping past and present-day distribution of red snapper in the northern Gulf.

Based on the above, we hypothesize that red snapper in the northern Gulf occur as a network (or metapopulation) of semi-isolated assemblages that are demographically independent over the short term, yet over the long term can influence each other's demographics via intermittent or periodic gene flow. Stated differently, each semi-isolated assemblage is, to varying degrees, self-replenishing but can be impacted by adjacent assemblages when sufficient gene flow occurs. This concept of metapopulation structure in red snapper closely follows metapopulation models discussed by Kritzer and Sale (2002), Hellberg et al. (2002), and Østergaard et al. (2003) that predict, respectively, that (*i*) populations may be asynchronous demographically but display homogeneity at selectively neutral (genetic) markers, (*ii*) populations may be independent in terms of recruitment events yet show no genetic differences due to sporadic gene flow, and (*iii*) temporal genetic divergence can exceed spatial genetic divergence. This type of metapopulation model may be common in marine systems, and, if not accounted for, could significantly impact assessments of critical fishery resource parameters such as population size, age structure, and recruitment.

The concept that different 'demographic' stocks of an exploited marine species may exist is not new but has not been employed widely relative to management planning and assessment and allocation of marine fish resources. Definitions of marine-fish stocks vary widely (Carvalho and Hauser 1995) and can depend on socio-economic and political as well as biological considerations. The most widely emphasized definition at present is 'genetic' in that discrete stocks are presumed to exist if heterogeneity in allele or genotype distributions occurs across a geographic surface. Carvalho and Hauser (1995), however, suggested that a 'stock' should have definable patterns of recruitment and mortality, raising the notion that geographic assemblages with different patterns of recruitment and mortality perhaps should be defined as different stocks. There is empirical evidence (Richards and Leberg 1996; Queney et al. 2000) that measures of genetic diversity can be insensitive to demographic variation, and there are a number of reports in exploited fishes of significant temporal variation in allele and genotype distributions (Hansen et al. 2002; Hauser et al. 2002; Turner et al. 2002; Shrimpton and Heath 2003; Lage and Kornfield 2006). There also are reports, including this paper, where the temporal variation appears to be significantly greater than spatial variation (Garant et al. 2000; Østergaard et al. 2003). The latter indicates that demographic differences in exploited species may not be uncommon.

The estimates of genetic effective size (N_{eV}) that revealed significant differences among red snapper across the northern Gulf were generated under the assumption that no migration into a locality occurred during the time interval when samples were obtained. This assumption would seem at odds with the absence of genetic divergence among samples. However, migration presumably can either increase or decrease the variance in allele frequency (hence generating under- or over-estimates of N_e, respectively), depending on whether the pattern of migration is periodic or continuous (Wang and Whitlock 2003). Consequently, the observed differences in N_{eV} among the geographic samples of red snapper could reflect differences in effective size, differences in patterns of migration, or both. The estimates of N_{eV} generated using the approach of Wang and Whitlock (2003) accounts for migration (estimated here to be 0.01 for the interval 1995–1997 and 0.02 for the interval 1995–2000) and yielded, for the sample from the north-central Gulf, N_{eV} estimates that were approximately 10–15 times smaller than the estimates generated assuming no migration. This finding is compatible with the occurrence of sustained migration over the long term (migration-drift equilibrium, Wang and Whitlock 2003) and is consistent with our metapopulation model. The lower estimates of N_{eV} generated when migration was included may indicate that red snapper in the northern Gulf are more genetically compromised than suggested by the estimates when no migration was assumed. More extensive sampling across the northern Gulf obviously is needed to place this finding into perspective and to generate estimates of N_{eV} and m for other localities across the northern Gulf.

Acknowledgments

We thank D. Nieland for providing life history and catch data, C. Bradfield for technical assistance, and J. Cowan, W. Patterson, and T. Turner for helpful discussions. Work was supported by the Marfin Program of the U.S. Department of Commerce (Grant NA87-FF-0426), the Gulf and South Atlantic Fisheries Development Foundation (Grant 70–04–20000/11824), and the Texas Agricultural Experiment Station (Project H-6703). Views expressed in this paper are those of the authors and do not necessarily reflect views of the sponsors. Figure 2 was reprinted from the following paper (Pruett, C. L., Saillant, E., and Gold, J. R., 2005; Historical population demography of red snapper (*Lutjanus campechanus*) from the northern Gulf of Mexico based on analysis of sequences of mitochondrial DNA, *Marine Biology*, 147, 593–602) with kind permission of Springer Science and Business Media. This paper is number 50 in the series "Genetics Studies in Marine Fishes" and Contribution No. 156 of the Center for Biosystematics and Biodiversity at Texas A&M University.

References

Arbogast, B. S., S. V. Edwards, J. Wakeley, P. Beerli, and J. B. Slowinski. 2002. Estimating divergence times from molecular data on phylogenetic and population genetic timescales. Annual review of Ecology and Systematics 33:707–740.

Camper, J. D., R. C. Barber, L. R. Richardson, and J. R. Gold. 1993. Mitochondrial DNA variation among red snapper (*Lutjanus campechanus*) from the Gulf of Mexico. Molecular Marine Biology and Biotechnology 3:154–161.

Carvalho, G. R., and L. Hauser. 1995. Molecular genetics and the stock concept in fisheries. Pages 55–80 *in* G. R. Carvalho and T. J. Pitcher, editors. Molecular genetics in fisheries. Chapman and Hall, London.

Conover, D. O., and S. B. Munch. 2002. Sustaining fisheries yields over evolutionary time scales. Science 297:94–96.

Conover, D. O., S. A. Arnott, M. R. Walsh, and S. B. Munch. 2005. Darwinian fishery science: lessons from the Atlantic silverside. Canadian Journal of Fisheries and Aquatic Sciences 62:730–737.

Fable, W. A. 1980. Tagging studies of red snapper (*Lutjanus campechanus*) and vermilion snapper (*Rhomboplites aurorubens*) off the south Texas coast. Contributions in Marine Science 23:115–121.

Fischer, A. J., M. S. Baker Jr., and C. A. Wilson. 2004. Red snapper, *Lutjanus campechanus*, demographic structure in the northern Gulf of Mexico based on spatial patterns in growth rates and morphometrics. Fishery Bulletin 102:593–603.

Fraser, D. J., C. Lippé, and L. Bernatchez. 2004. Consequences of unequal population size, aymmetric gene flow and sex-biased dispersal on population structure in brook trout (*Salvelinus fontinalis*). Molecular Ecology 13:67–80.

Garant, D., J. J. Dodson, and L. Bernatchez. 2000. Ecological determinants and temporal stability of the within-river population structure in Atlantic salmon (*Salmo salar* L.). Molecular Ecology 9:615–628.

Garber, A. F., M. D. Tringali, and K. C. Stuck. 2004. Population structure and variation in red snapper (*Lutjanus campechanus*) from the Gulf of Mexico and Atlantic Coast of Florida as determined from mitochondrial DNA control region sequence. Marine Biotechnology 6:175–185.

Gold, J. R., F. Sun, and L. R. Richardson. 1997. Population structure of red snapper from the Gulf of Mexico as inferred from analysis of mitochondrial DNA. Transactions of the American Fisheries Society 126:386–396.

Gold, J. R., E. Pak, and L. R. Richardson. 2001. Microsatellite variation among red snapper (*Lutjanus campechanus*) from the Gulf of Mexico. Marine Biotechnology 3:293–304.

Goodyear, C. P., and P. Phares. 1990. Status of red snapper stocks of the Gulf of Mexico: report for 1990. National Marine Fisheries Service, SE Fisheries Center, Miami Laboratory, CRD 89/90–05, Miami, Florida.

Hartl, D. L., and A. G. Clark. 1989. Principles of population genetics, 2nd edition. Sinauer Assoc., Inc., Sunderland, Massachusetts.

Hansen, M. M., D. E. Ruzzante, E. E. Nielsen, D. Bekkevold, and K.-L. D. Mensberg. 2002. Long-term effective population sizes, temporal stability of gentic composition and potential for local adaptation in anadromous brown trout (*Salmo trutta*) populations. Molecular Ecology 11:2523–2535.

Hauser, L., G. J. Adcock, P. J. Smith, J. H. B. Ramirez, G. R. Carvalho. 2002. Loss of microsatellite diversity and low effective population size in an overexploited population of New Zealand snapper (*Pagrus auratus*). Proceedings of the National Academy of Sciences USA 99:11742–11747.

Hellberg, M. E., R. S. Burton, J. E. Neigel, and S. R. Palumbi. 2002. Genetic assessment of connectivity among marine populations. Bulletin of Marine Science 70 (Supplement):273–290.

Hoese, H. D., and R. H. Moore. 1998. Fishes of the Gulf of Mexico, 2nd edition. Texas A&M University Press, College Station, Texas.

Johnson, A. G. 1987. An investigation of the biochemical and morphometric characteristics of red snapper (*Lutjanus campechanus*) from the southern United States. Unpublished manuscript, National Marine Fisheries Service, Panama City Laboratory, Panama City, Florida.

Jorde, P. E., and N. Ryman. 1995. Temporal allele frequency change and estimation of effective size in populations with overlapping generations. Genetics 139:1077–1090.

Jorde, P. E., and N. Ryman. 1996. Demographic genetics of brown trout (*Salmo trutta*) and estimates of effective population size from temporal change of allele frequencies. Genetics 143:1369–1381.

Kinlan, B. P., and S. Gaines. 2003. Propagule dispersal in marine and terrestrial environments: a community perspective. Ecology 84:2007–2020.

Kritzer, J. P., and P. F. Sale. 2004. Metapopulation ecology in the sea: from Levin's model to marine ecology and fisheries science. Fish and Fisheries 5:131–140.

Lage, C., and I. Kornfield. 2006. Reduced genetic diversity and effective population size in an endangered

Atlantic salmon (*Salmo salar*) population from Maine, USA. Conservation Genetics 7:91–104.

McKay, J. K., and R. G. Latta. 2002. Adaptive population divergence: markers, QTL and traits. Trends in Ecology and Evolution 17:285–291.

Neigel, J. 1997. A comparison of alternate strategies for estimating gene flow from genetic markers. Annual Review of Ecology and Systematics 28:105–128.

Østergaard, S., M. M. Hansen, V. Loeschcke, and E. E. Nielsen. 2003. Long-term temporal changes of genetic composition in brown trout (*Salmo trutta* L.) populations inhabiting an unstable environment. Molecular Ecology 12:3123–3135.

Patterson, W. F., III, J. C. Watterson, R. L. Shipp, and J. H. Cowan, Jr. 2001. Movement of tagged red snapper in the northern Gulf of Mexico. Transactions of the American Fisheries Society 130:533–545.

Pruett, C. L., E. Saillant, and J. R. Gold. 2005. Historical population demography of red snapper (*Lutjanus campechanus*) from the northern Gulf of Mexico based on analysis of sequences of mitochondrial DNA. Marine Biology 147:593–602.

Queney, G., N. Ferand, S. Marchandeau, M. Azevedo, F. Mougel, M. Branco, and M. Monnerot. 2000. Absence of a genetic bottleneck in a wild rabbit (*Oryctolagus cuniculus*) population exposed to a severe viral epizootic. Molecular Ecology 9:1253–1264.

Rice, W. R. 1989. Analyzing tables of statistical tests. Evolution 43:223–225.

Richards, C., and P. L. Leberg. 1996. Temporal changes in allele frequencies and a population's history of severe bottlenecks. Conservation Biology 10:832–839.

Saillant, E., and J. R. Gold. 2006. Population structure and variance effective size of red snapper (*Lutjanus campechanus*) in the northern Gulf of Mexico. Fishery Bulletin U.S. 104:136–148.

Shrimpton, J. M., and D. E. Heath. 2003. Census vs. effective population size in chinook salmon: large- and small-scale environmental perturbation effects. Molecular Ecology 12:2571–2583.

Smith, C. L. 1997. National Audubon Society field guide to tropical marine fishes of the Caribbean, the Gulf of Mexico, Florida, the Bahamas, and Bermuda. Knopf, Inc., New York.

Stepien, C. 1995. Population genetic divergence and geographic patterns from DNA sequences: examples from marine and freshwater fishes. Pages 263–287 *in* J. Nielson, editor. Evolution and the aquatic ecosystem: defining unique units in population conservation. American Fisheries Society, Symposium 17, Bethesda, Maryland.

Szedlmayer, S. T. 1997. Ultrasonic telemetry of red snapper, *Lutjanus campechanus*, at artificial reef sites in the northeast Gulf of Mexico. Copeia 1997:846–850.

Szedlmayer, S. T., and R. L. Shipp. 1994. Movement and growth of red snapper, *Lutjanus campechanus*, from an artificial reef area in the northeastern Gulf of Mexico. Bulletin of Marine Science 55:887–896.

Templeton, A. R. 1998. Nested clade analysis of phylogeographic data: testing hypotheses about gene flow and population history. Molecular Ecology 7:381–397.

Templeton, A. R., E. Routman, and C. A. Phillips. 1995. Separating population structure from population history: a cladistic analysis of the geographical distribution of mitochondrial DNA haplotypes in the tiger salamander, Ambystoma tigrinum. Genetics 140:767–782.

Trippel, E. A. 1995. Age at maturity as a stress indicator of fisheries. BioScience 45:759–771.

Turner, T. F., J. P. Wares, and J. R. Gold. 2002. Genetic effective size is three orders of magnitude smaller than adult census size in an abundant, estuarine-dependent marine fish (*Sciaenops ocellatus*). Genetics 162:1329–1339.

Wang, J. 2001. A pseudo-likelihood method for estimating effective population size from temporally spaced samples. Genetical Research 78:243–257.

Wang, J., and M. C. Whitlock. 2003. Estimating effective population size and migration rates from genetic samples over space and time. Genetics 163:429–446.

Waples, R. S. 1989. A generalized approach for estimating effective population size from temporal changes in allele frequency. Genetics 121:379–391.

Weir, B. S., and C. C. Cockerham. 1984. Estimating *F*-statistics for the analysis of population structure. Evolution 38:1358–1370.

Wilson, C. A., and D. L. Nieland. 2001. Age and growth of red snapper, *Lutjanus campechanus*, from the northern Gulf of Mexico off Louisiana. Fishery Bulletin U.S. 99:653–664.

Woods, M. K., A. J. Fischer, J. H. Cowan Jr., and D. L. Nieland. 2003. Size and age at maturity of female red snapper *Lutjanus campechanus* in the northern Gulf of Mexico. Proceedings of the Gulf and Caribbean Research Institute 54:526.537.

Appendix Table 1. Summary statistics at 19 nuclear-encoded microsatellite loci for the 1999 cohort of red snapper *Lutjanus campechanus* sampled at three localities in the northern Gulf of Mexico. *n* is sample size, *#A* is number of alleles, A_R is allelic richness, H_E is gene diversity (expected heterozygosity), P_{HW} is probability of conforming to expected Hardy-Weinberg genotypic proportions, and F_{IS} is an inbreeding coefficient measured as Weir and Cockerham's (1984) *f*. Boldface indicates significant departures from HW equilibrium following (sequential) Bonferroni correction.

Locus	Texas	Louisiana	Alabama	Locus	Texas	Louisiana	Alabama
*Lca*20				*Prs*240			
n	95	77	63	*n*	95	77	61
#A	3	4	5	#A	18	18	18
A_R	2.59	3.17	3.76	A_R	14.27	14.55	16.22
H_E	0.12	0.20	0.17	H_E	0.90	0.89	0.91
P_{HW}	1.000	0.181	1.000	P_{HW}	0.757	0.282	0.125
F_{IS}	–0.049	0.170	–0.064	F_{IS}	–0.025	0.032	0.024
*Lca*22				*Prs*248			
n	94	77	63	*n*	96	77	63
#A	11	10	10	#A	20	20	15
A_R	8.50	8.39	8.43	A_R	14.03	14.34	11.97
H_E	0.72	0.74	0.73	H_E	0.88	0.90	0.86
P_{HW}	0.081	0.818	0.964	P_{HW}	0.039	**0.000**	0.616
F_{IS}	0.115	–0.067	–0.121	F_{IS}	–0.043	0.087	0.003
*Lca*43				*Prs*257			
n	96	76	63	*n*	93	77	63
#A	9	6	8	#A	14	14	13
A_R	6.51	5.15	6.48	A_R	12.54	13.25	12.18
H_E	0.59	0.53	0.63	H_E	0.89	0.92	0.90
P_{HW}	0.387	0.204	0.014	P_{HW}	**0.001**	0.028	0.683
F_{IS}	0.070	0.086	0.175	F_{IS}	0.021	0.139	–0.033
*Lca*64				*Prs*260			
n	96	77	63	*n*	95	77	63
#A	10	11	9	#A	5	4	5
A_R	6.84	7.64	7.40	A_R	3.54	2.93	4.07
H_E	0.77	0.77	0.78	H_E	0.40	0.28	0.39
P_{HW}	0.581	0.681	0.917	P_H	0.281	1.000	0.945
F_{IS}	–0.009	0.019	–0.010	F_{IS}	0.050	–0.069	–0.093
*Lca*91				*Prs*275			
n	95	77	62	*n*	97	77	63
#A	5	5	7	#A	6	6	7
A_R	4.29	4.15	5.47	A_R	4.42	4.79	5.88
H_E	0.59	0.59	0.60	H_E	0.59	0.56	0.61
P_{HW}	0.912	0.602	0.499	P_{HW}	0.304	0.183	0.230
F_{IS}	–0.056	0.031	0.003	F_{IS}	0.026	–0.059	0.117
*Lca*107				*Prs*282			
n	94	77	61	*n*	96	77	63
#A	11	11	10	#A	13	12	12
A_R	9.05	8.57	9.05	A_R	7.65	8.92	7.39
H_E	0.82	0.80	0.82	H_E	0.62	0.66	0.62
P_{HW}	0.197	0.655	0.040	P_{HW}	0.859	0.314	0.511
F_{IS}	0.043	–0.002	0.086	F_{IS}	0.029	0.136	0.109

Appendix Table 1. (Continued)

Locus	Texas	Louisiana	Alabama
*Prs*55			
n	94	77	63
#A	8	7	3
A_R	3.93	4.09	2.43
H_E	0.25	0.14	0.17
P_{HW}	0.697	0.204	1.000
F_{IS}	0.058	0.151	–0.089
*Prs*137			
n	96	77	63
#A	10	11	11
A_R	7.70	7.88	9.08
H_E	0.69	0.72	0.73
P_{HW}	0.029	**0.001**	0.308
F_{IS}	–0.015	0.154	0.111
*Prs*221			
n	96	77	63
#A	14	8	13
A_R	9.72	8.91	8.57
H_E	0.80	0.79	0.75
P_{HW}	0.610	0.474	0.594
F_{IS}	0.043	0.003	–0.088
*Prs*229			
n	95	77	63
#A	6	7	6
A_R	5.16	5.88	5.73
H_E	0.54	0.59	0.56
P_{HW}	0.522	0.640	0.499
F_{IS}	0.119	0.002	–0.014
*Prs*303			
n	95	76	63
#A	9	8	6
A_R	5.31	5.38	4.28
H_E	0.42	0.39	0.43
P_{HW}	0.261	0.817	0.245
F_{IS}	0.100	–0.038	0.108
*Prs*328			
n	97	77	63
#A	7	5	4
A_R	4.02	3.53	3.42
H_E	0.56	0.54	0.57
P_{HW}	0.558	0.234	0.245
F_{IS}	0.054	0.158	0.022
*Prs*333			
n	94	77	61
#A	6	6	6
A_R	4.27	4.48	4.66
H_E	0.27	0.28	0.34
P_{HW}	0.903	0.885	0.314
F_{IS}	–0.113	–0.064	–0.022

Appendix Table 2. Summary statistics at 19 nuclear-encoded microsatellite loci for the 2000 cohort of red snapper *Lutjanus campechanus* sampled at three localities in the northern Gulf of Mexico. *n* is sample size, #*A* is number of alleles, A_R is allelic richness, H_E *is* gene diversity (expected heterozygosity), P_{HW} is probability of conforming to expected Hardy-Weinberg genotypic proportions, and F_{IS} is an inbreeding coefficient measured as Weir and Cockerham's (1984) *f*. Boldface indicates significant departures from HW equilibrium following (sequential) Bonferroni correction.

Locus	Texas	Louisiana	Alabama	Locus	Texas	Louisiana	Alabama
Lca20				*Prs*240			
n	65	32	44	*n*	63	27	39
#A	3	3	4	#A	17	15	16
A_R	2.61	2.95	3.59	A_R	14.40	15.00	14.45
H_E	0.09	0.12	0.23	H_E	0.88	0.92	0.88
P_{HW}	1.000	1.000	1.000	P_{HW}	0.804	0.011	0.186
F_{IS}	–0.028	–0.033	–0.087	F_{IS}	–0.082	0.098	–0.005
Lca 22				*Prs*248			
n	62	31	43	*n*	65	32	44
#A	11	8	9	#A	18	12	12
A_R	9.44	7.61	7.73	A_R	14.36	11.35	10.74
H_E	0.73	0.67	0.71	H_E	0.90	0.86	0.83
P_{HW}	0.026	0.533	0.933	P_{HW}	0.626	0.275	0.806
F_{IS}	–0.079	–0.201	–0.118	F_{IS}	–0.006	–0.016	–0.119
*Lca*43				*Prs*257			
n	65	32	44	*n*	58	31	40
#A	8	6	7	#A	15	13	14
A_R	6.77	5.95	6.44	A_R	13.30	12.85	13.18
H_E	0.57	0.55	0.59	H_E	0.91	0.90	0.91
P_{HW}	0.852	0.674	0.312	P_{HW}	0.206	0.529	0.607
F_{IS}	0.003	–0.072	–0.028	F_{IS}	–0.007	0.039	–0.041
*Lca*64				*Prs*260			
n	65	32	44	*n*	65	32	44
#A	11	7	7	#A	5	4	4
A_R	8.21	6.82	6.23	A_R	3.72	3.84	3.85
H_E	0.79	0.80	0.78	H_E	0.34	0.40	0.46
P_{HW}	0.935	0.168	0.715	P_{HW}	0.714	0.010	0.814
F_{IS}	0.022	–0.014	0.064	F_{IS}	0.044	0.371	0.016
*Lca*91				*Prs*275			
n	65	30	44	*n*	65	32	44
#A	7	6	5	#A	7	5	5
A_R	4.62	5.70	4.91	A_R	5.38	4.69	4.80
H_E	0.57	0.56	0.62	H_E	0.65	0.59	0.61
P_{HW}	0.088	0.094	0.143	P_{HW}	0.227	0.313	0.079
F_{IS}	–0.072	0.115	0.230	F_{IS}	0.028	0.094	0.215
*Lca*107				*Prs*282			
n	65	31	44	*n*	65	32	44
#A	10	7	9	#A	11	10	10
A_R	8.93	6.87	8.39	A_R	8.48	9.33	9.15
H_E	0.83	0.79	0.79	H_E	0.67	0.67	0.70
P_{HW}	0.774	0.213	0.545	P_{HW}	0.885	0.480	0.340
F_{IS}	0.015	0.027	–0.136	F_{IS}	0.059	0.167	0.062

Appendix Table 2. (Continued)

Locus	Texas	Louisiana	Alabama	Locus	Texas	Louisiana	Alabama
*Prs*55				*Prs*303			
n	65	32	43	*n*	65	32	44
#A	6	2	4	#A	8	6	4
A_R	3.87	2.00	3.26	A_R	6.19	5.53	3.94
H_E	0.15	0.12	0.21	H_E	0.47	0.48	0.39
P_{HW}	1.000	1.000	0.197	P_{HW}	0.479	0.378	0.853
F_{IS}	–0.047	–0.051	0.126	F_{IS}	0.043	0.150	–0.057
*Prs*137				*Prs*328			
n	65	32	44	*n*	65	32	44
#A	10	10	11	#A	5	4	5
A_R	7.52	9.33	9.67	A_R	3.63	3.69	4.22
H_E	0.68	0.76	0.69	H_E	0.54	0.53	0.59
P_{HW}	0.387	0.246	0.820	P_{HW}	0.946	0.874	1.000
F_{IS}	0.057	0.016	0.135	F_{IS}	–0.024	–0.004	–0.001
*Prs*221				*Prs*333			
n	65	32	44	*n*	65	32	44
#A	14	8	13	#A	4	5	4
A_R	10.43	7.77	10.43	A_R	3.46	4.82	3.47
H_E	0.79	0.75	0.80	H_E	0.24	0.36	0.28
P_{HW}	0.654	0.274	0.253	P_{HW}	1.000	1.000	0.197
F_{IS}	–0.012	0.129	0.034	F_{IS}	–0.105	–0.142	–0.057
*Prs*229							
n	65	32	44				
#A	5	5	6				
A_R	4.35	4.69	5.57				
H_E	0.53	0.44	0.55				
P_{HW}	0.269	1.000	0.084				
F_{IS}	0.101	–0.002	0.215				

American Fisheries Society Symposium 60:217–227, 2007

Demographic Differences in Northern Gulf of Mexico Red Snapper Reproductive Maturation: Implications for the Unit Stock Hypothesis

MELISSA W. JACKSON[1, 2]
Department of Marine Sciences
University of South Alabama
Mobile, Alabama 36688 USA

JAMES H. COWAN, JR.
Coastal Fisheries Institute and
Department of Oceanography and Coastal Sciences
School of the Coast and Environment
Louisiana State University
Baton Rouge, Louisiana 70803 USA

DAVID L. NIELAND
Louisiana Sea Grant College Program
Louisiana State University
Baton Rouge, Louisiana 70803 USA

Abstract.—Red snapper, *Lutjanus campechanus*, has been fished for over a century, with management beginning in the Gulf of Mexico (Gulf) in the early 1990s when perceptions of a declining population size surfaced. Red snapper are managed as a unit stock and the fishery management plan is based upon minimal data regarding reproductive output, and size and age at sexual maturation is not certain. Differences in size and age at sexual maturity of red snapper between the northeast and north-central Gulf were evaluated to test whether the population conforms to the unit stock hypothesis. Red snapper were collected during the spawning season in 1999, 2000, and 2001 from the Gulf off Alabama and Louisiana and were used to describe maturation schedules. Progression of oocyte maturation to vitellogenesis was used to define and identify sexually mature females. Combined data showed the smallest mature red snapper was 267 mm fork length (FL) and was two years old. The smallest with hydrated oocytes, indicative of imminent spawning, or postovulatory follicles, indicative of recent spawning, were 285 mm and 297 mm FL respectively, and both were two years old. Red snapper off Alabama reached maturation at smaller sizes and younger ages than those sampled off Louisiana. Growth rates did not differ between the regions. Such differences in maturation schedules may document an important stock response to reductions in population size.

[1] Corresponding author: melissa_woods@hotmail.com.
[2] Current address: P.O. Box 663, Cedar Key, Florida 32625

Introduction

Over decades of exploitation, strong size-selective fishing mortality has resulted in the removal of the largest, most fecund females (Trippel 1995) and in decreases of age- and/or size-at-sexual maturity in a number of fish populations worldwide (Haug and Tjemsland 1986; Jørgensen 1990, Bowering and Brodie 1991; Rijnsdorp 1991, Harris and McGovern 1997; Zhao and McGovern 1997, McGovern et al. 1998, see Sadovy 2001). Such changes in maturation schedules may evidence an important stock response to reductions in population size (Trippel 1995). Although flexibilities in life history traits such as age-at-maturation may help both to increase egg production and to compensate for reduced population numbers, it has been demonstrated that younger fish of some species produce eggs and larvae that are less likely to survive than those produced by older conspecifics (Knutsen and Tilseth 1985; Hislop 1988, Zastrow et al. 1989). This, in addition to reduced population egg production attributable to removal of large females, engenders concerns about whether young females are sufficient in their reproductive capacity to maintain population numbers at sustainable levels while fishing mortality remains high. Thus, effective management of fish populations depends in part upon an understanding of the age and size at which a species becomes sexually mature.

The red snapper *Lutjanus campechanus* has supported important recreational and commercial fisheries in the Gulf of Mexico (Gulf) for over a century. The Gulf red snapper fishery has been intensely managed since the early 1990s under the assumption of a unit stock, which implies that age- and size-at-maturity and maturation rates should vary little Gulf-wide. Despite management efforts, red snapper in the Gulf remain overfished (Goodyear 1995; Schirripa and Legault 1999).

The red snapper is a long-lived reef fish that can exceed 900 mm total length (TL) and can live more than 50 years (Wilson and Nieland 2001). The species is restricted to the South Atlantic Bight and the Gulf, including Mexican waters but not the Caribbean Sea (Hoese and Moore 1977). This gonochoristic species spawns multiple times during a prolonged season. Though much “gray” literature on red snapper exists, little addressing spawning and reproductive biology of the species in the wild has been published in the peer-reviewed literature. Camber (1955), Moseley (1966), and Futch and Bruger (1976) all presented limited data on red snapper length- or age-at-maturity and the duration of the spawning season in the Gulf. Length-at-maturity has been reported to be 320 mm fork length (FL; Camber 1955) and between 190 and 300 mm standard length (SL; Moseley 1966). Wilson et al. (1994) presented preliminary estimates of maturation at 290 mm FL. Age-at-maturity has been reported as greater than 2 years old (Futch and Bruger 1976).

Ovary analyses have also varied among studies. While authors prior to 1976 (Camber 1955; Moseley 1966, Bradley and Bryan 1975) used macro-assessment of gonad condition and maturation, later authors (Futch and Bruger 1976; Wilson and Nieland 1994; Collins et al. 1996) used histological techniques to determine stages of oocyte maturation. Histology is a more effective method for staging oocytes because it allows a detailed view of cross-sectioned and stained oocytes and thus, precise descriptions of oocyte maturation stages can be defined more clearly (West 1990). Despite this precision, problems still exist in the histological approach due to differences in terminology among authors, as well as subtle differences in the progress of oocyte maturation among species.

The purpose of this research was to gain a better understanding of red snapper population dynamics for management purposes. The objectives were to provide data on age- and size-at-reproductive maturity of red snapper in the northern Gulf and to determine if differences in these reproductive variables existed east and west of the Mississippi River over a time span of three years. Our goal was to sample over greater spatial and longer temporal scale than previous studies of red snapper reproductive biology.

Methods

During the recreational fishing seasons (April–October) of 1999, 2000, and 2001, red snapper were sampled by hook and line and from charter boat docks from the Gulf west of the Mississippi River off Fourchon, Louisiana, and east of the Mississippi River off Dauphin Island, Alabama (Figure 1). A minimum total of 600 males and females were annually targeted per region in 1999 and again in 2000, and 300 males and females per region were targeted in 2001. The minimum size of individuals for recreational harvest was initially 381 mm TL, but was increased to 457 mm TL for the last half of the 1999 season. The minimum size was 406 mm TL during the 2000 and 2001 seasons. In addition to these targeted fish from the recreational fishery, a National Marine Fisheries Service (NMFS) permit allowed us to collect undersized red snapper; however, we were able to exert more effort in collecting undersized fish off Alabama than Louisiana. In this study only female individuals are considered.

Fork length (mm) and total weight (TW; to the nearest 0.01 kg) were recorded for each fish sampled. Sagittal otoliths were removed and processed for age analysis following Cowan et al. (1995). Gonads were excised, placed in individually labeled plastic resealable bags, and transported on ice to the laboratory. After the gonads were weighed (nearest 0.1 g), the ovaries were stored in 10% formalin until further analysis.

Red snapper ovaries were examined microscopically to determine age- and size-at-maturity. Oocyte maturation stages were determined using histological analysis. The

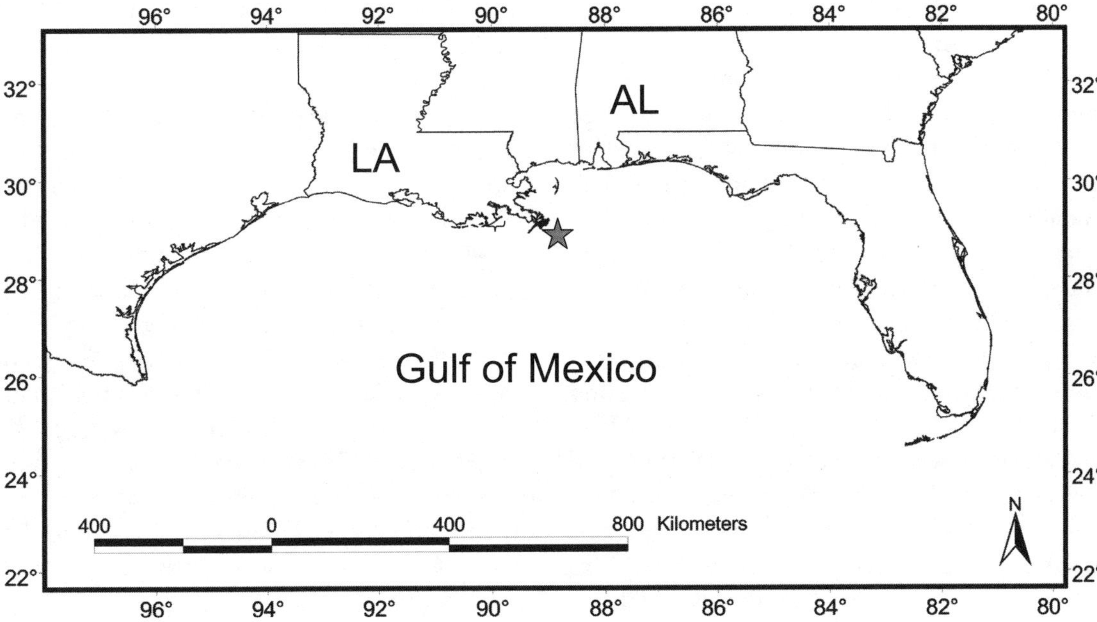

Figure 1. Red snapper *Lutjanus campechanus* were collected east of the Mississippi River outflow (star) plume off Alabama (AL) and west of the river off southeast Louisiana (LA).

lobes of red snapper ovaries are symmetrical (Collins et al. 1996); therefore, a subsample of formalin-fixed ovarian tissue (30–50 g) was dissected from one randomly chosen region of the two lobes (three regions per lobe: anterior, medial and posterior). Each subsample was embedded in Paraplast (Sherwood Medical Industries) and sectioned to 3 μm thickness. Sections were mounted on microscope slides, stained in Gill hematoxylin, and counter-stained in eosin Y.

Oocytes were categorized into one of four oocyte stages by microscopic examination of the prepared histology slides at 40x to 100x magnification. The four stages of oocyte maturation described by Wallace and Selman (1981) are primary growth (PG), cortical alveoli (CA), vitellogenic (V), and hydrated (H). Progression of oocyte maturation to vitellogenesis was used to define and identify mature females (Hunter and Goldberg 1980; Brown-Peterson et al. 1988; Nieland and Wilson 1993). Immature ovaries contained only PG and CA oocytes. Each histological section was also scanned for the presence of post-ovulatory follicles (POF), indicative of recent spawning activity.

Elevated gonadosomatic indices, or the gonad weight as the percentage of ovary free body weight, indicated that spawning began in May and ended by late September; the peak months of the spawning season appeared to be May, June, and July and spawning continued through September (Woods 2003). Thus, maturity analyses were restricted to fish collected in June, July, and August to minimize errors in differentiating between immature and resting or spent ovaries.

Females were classified as mature by size and age. Females were grouped into size classes of 50 mm FL, and into age classes. Statistical Analysis System (SAS Institute 1999) was used for statistical tests. Differences in defined reproductive characteristics between regions were tested by chisquare analysis. The Bonferroni technique was used reduce type I error.

Results

A total of 1,682 female red snapper was collected from the northern Gulf between June and early August from 1999 to 2001; 1,029 were from Alabama, 169 of those were under the legal size limit and collected with permission through a NMFS permit; 653 were from Louisiana, 111 of those were under the size limit and collected through a permit. Alabama females ranged from 237 to 916 mm FL, Louisiana females ranged from 292 to 910 mm FL. Ages of Alabama and Louisiana females ranged from 1 to 34 years and from 2 to 37 years, respectively. A total sample size of 903 females less than 625 mm FL was used for comparisons of length-at-maturity, and 806 females less than 7.5 years for comparisons of age-at-maturity.

The smallest mature Alabama red snapper was 267 mm FL; the smallest with hydrated oocytes, indicative of imminent spawning, and POF were 285 mm and 297 mm FL, respectively. The smallest mature Louisiana red snapper was 292 mm FL; the smallest with hydrated oocytes and POF were 304 and 306 mm FL, respectively. All of these fish were 2 years old.

Combined data generally showed that red snapper greater than 575 mm FL and greater than 5.5 years were 100% mature, with the exception of a 590 mm FL immature individual captured off Louisiana and a 720 mm FL, 7 year old immature individual captured off Alabama. A comparison of the maturity schedules for each region indicates that red snapper captured east of the Mississippi River off Alabama reached 50% maturity before 275 mm FL and before they are 2.5 years old and raw data showed 100% maturity by 575 mm FL and 4.5 years old. Females captured to the west, off Louisiana, reached 50% maturity before 325 mm FL and before they are 2.5 years old and raw data showed 100% maturity by 625 mm FL and 6.5 years old (Figures 2 and 3). Compared to Alabama, Louisiana female

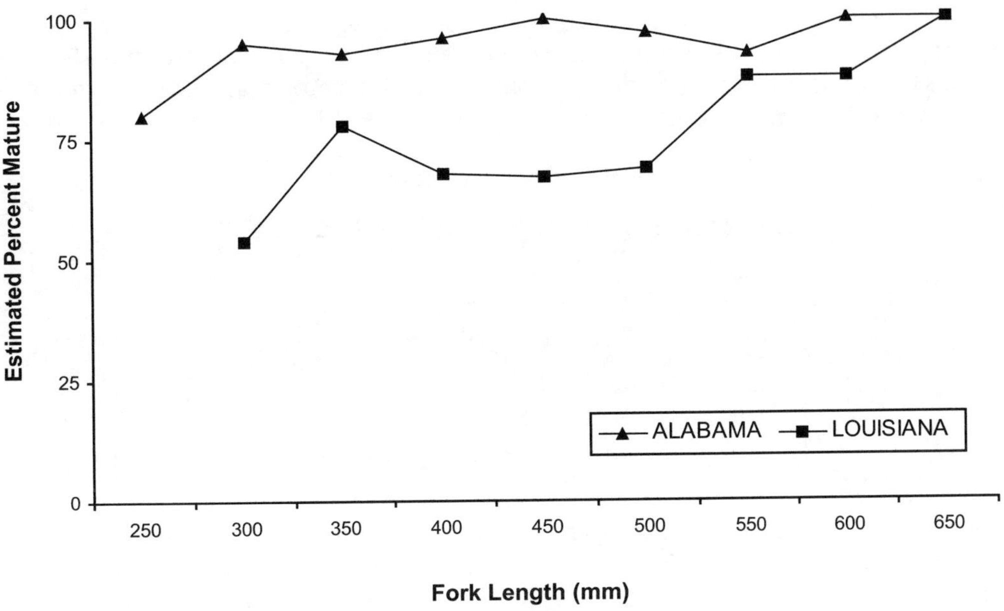

Figure 2. Percent of mature female red snapper *Lutjanus campechanus* sampled off the coasts of Alabama and Louisiana by fork length (mm).

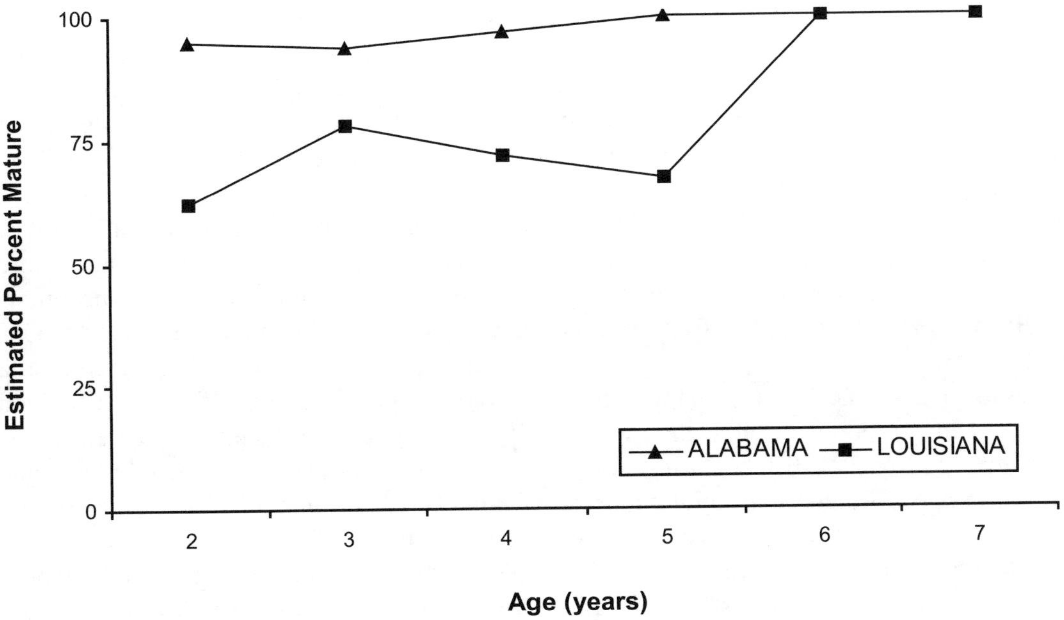

Figure 3. Percent of mature female red snapper *Lutjanus campechanus* sampled off the coasts of Alabama and Louisiana by age (years).

Table 1. Percent of mature female red snapper *Lutjanus campechanus* and sample size (*N*) by size class in 50 mm increments for Alabama and Louisiana. Size classes having significantly different proportions of mature females between states are indicated by (*). For Alabama, the 250 mm size class' smallest fish is 237; the size class is from 237–274. For Louisiana, the 300 mm size class' smallest fish is 292; the size class is from 292–324.

Size Class	Alabama		Louisiana	
FL (mm)	% Mature	*N*	% Mature	*N*
250	77	13		0
300*	89	64	54	13
350*	93	75	75	36
400*	97	133	66	56
450*	100	151	62	89
500*	98	96	63	46
550	94	64	88	17
600	100	39	88	8

red snapper progressed to 100% maturity over size- and age-classes more slowly. Chi-square analyses indicated that Alabama size classes between 300 and 500 mm FL and age classes between 3 and 5 years had significantly more mature females than those Louisiana classes (Tables 1 and 2).

Discussion

While estimates of size- and age-at-50% maturity fall within the broad range of estimates of past studies in different Gulf locations (Camber 1955; Moseley 1966; Futch and Bruger 1976; Wilson and Nieland 1994), only one other study has reported a maturation schedule to 100% maturity. Wilson and Nieland (1994) reported that females sampled from Louisiana reached 100% maturity by 420 mm FL, which differs from results reported here.

Regional differences in red snapper age- and size-at-maturation, as well as the rates at which they approach 100% maturity, could represent a real difference in population demographics of the red snapper east and west of the Mississippi River. Maturity schedules for a species generally are not static and it is rare for all individuals in a fish population to mature at the same age. However, a decrease in age-at-maturity can be an indication of a stressed population caused by compensatory responses to waning population size, or by genetic selection (Trippel 1995). More specifically, such a compensatory response may be due to fishing pressure, predator and prey abundance, stock composition and other biotic and abiotic environmental factors (Wootton 1990).

Although fishing mortality and natural mortality have been estimated east and west of the Mississippi River, these estimates are few and are based upon out-dated knowledge of longevity, and thus must be interpreted with caution. Goodyear (1995) provided revised estimates of Nelson and Manooch's (1982) red snapper natural mortality rates, concluding that both regional estimates were too high to be accurate and that natural mortality is probably less than 0.18 $years^{-1}$. This estimate also may be too high given new data on red snapper longevity (Wilson and Nieland 2001). In regards to fishing mortality, Schirripa and Legault (1999) estimated Gulf-wide fishing mortality to be 2.5–4.5 times higher than that believed to be sustainable. Nelson and Manooch (1982) estimated red snapper fishing mortality off Louisiana to be between 0.58 and 0.74 $year^{-1}$ and for west Florida, between 0.23 and 0.25 $year^{-1}$. More than 10 years later, Watterson (1998) estimated fishing mortality to be 1.14 $year^{-1}$off Alabama when natural mortality was assumed to be 0.20 $year^{-1}$.

Maturity schedules of red snapper populations may vary by region due to differences in compensatory responses. If mortality rates are different, or if the environments in areas off the Alabama and Louisiana coastlines differ, then population demographics could also differ. Greater fishing pressure off Alabama could decrease the density of fish per unit area, thereby decreasing intra-specific competition and allowing for an increase in resource availability. However, the red snapper is believed to be overfished Gulf-wide (Schirripa and Legault 1999) and may have reached its compensatory maximum physiological growth potential, resulting in no observed difference in growth between the regions. As such, increased available energy in Alabama waters may be allocated to reproduction (Wootton 1990; Roff 1992). This would allow fish to reproduce at a smaller size and a younger age in Alabama waters.

Environmental differences also could contribute to demographic dissimilarity between regions. While both regions have a great number of artificial reefs that red snapper are known to inhabit, Alabama's artificial reefs are predominantly small, low vertical relief structures. In contrast, most of Louisiana's artificial reefs are oil and gas platforms that extend vertically to the surface, perhaps increasing their value as habitat. To date, no direct comparisons of oil rigs and lower relief artificial reefs as red snapper habitat have been made. Perhaps more importantly there is a large amount of natural low relief hard bottom off each of the coastlines (Gore 1992) and differences in the amount of structural habitat provided by either type of artificial reef is likely to be negligible.

There also may be differences in the amount of primary and secondary production available to higher trophic levels between the two regions. While coastal waters of both regions are production enhanced by river-dominated estuaries (Mobile Bay and the Mississippi-Atchafalaya Rivers for Alabama and Louisiana, respectively), the Mississippi River system drains 41% of the contiguous United States, discharging a great amount of nutrient-enriched freshwater and sediment into the Gulf (Milliman and Meade 1983); thus, the area of the continental shelf between the Mississippi River and the Yucatan Peninsula is known historically to be extremely high in primary and secondary production due to comparatively high amounts of nutrients discharged by the Mississippi River. Whether higher secondary production off Louisiana acts to affect the maturation schedule of red snapper is

Table 2. Percent of mature female red snapper *Lutjanus campechanus* and sample size (*N*) by age classes for Alabama and Louisiana. Age classes having significantly different proportions of mature females between states are indicated by (*).

Age Class	Alabama		Louisiana	
years	% Mature	*N*	% Mature	*N*
1	0	8		
2	87	55	63	16
3*	93	75	77	117
4*	97	200	65	99
5*	100	86	65	54
6	100	29	97	34
7	96	25	100	8

unknown. However, experiments with other species have shown that differences in food supply can alter the age at which fish become mature. In Pacific herring *Clupea pallasii*, brown trout *Salmo trutta*, and threespine sticklebacks *Gasterosteus aculeatus*, fish receiving higher rations of food matured earlier (Bagenal 1969; Wootton 1973; Hay et al. 1988). It may or may not be the case that red snapper off Louisiana receive greater dietary nutrients due to influence of the Mississippi River. Regardless of their high nutrient environment, they do not appear to be benefiting in a manner that allows them to reproduce at a younger age as do other fishes receiving higher rations of food. Furthermore, recent studies show that over a ten year period, size-at-age of Louisiana red snapper is decreasing (Nieland et al. 2007, this volume).

Shifts in maturation schedules in red snapper may be attributable to genetic selection coincident with increased mortality (Roff 1992). Tagging studies have inferred that red snapper movement may be sufficient to facilitate stock mixing in the northern Gulf (Patterson et al. 2001), but most movement of red snapper off Alabama is to the east and very little movement and mixing occurs west of the Mississippi River (Patterson et al. 2001). Furthermore, genetic studies of red snapper stock in the northern Gulf have not strongly supported genetic differences among regions (Camper et al. 1993; Gold et al. 1997, Heist and Gold 2000; but see Chapman et al. 1995). However, these genetic tests would not indicate differences in a maturity genotype. A genotype for smaller size and younger age-at-maturity can be selected for in some species (Trippel 1995). Early maturing genotypes reproduce before being fully recruited to the fishery. Genotypes that mature at larger sizes or older ages are more likely to be removed before reproduction. In contrast, fish that mature early may participate in one or more spawning seasons before being captured. The progeny for later maturing fish would be selected out of the population over time. Therefore, it may be that the late-maturing genotypes have been removed from the Alabama population due to high fishing pressure along this small coastline in the 1980s and 1990s; this still may be occurring today. This process also may account for differences in size- and age-at-maturity.

Other fish populations have evidenced changes in maturity schedules in response to reductions in population size. Zhao and McGovern (1997) identified changes through time in age- and length-at-maturity, and in growth rate of vermilion snapper *Rhomboplites aurorubens* in the South Atlantic Bight; these changes were probably due to gradual fishing pressure increases that ultimately resulted in growth overfishing. A preliminary study indicated that, compared to specimens sampled from the region five years earlier (Grimes and Huntsman 1980), vermilion snapper decreased in size- and age-at-maturation; however, length-at-age for one and two year old fish changed little (Collins and Pinckney 1988). A temporal comparison of the stock in 1979–1980 to the stock in 1985–1987 also concluded that vermilion snapper declined in size- and age-at-maturity through time (Zhao and McGovern 1997) and that growth rate decreased through time (Zhao et al. 1997). Because vermillion snapper growth rates decreased with time, Zhao and McGovern (1997) suggested that the decrease in size-at-maturity probably was not part of a density-dependent compensatory response to harvesting. More likely, it was a response to growth overfishing caused by the selective removal and incomplete replacement of faster-growing, later-maturing individuals (Zhao and McGovern 1997).

In contrast, Jørgensen (1990) suggested that changes through time in age-at-maturity and growth rate of Atlantic cod *Gadus morhua* were a density-dependent compensatory response to harvesting. The cod population has been intensely exploited for more than 50 years and cod stocks have undergone severe population declines. Jørgensen (1990) determined that despite high fishing pressures, it was the age distribution—and not the size distribution—of the population that changed. It is likely that in response to increased mortality, fish grow faster and mature at a younger age; implicitly, there is a minimum threshold for size-at-maturity (Jørgensen 1990). If this idea is correct, then declines in size- and in age-at-maturity should not necessarily occur simultaneously.

Reznick et al. (2001) presented a case in which populations of guppies, *Poecilia reticulata*, experiencing high and low predation pressures had different population demographics. Guppies experiencing high predation had a smaller size-and younger age-at-maturity, and a faster growth rate than fish experiencing low predation pressure because higher levels of resource availability existed as an indirect consequence of high predation. The populations experienced different mortality, lived in different environments, and due to mortality and environmental differences, food availability differed. Guppies at sites of low-predation did not have a greater density per unit area but had more large, old fish and fewer small, young fish, and thus, greater biomass.

Although a difference in mortality due to fishing is likely to be the cause of differences in size- and age-at-maturity in regions east of the Mississippi River off Alabama and west of the River off Louisiana, no definitive conclusion can be made at this time. It is evident that differences in population demographics do exist between the two regions based upon maturation schedules. Furthermore, studies off the Louisiana coast over a 10 year time span show a decrease in total length of red snapper at ages 4, 5, and 6 years (Nieland et al. 2007). Implications of these differences in future stock assessments should be considered, along with other characteristics of red snapper growth and reproductive biology. To prevent regional overfishing, and to learn more about stock dynamics, an adaptive management approach should be considered.

For example, if red snapper reproductive characteristics differ east and west of the Mississippi River due to unsustainable fishing mortality rates off Alabama and northwest Florida, then an adaptive management approach decreasing fishing mortality off coastal Alabama should be enforced to prevent regional overfishing. Moreover, if red snapper reproductive characteristics respond to become more like those of the females off Louisiana through time, then much can be learned about how fish stocks respond to changes in fishing pressure. In contrast, if reproductive characteristics remain unchanged, then the environment may be driving differences in inherent reproductive capacity. Should this be true, a more environmental-based ecosystem approach to management could be indicated. Either way, our understanding of population dynamics would be enormously improved.

Acknowledgments

This research was a partial fulfillment of M. W. Jackson's thesis research at the University of South Alabama. It was funded through the Department of Commerce, Marine Fisheries Initiative Grant No. NA87FF0424 and Environmental Defense. Yvonne Allen provided Figure 1. We thank R. Shipp, A. Fischer, S. Baker, J. McCawley, A. Strelcheck, W. Patterson, W. Ingram and many students and staff at Dauphin Island Sea Laboratory and LSU for their help in the field and the laboratory. We are grateful to Captains Mike Thierry and Steve Tomeny and crews and the many anglers for their help in collecting fish.

References

Bagenal, T. B. 1969. The relationship between food supply and fecundity in brown trout *Salmo trutta* L. Journal of Fish Biology 1:169–182.

Bowering, W. R., and W. B. Brodie. 1991. Distribution of commercial flatfishes in the Newfoundland-Labrador region of the Canadian northwest Atlantic and changes in certain biological parameters since exploitation. Netherlands Journal of Sea Research 27 (3–4): 407–422.

Bradley, E., and C. E. Bryan. 1975. Life history of the red snapper (*Lutjanus campechanus*) in the northwestern Gulf of Mexico: 1970–1974. Proceedings of the 27th Annual Gulf and Caribbean Fisheries Institute and the 17th Annual International Game Fish Research Conference, Miami Beach, Florida.

Brown-Peterson, N., P. Thomas, and C. R. Arnold. 1988. Reproductive biology of the spotted sea trout, *Cynoscion nebulosus*, in South Texas. Fishery Bulletin 86:373–388.

Camber, C. I. 1955. A survey of the red snapper fishery of the Gulf of Mexico, with special reference to the Campeche Banks. State of Florida Board of Conservation Technical Series No. 12.

Camper, J. C., R. C. Barner, L. R. Richardson, and J. R. Gold. 1993. Mitochondrial DNA variation among red snapper (*Lutjanus campechanus*) from the Gulf of Mexico. Molecular Marine Biology and Biotechnology 2(3):154–161.

Chapman, R. W., S. A. Bortone, and C. M. Woodley. 1995. A molecular approach to stock identifiacation and recruitment patterns in red snapper, *Lutjanus campechanus.* Final report for Cooperative Agreement NA17FF0379–03 Marine Fisheries Initiative (MARFIN) Program. Institute for Coastal and Estuarine Research, University of West Florida, Pensacola, Florida, and Marine Resources Research Institute, South Carolina Wildlife and Marine Resources Department, Charleston, South Carolina.

Collins, L. A., A. G. Johnson, and C. P. Keim. 1996. Spawning and annual fecundity of the red snapper (*Lutjanus campechanus*) from the northern Gulf of Mexico. Pages 174-188 *in* F. Arreguin-Sanchez, J. L. Munro, M. C. Balgos, and D. Pauly editors. Biology, fisheries and culture of tropical groupers and snappers. ICLARM Conference Proceedings 48.

Collins, M. R., and J. L Pinckney. 1988. Size and age at maturity for vermilion snapper (*Rhomboplites aurorubens*) in the South Atlantic Bight. Gulf of Mexico Science 10:51–53.

Cowan, J. H., Jr., R. L. Shipp, K. B. Bailey, IV, and D. W. Haywick. 1995. Procedure for rapid processing of large otoliths. Transactions of American Fisheries Society 124:280–222.

Futch, R. B., and G. E. Bruger. 1976. Age, growth, and reproduction of red snapper in Florida waters. Pages 165–184 *in* H. R. Bullis and A. C. Jones, editors. Proceedings: Colloquium on snapper-grouper fishery resources for the western central Atlantic Ocean. Florida Sea Grant College Report 17.

Gold, J. R., F. Sun, and L. R. Richardson. 1997. Population structure of red snapper from the Gulf of Mexico as inferred from analysis of mitochondrial DNA. Transactions of American Fisheries Society 126:386–396.

Goodyear, C. P. 1995. Red snapper in U.S. waters of the Gulf of Mexico. Contribution MIA-95/96–05. Southeast Fisheries Center, Miami Laboratory, Miami, Florida.

Gore, R. 1992. The Gulf of Mexico. Pineapple Press, Inc., Sarasota, Florida.

Grimes, C. B., and G. R. Huntsman. 1980. Reproductive biology of the vermilion snapper, *Rhomboplites aurorubens*, from North Carolina and South Carolina. Fishery Bulletin 78:137–146.

Harris, P. J., and J. C. McGovern. 1997. Changes in the life history of porgy, *Pagrus pagrus*, from the southeastern United States, 1972–1994. Fishery Bulletin 95:732-747.

Hay, D. E., J. R. Brett, E. Billinsky, D. T. Smith, E. M. Donaldson, G. A. Hunter, and A. V. Solemie. 1988. Experimental impoundments of prespawning Pacific herring (*Clupea harengus pallasi*): effects of feeding and density on maturation, growth, and approximate analysis. Canadian Journal of Fisheries and Aquatic Sciences 45:388–398.

Heist, E. J. and J. R. Gold. 2000. DNA microsatellite loci and genetic structure of red snapper in the Gulf of Mexico. Transactions of American Fisheries Society 129:469-475.

Hislop, J. R. G. 1988. The influence of maternal length and age on the size and weight of the eggs and relative fecundity of haddock, *Melanogrammus aeglefinus*, in British waters. Journal of Fish Biology 32(6):923–930.

Hoese, H. D. and Moore R. H. 1977. Fishes of the Gulf of Mexico. Texas A&M University Press, College Station, Texas.

Hunter, J. R., and S. R. Goldberg. 1980. Spawning incidence and batch fecundity in northern anchovey, *Engraulis mordax.* Fisheries Bulletin 77:641–652.

Jørgensen, T. 1990. Long-term changes in age at sexual maturity of northeast Arctic cod (*Gadus morhua*). Journal du Conseil International pour l'Exploration de la Mer 46:235–248.

Knutsen, G. M., and S. Tilseth. 1985. Growth, development and feeding success of Atlantic cod larvae *Gadus morhua* related to egg size. Transactions of American Fisheries Society 114:507–511.

Mace, P. M. 1994. Relationships between common biological reference points used as thresholds and targets of fisheries management strategies. Canadian Journal of Fisheries and Aquatic Sciences 51:110–122.

McGovern, J. C., D. Wyanski, O. Pashuk, C. S. Manooch, G. R. Sedberry. 1998. Changes in the sex ratio size at maturity of gag, *Mcyteroperca microlepis*, from the Atlantic coast of the southern U.S. during 1976–1995. Fishery Bulletin 96:797–807.

Milliman, J. D., and Meade, R. H. 1983. World-wide delivery of river sediment to the ocean. Journal of Geology 91:121.

Moseley, F. N. 1966. Biology of red snapper, *Lutjanus aya*, of the northwestern Gulf of Mexico. Publications of the Institute of Marine Science 11:90–101.

Nelson, R. S. and C. S. Manooch. 1982. Growth and mortality of red snappers in the west-central Atlantic Ocean and northern Gulf of Mexico. Transactions of American Fisheries Society 111:465–475.

Nieland, D. L., C. A. Wilson, III, and A. J. Fischer. 2007. Declining size at age among red snapper in the northern Gulf of Mexico off Louisiana, USA: recovery or collapse? Pages 329–335 *in* W. F. Patterson, III, J. H. Cowan, Jr., G. R. Fitzhugh, and D. L. Nieland, editors. Red snapper ecology and fisheries in the U.S. Gulf of Mexico. American Fisheries Society, Symposium 60, Bethesda, Maryland.

Nieland, D. L. and C. A. Wilson. 1993. Reproductive biology and annual variation of reproductive variables of black drum in the northern Gulf of Mexico. Transactions of American Fisheries Society 122:318–327.

Patterson, W. P., J. C. Watterson, and R. L. Shipp, J. H. Cowan, Jr. 2001. Movement of tagged red snapper in the northern Gulf of Mexico. Transactions of American Fisheries Society 130:533–545.

Reznick, D., M. J. Butler, and H. Rodd. 2001. Life-history evolution in guppies, the comparative ecology of high- and low-predation environments. American Naturalist 157:126–140.

Rijnsdorp, A. D. 1991. Reproductive variability in North Sea plaice, sole, and cod. Journal du Conseil International pour l'Exploration de la Mer 47:352–375.

Roff, D. A. 1992. The Evolution of life histories: theory and analysis. Chapman and Hall, London.

SAS Institute. 1999. Statsistical Analysis System. Software package: Version 8. SAS Institute Inc. Cary, North Carolina.

Sadovy, Y. 2001. The threat of fishing to highly fecund fishes. Journal of Fish Biology 59(A):90–108.

Schirripa, M. J. and C. M. Legault. 1999. Status of the red snapper in U.S. waters of the Gulf of Mexico: updated through 1998. National Marine Fisheries Service. Southeast fisheries Science Center, Miami Laboratory, Miami SFD-99/00–75.

Trippel, E. A. 1995. Age at maturity as a stress indicator in fisheries. BioScience 45:759-771.

Wallace, R. A. and K. Selman. 1981. Cellular and dynamic aspects of oocyte growth in teleosts. American Zoologist 21:325–343.

Watterson, C. 1998. Movement of red snapper, *Lutjanus campechanus*, in the north central Gulf of Mexico: potential effects of hurricanes. Master's thesis. University of South Alabama, Mobile, Alabama.

West, G. 1990. Methods of assessing ovarian development in fishes: a review. Australian Journal of Marine and Freshwater Resources 41:199–222.

Wilson, C. A. and D. L. Nieland. 1994. Reproductive biology of red drum, *Sciaenops ocellatus*, from the neritic waters of the Gulf of Mexico. Fishery Bulletin. 92:841-850.

Wilson, C. A. and D. L. Nieland. 2001. Age and growth of red snapper, *Lutjanus campechanus*, from the northern Gulf of Mexico off Louisiana. Fishery Bulletin 99:653–664.

Woods, M. K. 2003. Demographic differences in reproductive biology of female red snapper in the northern Gulf of Mexico. Master's thesis. University of South Alabama. Mobile, Alabama.

Wootton, R. J. 1973. Effect of size of food ration on egg production in the female threespined stickleback, *Gasterosteus aculeatus* L. Journal of Fisheries Biology 5:683–688.

Wootton, R. J. 1990. Ecology of teleost fishes. Chapman and Hall, New York.

Zastrow, C. E., E. D. Houde, and E. H. Saunders. 1989. Quality of striped bass (*Morone saxatillis*) eggs in relation to river source and female weight. Rapports et Proces-Verbaux des Reunions de la Commission Internationale pour l'Exploration Scientifique de la Mer. 191:34–42.

Zhao, B. and J. C. McGovern. 1997. Temporal variation in sexual maturity and gear-specific sex ratio of the vermilion snapper, *Rhomboplites aurorubens* in the South Atlantic Bight. Fishery Bulletin 95:837–848.

Zhao, B., J. C. McGovern, and P. J. Harris. 1997. Age, growth, and temporal change in size at age of the vermilion snapper from the South Atlantic Bight. Transactions of the American Fisheries Society 126:181–193.

American Fisheries Society Symposium 60:229–243

Modeling the Dependence of Batch Fecundity on Size and Age for Use in Stock Assessments of Red Snapper in U.S. Gulf of Mexico Waters

Clay E. Porch[1]
National Marine Fisheries Service
Southeast Fisheries Science Center, Sustainable Fisheries Division
75 Virginia Beach Drive, Miami, Florida 33149 USA

Gary R. Fitzhugh, Michelle S. Duncan, and L. A. Collins
National Marine Fisheries Service
Panama City Laboratory, 3500 Delwood Beach Road
Panama City, Florida 32408 USA

Melissa W. Jackson[2]
Department of Marine Sciences
University of South Alabama, Mobile, Alabama 36688 USA

Abstract.—The red snapper *Lutjanus campechanus* exhibits rapid growth early in life, yet is a relatively long-lived species with an indeterminate spawning pattern. Batch fecundity (*BF*), a principal determinant of the reproductive potential of indeterminate spawners, appears to increase geometrically with length and asymptotically with age based upon combined data from National Marine Fisheries Service (NMFS) and Marine Fisheries Initiative (MARFIN) reproductive sampling programs. As the life history pattern sets up a potential disconnection between size and age, we modeled their interaction as explanatory variables of *BF*. Visual analysis suggests a dome-shaped relationship between *BF* and age within a given size-class. To test this, a general log-linear model and a new "standardized-age" model (where the effect of age depends explicitly on the size of the fish) were fitted to data. These analyses suggest that the effect of age, while statistically significant, was relatively small (the models with age and length terms explaining only slightly more of the variation in *BF* than a simple power function of length). The age effect seems to be most pronounced for fish that are exceptionally small or exceptionally large given their age, which constitute a small fraction of the sample and presumably also a small fraction of the population at large. Hence, it seems unlikely that including this interaction would have any important ramifications for stock assessments of red snapper. Nevertheless, the effect of age on *BF* remains open to the degree that old fish (i.e., greater than age 13) were rare in the combined data set and tended to exhibit lower *BF* values than predicted. Age effects of other reproductive determinants remain to be evaluated but we found hydrated females to be of greater size-at-age than nonhydrated red snapper suggesting an age-length interaction on batch frequency.

[1] Corresponding author: clay.porch@noaa.gov
[2] Current address: P.O. Box 663, Cedar Key, Florida 32625

Introduction

Many fish species with prolonged spawning periods exhibit an indeterminate spawning pattern, where annual fecundity is not fixed at the onset of spawning and previtellogenic oocytes can continue to mature and be spawned during the spawning season (Hunter et al. 1985; Murua et al. 2003). For these species, total fecundity may be modeled on a per capita basis as the product of the proportion female, proportion mature, batch fecundity, batch frequency and perhaps some measure of the quality of the eggs. Each of these variables is likely to change in some way over the lifetime of an individual. Batch fecundity (*BF*), for example, is known to be highly correlated with fish size (Bagenal 1978; Hunter et al. 1985) and is commonly modeled as a linear or power function of length or weight.

Recent stock assessments of red snapper *Lutjanus campechanus*, like most stock assessments, are structured in terms of age rather than size. Accordingly, the functional dependence on size must be extrapolated to age by use of a growth curve or age-length key. Of course such an approach implicitly assumes *BF* and age are correlated only to the extent that size and age are correlated; i.e., it assumes age has no direct bearing on the number of eggs produced. However, it has sometimes been observed that the extrapolated fecundity-age relationship differs from the relationship inferred when fecundity and age are compared directly. In some cases the differences can be attributed to the method used to convert size to age, but in others the effects of size and age appear genuinely disconnected (Goeman 1983; Koslow et al. 1995; Eenennaam and Doroshov 1998; Holmgren 2003).

The red snapper would seem to be a good candidate for detecting age and size effects on fecundity. From a life history standpoint, red snapper, like some other lutjanids, mature at a young age (Jackson et al. 2007, this volume), grow rapidly during the first few years of life, grow slowly thereafter and can live for more than 50 years (Wilson and Nieland 2001; Newman 2001; Kritzer 2004). Red snapper have a fairly prolonged spawning period (several months) and oocyte development is consistent with an indeterminate fecundity pattern (Collins et al. 1996) thus necessitating a *BF* approach (see Murua et al. 2003). We have combined the results of two reproductive studies to obtain a relatively large data set with over 500 estimates of *BF* at age and size, which is not common in the fish literature. Our objective was to examine the interaction of size and age as explanatory variables of *BF* and ultimately gauge how this might affect estimates of red snapper reproductive output.

Methods

Data

Red snapper reproductive samples were obtained during a National Marine Fisheries Service (NMFS) long-term study of reproduction (Fitzhugh et al. 2004) and from a Marine Fisheries Initiative Program (MARFIN) study of stock differentiation in the northern Gulf of Mexico (Cowan et al. 2002; Woods 2003; Jackson et al. 2007). During both studies, ovaries were processed for histological observation to identify hydrated oocytes for *BF* estimates following methods used by Nieland and Wilson (1993) and Collins et al. (1996). Histological slides were prepared by Louisiana State University School of Veterinary Medicine, Department of Pathobiological Sciences.

Histological slides were examined microscopically at 32x to 800x magnification to determine oocyte maturation. Using the oocyte maturation characteristics described by Wallace and Selman (1981), oocytes were staged accordingly to determine the leading oocyte stage. Females displaying vitellogenic or more advanced oocytes (yolked oocytes) during the peak spawning months of June–August were defined as mature. Females were classified as "spawning" depending upon the presence of hydrated oocytes, indicative of imminent spawning, or postovulatory follicles (POF), indicative of recent spawning (Hunter and Macewicz 1985). Batch fecundity was determined using the hydrated oocyte method described by Hunter et al. (1985). Any sections showing recent postovulatory follicles, suggesting the female had partially spent her current batch, were eliminated from the fecundity estimates.

The data are summarized by age and size in Table 1.

Analyses

The relationship between *BF* and total length (l) has often been modeled with the power function

$$(1)\quad BF = al^b e^{\varepsilon}$$

where e^{ε} represents a multiplicative model error (in which case the parameters a and b are usually estimated by a linear regression of the logarithm of *BF* on the logarithm of l). For stock assessment purposes, this length-based model has been converted into an age-based function by use of a von Bertalanffy growth equation

$$(2)\quad BF = a\left(l_{\infty}(1 - e^{-k(t-t_0)})\right)^b$$

where t is age, l_{∞} is the asymptotic length, k is the growth rate coefficient and t_0 is the intercept.

The fecundity and growth parameters in equation (2) have usually been estimated independently through separate regressions of *BF* on length and length on age, respectively. In principle however, equation (2) may be collapsed into the simpler form

$$(3)\quad BF = c\left(1 - e^{-kt}\right)^b e^{\varepsilon}$$

and the parameters estimated through a single regression on age. The variables a and l_{∞} in equation (2) have been combined into a single variable c because they would otherwise be completely confounded. Similarly, the parameter t_0 is absent because b and t_0 are highly correlated and nearly identical curve shapes can be achieved with t_0 fixed to zero. Note that if fecundity were primarily a function of length, then equation (3) would not be expected to fit the observations of *BF* as well as equation (1) except in the unlikely event that the variation in growth among individuals is negligible. Conversely, if fecundity were primarily an increasing or asymptotic function of age, then equation (3) might be expected to fit the data better than equation (1).

It is possible, of course, that *BF* depends to an important extent on both length and age. One method that has been suggested to test this possibility is to regress a measure of *BF* that has been corrected for size against age. For example, let us suppose that the primary factor controlling the number of oocytes that can be hydrated at any given time is the size of the body cavity, then to the extent that the size of the body cavity is correlated with body weight (total weight minus gonad weight, w), one might expect $BF = awe^{\varepsilon}$ (ε = a random error term). Suppose further that age somehow impacts *BF* in a way that is independent of body weight such that $BF = awf_t e^{\varepsilon}$ (f_t = some arbitrary function of age). In that case the effect of age should be clear from a plot of BF/w against t (because $BF/w = af_t e^{\varepsilon}$). Unfortunately, the clarity of this approach is muddled considerably when *BF* is a nonlinear function of w or when the effects of t and w are not independent. For example, suppose the true relationship were $BF = aw^b f(t)\, e^{\varepsilon}$, then to the extent that b differs from unity, the trend re-

Table 1a. Average red snapper *Lutjanus campechanus* batch fecundity in thousands of eggs by age and length (midpoint of 2 cm size intervals).

	Age												
TL (cm)	2	3	4	5	6	7	8	9	10	11	12	13	16+
30	14												
32	93												
34	124		7										
36	180	82											
38	299	459											
40	498	1621	688										
42	385	1356	1229	33									
44	517	1578	1183		67								
46	221	2113	3388										
48	318	2171	3015	602	302								
50	306	787	3354	517									
52		602	2795	844	644								
54		375	1934	301									
56		1024	2954	1108									
58		280	1805	1121									
60			1860	743	236								
62			2458	1845	478								
64		240	513	642	579								
66		224	375	3426	475	204							
68				1162	1544		441						
70			1674	2433	1247	1574			84				
72			1327	635	4314								
74				2546	1238	888	2699	864					
76			1084	4056	6004	4560		300		1190		744	
78				878	4996	2681	2563	551					
80					1271	3799	161	2237	1824	1423	512		
82				550	3439	60	1012		1894	1200			
84						2334			3205	2131	2621		
86					135	260	295	999	3207	3387			
88						860	1298			2425		7980	1642
90						1018		1141		2237			1280
96													3442

vealed by plotting BF/w on t will be contaminated by the effect of weight (because $BF/w = aw^{b-1}f_t e^{\varepsilon}$).

An approach that is perhaps more helpful is to view the data by age and size simultaneously, either by use of summary tables such as Table 1 or by constructing three-dimensional plots as illustrated in Figure 1. From this vantage the relationship between red snapper BF and length appears to be geometric and the relationship between BF and age is roughly asymptotic, just as one might expect when fecundity is dependent primarily on length and length increases asymptotically with age (owing to growth). Within a given size-class, however, the relationship between BF and age appears to be dome-shaped. Older fish seem to be more productive than younger fish of the same size up to a point, after which they are much less productive. In other words, fish that are unusually old for their size (small for their age) are relatively unproductive.

Table 1b. Number of red snapper sampled by age and length (midpoint of 2 cm size intervals).

	Age												
TL (cm)	2	3	4	5	6	7	8	9	10	11	12	13	16+
30	1												
32	4												
34	4		2										
36	9	2											
38	6	7											
40	8	26	4										
42	11	16	8	1									
44	4	20	8		1								
46	3	19	14										
48	3	21	21	3	1								
50	2	5	20	3									
52		5	12	3	2								
54		6	17	2									
56		4	13	5									
58		2	7	3									
60			7	2	1								
62			9	7	1								
64		1	1	2	1								
66		1	2	8	2	1							
68				3	4		1						
70			4	5	2	2			1				
72			2	2	7								
74				4	3	1	2	2					
76			1	5	8	4		1		1		1	
78				1	6	4	3	1					
80					2	4	1	1	1	1	1		
82				1	3	1	1		1	1			
84						2			2	1	2		
86					1	1	1	1	2	1			
88						1	1			1		1	2
90						1		1		1			5
96													2

In the absence of a theoretical model with a physiological basis, a number of authors have tested for suspected age effects by use of generalized log-linear models of the form

$$(4)\quad BF = e^{\alpha + \sum_{j,k} \beta_{jk} t^j l^k + \varepsilon}$$

where j and k index the order of the various polynomial terms for age, length or weight (e.g., Messieh 1976; Bowering 1980; Eldridge and Jarvis 1995; Nitschke et al. 2001). The parameters in (4) may be estimated by a multivariate regression on age and length, in which case important age effects should be evidenced by statistically significant estimates for the β_j coefficients. The primary weaknesses of this approach are the limited ability to model complex age-length interactions and lack of any physical interpretation.

The dome-shaped patterns evident in Figure 1 suggest the effect of age and its interaction with size might be better tested by

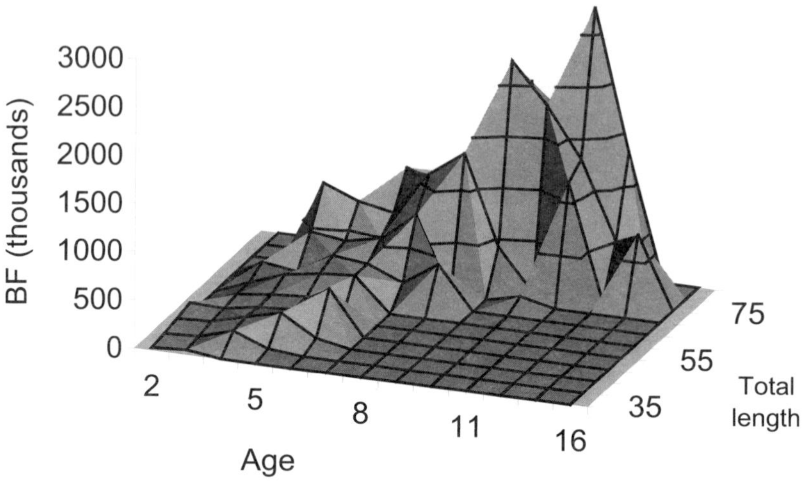

Figure 1. Plot of red snapper *Lutjanus campechanus* mean batch fecundity (*BF*) in thousands of hydrated oocytes against age (years) and total length (in 5 cm categories).

a model that takes explicit account of the degree to which the age of an individual departs from the expectation given its size. With this in mind, we suggest a model of the form

$$(5) \quad BF = al^b\left(1+\sum_{k=1}^{3}\beta_k\left(\frac{t-\bar{t}_l}{\sigma_t}\right)^k\right)e^{\varepsilon}$$

$$t = \bar{t}_l e^{\eta}$$

Here $\bar{t}_l$ is the age expected for a given length (modeled by the inverted von Bertalanffy equation) and σ_t is the standard deviation of age with respect to length (here modeled as a constant coefficient of variation $\sigma_t = \phi\, \bar{t}_l$). The interaction between age and length is implicit in the term $(t - \bar{t}_l)/\sigma_t$. The polynomial function allows for the possibility of a dome-shaped relationship. Hereafter, the term $(t - \bar{t}_l)/\sigma_t$ will be referred to as the standardized age inasmuch as it reflects the deviation in age from the mean measured in standard deviation units. It would be expected to be approximately normal distributed with mean 0 and standard deviation 1.

The parameters in (5) were estimated with the two-step approach that is often used to extrapolate length-based models to age. The growth parameters, $\bar{t}(l)$ and σ_t , were first estimated with a nonlinear regression of ln*t* on *l* (assuming age is an inverted von Bertalanffy function of length). The fecundity parameters (*a,b,β*) were then estimated by a nonlinear regression of ln*BF* on age and length using the previously estimated values for $\bar{t}(l)$ and σ_t.

Statistical comparisons among alternative models were made using Akaike's information criterion (*AIC*; Akaike 1973):

$$AIC = -2lnL + 2p ,$$

where *p* is the total number of parameters estimated and *L* is the measure of goodness of fit (e.g., likelihood function) being maximized. The *AIC* attempts to identify the most parsimonious explanation of the data by balancing the relative improvement in model fit against the number of parameters required to achieve that fit. The 'best' model is considered to be the one with the lowest *AIC*. A rule of thumb is that differences in *AIC* of less than 2 constitute weak evidence that one model is better than another, differences between 3 and 10 are regarded as moderate evidence, and differences greater than 10 are

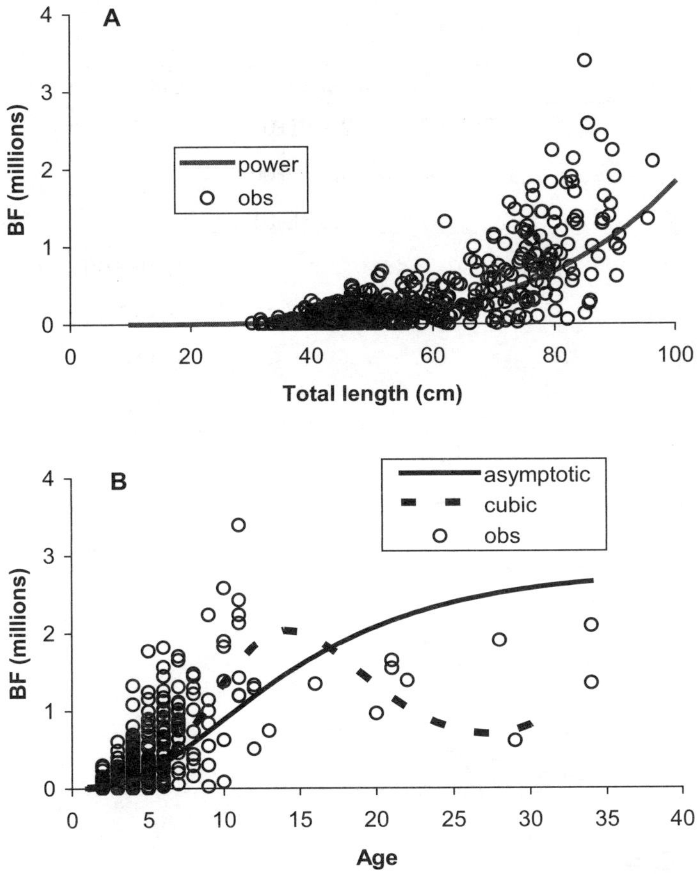

Figure 2. Observed and predicted values of red snapper *Lutjanus campechanus* batch fecundity (*BF*), in millions of hydrated oocytes, expressed as a function of A) total length (cm) and B) age (years). For convenience of scale, an 84 cm, age 13 female exhibiting a high *BF* estimate of 8 million ova is not shown on the figures but was included in the equation fits. Equation parameters are indicated in Table 1.

regarded as strong evidence (Burnham and Anderson 2002).

Results

Plots of red snapper *BF* against total length and age are shown in Figures 2a and 2b, respectively. Fecundity appears to increase geometrically with length and more or less asymptotically with age. The length-based power function (equation 1) fit the data significantly better than the age-based exponential function (equation 3) according to the Akaike's information criteria (see Table 1), but explained only slightly more of the variation in *BF* (r^2 = 42% versus 37%). Interestingly, each *BF* for red snapper over 15 years of age was consistently less than expected (below the fitted curve) and generally less than the average *BF* for 11 year-old fish. This suggested the possibility of a dome-shaped relationship between *BF* and age, which might be modeled better by a polynomial function analogous to equation (4) (but without the length terms). Indeed, a cubic polynomial was found that fit the

Table 2. Fitted length-based power function compared with age-based asymptotic and cubic models for red snapper. 'RMSE' is root mean square error.

Statistic	power	asymptotic	cubic
a, α	2.78E-03	2.80E+06	8.08E+00
b, β	4.41E+00	3.33E+00	
K		1.23E-01	
β(t)			1.10E+00
$\beta(t^2)$			−5.86E-02
$\beta(t^3)$			9.40E-04
RMSE	1.34	1.39	1.38
R^2	0.42	0.37	0.38
AIC	821.3	861.8	856.9

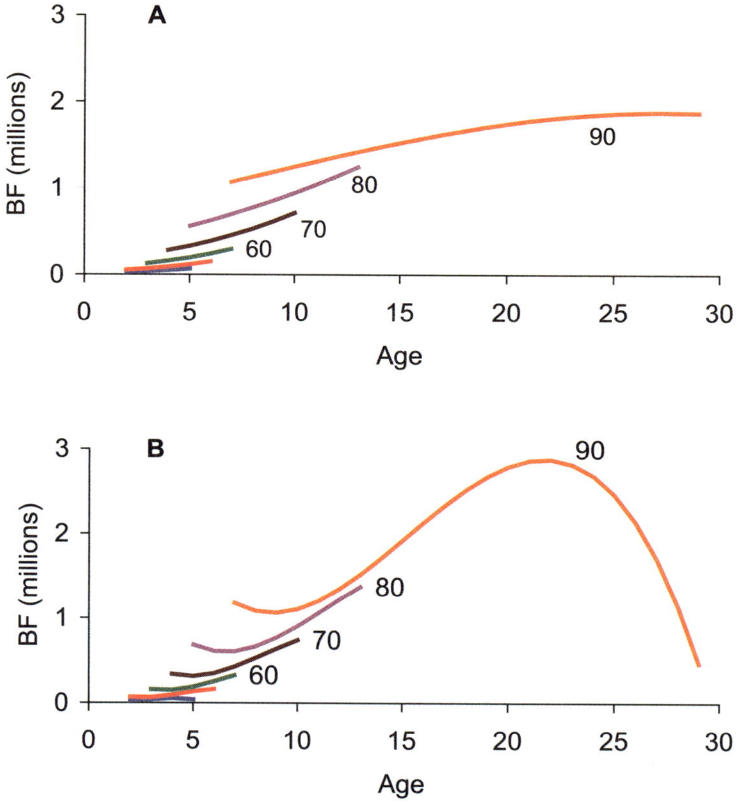

Figure 3. Model predictions of red snapper *Lutjanus campechaunus* batch fecundity (*BF*) in millions of hydrated oocytes, by age for selected total lengths ranging from 40 to 90 cm (inset numbers). Panel (A) refers to the results from the log-linear model and panel (B) refers to the results from the standard age model. The range of ages shown for each length category corresponds to the range of ages available in the data.

data significantly better than the asymptotic model (see Table 1); however, it explained only slightly more of the variation in *BF* (r^2 = 38.3%) and behaved unreasonably for the oldest ages (Figure 2b). The quadratic model was more stable than the cubic and captured the desired dome shape, but did not fit as well as the asymptotic model.

The relative importance of age and length, and the potential importance of an age-length interaction, was examined further by building a log-linear model of the form given in equation (4). The model building exercise proceeded in stepwise fashion, starting with the intercept and adding successively higher terms if they contributed significantly to the model's ability to fit the data (as judged by *AIC*). Important age-length interactions, if they exist, should be evidenced by nonzero $\beta_{j,k}$ ($j,k > 0$) coefficients. The results for select models are shown in Table 2. The simple log-linear function of length provided a fit to the data that was comparable to the power function discussed earlier. The addition of several age and length coefficients resulted in modest decreases in the *AIC*. The most parsimonious log-linear model (quadratic terms for age, linear terms for length, and a linear age/length interaction) produced an *AIC* value of 817, indicating moderate statistical evidence for an age effect. However, the r^2 for this model (43%) was only slightly better than the 42% achieved by the simple power function of length. The estimated impact of age (over and above length) is demonstrated graphically in Figure 3a.

The standard-age model (Table 3) proved to be the most parsimonious of all the models examined (*AIC* = 810). It was better able to

Table 3. Selected log-linear models for red snapper *Lutjanus campechaunus* developed during stepwise model building procedure (based on the log-linear model in equation 4). Shaded region highlights the most parsimonious log-linear model. The abbreviation 'quad.' refers to a model with only quadratic or lower terms. The notation 'l+a' refers to a model with both length and age terms. The labels in the statistics column (such as t^2 or tl^2) identify the β coefficient corresponding to that combination of age and length terms. The notation 'RMSE' refers to the root mean square error.

statistic	linear l	linear l+a	quad. 1	Quad. l+a	final	cubic	Full
intercept	7.46E+00	7.53E+00	6.23E+00	5.66E+00	6.41E+00	-1.39E+00	-1.22E+00
t		1.76E-02		2.41E-01	5.24E-01	5.05E-01	5.24E-01
l	7.59E-02	7.30E-02	1.19E-01	1.27E-01	7.80E-02	4.98E-01	4.88E-01
t^2				-5.94E-03	-1.35E-03	-1.64E-02	-3.30E-02
l^2			-3.55E-04	-5.95E-04		-7.31E-03	-7.26E-03
tl					-5.00E-03	-2.36E-03	1.29E-04
t^3						2.47E-04	3.85E-05
l^3						3.96E-05	4.07E-05
t^2l							2.91E-04
Tl^3							-4.55E-05
RMSE	1.34	1.34	1.34	1.33	1.33	1.32	1.32
R^2	0.42	0.42	0.42	0.43	0.43	0.44	0.44
AIC	823.8	825.3	824.2	817.6	817.0	817.6	821.5

Table 4. Parameter estimates for the standard-age model (equation 5) applied to the red snapper *BF* data. The parameters *a* and *b* relate *BF* to length. The parameters l_∞, *k*, t_0 and ϕ (coefficient of variation of age with length) relate standard-age to length. The β parameters relate *BF* to standard age. The notation 'RMSE' refers to the root mean square error.

statistic	estimate
a	4.24E–04
b	4.87E+00
l_∞	1.11E+02
k	1.23E–01
t_0	–1.29E+00
ϕ	2.70E–01
β_1	4.01E–01
β_2	1.17E–01
β_3	–4.47E–02
RMSE	1.31
R^2	0.45
AIC	810.2

capture the dome-shaped relationship between *BF* and age within size classes that was evident in Figure 1. Batch fecundity was estimated to increase with age (in a given size-class) until a standard age of about 2.8, after which it was estimated to decrease rapidly with age (Figure 3b). Nevertheless the standard age model explained only slightly more of the total variation in *BF* (r^2 = 45%) than the simple power function of length (r^2 = 42%).

The fact that the age effect is statistically significant, but small in magnitude, raises concerns that it may simply be an artifact of small sample sizes for older ages or some other sampling-related issue. We examined the former by removing fish older than age 13 from the analyses (a total of 9 individuals). The results were qualitatively identical in terms of *AIC* and r^2 rankings, indicating that the oldest ages were not unduly influential. A broader perspective was gained when the size-at-age of the red snapper obtained for our fecundity samples were inspected against a larger data set of red snapper age and length obtained from the landed catch (Figure 4; see Allman and Fitzhugh 2007, this volume). The comparison in Figure 4a shows that most of the fecundity samples were obtained from red snapper during their earlier and relatively rapid growth phase. The contrast in Figure 4b demonstrates that the red snapper sampled for fecundity were notably larger at age than those typically sampled from the landings. The mean length at age for the NMFS and MARFIN samples were similar.

Discussion

The fecundity models used in the assessments of Gulf red snapper were variations of equations (2) and (3), which do not admit strong departures from the relationship one would expect if *BF* were primarily length-based and growth followed the von Bertalanffy equation. Accordingly, the existence of substantial, un-modeled age-dependencies would have important implications for the management of this species. This study examined the issue by focusing on the relationship between the quantities for which we have the most complete data set: *BF*, length, and age. The results are consistent with the hypothesis of Bagenal (1978) and others that length is more important than age in determining *BF*.

It may be that the effect of age is of a more indirect nature. The observation that older individuals are more productive than younger individuals of the same size until some critical age is reached was nearly ubiq-

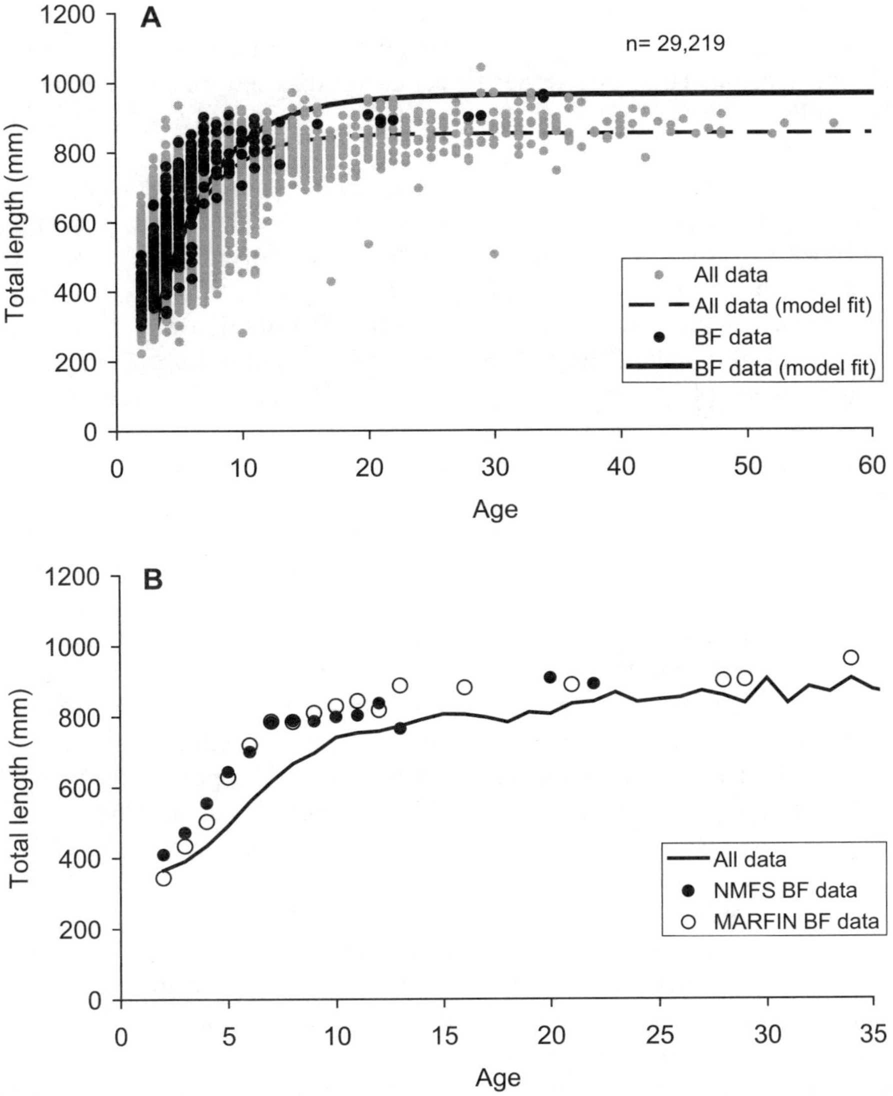

Figure 4. Red snapper *Lutjanus campechaunus* length as a function of age. Panel (A) compares the distribution of lengths from our *BF* data against a much larger data set that was collected from similar sources. The lines represent von Bertalanffy growth curves that were fitted to the two respective data sets. Panel (B) compares the average size at age from the MARFIN *BF* data, NMFS *BF* data, and the larger data set.

uitous across size classes (Table 1; Figure 1). This was confirmed by the statistically superior fit of the standardized-age model, which explicitly accounts for the degree to which the age of a fish departs from the expectation given its size. The effect seems to be most pronounced for fish that are exceptionally small or exceptionally large given their age, which constitute a small fraction of the sample (Table 1) and presumably also a small fraction of the population at large. As a result, the standard age model explained only a few percent more of the total variance in *BF* than did the simple power function of length alone. Hence, it seems unlikely that including this interaction would have any important ramifications for stock assessments of red snapper.

It was also demonstrated that the red snapper in the MARFIN and NMFS *BF* data sets were longer on average than the red snapper from a larger random sample of the landings (Figure 4). Initially we suspected size-biased sampling inasmuch as samplers were directed to target larger fish during the last year of the MARFIN study, but the average size at age was nearly identical to that of the NMFS samples (where most sampling was intentionally random). In fact, we found the similarity in average size at age to be rather remarkable considering the MARFIN samples were obtained primarily from charter boats operating off Alabama, whereas the NMFS samples were obtained mostly from recreational and commercial landings distributed over a much broader area. This suggests that the difference in size between the *BF* samples and the larger data set is likely not the result of differences in sampling, but reflective of a fundamental difference between reproductively active females and the general population. There is no evidence for sexually dimorphic growth in red snapper in the larger data set, confirming Goodyear (1995). Therefore we think a more likely explanation is that the frequency of spawning increases with size, therefore the probability of sampling a fish with hydrated oocytes (and including it in the *BF* data set) also increases with size. Unfortunately, our data are insufficient to establish a clear relationship between spawning frequency and size. Whatever the cause, the size-bias effect can be corrected for during the conversion to age by applying the size-specific function to the distribution of length expected for each age-class rather than to the mean length predicted by the growth curve. In practice, however, this is not expected to have a strong impact on the assessment because the size bias appears to be similar for most of the age classes that contribute substantially to the reproductive potential of the stock.

On balance, the above results suggest that the length-based (age-converted) models currently used for red snapper stock assessments adequately account for the relative change in *BF* with age. The importance of age to *BF* appears to be limited to an interaction with length that pertains primarily to those individuals that are unusually large or small for their age. Of course there are other factors that may contribute to the reproductive potential of a given age-class besides *BF*. The latest red snapper stock assessment (SEDAR 2004) modeled the fraction of the population that is mature as an increasing function of age (see also Jackson et al. 2007). Fitzhugh et al. (2004) suggested the possibility that spawning frequency also increased with age, although their results were not statistically significant. Several recent studies of other species have found that older fish produce more eggs of better quality than their younger counterparts (Trippel et al. 1997; Trippel 1998; Berkeley et al. 2004).

Another caveat to consider is that the *BF* data set examined here, while the largest for red snapper to date, does not include many observations of females older than age 13 ($n = 9$). This low sample size, coupled with the naturally high variability in *BF*, severely limits the ability to detect important effects such as senescence, which may occur at much older ages. Thus, while the *BF* values for these 9 older females may seem to be somewhat lower than expected, the trend was not statistically significant. In this light, one wonders whether more examples of reproductive senescence would be detected if more aging were conducted in concert with reproductive studies. Physiological changes known to occur with age in other species (e.g., disproportionate increases in ovarian connective tissue, thickening of the ovarian wall, declines in pituitary function) would suggest the possibility of eventual reproductive senescence. Certainly captive fishes often live beyond their reproductive years (Woodhead 1979) and there is some evidence that this may be true in wild populations as well (Koslow et al. 1995; Eenennaam and Doroshov 1998; and Reznick et al. 2004). Accordingly, we do not regard as closed the question of changes in reproductive potential with age over and above the effect of length. In particular, we recommend additional reproductive sampling, particularly of older individuals, and further research to examine the relationship of spawning frequency and egg quality to age.

Finally, the models used to relate *BF* to age, both here and in previous studies, typically treat age as a covariate measured without error. In fact, age determination is often imprecise and the degree of imprecision generally increases with age. In the case of equation (4), this is a classic log-linear random effects problem that is easily handled in many familiar statistical packages. The standardized-age model represented by equation (5) is also a random effects model of sorts, except in this case the random variable t is expressed as a function of the independent covariate l, which is not so easily handled by most statistical packages (but could be programmed into more general random effects software). In practice, it is often observed that the parameter estimates do not change much when random effects are included. However the degree of uncertainty is better characterized, which can affect the outcome of hypothesis tests. In the present case, the models defined by equations (4) and (5) indicated that the influence of age on *BF* was rather small once length was accounted for. Thus, the question of statistical significance was rendered moot and the random effects models were not pursued further. Nevertheless, we recommend the use of random effects models in the future should the number of older red snapper in the samples be increased.

Acknowledgments

We thank the many port agents, biologists, and anglers for assistance in sampling red snapper over the years. Several people were involved with the preparatory work, aging and data assembly. We thank Robert Allman, Linda Lombardi-Carlson, Bill Fable and Bill Walling at the NMFS Panama City, FL laboratory. Faculty, staff and students at the University of South Alabama and Louisiana State University also were involved in this research. In particular, we thank Jim Cowan, Chuck Wilson, Dave Nieland, Andy Fischer, and Scott Baker. The NMFS project was supported in part by the MARFIN (Marine Fisheries Initiative) and Red Snapper Research programs. Data supplied by Melissa Woods Jackson was funded through MARFIN Grant # NA87FF0424 to the University of South Alabama. Thanks are also due to the participants of the SEDAR 7 data workshop who offered helpful comments on a related, unpublished manuscript.

References

Akaike, M. 1973. Information theory and an extension of the maximum likelihood principle. Pages 267–281 *in* B. Petrov and F. Csaki, editors, Proceedings of the second International Symposium on Information Theory. Akademia Kiado, Budapest, Hungary.

Allman, R. J., and G. R. Fitzhugh. 2007. Temporal age progressions and relative year-class strength of Gulf of Mexico red snapper. Pages 311–328 *in* W. F. Patterson, III, J. H. Cowan, Jr., G. R. Fitzhugh, and D. L. Nieland, editors. Red snapper ecology and fisheries in the U.S. Gulf of Mexico. American Fisheries Society, Symposium 60, Bethesda, Maryland.

Bagenal, T. B. 1978. Aspects of fish fecundity in ecology of freshwater fish production. Blackwell Scientific Publications: 75–102.

Berkeley, S. A., M. A. Hixon, R. L. Larson, and M. S. Love. 2004. Fisheries sustainability via protection of age structure and spatial distribution of fish populations. Fisheries 29(8):23–32.

Bowering, W. R. 1980. Fecundity of Greenland halibut,

Reinhardtius hippoglossoides (Walbaum), from southern Labrador and southeastern Gulf of St. Lawrence. Journal of Northwest Atlantic Fishery Science 1:39–43.

Burnham, K. P., and D. R. Anderson 2002. Model selection and Multimodal Inference: a practical information-theoretic approach, 2nd Edition. Springer-Verlag, New York.

Collins, L. A., A. G. Johnson, C. P. Keim. 1996. Spawning and annual fecundity of the red snapper (*Lutjanus campechanus*) from the northeastern Gulf of Mexico. Pages 174–188 *in* F. Arreguin-Sanchez, J. L. Munro, M. C. Balgos, and D. Pauly, editors. Biology, fisheries and culture of tropical groupers and snappers. ICLARM Conference Proceedings Number 48.

Cowan, J. H., Jr., M. Woods, W. Patterson, III, and D. Nieland. 2002. Otolith microchemistry (and reproductive biology) portion—stock structure of red snapper in the northern Gulf of Mexico: is their management as a single stock justified based on spatial and temporal patterns of genetic variation, otolith microchemistry, and growth rates? Final Report to the U.S. Department of Commerce, National Marine Fisheries Service, Marine Fisheries Initiative (MARFIN) Cooperative Agreement NA87FF0425. St. Petersburg, Florida.

Eldridge, M. B., and B. M. Jarvis. 1995. Temporal and spatial variation in fecundity of yellowtail rockfish. Transactions of the American Fisheries Society 124:16–25.

Eenennaam, J. P., and S. I. Doroshov. 1998. Effects of age and body size on gonad development of Atlantic sturgeon. Journal of Fish Biology 53:624–637.

Fitzhugh, G. R., M. S. Duncan, L. A. Collins, W. T. Walling and D. W. Oliver. 2004. Characterization of red snapper (*Lutjanus campechanus*) reproduction: for the 2004 Gulf of Mexico SEDAR. National Marine Fisheries Service, Panama City Laboratory Contribution Series:04–01. Panama City, Florida.

Goeman, T. J. 1983. Freshwater drum spawning and fecundity in the upper Mississippi River. Procedings of the Iowa Academy of Sciences 90(4):132–133.

Goodyear, C. P. 1995. Red snapper in U.S. waters of the Gulf of Mexico. Southeast Fisheries Science Center, Coastal Resources Division, MIA-95/96–05.

Holmgren, K. 2003. Omitted spawning in compensatory-growing perch. Journal of Fish Biology 62:918–927.

Hunter, J. R., N. C. H. Lo, and R. J. H. Leong. 1985. Batch fecundity in multiple spawning fishes. Pages 67–77 *in* R. Lasker, editor. An egg production method for estimating spawning biomass of pelagic fish: application to the northern anchovy, *Engraulis mordax*. NOAA/NMFS Technical Report 36.

Hunter, J. R., and B. J. Macewicz. 1985. Measurement of spawning frequency in multiple spawning fishes. Pages 79–94 *in* R. Lasker, editor. An egg production method for estimating spawning biomass of pelagic fish: an application to the northern anchovy, *Engraulis mordax*. NOAA/NMFS Technical Report 36.

Jackson, M. W., J. H. Cowan, Jr., and D. L. Nieland. 2007. Demographic differences in northern Gulf of Mexico red snapper reproductive maturation: implications for the unit stock hypothesis. Pages 217–227 *in* W. F. Patterson, III, J. H. Cowan, Jr., G. R. Fitzhugh, and D. L. Nieland, editors. Red snapper ecology and fisheries in the U.S. Gulf of Mexico. American Fisheries Society, Symposium 60, Bethesda, Maryland.

Koslow, J. A., J. Bell, P. Virtue, and D. C. Smith. 1995. Fecundity and its variability in orange roughy: effects of population density, condition, egg size, and senescence. Journal of Fish Biology 47:1063–1080.

Kritzer, J. P. 2004. Sex-specific growth and mortality, spawning season, and female maturation of the stripey bass (*Lutjanus carponotatus*) on the Great Barrier Reef. Fishery Bulletin 102:94–107.

Messieh, S. N. 1976. Fecundity studies on Atlantic herring from the southern Gulf of St. Lawrence and along the Nova Scotia Coast. Transactions of the American Fisheries Society 105(3):384–394.

Murua, H., G. Kraus, F. Saborido-Rey, P. R. Witthames, A. Thorsen, and J. Junquera. 2003. Procedures to estimate fecundity of marine fish species in relation to their reproductive strategy. Journal of Northwest Atlantic Fishery Science 33:33–53.

Newman, S. J. 2002. Growth rate, age determination, natural mortality and production potential of the scarlet seaperch, *Lutjanus malabaricus* Schneider 1801, off the Pilbara coast of north-western Australia. Fisheries Research 58:215–225.

Nieland, D. L., and C. A. Wilson. 1993. Reproductive biology and annual variation of reproductive variables of black drum in the northern Gulf of Mexico. Transactions of the American Fisheries Society 122:318–327.

Nitschke, P., M. Mather, and F. Juanes. 2001. A comparison of length-, weight-, and age-specific fecundity relationships for cunner in Cape Cod Bay. North American Journal of Fisheries Management 21:86–95.

Reznick, D. N., M. J. Bryant, D. Roff, C. K. Ghalambor, and D. E. Ghalambor. 2004. Effect of extrinsic mortality on the evolution of senescence in guppies. Nature (London) 431:1095–1099.

SEDAR, 2004. Southeast Data Assessment and Review (SEDAR) 7: Gulf of Mexico red snapper workshop report. National Marine Fisheries Service, Southeast Fisheries Science Center, Miami, Florida. <www.sefsc.noaa.gov/sedar/>

Trippel, E. A., O. S. Kjesbu, and P. Solemdal. 1997. Effects of adult age and size structure on reproductive output in marine fishes. Pages 31–62 *in* R. C. Chambers and E. A. Trippel, editors. Early life history and recruitment in fish populations. Fish and Fisheries Series 21. Chapman and Hall, London.

Trippel, E. A. 1998. Estimation of stock reproductive potential: history and challenges for Canadian Atlantic gadoid stock assessments. Journal of Northwest Atlantic Fishery Science 25:61–81.

Wallace, R. A., and K. Selman. 1981. Cellular and dynamic aspects of oocyte growth in teleosts. American Zoologist 21:325–343.

Wilson, C. A. and D. L. Nieland. 2001. Age and growth of red snapper, *Lutjanus campechanus*, from the northern Gulf of Mexico off Louisiana. Fishery Bulletin 99:653–664.

Woodhead, A. D. 1979. Senescence in fishes. Symposia of the Zoological Society of London 44:179–205.

Woods, M. K. 2003. Demographic differences in reproductive biology of female red snapper (*Lutjanus campechanus*) in the northern Gulf of Mexico. Master's thesis. University of South Alabama, Mobile, Alabama.

American Fisheries Society Symposium 60:245–261, 2007

A Review of Movement in Gulf of Mexico Red Snapper: Implications for Population Structure

William F. Patterson, III[1]
Department of Biology
University of West Florida
11000 University Parkway, Pensacola, Florida 32514 USA

Abstract.—A review of studies examining stage-specific distribution and movement of various life stages of red snapper, *Lutjanus campechanus*, in U.S. waters of the Gulf of Mexico (GOM) was conducted to draw inference about population structure. Hydrodynamic modeling of neither egg nor larval transport has been conducted for GOM red snapper; thus, the potential for planktonic dispersal among regions is currently unknown. However, recent studies of other reef fishes have demonstrated that larval fishes may not act as passive particles. Postsettlement movement, or the lack thereof, may be just as important for describing population connectivity and structure as planktonic transport. Red snapper juveniles display thigmotaxis and have been shown to undergo an ontogentic shift in which the dimension and complexity of their habitat increases with fish size. Tagging data demonstrate that while a substantial percentage of tagged fish were recaptured near their release sites, movement on the scale of hundreds of km also has been reported. Direct estimates of movement and population mixing from ultrasonic tagging, conventional tagging, and otolith chemistry studies indicate movement of some individuals may be sufficient to promote genetic exchange among regions, but overall movement is likely insufficient to affect population demographic differences observed among regions. Therefore, GOM red snapper meet criteria for consideration as a metapopulation: subpopulations are distinct, dispersal mechanisms exist among subpopulations, and asynchrony in population demographics is apparent among subpopulations.

Introduction

Red snapper, *Lutjanus campechanus*, have been intensively managed in U.S. waters of the Gulf of Mexico (GOM) since the late 1980s when they were first estimated to be overfished and undergoing overfishing (Goodyear 1988; reviewed in Hood et al. 2007, this volume). Despite increasingly stringent harvest regulations placed on the directed fishery since the early 1990s, the stock has failed to recover, or even show significant signs of moving toward recovery (Hood et al. 2007; Porch 2007, this volume). Goodyear (1995) estimated shrimp trawl bycatch was the most significant source of mortality for GOM red snapper, the implications of which were the directed fishery would have to be either severely curtailed or closed in order to recover the stock if bycatch could not be minimized. Although more restrictive size

[1] Corresponding author: wpatterson@uwf.edu.

and daily catch limits were put in place, other management options also have been explored in hopes of lessening the impact of regulations on the directed fisheries. Bycatch reduction devices (BRDs) were required in 1998 for shrimp trawlers operarting west of Cape San Blas, Florida, and in the entire U.S. GOM in 2004 (Hood et al. 2007). Stock enhancement also has been suggested as a means to recover red snapper without restricting the directed fishery (Papanikos et al. 2003; Ogle and Lotz 2006). Last, marine protected areas (MPAs) have been recommended as another alternative to increase red snapper biomass, although MPAs may provide additional conservation benefits that extend well beyond the fishery (Holland and Brazee 1993; Trexler and Travis 2000; Baskett et al. 2005).

Nearly all of the significant conservation questions concerning GOM red snapper have at their core the issues of population structure and the stage-specific distribution and movement of individuals. Clearly, red snapper are not unique in that respect as the importance of understanding population structure and connectivity has been stressed since early in the 20th century (Hjort 1914; Secor 2002, 2006). In reef fishes, eggs and larvae traditionally have been viewed as the most likely life history stages during which population mixing may occur given the potential for long-distance dispersal of planktonic early life stages and the often sedentary nature of adults (Jones et al. 1999; Swearer et al. 2002). Following that logic, several authors invoked planktonic transport of red snapper early life stages to explain the lack of genetic divergence reported among northern GOM regions (Goodyear 1995; Gold et al. 1997, 2001; Saillant and Gold 2006). Recent studies of other reef fishes, however, have indicated larvae may not act as passive particles in the sea and that self-recruitment mechanisms are prevalent (Cowen and Castro 1994; Jones et al. 1999, 2005; Swearer et al. 2002; Cowen et al. 2006; Almany et al. 2007). As a corollary, postsettlement movement may be more important than previously realized in facilitating population mixing in reef fishes, especially in large reef fishes, such as snappers and groupers, that may at times move great distances (Patterson et al. 2001; Lindberg et al. 2006).

The objective of this paper was to review the literature on GOM red snapper movement in order to draw inference about the implications observed movement has for red snapper population structure. Little is known about oceanographic transport of red snapper eggs and larvae, but what has been described about the occurrence of these life stages and their distributions is discussed briefly. The bulk of the paper reviews studies of postsettlement habitat, site fidelity, and movement of red snapper. Much research effort has been expended in recent years describing ontogentic shifts in red snapper habitat-specific distribution, as well as the potential for postsettlement movement to affect population connectivity, or the lack of movement, to shape localized population demographics. Overall, this review is aimed at facilitating a better understanding of GOM red snapper population structure and connectivity.

Methods

A literature search was conducted for GOM red snapper within Cambridge Scientific Abstracts' Natural Sciences Database (www.csa.com). Separate searches were conducted for "*Lutjanus campechanus*" or "red snapper" appearing anywhere within citations published between 1980 and 2006. Unique citations of peer-reviewed publications were placed into one of ten categories: age and growth, bycatch, culture, diet/bioenergetics, fisheries management, genetics, habitat, movement, MPAs, reproduction and early life history, and miscellaneous. Papers were evaluated in the context of red snapper movement, population connectivity, and population structure. Additional papers reviewed include technical documents presented at the 2004 Southeast Data Assessment and Review (SEDAR) workshops for GOM red snapper (SEDAR7), technical reports featuring red snapper movement that were cited in other studies, and peer-reviewed publications published prior to 1980 and cited in subsequent papers.

Results and Discussion

The literature search within Cambridge Scientific Abstracts' Natural Sciences Database yielded 149 GOM red snapper papers that appeared in the literature between 1980 and 2006. More papers were published in the genetics (n = 23) category than any other. Nine of those manuscripts detailed studies examining molecular markers in red snapper fillets in order to distinguish them from mislabeled species in the marketplace (e.g., Marko et al. 2004), but the subject of the majority of the genetics papers (n = 13) was estimating genetic population structure in GOM red snapper (e.g., Pruett et al. 2005; Saillant and Gold 2006). Papers examining red snapper fisheries management (n = 15) ranged from examining the effect of regulatory discards in the directed fisheries to estimating the value of the recreational fishery to an assessment of implementing an individual transferable quota system in the commercial fishery. Several papers (n = 19) described red snapper habitat affinity and ontogentic shifts in habitat utilization, with papers split among juvenile habitat studies (n = 9), natural hardbottom habitat of adults (n = 1, but 2 others in the MPA category), and artificial reefs (n = 9). Diet and bioenergetics studies (n = 7) also tended to emphasize ontogentic shifts and habitat-specific differences in diet. Age and growth papers were prevalent (n = 16), but only in one were differences in red snapper size at age tested among GOM regions (Fischer et al. 2004). Movement studies (n = 13) examined life stage specific site fidelity and movement, as well as the residency of adult red snapper at artificial reef sites. Twenty-one percent (n = 32) of the studies examined alternative management strategies for rebuilding red snapper, including stock enhancement (culture; n = 15), bycatch reduction (n = 11), and the efficacy of MPAs for rebuilding red snapper spawning stock biomass (n = 6). Relatively few studies were directed at reproductive biology (n = 4) or early life stages (n = 4), the latter result highlighting the paucity of information available on red snapper eggs and larvae in the wild.

Dispersal of Early Life Stages

Authors of early studies of GOM red snapper reproductive biology concluded that fish began reaching sexual maturity at small size (<300 mm total length TL) and had protracted spawning seasons extending throughout summer months (Bradley and Bryan 1975; Futch and Bruger 1976; Moseley 1966). Collins et al. (1996) were the first to estimate batch fecundity in red snapper and to establish that the spawning season extended from April through October in both the eastern and western GOM. The protracted spawning season for red snapper, combined with a larval stage duration of approximately 20 d (Szedlmayer and Conti 1999; Drass et al. 2000; Rooker et al. 2004), provides the potential for significant planktonic dispersal, and several authors have hypothesized that oceanographic transport of eggs and larvae is at least partially responsible for the lack of significant differences reported in selectively neutral genetic markers among GOM regions (Goodyear 1995; Gold et al. 1997; Saillant and Gold 2006). Despite the lack of significant genetic differences, Jackson et al. (2007) reported maturity schedules were significantly different between red snapper populations east and west of the mouth of the

Mississippi River. They suggested early maturity at smaller size in the eastern GOM may be a genotypic response to high fishing mortality having selectively removed later maturing genotypes, a response that would not be apparent in selectively neutral genetic markers such as mitochondrial DNA (mtDNA) or nuclear DNA microsatellites (Pruett et al. 2005; Saillant and Gold 2006). Regardless of the causative factor of differences in reproductive biology parameters between the eastern and western GOM, the fact that regional population demographic differences exist implies some degree of isolation between the eastern and western GOM. Regional differences in size at age reported by Fischer et al. (2004) further support that population structure exists in GOM red snapper, which has not been revealed by traditional fisheries genetics applications (Pruett et al. 2005; Gold and Saillant 2007, this volume).

Relatively little was known until recently about the distribution of red snapper eggs and larvae in the GOM, and the extent to which interregional mixing may occur in the plankton remains unresolved (Hanisko et al. 2007, this volume; Lyczkowski-Shultz and Hanisko 2007, this volume). Collins et al. (1980) described morphometric and meristic characteristics of larvae and Potthoff et al. (1988) described their osteological development. Drass et al. (2000) were the first to describe characters that distinguished larval red snapper as small as 3.5 mm (mid-flexion) from potentially co-occurring congeners and confamilials. Based on those characters, Lyczkowski-Shultz and Hanisko (2007) reported a total of 1,692 red snapper larvae were identified in >14,000 bongo and neuston net samples collected on National Marine Fisheries Service (NMFS) research surveys between 1982 and 2003. Those data were used to compute fishery-independent indices of spawning stock biomass (Hanisko et al. 2007), but as yet no modeling exercise has been conducted to estimate the effect of oceanographic processes on the distribution and potential dispersal of larvae (e.g., Hanisko and Lyczkowski-Shultz 2003; Fitzhugh et al. 2005; Cowen et al. 2006).

Several authors hypothesized that oceanographic transport of eggs and larvae may be sufficient to facilitate population mixing despite the lack of hydrodynamic modeling of egg or larval transport in the northern GOM. Gold et al. (1997) reported mtDNA haplotype frequencies were not significantly different among northern GOM regions; thus, the authors failed to reject the null hypothesis that GOM red snapper constitute a single panmictic stock. They suggested genetic mixing among regions, or populations, may occur during planktonic egg and larval stages due to the preponderance of evidence, at that time, that red snapper adults were sedentary. However, the authors also suggested, based on intrapopulational mtDNA diversity differences, that GOM red snapper might include recently derived, but as yet not genetically distinct, population subunits. Pruett et al. (2005) conducted nested clade analysis of mtDNA haplotypes and concluded the genetic history of GOM red snapper was complex, as mtDNA frequencies suggested periods of both range expansion and ones of restricted flow resulting from isolation by distance. They hypothesized that apparently restricted gene flow among contemporary red snapper populations may yield metapopulation structure, but that hypothesis likely is not testable with selectively neutral genetic markers. Nevertheless, Pruett et al. (2005) suggested asynchrony observed in red snapper population demographics among northern GOM regions (Fischer et al. 2004; Jackson et al. 2007) was evidence that metapopulation structure may exist within GOM red snapper, despite the lack of divergence in selectively neutral genetic markers. Last, the authors concluded that precise estimates of exchange (i.e., movement) among regions was required to further examine the existence of metapopulation structure.

Discussions of interregional or interpopulational connectivity in GOM red snapper, as presented above, typically have centered on hypothesized, but as yet untested, oceanographic transport of eggs and larvae. Recent studies of other reef fishes, however, have indicated that larvae may not behave as passive particles in the sea and that self-recruitment mechanisms, including ones promoting endemism, are prevalent (Cowen and Castro 1994; Jones et al. 1999, 2005; Swearer et al. 2002; Cowen et al. 2006; Almany et al. 2007). Cowen et al. (2000) reported hydrographic model simulations of larval fish transport within the Caribbean Basin tended to overestimate dispersal when simple advection was assumed, thus demonstrating the importance of local retention in maintaining population structure. Other authors have demonstrated retention mechanisms in meroplanktonic invertebrates that metamorphose into sessile adults (e.g., Ayre and Hughes 2000; Johnson and Black 2006; Gilg and Hilbish 2003), as well as in reef fishes that display limited postsettlement home ranges (e.g., Doherty et al. 1995; James et al. 2002; Almany et al. 2007). However, the potential for interpopulational mixing clearly is greater, postsettlement, for reef-associated fishes that do not demonstrate high long-term site fidelity (Ingram and Patterson 2001; Patterson et al. 2001; Meyer et al. 2007) or that display ontogentic habitat shifts that occur over significant distances (Bryant et al. 1989; Lindberg et al. 2006).

Ontogenetic Shifts in Red Snapper Habitat

Postsettlement movement in GOM red snapper has been investigated with several different approaches to address various questions about red snapper population ecology. Several authors have examined ontogenetic habitat shifts (e.g., Bradley and Bryan 1975; Szedlmayer and Howe 1997; Szedlmayer and Conti 1999; Rooker et al. 2004), with recent studies aimed at defining essential fish habitat (EFH) of juveniles in order to mitigate shrimp trawl bycatch (e.g., Patterson et al. 2005; Wells 2007). When results of habitat studies are considered in totality, some consistent themes begin to emerge relative to red snapper habitat requirements and how they shift ontogenetically. First, like most reef fishes, juvenile red snapper display a strong thigmotaxis, thus they seek structured environments (Workman et al. 1994; Szedlmayer and Howe 1997; Bailey et al. 2001; Franks et al. 2004). Szedlmayer and Howe (1997) reported juvenile red snapper selected oyster shell versus sand habitat in tank trials, while Patterson et al. (2005) reported highest juvenile red snapper densities in trawl samples off Alabama and Mississippi came from high-relief (2–3 m) shell rubble ridge habitats. Seemingly contrary results were presented by Rooker et al. (2004) and Geary et al. (2007, this volume) that demonstrated juvenile red snapper associated with Texas bank systems were found in high densities in relatively unstructured mud habitats, and Patterson et al. (2005) reported moderately high juvenile densities occurred in sand habitats off Alabama and Mississippi. However, sampling trawls deployed by Rooker et al. (2004) and Geary et al. (2007) contained small-mesh bags that retained recently-settled juveniles for which biogenic structures such as worm tubes may provide sufficient structure in soft sediments (Workman et al. 2002). Furthermore, mud habitats were immediately adjacent to shell rubble habitats and may have served as foraging areas. Patterson et al. (2005) reported red snapper juvenile density in sand habitats was significantly correlated with sponge biomass. Hence, they concluded sponges provided habitat complexity at a scale sufficient for juvenile snapper. Similarly, Workman and Foster (1994) reported juvenile red snapper encountered in sand habitats typically were associated with objects, such as squid egg cases, woody debris, or discarded drink cans, that fish used for either refuge or orientation.

Much evidence suggests that as red snapper age they recruit to habitats characterized by increasing vertical dimension. Patterson et al. (2005) demonstrated age 0 red snapper were found in low-relief shell rubble and sand (interspersed with sponge) habitats, but age 1 fish were found at deeper (40 versus 20 m depth) sites that had greater vertical relief and complexity. Bailey et al. (2001) conducted tank trials in which they tested the effect of structure (concrete blocks) and the presence of adult conspecifics on the location of juvenile red snapper within tanks. When adults were not present, juveniles oriented to experimental reefs, but when present, adults displayed agonistic behavior in excluding juveniles from the preferred habitat. Wells (2007) reported that age 0 fish off Alabama were abundant in shell rubble habitats, but larger, older (age 2+) fish were concentrated in natural reef habitats. Bradley and Bryan (1975) reported ontogenetic movement of red snapper to structured habitats of increasing dimension occurred as an onshore to offshore migration throughout the juvenile stage. [It should be noted that natural hardbottoms and banks that constitute the most significant natural reef areas in both the western and eastern GOM are found predominantly on the outer shelf, while lower-relief shell rubble ridges and banks are found in relatively shallow (<20 m) nearshore waters (Parker et al. 1983; Schroeder et al. 1988; Laswell et al. 1990; Dufrene et al. 2003; Gledhill and David 2004; Rooker et al. 2004; Kraus et al. 2006)]. Mitchell et al. (2004) reported larger (median TL ranged from 545 to 815 mm among surveys), older red snapper were captured at higher rates in outer shelf habitats during experimental longline surveys in the eastern (off Alabama-Mississippi) and western (off Texas) GOM. Fishery-dependent data confirm that larger, older fish are captured much more frequently in commercial fishery sectors operating farther from shore (Allman and Fitzhugh 2007, this volume).

Analysis of natural ontogenetic shifts in red snapper habitat utilization is complicated due to the proliferation of artificial reefs deployed in the north-central GOM and the vast number of petroleum platforms, which function as artificial reefs, erected in the northwestern GOM (Wilson and Nieland 2004). However, examination of the literature on red snapper recruitment to and residency at artificial reefs and platforms reveals the same pattern of increasing dimension of utilized habitats with increasing fish size. Szedlmayer and Lee (2004) reported juvenile red snapper as small as 18 mm settled in relatively unstructured open shelf habitats in summer, but by winter age 0 fish had recruited to experimental artificial reefs that provided greater habitat complexity and relief. Nieland and Wilson (2003) randomly sampled red snapper (n = 300) killed during the explosive removal of an obsolete oil platform off Louisiana. Otolith-based aging revealed the majority of fish were 2 and 3 year olds (53% and 37%, respectively), while virtually no (n = 2) age 1 fish were present in their sample. Similarly, size frequency data from small-scale (<5 m^3) artificial reefs off Alabama and northwest Florida indicate the majority of red snapper present are 2 and 3 year old fish (Strelcheck et al. 2005; Patterson, unpublished data). The lack of older red snapper at both platforms and artificial reef sites may indicate thigmotaxis or the threat of predation subsides with age and size; thus, larger, older fish display lower site fidelity and greater movement (Patterson et al. 2001; Patterson and Cowan 2003; Stelcheck et al. 2007, this volume). Alternatively, high fishing mortality rates at platforms and artificial reefs may remove snapper very quickly from the population once fish recruit to the commercial or recreational fisheries (Nieland and Wilson 2003).

Direct Estimates of Post Settlement Movement

More important to population connectivity than the distribution of fish at single points

Table 1. Results from ultrasonic tagging studies of sub-adult and adult red snapper *Lutjanus campechaunus* in the northern Gulf of Mexico.

Study	Location and Habitat	Area of Detection per Hydrophone	Hydrophones per Site	Number Tagged	Mean TL at Tagging mm	Mean Days Detected in Study Area	Max Days Detected in Study Area
Szedlmayer (1997)	Alabama; artificial reefs	3.1 km^2	1; roving	23	349	150	597
Szedlmayer and Schroepfer (2005)	Alabama; artificial reefs	≤ 8.0 km^2	3–4; fixed	54	589	212	595
Schroepfer and Szedelmayer (2006)	Alabama; artificial reefs	≤8.0 km^2	3–4; fixed	77	542	179	597
Westmeyer et al. (2007, this volume)	Louisiana; petroleum platforms	0.02 km^2	7 within a 35 km^2 area	125	360	64	202

[a] Area of detection was estimated based on maximum detection radii from hydrophones reported in each study.

in space or time is the degree of site fidelity (philopatry) individuals display and the spatial scale over which movement occurs. Some inference can be drawn about red snapper movement due to seasonal occurrence of fish in certain habitats and ontogenetic habitat shifts described above, but movement on multiple temporal and spatial scales has been estimated directly in several studies. The two main approaches that have been applied to estimate red snapper site fidelity and movement are conventional and, more recently, ultrasonic tagging. Benefits of conventional tagging include tags being inexpensive and relatively unobtrusive to fish, individual tagged fish can be identified, and nonscientists can be trained to apply tags (Patterson et al. 2001; Diamond et al. 2007, this volume). However, movement can only be estimated as straight-line distances between release and recapture locations and reporting rates by fishers often are low in heavily regulated fisheries (Fable 1990). Tag loss also can be problematic (Patterson et al. 2001). With ultrasonic tagging, individuals can be tracked nearly continuously within the range of receivers. Depending on the types of tags deployed, individual tags (fish) can be identified based on their frequency or ping rate, but the ability to track individuals is affected by receiver range and tag battery life. If functionality of tags is compromised, then a fish present but not detected would be perceived as having left the study area (Westmeyer et al. 2007, this volume).

Szedlmayer (1997) conducted the first ultrasonic tagging experiment on red snapper at artificial reef sites off Alabama (Table 1). He concluded from study results that red snapper displayed "high" site fidelity to artificial reefs, yet the mean time fish were detected in his study area was only 150 d for a species with maximum longevity >50 years (Wilson and Nieland 2001). While one tagged individual was detected for 597 d, several others (n = 6 of 23) were lost from the study area. Both "stayers" and "movers" (from Diamond et al.

2007) have important implications for population connectivity and structure (Dieckmann et al. 1999; Doebeli and Ruxton 1997; Fraser et al. 2001), but movers can no longer be tracked with ultrasonic receivers once they move beyond the range of receivers. Hence, emphasis in red snapper ultrasonic tagging analysis and interpretation has tended to be weighted toward the stayers. For example, Schroepfer and Szedlmayer (2006) concluded that ultrasonically tagged red snapper displayed high site fidelity to artificial reef sites because 87% (13 of 15) of fish were detected within 200 m of study sites 99% of the time they were detected. However, the probability that fish remained resident at reef sites after a year was only approximately 50% (i.e., 50% annual site fidelity). Westmeyer et al. (2007) reported even lower probability of detection at petroleum platforms off Louisiana after one year, but tag battery failure and thermocline effects on receivers likely negatively affected their estimates of site fidelity.

Ultrasonic tagging data are useful for examining short-term movements in resident individuals (stayers), but understanding population connectivity and structure is perhaps more dependent on estimating dispersion distances and rates of movers (Doebeli and Ruxton 1997). Conventional tagging studies are better suited for that purpose, despite the limitations cited above, and several large-scale conventional tagging studies have been conducted on GOM red snapper since the 1960s (Table 2). Among the various studies, fish were captured at natural reefs, artificial reefs (including petroleum platforms), or both. However, most of the existing movement data available from conventional tagging studies are from studies conducted over

Table 2. Movement and site fidelity estimates from conventional tagging studies of sub-adult and adult red snapper *Lutjanus campechanus* conducted in the northern Gulf of Mexico.

Study	Location and Habitat	Number Tagged	Mean TL at Tagging mm	Number Recaptured	Mean/Max Days Free	Mean/Max km Moved	Site Fidelity[a]
Beaumariage (1969)	West Florida; natural reefs	1,126	NA	384	113/ 2,049	NA/ 279	90% recaptured within 5 km of release site
Fable (1980)	Texas; natural reefs and oil platforms	299	286	17	112/ 253	0.3/ 5	94% recaptured at release site
Szedlmayer and Shipp (1994)	Alabama; artificial reefs	1,155	287	146	137/ 430	4.6/ 32	74% recaptured within 2 km of release site
Patterson and Cowan (2003)[b]	Alabama; artificial reefs	2,932	335	599	404/ 1,501	30.9/ 558	25–27% per year
Strelcheck et al. (2007, this volume)	Alabama; artificial reefs	4,317	335	629	401/ 1,587	2.1/ 202	48–50% per year
Diamond et al. (2007, this volume) (TTU Tagging)	Texas; artificial and natural reefs	5,614	363	130	166/ 564	9.8/ 58.3	52.4% recaptured at release site

[a]Site fidelity to release site was directly estimated as an annual rate by Patterson and Cowan (2003) and Strelcheck et al. (2007). from the decline in recaptures made by researchers at unreported study sites over time.
[b]Data presented in Patterson and Cowan (2003) include data presented in Patterson et al. (2001) plus additional recaptures.

artificial reef sites in the north-central GOM (e.g., Szedlmayer and Shipp 1994; Patterson and Cowan 2003; Strelcheck et al. 2007). An exception to that is data from the Schlitz Tagging Program conducted off Florida in the 1960s by Florida Department of Natural Resources personnel. Fish in that program were captured and tagged over natural reef sites, most of which occurred off northwest Florida (Beaumariage 1969). Fable (1980) also reported movement data from fish captured and tagged over a variety of natural and artificial habitats off Texas, as did Diamond et al. (2007).

Several consistent trends exist in the movement data among conventional tagging studies, although considerable variability also exists in results among them. Most tagged individuals have been small, young fish, with mean TL between 299 and 363 mm across studies (Table 2). Patterson et al. (2001) reported fish size significantly affected the likelihood and distance of movement away from release sites, and Diamond et al. (2007) reported larger fish had a higher probability of movement than had smaller ones. Red snapper can attain sizes of nearly 1 m TL (Wilson and Nieland 2001); thus, movement estimates based on a sample of small, young individuals may be conservative when applied to larger, older fish in the population.

The scale of observed movement generally increased with sample size and the temporal scale of tagging studies. Movement data presented by Diamond et al. (2007) from tag returns of fish tagged off Texas represents one departure from that trend. Substantially more fish were tagged in their study (n = 5,614) than in any other, yet only modest levels of movement were observed. However, some of their samples came from deep water (to 100 m); thus, acute or chronic effects of barotrauma may have affected their functional sample size (Patterson et al. 2002; Rummer 2007, this volume) and overall tag return rate (2.8% versus ~6–35% among other studies). Furthermore, the mean and maximum times that recaptured fish were free in their study (564 d) were only about a third of those reported by others (Table 2).

Perhaps the most striking characteristic of red snapper movement data that is consistent among tagging studies is that data tend to be positively skewed and are characterized by a negative binomial distribution (Patterson et al. 2001). That consistent pattern results from a high percentage of zeros in the data (stayers) and the fact that most movers moved only small (<10 km) rather than large (>50 km) distances prior to recapture [e.g., Table 8 in Beaumariage (1969); Figure 5 in Szedlmayer and Shipp (1994); Figure 5 in Patterson and Cowan (2003)]. Fraser et al. (2001) reported similar movement distributions are common across many taxa, and sought to explain the ecological and evolutionary significance of dispersing phenotypes in populations. They demonstrated movers (their "dispersers") within populations of giant rivulus, *Rivulus hartii*, in Trinidadian streams were individuals who displayed boldness versus fearfulness in traversing open spaces in test tanks prior to tagging. Once tagged and released back into the wild, bold individuals not only moved greater distances in streams, but also had higher individual growth rates. Diamond et al. (2007) also reported tagged red snapper that moved away from tagging sites off Texas grew at faster rates than ones that stayed. Fraser et al. (2001) concluded that bold behavior traits contributed to greater fitness of surviving movers versus stayers, although the cost of boldness, hence movement, was greater exposure to predation risk.

Currently, it is unknown what the cost of movement away from reef structure is for red snapper. Observed postsettlement movement has been lower in juveniles than in sub-adults, and the scale of movement generally increased with size in the adult life stage (Patterson et al. 2001; Workman et al. 2002; Diamond et al. 2007). Lorenzen (2000,

2006) demonstrated the existence of an allometric relationship between natural mortality (*M*) and fish body weight. He demonstrated *M* declines with increasing body size, which in turn most likely results from a concomitant decrease in predation risk (Sogard 1997; Murakami et al. 2005). Therefore, adult red snapper site fidelity to reefs may decrease, and their movement distances increase, with increasing size and/or age due to a relaxation of predation risk as fish attain larger sizes. But despite the potential for interpopulational genetic mixing resulting from extreme (>100 km) movement observed in some tagged fish (e.g., Beaumariage 1969; Patterson et al. 2002; Strelcheck et al. 2007), Pruett et al. (2005) reported evidence of evolutionarily recent isolation by distance was apparent in mtDNA haplotype frequencies, and Saillant and Gold (2006) reported significant differences in red snapper genetic effective population size among northern GOM regions. Therefore, although some selection pressure may exist to maintain mover phenotypes in red snapper populations (Fraser et al. 2001), competing pressures must also exist else selection likely would have driven red snapper populations to display even greater movement than what has been observed (Jonsson and Jonsson 1994; Doebeli and Ruxton 1997). Perhaps selection pressures for movement versus philopatry are stage-dependent in red snapper, as has been demonstrated in other marine and anadromous species that maintain divergent migratory tactics (reviewed in Secor 1999). For small, young fish, fitness tradeoffs existing between growth and defense (Sibly et al. 1985) likely favor high site fidelity to reef structure (thigmotaxis) as a refuge from predation (Overholtzer-McLeod 2005), even when high fish density compromises growth (Strelcheck et al. 2005; Lindberg et al. 2006). Predation pressure likely is lower for larger, older fish, for which large size alone may convey fitness, especially in females (Munch and Conover 2003; Andersen et al. 2007); therefore, reef dependency is relaxed (Patterson et al. 2001; McCawley et al. 2007, this volume). It is unknown, however, what factors may contribute to extreme (>100 km) movement observed in some red snapper. Patterson et al. (2001) reported tagged fish at liberty during hurricanes moved significantly farther than ones that were not exposed to storms. However, Beaumariage (1969) did not report storm effects and several fish recently tagged off northwest Florida have moved extreme distances in the absence of storms (Patterson, unpublished data).

Implications for Population Structure

Postsettlement movement observed in red snapper has significant implications for population structure. Traditionally, population, or stock, structure in GOM red snapper has been evaluated with population genetics techniques. Results of studies designed to examine genetic population structure consistently have shown that interregional variability in selectively-neutral genetics markers, such as mtDNA and nuclear DNA microsatellites, is low. Thus, significant differences among regions in haplotype frequencies have not been found (Gold et al. 1997; Saillant and Gold 2006; Gold and Saillant 2007). However, Saillant and Gold (2006) reported 10-fold differences in genetic effective population size estimates among southwest, northwest, and north-central regions of the U.S. GOM, which they inferred likely reflected interregional differences in patterns and intensity of migration. Pruett et al. (2005) reported results of nested clade analysis performed on region-specific mtDNA haplotype frequencies were consistent with the hypothesis that red snapper populations were semi-isolated within regions, despite the lack of significant interregional genetic heterogeneity found in selectively neutral markers. Even in the absence of significant gene flow due to oceanographic transport of eggs and larvae, it is possible that extreme (>100 km) movement observed in some adults is sufficient to facilitate genetic mixing among regions (Nolan et al. 1991), yet also so rare as to be inconsequential to the maintenance of persistent interregional differences in population demographics (Policansky and Magnuson 1998). This may be especially true currently as relatively few large fish that are more likely to be movers exist in the truncated age distribution of the overfished stock (Porch 2007).

Previously narrow definitions of genetic stock structure in marine fishes have been replaced in recent years with metapopulation concepts due to issues similar to those raised above for red snapper (Thorrold et al. 2001; Kritzer and Sale 2004, 2005). Kritzer and Sale (2005) stated that in order to invoke metapopulation dynamics, subpopulations must be distinct, have dispersal mechanisms among them, and display asynchrony in population dynamics; Levins' (1969) earlier emphasis on extinction risk was abandoned. Pruett et al. (2005) concluded that results of nested clade analysis, in light of asynchronous population dynamics parameters among GOM regions, were consistent with the interpretation that GOM red snapper constituted a metapopulation. I submit that red snapper movement data, presented here in their various forms, also are consistent with that interpretation. Movement sufficient to affect mixing of genetic resources among regions has been demonstrated, but not on scales that would be likely to diminish regional differences in population demographics, such as those demonstrated by Fischer et al. (2004) and Jackson et al. (2007).

Pruett et al. (2005) indicated precise estimates of movement among regions are required in order to test whether GOM red snapper constitute a metapopulation. Other authors also have recognized that estimates of interpopulational mixing rates on ecological versus evolutionary time scales are required to assess metapopulation structure in fishes (Secor 1999; reviewed in Sale et al. 2005). Perhaps the most powerful tool yet found for that purpose is the use of otolith chemistry as a natural tag to track movement of fish among regions (Begg et al. 2005; Campana 2005). Since Thorrold et al.'s (2001) groundbreaking work employing otolith chemistry as a natural tag to examine weakfish, *Cynoscion regalis*, natal homing and population connectivity, several authors have likewise drawn inferences about population structure in marine fishes via natural tags based on otolith elemental signatures (e.g., Geffen et al. 2003; Miller and Shanks 2004; Hamer et al. 2005; Jonsdottir et al. 2006).

Application of otolith chemistry as a natural tag also has been applied to examine population structure in GOM red snapper. Patterson et al. (1998) reported otolith elemental signatures of age 0 red snapper were significantly different among north-central, northwestern, and southwestern regions of the U.S. GOM. Patterson et al. (in press) reported region-specific otolith elemental signatures (Ba, Mg, Mn, and Sr concentrations) of age 0 fish were significantly different among five successive (1996–2000) year classes, and that classification success in most (n = 4 of 5) years was sufficient to employ signatures as natural tags (e.g., jackknifed classification success of liner discriminant function models approached 80%). Analysis of the core elemental chemistry of otoliths collected from members of the 1996–2000 year classes sampled among study regions in 2001 revealed red snapper displayed strong intraregional philopatry in the first year of life (Cowan et al. 2002). Almost no mixing was estimated to have occurred between the north-central and northwestern GOM as fish aged, which is consistent with data from conventional tagging studies that have not demonstrated mixing between areas east and west of the Mississippi River. Mixing between the northwestern and southwestern regions, however, was greater, with a net subsidy of recruits apparently provided to the southwestern region from the northwestern region.

Overall, postsettlement movement data presented herein are consistent with the inference that GOM red snapper constitute a metapopulation. However, as suggested by Pruett et al. (2005) and despite the resources already invested in estimating movement in red snapper, more precise estimates of movement and exchange rates are required to assess interpopulational connectivity. For example, conventional tagging studies conducted to date have been designed to estimate movement away from tagging sites in a given GOM region, but not necessarily to estimate connectivity among regions. Coordination among ongoing tagging programs would prove beneficial for that purpose. Otolith chemistry has shown great promise as a tool to examine population structure and connectivity in GOM red snapper, but misclassification error (~20%) of age 0 fish is problematic. Furthermore, not all regions of the GOM where red snapper occur have been studied. Future otolith elemental chemistry research should examine elemental signatures from age 0 fish across the entire GOM Basin, as well as examine the potential for increased region-specific classification accuracies of age 0 fish by addition of other elements and stable isotope values of C and O to otolith signatures.

Acknowledgments

The author acknowledges the significant contributions reviewed here that others have made to understanding GOM red snapper population structure and the selection pressures that may affect movement and population connectivity. The idea for this paper originated when the author was a postdoctoral researcher working with C. Wilson, J. Cowan, and J. Gold. I thank them for stimulating discussions related to red snapper population ecology and population structure during that time and since. An earlier draft of this manuscript was improved based on critiques provided by D. Nieland and two anonymous reviewers.

References

Allman, R. J., and G. R. Fitzhugh. 2007. Temporal age progressions and relative year-class strength of Gulf of Mexico red snapper. Pages 311–328 *in* W. F. Patterson, III, J. H. Cowan, Jr., G. R. Fitzhugh, and D. L. Nieland, editors. Red snapper ecology and fisheries in the U.S. Gulf of Mexico. American Fisheries Society, Symposium 60, Bethesda, Maryland.

Almany, G. R., M. L. Berumen, S. R. Thorrold, S. Planes, and G. P. Jones. 2007. Local replenishment of coral reef fish populations in a marine reserve. Science 316:742–744.

Andersen, K. H., K. D. Farnsworth, U. H. Thygesen, and J. E. Beyer. 2007. The evolutionary pressure from fishing on size at maturation of Baltic cod. Ecological Modeling 204:246–252.

Ayre, D. J., and T. P. Hughes. 2000. Genotypic diversity and gene flow in brooding and spawning corals along the Great Barrier Reef, Australia. Evolution 54:1590–1605.

Bailey, H. K., IV, J. H. Cowan, Jr., and R. L. Shipp. 2001. Experimental evaluation of potential effects of habitat size and presence of conspecifics on habitat association by young-of- the-year red snapper. Gulf of Mexico Science 19:119–131.

Baskett, M. L., S. A. Levin, S. D. Gaines, and J. Dushoff. 2005. Marine reserve design and the evolution of size at maturation in harvested fish. Ecological Applications 15:882–901.

Beaumariage, D. S. 1969. Returns from the 1965 Schlitz tagging program including a cumulative analysis of previous results. Florida Department of Natural Resources Technical Series 59:1–38.

Begg, G. A., S. E. Campana, A. J. Fowler, and I. A. Suthers. 2005. Otolith research and application: current directions in innovation and implementation. Marine and Freshwater Research 56:477–484.

Bradley, E., and C. E. Bryan. 1975. Life history and fishery of the red snapper (*Lutjanus campechanus*) in the northwestern Gulf of Mexico: 1970–1974. Proceedings of the Gulf and Caribbean Fisheries Institute 27:77–106.

Bryant, H. E., M. R. Dewey, N. A. Funicelli, G. M. Ludwig, D. A. Meineke, and L. J. Mengal. 1989. Movement of five selected sportfish species of fish in Everglades National Park. Bulletin of Marine Science 44:515.

Campana, S. E. 2005. Otolith science entering the 21st century. Marine and Freshwater Research 56:485–496.

Collins, L. A., J. H. Finucane, and L. E. Barger. 1980. Description of larval and juvenile red snapper, *Lutjanus campechanus*. Fishery Bulletin 77:965–974.

Collins, L. A., A. G. Johnson, and C. P. Keim. 1996. Spawning and annual fecundity of the red snapper (*Lutjanus campechanus*) from the northeastern Gulf of Mexico. Pages 174–188 *in* F. Arreguin-Sanchez, J. L. Munro, M. C. Balgos, and D. Pauly, editors. Biology, fisheries and culture of tropical groupers and snappers. ICLARM Conference Proceedings Number 48.

Cowan, J. H., Jr., M. Woods, W. Patterson, III, and D. Nieland. 2002. Otolith microchemistry (and reproductive biology) portion- Stock structure of red snapper in the northern Gulf of Mexico: is their management as a single stock justified based on spatial and temporal patterns of genetic variation, otolith microchemistry, and growth rates. National Marine Fisheries Service, Marine Fisheries Initiative (MARFIN) Grant NA87FF0425.

Cowen, R. K., and L. R. Castro. 1994. Relation of coral reef fish larval distributions to island scale circulation around Barbados, West Indies. Bulletin of Marine Science. 54:228–244.

Cowen, R. K., C. B. Paris, and A. Srinivasan. 2006. Scaling of connectivity in marine populations. Science 311:522–527.

Cowen, R. K., K. M. M. Lwiza, S. Sponaugle, C. B. Paris, and D. B. Olson. 2000. Connectivity of marine populations: open or closed? Science 287:857–859.

Diamond, S. L., M. D. Campbell, D. Olson, Y. Wang, J. Zeplin, and S. Qualia. 2007. Movers and stayers: individual variability in site fidelity and movements of red snapper off Texas. Pages 163–187 *in* W. F. Patterson, III, J. H. Cowan, Jr., G. R. Fitzhugh, and D. L. Nieland, editors. Red snapper ecology and fisheries in the U.S. Gulf of Mexico. American Fisheries Society, Symposium 60, Bethesda, Maryland.

Dieckmann, U., B. O'Hara, and W. Weisser. 1999. The ecology of dispersal. Trends in Ecology & Evolution 14:88–90.

Doebeli, M., and G. D. Ruxton. 1997. Evolution of dispersal rates in metapopulation models: branching and cyclic dynamics in phenotype space. Evolution 51:1730–1741.

Doherty, P. J., Planes, S., and P. Mather. 1995. Gene flow and larval duration in seven species of fish from the Great Barrier Reef. Ecology 76:2373–2391.

Drass, D. M., K. L. Bootes, J. Lyczkowski-Shultz, B. H. Comyns, G. J. Holt, C. M. Riley, and R. P. Phelps. 2000. Larval development of red snapper, *Lutjanus campechanus*, with comparisons to co-occurring snapper species. Fishery Bulletin 98:507–527.

Dufrene, T. A., S. J. Bentley, and Y. C. Allen. 2003. Geologic and geoacoustic study of surficial deposits, north-central Gulf of Mexico continental shelf. Transactions of the Gulf Coast Association of Geological Societies 53:210–216.

Fable, W. A., Jr. 1980. Tagging studies of red snapper (*Lutjanus campechanus*) and vermilion snapper (*Rhomboplites aurorubens*) off the south Texas coast. Contributions in Marine Science 23:115–121.

Fable, W. A., Jr. 1990. Summary of king mackerel tagging in southeastern USA: mark-recapture techniques and factors influencing tag returns. Pages 161–167 *in* N. C. Parker, editor. Fish Marking Techniques. American Fisheries Society, Symposium 7, Bethesda, Maryland.

Fischer, A. J., M. S. Baker, C. A. Wilson. 2004. Red snapper (*Lutjanus campechanus*) demographic structure in the northern Gulf of Mexico based on spatial patterns in growth rates and morphometrics. Fishery Bulletin 102:593–603.

Fitzhugh, G. R., C. C. Koenig, F. C. Coleman, C. B. Grimes, and W. Sturges. 2005. Spatial and temporal patterns in fertilization and settlement of young gag (*Mycteroperca microlepis*) along the West Florida Shelf. Bulletin of Marine Science 77:377–396.

Franks, J. S., J. R. Hendon, and N. M. Garber. 2004. Red snapper (*Lutjanus campechanus*) associated with a small artificial structure in the Mississippi Sound, a northern Gulf of Mexico estuary. Proceedings of the Gulf and Caribbean Fisheries Institute 55:853–864.

Fraser, D. F., J. F. Gilliam, M. J. Daley, A. N. Le, and G. T. Skalski. 2001. Explaining leptokurtic movement distributions: Intrapopulation variation in boldness and exploration. American Naturalist 158:124–135.

Futch, R. B., and G. E. Bruger. 1976. Age, growth, and reproduction of red snapper in Florida waters. Pages 165–184 *in* H. R. Bullis and A. C. Jones, editors. Proceedings: Colloquium on snapper-grouper fishery resources for the western central Atlantic Ocean. Florida Sea Grant College Report 17.

Geary, B. W., J. J. Mikulas, Jr., J. R. Rooker, A. M. Landry, Jr. and T. M. Dellapenna. 2007. Patterns of habitat used by newly settled red snapper in the northwestern Gulf of Mexico. Pages 25–38 *in* W.

F. Patterson, III, J. H. Cowan, Jr., G. R. Fitzhugh, and D. L. Nieland, editors. Red snapper ecology and fisheries in the U.S. Gulf of Mexico. American Fisheries Society, Symposium 60, Bethesda, Maryland.

Geffen, A. J., K. Jarvis, J. P. Thorpe, R. T. Leah, and R. D. M. Nash. 2003. Spatial differences in the trace element concentrations of Irish Sea plaice *Pleuronectes platessa* and whiting *Merlangius merlangus* otoliths. Journal of Sea Research 50:247–256.

Gilg, M. R., and T. J. Hilbish. 2003. Patterns of larval dispersal and their effect on the maintenance of a blue mussel hybrid zone in southwestern England. Evolution 57:1061–1077.

Gledhill, C., and A. David. 2004. Survey of fish assemblages and habitat within two marine protected areas on the west Florida shelf. Proceedings of the Gulf and Caribbean Fisheries Institute 55:614–625.

Gold, J. R., E. Pak, and L. R. Richardson. 2001. Microsatellite variation among red snapper (*Lutjanus campechanus*) from the Gulf of Mexico. Marine Biotechnology 3:293–304.

Gold, J. R., and E. Saillant. 2007. Population structure of red snapper in the northern Gulf of Mexico. Pages 201–216 *in* W. F. Patterson, III, J. H. Cowan, Jr., G. R. Fitzhugh, and D. L. Nieland, editors. Red snapper ecology and fisheries in the U.S. Gulf of Mexico. American Fisheries Society, Symposium 60, Bethesda, Maryland.

Gold, J. R., F. Sun, and L. R. Richardson. 1997. Population structure of red snapper from the Gulf of Mexico as inferred from analysis of mitochondrial DNA. Transactions of the American Fisheries Society 126:386–396.

Goodyear, C. P. 1988. Recent trends in the red snapper fishery of the Gulf of Mexico. National Marine Fisheries Service, Southeast Fisheries Science Center, Miami Laboratory Report CRD-87/88–16, Miami, Florida.

Goodyear, C. P. 1995. Red snapper stocks in U.S. waters of the Gulf of Mexico. National Marine Fisheries Service, Southeast Fisheries Center, Miami Laboratory, CRD 95/96–05, Miami, Florida.

Hamer, P. A., G. P. Jenkins, and B. M. Gillanders. 2005. Chemical tags in otoliths indicate the importance of local and distant settlement areas to populations of a temperate sparid, *Pagrus auratus*. Canadian Journal of Fisheries and Aquatic Sciences 62:623–641.

Hanisko, D. S., J. Lyczkowski-Shultz. 2003. Occurrence and pelagic habitat of reef fish larvae in the Gulf of Mexico. Pages 205–224 *in* D. R. Stanley and A. Scarborough-Bull, editors. Fisheries, reefs, and offshore development. American Fisheries Society, Symposium 36, Bethesda, Maryland.

Hanisko, D. S., J. Lyczkowski-Shultz, and G. W. Ingram. 2007. Indices of larval red snapper occurrence and abundance for use in stock assessment. Pages 285–300 *in* W. F. Patterson, III, J. H. Cowan, Jr., G. R. Fitzhugh, and D. L. Nieland, editors. Red snapper ecology and fisheries in the U.S. Gulf of Mexico. American Fisheries Society, Symposium 60, Bethesda, Maryland.

Hjort, J. 1914. Fluctuations in the great fisheries of Northern Europe. Conseil Permanent International pour L'Explorations de la Mer 20:1–228.

Holland, D. S., and R. J. Brazee. 1993. Marine reserves for fisheries management. Marine Resource Economics 11:157–171.

Hood, P. B., A. J. Strelcheck, and P. Steele. 2007. A history of red snapper management in the Gulf of Mexico. Pages 267–284 *in* W. F. Patterson, III, J. H. Cowan, Jr., G. R. Fitzhugh, and D. L. Nieland, editors. Red snapper ecology and fisheries in the U.S. Gulf of Mexico. American Fisheries Society, Symposium 60, Bethesda, Maryland.

Ingram, G. W., and W. F. Patterson, III. 2001. Movement patterns of red snapper (*Lutjanus campechanus*), greater amberjack (*Seriola dumerili*), and gray triggerfish (*Balistes capriscus*) in the Gulf of Mexico and the utility of marine reserves as management tools. Proceedings of the Gulf and Caribbean Fisheries Institute 52:686–699.

Jackson, M. W., J. H. Cowan, Jr., and D. A. Nieland. 2007. Demographic differences in northern Gulf of Mexico red snapper reproductive maturation: implications for the unit stock hypothesis. Pages 217–227 *in* W. F. Patterson, III, J. H. Cowan, Jr., G. R. Fitzhugh, and D. L. Nieland, editors. Red snapper ecology and fisheries in the U.S. Gulf of Mexico. American Fisheries Society, Symposium 60, Bethesda, Maryland.

James, M. K., P. R. Armsworth, L. B. Mason, and L. Bode. 2002. The structure of reef fish metapopulations: modelling larval dispersal and retention patterns. Proceedings of the Royal Society of London, Series B: Biological Sciences 269:2079–2086.

Johnson, M. S., and R. Black. 2006. Islands increase genetic subdivision and disrupt patterns of connectivity of intertidal snails in a complex archipelago. Evolution 60:2498–2506.

Jones, G. P., S. Planes, and S. R. Thorrold. 2005. Coral reef fish larvae settle close to home. Current Biology 15:1314–1318.

Jones, G. P., M. J. Milicich, M. J. Emslie, and C. Lunow. 1999. Self-recruitment in a coral reef fish population. Nature (London) 402:802–804.

Jonsdottir, I. G., S. E. Campana, and G. Mareinsdottir. 2006. Stock structure of Icelandic cod *Gadus morhua* L. based on otolith chemistry. Journal of Fish Biology 69:136–150.

Jonsson, B., and J. Jonsson. 1993. Partial migration: niche shift versus sexual maturation in fishes. Reviews in Fish Biology and Fisheries 3:348–365.

Kraus, R. T., R. L. Hill, J. R. Rooker, and T. M. Dellapenna. 2006. Preliminary characterization of a mid-shelf bank in the northwestern Gulf of Mexico as essential habitat of reef fishes. Proceedings of the Gulf and Caribbean Fisheries Institute 57:621–632.

Kritzer, J. P., and P. F. Sale. 2004. Metapopulation ecology in the sea: From Levins' model to marine ecology and fisheries science. Fish and Fisheries 5:131–140.

Kritzer, J. P., and P. F. Sale. 2005. The metapopulation ecology of coral reef fishes. Pages 31–68 *in* J. P. Kritzer and P. F. Sale, editors. Marine Metapopulations. Academic Press, Boston, Massachusetts.

Laswell, J. S., W. W. Sager, W. W. Schroeder, R. Rezak, K. S. Davis, and E. G. Garrison. 1990. Atlas of high-resolution geophysical data, Mississippi-Alabama marine ecosystem study. OCS Study/MMS 90–0000. U.S. Department of the Interior, Minerals Management Service, Gulf of Mexico OCS Regional Office, New Orleans, Louisiana.

Levins, R. 1969. Some demographic and genetic consequences of environmental heterogeneity for biological control. Bulletin of the Entomological Society of America 15:237–240.

Lindberg, W. J., T. K. Frazer, K.M. Portier, F. Vose, J. Loftin, D. J. Murie, D. M. Mason, B. Nagy, and M. K. Hart. 2006. Density-dependent habitat selection and performance by a large mobile reef fish. Ecological Applications 16:731–746.

Lorenzen, K. 2000. Allometry of natural mortality as a basis for assessing optimal release size in fish-stocking programmes. Canadian Journal of Fisheries and Aquatic Sciences 57:2374–2381.

Lorenzen, K. 2006. Population management in fisheries enhancement: Gaining key information from release experiments through use of a size-dependent mortality model. Fisheries Research 80:19–27.

Lyczkowski-Shultz, J., and D. S. Hanisko. 2007. A time series of observations on red snapper larvae from SEAMAP surveys, 1982–2003: seasonal occurrence, distribution, abundance and size. Pages 3–23 *in* W. F. Patterson, III, J. H. Cowan, Jr., G. R. Fitzhugh, and D. L. Nieland, editors. Red snapper ecology and fisheries in the U.S. Gulf of Mexico. American Fisheries Society, Symposium 60, Bethesda, Maryland.

Marko, P. B., S. C. Lee, A. M. Rice, J. M. Gramling, T. M. Fitzhenry, J. S. Mcalister, G. R. Harper, and A. L. Moran. 2004. Fisheries: mislabelling of a depleted reef fish. Nature (London) 430:309–310.

McCawley, J. R., and J. H. Cowan, Jr. 2007. Seasonal and size specific diet and prey demand of red snapper on Alabama artificial reefs. Pages 77–104 *in* W. F. Patterson, III, J. H. Cowan, Jr., G. R. Fitzhugh, and D. L. Nieland, editors. Red snapper ecology and fisheries in the U.S. Gulf of Mexico. American Fisheries Society, Symposium 60, Bethesda, Maryland.

Meyer, C. G., K. N. Holland, and Y. P. Papastamatiou. 2007. Seasonal and diel movements of giant trevally *Caranx ignobilis* at remote Hawaiian atolls: implications for the design of marine protected areas. Marine Ecology Progress Series 333:13–25.

Miller, J. A., and A. L. Shanks. 2004. Evidence for limited larval dispersal in black rockfish (*Sebastes melanops*): implications for population structure and marine-reserve design. Canadian Journal of Fisheries and Aquatic Sciences 61:1723–1735.

Mitchell, K. M., T. Henwood, G. R. Fitzhugh, and R. J. Allman. 2004. Distribution, abundance, and age structure of red snapper (*Lutjanus campechanus*) caught on research longlines in U.S. Gulf of Mexico. Gulf of Mexico Science 22:164–172.

Moseley, F. N. 1966. Biology of the red snapper, *Lutjanus aya Bloch*, of the northwestern Gulf of Mexico. Publication of the Institute of Marine Science 11:90–101.

Munch, S. B., and D. Conover. 2003. Rapid growth results in increased susceptibility to predation in *Menidia menidia.* Evolution 57:2119–2127.

Murakami, T., K. Yoshioka, S. Aida, T. Unimo, and H. Nakagawa. 2005. Relation of size at release of hatchery-reared black rockfish *Sebastes inermis* to survival in a *Zostera* belt around Ikuno Island, Hiroshima Prefecture, Japan. Bulletin of the Japanese Society of Scientific Fisheries 71:354–362.

Nieland, D. L., and C. A. Wilson. 2003. Red snapper recruitment to and disappearance from oil and gas platforms in the northern Gulf of Mexico. Pages 73–82 *in* D. R. Stanley and A. Scarborough-Bull, editors. Fisheries, reefs, and offshore development. American Fisheries Society, Symposium 36, Bethesda, Maryland.

Nolan, K., J. Grossfield, and I. Wirgin. 1991. Discrimination among Atlantic coast populations of American shad (*Alosa sapidissima*) using mitochondrial DNA. Canadian Journal of Fisheries and Aquatic Sciences 48:1724–1734.

Ogle, J. T., and J. M. Lotz. 2006. Characterization of an experimental indoor larval production system for red snapper. North American Journal of Aquaculture 68:86–91.

Overholtzer-McLeod, K. L. 2005. Post-settlement emigration affects mortality estimates for two Bahamian wrasses. Coral Reefs 24:283–291.

Papanikos, N., R. Phelps, K. Williams, A. Ferry, and D. Maus. 2003. Egg and larval quality of natural and induced spawns of red snapper, *Lutjanus campechanus*. Fish Physiology and Biochemistry 28:487–488.

Parker, R. O., Jr., D. R. Colby, and T. D. Willis. 1983. Estimated amount of reef habitat on a portion of the U.S. south Atlantic and Gulf of Mexico continental shelf. Bulletin of Marine Science 33:935–940.

Patterson, W. F., III and J. H. Cowan, Jr. 2003. Site fidelity and dispersion of tagged red snapper, *Lutjanus campechanus*, in the northern Gulf of Mexico. Pages 181–194 *in* D. R. Stanley and A. Scarborough-Bull, editors. Fisheries, Reefs, and Offshore Development. American Fisheries Society, Symposium 36, Bethesda, Maryland.

Patterson, W. F., III, J. H. Cowan, Jr., E. Y. Graham, and W. B. Lyons. 1998. Otolith microchemical fingerprints of age-0 red snapper, *Lutjanus campechanus*, from the northern Gulf of Mexico. Gulf of Mexico Science 16:83–91.

Patterson, W. F., III, G. W. Ingram, Jr., R. L. Shipp, and J. H. Cowan, Jr. 2002. Indirect estimation of red snapper (*Lutjanus campechanus*) and gray triggerfish (*Balistes capriscus*) release mortality. Proceedings of the Gulf and Caribbean Fisheries Institute 53:526–536.

Patterson, W. F., III, J. H. Cowan, Jr., C. A. Wilson, and Z. Chen. In press. Temporal and spatial variability in juvenile Gulf of Mexico red snapper otolith elemental signatures. Transactions of the American Fisheries Society.

Patterson, W. F., III, J. C. Watterson, R. L. Shipp, and J. H. Cowan, Jr. 2001. Movement of tagged red snapper in the northern Gulf of Mexico. Transactions of the American Fisheries Society 130:533–545.

Patterson, W. F., III, C. A. Wilson, J. H. Cowan, Jr., S. J. Bentley, T. A. Henwood, Y. C. Allen, and T. A. Dufrene. 2005. Delineating essential juvenile red snapper habitat in the north central Gulf of Mexico. Pages 277–288 *in* P. W. Barnes and J. P. Thomas, editors. Effects of fishing on benthic habitats. American Fisheries Society, Symposium 41, Bethesda, Maryland.

Policansky, D. and J. J. Magnuson. 1998. Genetics, metapopulations, and ecosystem management of fisheries. Ecological Applications 8(1):S119–S123.

Porch, C. E. 2007. An assessment of the red snapper fishery in the U.S. Gulf of Mexico using a spatially-explicit age-structured model. Pages 355–384 *in* W. F. Patterson, III, J. H. Cowan, Jr., G. R. Fitzhugh, and D. L. Nieland, editors. Red snapper ecology and fisheries in the U.S. Gulf of Mexico. American Fisheries Society, Symposium 60, Bethesda, Maryland.

Potthoff, T., S. Kelly, and L. A. Collins. 1988. Osteological development of the red snapper, *Lutjanus campechanus*. Bulletin of Marine Science 43:1–40.

Pruett, C. L., E. Saillant, and J. R. Gold. 2005. Historical population demography of red snapper (*Lutjanus campechanus*) from the northern Gulf of Mexico based on analysis of sequences of mitochondrial DNA. Marine Biology 147: 593–602.

Rooker, J. R., A. Landry, Jr., B. W. Geary, and J. A. Harper. 2004. Assessment of a shell bank and associated substrates as nursery habitat of postsettlement red snapper. Estuarine, Coastal and Shelf Science 59:653–661.

Rummer, J. L. 2007. Factors affecting catch and release (CAR) mortality in fish: insight into CAR mortality in red snapper and the influence of catastrophic decompression. Pages 123–144 *in* W. F. Patterson, III, J. H. Cowan, Jr., G. R. Fitzhugh, and D. L. Nieland, editors. Red snapper ecology and fisheries in the U.S. Gulf of Mexico. American Fisheries Society, Symposium 60, Bethesda, Maryland.

Saillant, E., and J. R. Gold. 2006. Population structure and variance effective size of red snapper (*Lutjanus campechanus*) in the northern Gulf of Mexico. Fishery Bulletin 104:136–148.

Sale, P. F., R. K. Cowen, B. S. Danilowicz, G. P. Jones, J. P. Kritzer, K. C. Lindeman, S. Planes, N. V. C. Polunin, G. R. Russ, Y. J. Sadovy, and R. S. Steneck. 2005. Critical science gaps impede use of no-take fishery reserves. Trends in Ecology & Evolution 20:74–80.

Schroeder, W. W., A. W. Shultz, and J. J. Dindo. 1988. Inner-shelf hardbottom areas, northeastern Gulf of Mexico. Transactions of the Gulf Coast Association of Geological Societies 38:535–541.

Schroepfer, R. L., and S. T. Szedlmayer. 2006. Estimates of residence and site fidelity for red snapper *Lutjanus campechanus* on artificial reefs in the northeastern Gulf of Mexico. Bulletin of Marine Science 78:93–101.

Secor, D. H. 1999. Specifying divergent migrations in the concept of stock: the contingent hypothesis. Fisheries Research 43:13–34.

Secor, D. H. 2002. Historical roots of the migration triangle. ICES Marine Science Symposia 215:229–335.

Secor, D. H. 2006. The year-class phenomenon and the storage effect in marine fishes. Journal of Sea Research 57:91–103.

Sibly, R. M., P. Calow, and N. Nichols. 1985. Are patterns of growth adaptive? Journal of Theoretical Biology 112:553–574.

Sogard, S. M. 1997. Size-selective mortality in the juvenile stage of teleost fishes: a review. Bulletin of Marine Science 60:1129–1157.

Strelcheck, A., J. H. Cowan, Jr., and W. F. Patterson, III. 2007. Site fidelity, movement, and growth of red snapper: Implications for artificial reef management. Pages 147–162 *in* W. F. Patterson, III, J. H. Cowan, Jr., G. R. Fitzhugh, and D. L. Nieland, editors. Red snapper ecology and fisheries in the U.S. Gulf of Mexico. American Fisheries Society, Symposium 60, Bethesda, Maryland.

Strelcheck, A. J., J. H. Cowan, Jr., and A. Shah. 2005. The influence of reef location on artificial reef fish assemblages in the northcentral Gulf of Mexico. Bulletin of Marine Science 77:425–440.

Swearer, S. E., J. S. Shima, M. E. Hellberg, S. R. Thorrold, G. P. Jones, D. R. Robertson, S. G. Morgan, K. A. Selkoe, G. M. Ruiz, and R. R. Warner. 2002. Evidence of self-recruitment in demersal marine populations. Bulletin of Marine Science 70:251–271.

Szedlmayer, S. T. 1997. Ultrasonic telemetry of red snapper, *Lutjanus campechanus*, at artificial reef sites in the northeast Gulf of Mexico. Copeia 1997:846–850.

Szedlmayer, S. T., and J. Conti. 1999. Nursery habitats, growth rates, and seasonality of age-0 red snapper, *Lutjanus campechanus*, in the northeast Gulf of Mexico. Fishery Bulletin 97:626–635.

Szedlmayer, S. T., and J. C. Howe. 1997. Substrate preference in age-0 red snapper, *Lutjanus campechanus*. Environmental Biology of Fishes 50:203–207.

Szedlmayer, S. T., and J. D. Lee. 2004. Diet shifts of juvenile red snapper (*Lutjanus campechanus*) with changes in habitat and fish size. Fishery Bulletin 102:366–375.

Szedlmayer, S. T., and R. L. Schroepfer. 2005. Long-term residence of red snapper on artificial reefs in the northeastern Gulf of Mexico. Transactions of the American Fisheries Society 134:315–325.

Szedlmayer, S. T., and R. L. Shipp. 1994. Movement and growth of red snapper, *Lutjanus campechanus*, from an artificial reef area in the northeastern Gulf of Mexico. Bulletin of Marine Science 55:887–896.

Thorrold, S. R., C. Latkoczy, P. K. Swart, and C. M. Jones. 2001. Natal homing in a marine fish metapopulation. Science 291:297–299.

Trexler, J. C., and J. Travis. 2000. Can marine protected areas restore and conserve stock attributes of reef fishes? Bulletin of Marine Science 66:853–873.

Wells, R. J. D. 2007. The effects of trawling and habitat use on red snapper and the associated community. Doctoral dissertation. Louisiana State University, Baton Rouge, Louisiana.

Westmeyer, M. P., C. A. Wilson, and D. L. Nieland. 2007. Fidelity of red snapper to petroleum platforms in the northern Gulf of Mexico. Pages 105–121 *in* W. F. Patterson, III, J. H. Cowan, Jr., G. R. Fitzhugh, and D. L. Nieland, editors. Red snapper ecology and fisheries in the U.S. Gulf of Mexico. American Fisheries Society, Symposium 60, Bethesda, Maryland.

Wilson, C. A., and D. L. Nieland. 2001. Age and growth of red snapper, *Lutjanus campechanus*, from the northern Gulf of Mexico off Louisiana. Fishery Bulletin 99:653–664.

Wilson, C. A., and D. L. Nieland. 2004. The role of oil and gas platforms in providing habitat for northern Gulf of Mexico red snapper, *Lutjanus campechanus*. Proceedings of the Gulf and Caribbean Fisheries Institute 55:757–764.

Workman, I. K., and D. G. Foster. 1994. Occurrence and behavior of juvenile red snapper, *Lutjanus campechanus*, on commercial shrimp fishing grounds in the northeastern Gulf of Mexico. Marine Fisheries Review 56:9–11.

Workman, I., A. Shah, D. Foster, and B. Hataway. 2002. Habitat preferences and site fidelity of juvenile red snapper (*Lutjanus campechanus*). ICES Journal of Marine Science 59:S43–S50.

Fisheries Management and Conservation Overview

James H. Cowan, Jr.

The third section of this book contains 8 papers devoted to red snapper Fisheries Management and Conservation, and well illustrates how far our understanding both of ecology and stock dynamics has progressed since 1989 when the first rebuilding plan was implemented. The first paper in this section is by Hood, who provides a detailed description of the controversial history of red snapper management in the U.S. Gulf of Mexico. The second paper, by Hanisko and coauthors, describes the development and use of indices of larval abundance; relative abundance estimates are derived separately for the eastern and western Gulf of Mexico and are used as tuning indices in stock assessments. Nieland and coauthors in paper 3 of this section report a significant decrease in mean total length at age for ages 2–6 years in commercial catches of red snapper off Louisiana, and provide contrasting interpretations of potential causes for their observations. Allman and Fitzhugh use otolith-aged red snapper obtained from landings by commercial hook and line, longline and the recreational sector over a 12-year period to examine differences in age-structure over time and space, relative year-class success, and the impacts of observed differences on each sector of the fishery. In a second paper in this section, Neiland and coauthors describe samples taken in cooperation with commercial fishers off Louisiana to show that both regulatory discard and mortality rates were high during their study and conclude that minimum size regulations appear to do little to protect juvenile red snapper from commercial fishing mortality. The historical reconstruction of landings data is the subject of a paper by Porch and coauthors, who conclude that a substantial fishery for red snapper existed in the Gulf of Mexico as early as 1872, and that by the turn of the century landings were comparable to those of recent times, but may have been more dependent upon fishing grounds located in Mexico. Porch contributes another paper to this section that illustrates the state of assessment science for red snapper in the Gulf of Mexico, describing a flexible age-structured assessment model that includes data back to the inception of the fishery as described above. Porch's results are consistent with the most recent NMFS findings, based upon a shorter time-series of landings data, that are being used to define the current biological bench marks for the red snapper stock (SEDAR 2005). Interpretation of results from both Porch and the NMFS assessments indicate that red snapper biomass and other biological benchmarks are well below levels in the Gulf of Mexico that are considered to be risk-averse from a conservation perspective, and that stocks are not likely to recover in the near-term without substantial reductions in fishing mortality relative to assessed levels. In the final paper in this section of the book, Strelcheck and Hood provide more support for the above assertion, reporting that red snapper remain overfished and are experiencing overfishing based upon the most recent NMFS assessment (SEDAR 2005), and cast further doubt about whether existing plans for rebuilding the stock will be sufficient to accomplish long-term goals, even given that recovery under the plan is not required until the year 2032.

As is demonstrated by the collection of papers in this book, information about the biology and ecology of this species far exceeds that for many species under management in U.S. waters, and methods of data gathering and processing, and stock assessment techniques have withstood numerous internal and external reviews over the period of record. In fact, I believe that the methods being used to assess the status of red snapper in the U.S. Gulf of Mexico are state-of-the-art. Yet, management of this species has been controversial since the beginning, despite the fact that data to inform the assessment process has increased enormously, and the message concerning stock status and the need to reduce fishing mortality rate has been remarkably consistent. I suggest that the controversy arose not because of real doubts about the quality of advice being given to fisheries governance by assessment scientists, but rather because recreational and commercial fishers almost equally split the directed harvest, while much of the mortality (>80%) from fishing on this population occurs as bycatch of juveniles in the shrimp fishery. As such, the red snapper is a prime example of the 'paradox of the commons' and a poster child for illustrating the paralyzing affects that user conflict can have on political will and U.S. fisheries governance as it exists today. I have been a direct observer of these conflicts, as I was a member and chair of the reef fish stock assessment panel, or a member of the standing scientific and statistical committee (panels of experts that serve in an advisory capacity to the Gulf of Mexico Fisheries Management Council) from 1992–2006.

I believe that because of political pressures, fisheries governance failed all of the users of this valuable resource by knowingly acting in a risk-prone manner with respect to controls on fishing mortality; however, I do recognize that user conflicts sometimes constrain management options. Very early in the history of management, however, stock assessment scientists determined that reductions in the level of juvenile mortality attributable to bycatch would be required to recover the stock in the absence of significant, some argued draconian, cuts in directed harvest. At the time, around 1990, stock assessments indicated that bycatch reductions on the order of 40–50%, along with significant reductions in the directed harvest would be necessary to ensure recovery. Assessment models were used to generate a range of "allowable biological catches" (ABC) for the directed fishery that expressed both the uncertainties in the assessment process, and the probability of recovery given some future, but critically important, technological solution to reduce bycatch. From that point forward, rather than reducing catch by the directed fishery to levels that increased the likelihood of recovery, fisheries governance chose to manage predicated upon the notion that bycatch reduction would occur, and selected "total allowable catch (TAC)" for the directed fishery from the high-risk range of ABCs. Others defended the choice of risk prone TACs by suggesting that red snapper stock productivity increased in response to the addition of new habitat in the form of artificial reefs and oil and gas platforms in the northern Gulf of Mexico, despite little or no evidence that habitat was limiting to stock size, and habitat area added by all artificial structures combined represents less that 5% of available natural habitats. So, we fished on in the face of increasing evidence, some of which appears in this volume, that things were going south.

Where are we today? Language in the most recent reauthorization of the Sustainable Fisheries Act (January 2007) mandates that overfishing in U.S. federal waters must end on all species by 2010. Red snapper are overfished and fishing mortality is still too high. A technological solution to bycatch reduction through use of "bycatch reduction devices" has been elusive (only a 10–15% reduction in CPUE of juveniles has been achieved to date), owing mostly to a quirk in juvenile red snapper behavior. Moreover, recent study (Wells 2007) suggests that natural mortality rates of age-0 red snapper are higher than previously thought, which if true, implies that attempts to further reduce bycatch via either seasonal or area closures of shrimp fishing in nursery areas would result in diminishing returns. The commercial sector of the directed fishery is now operating under a dedicated access program that has successfully eliminated derby conditions, but which also may allow fishers to prospect for better fishing grounds on offshore natural habitats that have been largely unfished for many years because of combined effects of regulation and distance from shore. Evidence is also mounting that, for red snapper, any increase in production potential attributable to artificial habitats is being more than offset by increased vulnerability to recreational and commercial fishing on these structures, and there is no evidence that red snapper can now, or have ever been able to, withstand harvest levels much in excess of current yields. In response to all of this, fisheries governance has recently accepted the advice of assessment scientists by reducing directed harvest levels by around 45%. Still, I think that risk-prone decisions in the past have made the future of this fishery and conservation of this resource uncertain. We shall see.

References

SEDAR 2005. Southeast Data, Assessment, and Review: Stock Assessment Report of SEDAR 7: Gulf of Mexico Red Snapper. SEDAR7. Charleston, South Carolina.

Wells, R. J. D. 2007. The effects of trawling and habitat use on red snapper and the associated community. Doctoral dissertation. Louisiana State University, Baton Rouge, Louisiana.

American Fisheries Society Symposium 60:267–284, 2007

A History of Red Snapper Management in the Gulf of Mexico

Peter B. Hood[1], Andrew J. Strelcheck, and Phil Steele
NOAA Fisheries Service
Southeast Regional Office
263 13th Avenue South, St. Petersburg, Florida 33701 USA

Abstract.—The red snapper *Lutjanus campechanus* fishery has been in existence in the Gulf of Mexico since the mid-1800s. However, management of this species did not begin until more than a century afterward. Federal management of the fisheries in the Gulf of Mexico began in 1976 with the passage of the Magnuson-Stevens Fishery Conservation and Management Act and the establishment of the Gulf of Mexico Fishery Management Council (Council). One of the first fishery management plans (FMP) developed by the Council was the Reef Fish FMP. This FMP was implemented in November 1984 and established the first red snapper size and bag limits. In 1988, the stock was determined to be overfished. Since then, the fishery has been managed to stay within total allowable catch levels in order to rebuild the stock. Management methods have included size limits, bag limits, season closures, trip limits, and license limitation programs. The success of these methods has been limited in part due to high levels of juvenile red snapper mortality associated with shrimp trawling, high rates of discard mortality from the directed fishery, and socioeconomic requirements of the directed fisheries to maintain some minimal level of harvest.

Introduction

The Gulf of Mexico red snapper *Lutjanus campechanus* fishery has been in existence in the Gulf of Mexico (GOM) since the 1840s. The development of the commercial fishery has been well described by Camber (1955) and Carpenter (1965). The fishery began in the northeastern GOM centering on Pensacola, Florida. Because of readily available ice and trains to transport the fish, landings grew into the millions of pounds. By the early 1900s, landings regularly exceeded 10 million pounds (mp). As local waters began to be depleted, the fishery expanded south into the Florida Middle Grounds off Tampa Bay, west into the Texas Lumps, and southwest to the Campeche Banks off Mexico (Camber 1955; Carpenter 1965). Technological developments such as motorized vessels and fathometers allowed the fishery to become more efficient (Camber 1955; Carpenter 1965). Landings peaked in the mid-1960s at 14 mp and have declined because of the closure of foreign waters and declines in the GOM red snapper population size (Waters 2003).

Reliable estimates of the recreational harvest of red snapper in the GOM were

[1] Corresponding author: peter.hood@noaa.gov.

not available prior to 1981. In general, the for-hire sector harvests more red snapper than the private-rental sector of the fishery (Schirripa and Legault 1999). Projected recreational harvests by the Southeast Data, Assessment, and Review (SEDAR) suggest recreational harvest did not exceed 1 mp until the mid-1960s and harvest peaked in 1980 at 4.5 mp (SEDAR 2005a). From 1981 to 1990, landings declined from 4.1 to 1.4 mp suggesting a decline in the GOM stock (Schirripa and Legault 1999).

Federal management of the red snapper fishery in the exclusive economic zone (EEZ) began in 1976 when the Magnuson-Stevens Fishery Management and Conservation Act (Magnuson-Stevens Act) was implemented. This act established the responsibility for federal fishery management decision-making between the Secretary of Commerce (Secretary) and the Gulf of Mexico Fishery Management Council (Council). The fishery management plan (FMP) for reef fish, in which red snapper is included in the management unit, was implemented in 1981 and noted that commercial and recreational red snapper landings were in decline (GMFMC 1981). The first assessment of the red snapper stock was conducted in 1988 and showed the stock was overfished and undergoing overfishing (Goodyear 1988). The assessment also noted the GOM shrimp trawl fishery contributed heavily to the red snapper fishing mortality rate (*F*) by harvesting juvenile red snapper as bycatch. Therefore, the Council and NOAA Fisheries Service were obligated to rebuild this stock.

The Council and NOAA Fisheries Service have faced and continue to face several challenges to rebuild the red snapper stock. The greatest is constraining harvest by the directed fishery and shrimp trawl bycatch of juvenile red snapper to levels allowing the stock to rebuild, while allowing enough fish to be caught to maintain the economic viability of both the directed and shrimp fisheries. The purpose of this paper is to describe the federal management of GOM red snapper after the implementation of the Magnuson-Stevens Act, discuss the type of actions taken, and discuss the effectiveness of the management measures employed to date. This information helps to put into context rationale for research described in this volume. The paper also provides background material necessary for Strelcheck and Hood (2007, this volume) to discuss recent actions and future management challenges.

Management Process

The Council is responsible for preparing, monitoring, and revising FMPs within the GOM and the Secretary is responsible for implementing proposed FMPs and amendments after ensuring management measures are consistent with the Magnuson-Stevens Act and other applicable laws (Wallace and Fletcher 2000). The Council is composed of 17 members including state fisheries officials, stakeholders, and the NOAA Fisheries Service Regional Administrator (Magnuson-Stevens Act 302). In developing regulations, the Council relies on input from the public through scoping and public testimony; input from various Council-established panels comprised of stakeholders, biologists, economists, and sociologists; and input from NOAA Fisheries Service and other state and federal agencies (Wallace and Fletcher 2000). Once the Council finishes an FMP or amendment, it is sent to the Secretary for approval, disapproval, or partial approval. If regulations are needed to fulfill the mandates of the FMP or amendment, the Secretary is also responsible for developing a final rule implementing the regulations.

FMPs and amendments often require more than a year to go from development to implementation of a regulation. Actions are occasionally required to either address emergencies within a fishery or to provide a stopgap until an FMP or amendment are implemented (Wallace and Fletcher 2000). In such cases, the Council can request an emergency action or interim measure. These remain in effect for only 186 d after the date of publication of the rule and may be extended for one additional period of not more than 186 d provided the public has had an opportunity to comment on the emergency actions and interim measures. However, the Magnuson-Stevens Act also states when a Council requests either an emergency action or an interim measure be taken, the Council should also be actively preparing regulations addressing the emergency on a permanent basis.

Regulation of GOM red snapper did not begin until 1984 with the implementation of the Reef Fish FMP. This plan included a 13-in total length (TL) minimum size limit for both the commercial and recreational fisheries (Table 1). The first stock assessment of the GOM red snapper stock occurred in 1988 and concluded the stock was overfished and overfishing was occurring (Goodyear 1988). The assessment also noted the GOM shrimp trawl fishery contributed heavily to the red snapper *F* by harvesting age-0 and age-1 juvenile red snapper as bycatch. Because of the overfishing and overfished status of the stock, it was necessary for the Council and NOAA Fisheries Service to develop and implement further regulations to improve the stock's status.

Stock rebuilding and setting total allowable catch (TAC)

The Council has developed and modified a rebuilding plan for the overfished GOM red snapper stock. Considerations for the plan include a target (the level the stock needs to be rebuilt to), a time period (the time needed to achieve the target), and a harvest strategy (the level of TAC set over time that allows the stock to rebuild). In reviewing information on GOM reef fish stocks, the Council, in Amendment 1, developed a framework for setting TAC with the goal of stabilizing the long term stock condition of all reef fish species to 20% spawning stock biomass per recruit (SSBR), with TAC recommendations based on rebuilding overfished stocks by 2000 (Table 1).

The 1988 stock assessment suggested *F* in the directed red snapper fishery would need to be reduced by approximately 75% for the stock to rebuild by 2000 (Goodyear 1988). The Council felt to achieve such a reduction would create severe negative economic impacts in the directed fishery. Therefore, the Council selected actions in Amendment 1 predicted to reduce harvest and *F* by approximately 20%, recognizing further regulations would be needed to rebuild the stock. These actions set the commercial quota at 3.1 mp and established a seven-fish daily bag limit for the recreational fishery (Tables 1–3). Amendment 1 also set the allocations of reef fish between commercial and recreational fisheries based on the historical averages during the base period of 1979–1987. For red snapper, this allocation ratio was 51% commercial and 49% recreational.

A new red snapper stock assessment was conducted in 1990 (Goodyear and Phares 1990). This assessment concluded the stock condition was less than one percent of the target 20% SSBR and the rebuilding time period ending in 2000 was unrealistic. To rebuild the stock by 2000, a complete closure of the directed fishery would be required and there would also need to be a 60% reduction in shrimp trawl bycatch of juvenile red snapper. Therefore, the time period for red snapper rebuilding was extended to 2007 through Amendment 3 in a revised framework (Table 1). Additionally, the amendment revised OY and overfishing definitions, replaced the 20% SSBR target with a target of 20% spawning potential ratio (SPR). The Council and NOAA Fisheries Service did implement through a

Table 1. Year implemented, rule-making vehicle, action, and rationale for red snapper *Lutjanus campechanus* management measures from 1984 to 2006.

Year	Rule-making Vehicle	Action	Rationale
1984	FMP[1]	• 13 inch minimum TL	• Estimated 18-25% increase in yield • Some at this size sexually mature and have spawned
1990	Amendment 1[1]	• 7-fish bag limit • 3.1 mp commercial quota • Rebuilding goal 20% SSBR	• Actions estimated to achieve a 20 percent reduction in harvest.
1991	Amendment 3[1]	• Revise TAC framework to be more flexible	• Improve the efficiency of the TAC setting process
1991	Regulatory amendment[1]	• 2.04 mp commercial quota • 1.96 mp recreational allocation • Effect 50% bycatch reduction by 1994 in the shrimp fishery • Projected to achieve 20% SPR by 2007	• Reduces TAC an additional 20 percent • Should allow stock to rebuild to 20 percent SPR by 2007 • Further control F
1992	Emergency rule[2]	• Open commercial red snapper fishery from April 3 – May 14 with 1,000 lbs trip limit due to the season closing in just 53 days	• Ameliorate adverse economic caused by a short season, an influx of non-traditional vessels in the fishery, and depressed ex-vessel prices
1992	Amendment 4[1]	• Moratorium on the issuance of new reef fish commercial permits for three years	• Limit participation in an overcapitalized fishery and allow time to develop a limited-access fishery
1992	Emergency rule[2]	• Create commercial red snapper 2,000 lbs and 200 lbs endorsement for 1993	• Limit effort primarily to those with a historical dependence in the fishery • Allow a bycatch provision • Extend the fishing year
1992	Emergency rule[2]	• Close the commercial fishery from December 1, 1992 to February 15, 1993	• Provide time to implement trip limit endorsement system
1993	Regulatory amendment[1]	• 3.06 mp commercial quota • 2.94 mp recreational allocation • Projected to achieve 20% SPR by 2009 • Change opening day of the 1994 commercial season to February 10 • Restrict commercial vessels to landing no more than one trip limit per day	• Continue rebuilding plan • Facilitate enforcement of the trip limits • Minimize fishing during hazardous winter weather • Ensure the commercial red snapper fishery is open during Lent
1993	Amendment 6[1]	• Extended commercial red snapper endorsements	• Limit effort primarily to those with a historical dependence in the fishery • Allow a bycatch provision • Extend the fishing year
1994	Amendment 5[1]	• Raise minimum size limit incrementally from 14 to 16 inches TL over a 5-year period • Establish Class 1 and Class 2 licenses • Create Alabama SMZs	• Increase yield per recruit and help rebuild the stock • Limit pulse and derby commercial fishery • Limit fishing on artificial reefs off Alabama
1994	Regulatory amendment[1]	• Change opening day of the commercial season to February 24, 1995 • Retain 6 million pound red snapper TAC and commercial trip limits • Reduced the daily bag limit from 7 fish to 5 fish • Increase the minimum size limit for recreational fishing from 14 inches to 15 inches a year ahead of the scheduled automatic increase.	• Ensure the commercial red snapper fishery is open during Lent • Continue rebuilding plan • Because the recreational sector exceeded its 2.94 million pound red snapper allocation each year since 1992, further restrict recreational F
1994	Amendment 7[1]	• Establish dealer reporting	• Improve accountability for landings
1995	Regulatory amendment[1]	• Raise TAC from 6 mp to 9.12 mp • Start commercial season February 28	• Revise rebuilding plan taking into account new information • Ensure the commercial red snapper fishery is open during Lent
1994	Amendment 9[1]	• Allow collection of commercial landings 1990-92 for ITQ • Extend the moratorium on the issuance of new reef fish permits	• Need for historical red snapper landings for commercial fishermen to establish baseline information for an IFQ program • Allow time for evaluation and development of a more comprehensive controlled access system
1995	Amendment 8[1]	• Attempted to establish ITQ system (Congress repealed it)	• Reduce overcapitalization of commercial fishery • End derby fishery • Reduce user conflicts
1996	Regulatory amendment[1]	• Increase TAC to 9.12 mp • Extend recovery date to 20% SPR to 2019 • Split commercial quota in a spring and fall season	• TAC recommendations based on a new stock assessment and recovery plan range from 6 to 10 mp • Provide commercial fishermen an income going into the fall holiday season

Table 1. (Continued)

Year	Rule-making Vehicle	Action	Rationale
1996	Amendment 13[1]	• Extend the red snapper endorsement system through the remainder of 1996 and, if necessary, through 1997, in order to give the Council time to develop a permanent limited access system	• Continue permit limitations to avoid open access to red snapper by all commercially permitted vessels
1997	Amendment 12[1]	• NMFS disapproved proposed provisions to cancel the automatic comm. red snapper size limit increases to 15 inches total length in 1996 and 16 inches total length in 1998	• Minimum size limit increase assumes a 33 % discard mortality rate, a rate thought to be too high.
1997	Regulatory amendment[1]	• Change start of fall season from 9/15 to 9/2 • Fall season first 15 days of each month until the quota is filled. • Change the recreational red snapper allocation to a quota • RA close recreational fishery in EEZ when landings projected to exceed its allocation	• Earlier opening of the season avoids bad weather and Labor Day weekend conflicts with anglers • Helps extend the season • Quota will better control angler harvest • Quota allows for quicker action by RA to close the fishery when needed
1997	Regulatory amendment[1]	• Cancel planned increase in the red snapper minimum size limit to 16 inches TL	• Gains to the fishery from size limit increase offset by decreases in yield per recruit
1998	Amendment 15[1]	• Establish a permanent two-tier red snapper license limitation system (Class 1 and Class 2) • The comm. season was split in two, with two thirds of the quota allocated to a February 1 opening and the remaining quota to a September 1 opening.	• Without transferability, the previous system was a closed-access system • Spread out landings over a longer period of time and give fishermen more options about when to fish
1998	Regulatory amendment[1]	• Maintain 9.12 mp TAC • Zero bag limit for the captain and crew of for-hire recreational vessels (not implemented)	• Rebuilding projected to continue to 20% SPR with current TAC • Zero bag limit for captain and crew projected to extend recreational season 1-2 weeks
1998	Regulatory amendment[1]	• 6 mp TAC, with release of all or part of the remaining 3.12 mp contingent upon the capability of BRDs to achieve better than a 50 percent reduction in juvenile red snapper shrimp trawl mortality • Reduce the bag limit to 4 fish and zero fish for captain and crew of for-hire vessels • Set the opening date of the rec fishing season to March 1 • Reduce the minimum size limit for red snapper to 14 inches total length for both directed fisheries • Change the opening of the fall fishing season from the first 15 days to the first 10 days of each month beginning September 1	• A 1998 NMFS study suggested BRDs could achieve bycatch mortality reductions of Age-0 and Age-1 red snapper by over 60 percent • Reduce recreational catch to avoid quota closures • Close the recreational fishery during the least favorable months for fishing to reduce effort • Previous size limits were based on a release mortality of less than 33%. New information suggested release mortality of greater than 33%
1998	Emergency rule[2]	• Reduce the recreational bag limit for red snapper from 5 to 4 fish per person • Reopen the recreational fishing season in January 1999	• Reduce recreational F to prevent the fishery from exceeding its quota
1999	Interim rule[2]	• Increase the minimum size of recreationally caught red snapper to 18 inches • Close the recreational red snapper fishery in the EEZ on August 19, 1999	• Extend the recreational season by 2 weeks
1999	Interim rule[2]	• Change 2000 recreational season from April 24 to October 31 • Reinstate 4-fish bag limit for captain and crew • Reduce opening of spring commercial seasons from 15 to 10 days	• Allow for a fall recreational fishery • Allow flexibility for charter fishermen to manage their catch • Extend the spring commercial season
2000	Amendment 17[1]	• Extend the reef fish permit moratorium for another five years, from the existing expiration date of December 31, 2000 to December 31, 2005, unless replaced sooner by a comprehensive controlled access system.	• Provide a stable environment for the fishery • Prevent the fishery from further overcapitalization • Allow time for evaluation and development of a more comprehensive controlled access system

Table 1. (Continued)

Year	Rule-making Vehicle	Action	Rationale
2000	Regulatory amendment[1]	• Maintain the TAC at 9.12 mp for the next two years • Increase the recreational minimum size limit from 15 inches to 16 inches TL • Set the red snapper recreational bag limit at 4 fish • Reinstate the for-hire captain and crew bag limit • Set the recreational red snapper season from April 15 to October 31, subject to revision by the RA to accommodate reinstating the bag limit for captain and crew • Set the commercial red snapper Spring season to open on February 1 and be open from noon on the 1st to noon on the 10th of each month until the Spring sub-quota is reached • Set the commercial red snapper Fall season to open on October 1 and be open from noon on the 1st to noon on the 10th of each month until the remaining commercial quota is reached • Retain the red snapper commercial minimum size limit at 15 inches TL • Allocate the red snapper commercial season sub-quota at 2/3 of the commercial quota, with the Fall season sub-quota as the remaining commercial quota.	• Maintain stability in the fishery by maintaining TAC • Reduce the recreational F • Extend the recreational season • Extend the commercial season • Maintain price stability for the commercial fishery • Delay the fall season to increase red snapper prices • Allow more flexibility in assigning the commercial spring and fall quotas should TAC change
2003	Amendment 20[1]	• Establish a 3-year moratorium on the issuance of any additional charter vessel/headboat permits for vessels fishing the EEZ of the Gulf of Mexico (Gulf) for Reef Fish or CMP fishes • Allow permits (except those issued to historical captains) to be transferable to other persons • Require vessel captains or vessel owners to participate in data collection surveys as a permit condition.	• Cap effort in the for-hire fishery
2005	Amendment 22[1]	• Establish status determination criteria and biological reference points • Establish red snapper rebuilding plan • Establish additional reef fish bycatch reporting methodologies	• Bring the red snapper fishery into compliance with requirements added to the MSFCMA through the SFA • Establish a schedule for rebuilding the overfished 1red snapper stock meets MSFCMA requirements • Document and reduce red snapper bycatch
2005	Amendment 24[1]	• Extend the commercial reef fish permit moratorium indefinitely from the existing expiration date of December 31, 2005, unless replaced by a comprehensive controlled access system.	• Provide a stable environment for the fishery • Prevent the fishery from further overcapitalization • Allow time for evaluation and development of a more comprehensive controlled access system
2006	Amendment 25[1]	• Extend the recreational for-hire reef fish permit moratorium indefinitely from the expiration date of June 16, 2006 and create a limited access system.	• Cap effort in the for-hire fishery
2006	Amendment 26[1]	• Establish an individual fishing quota program for the commercial red snapper fishery	• Reduce overcapacity in the commercial red snapper fishery • Eliminate, to the extent possible, the problems associated with derby fishing

[1]Copies of the FMP/amendment can be obtained from the Gulf of Mexico Fishery Management Council, 2203 N. Lois Ave., Tampa, FL 33607
[2]Copies of the rule can be obtained from the Southeast Regional Office, 263 13th Avenue South, St. Petersburg, FL 33701

regulatory amendment a 4 mp TAC for the 1991 fishing year, of which 2.04 mp was the commercial quota and 1.96 mp was for the recreational allocation (Tables 1–3). Hence the commercial quota was reduced. Additionally, the Council proposed a 50% reduction of juvenile red snapper in shrimp trawl bycatch occur by 1994. However, Congress placed a 3-year moratorium on bycatch measures so NOAA Fisheries Service could conduct a research program to assess the effect of shrimp fishery on federally managed species.

In 1992, the Council's Reef Fish Stock Assessment Panel (RFSAP) reviewed a new red snapper stock assessment (Goodyear 1992). The panel looked at stock rebuilding scenarios using shrimp bycatch reduction levels ranging from 40 to 60% and rebuilding time periods ending in 2007 to 2009 (RFSAP 1992). Depending on the rebuilding target and bycatch, the RFSAP recommended TAC be set between 4 mp and 6 mp. A rebuilding time period through 2007 had been established in Amendment 3 because this seemed a reasonable time period to rebuild the stock. However, Amendment 3 also established that rebuilding periods could be changed through framework actions and that rebuilding periods could not exceed 1.5 times the biological generation time for a managed species. The 1992 stock assessment estimated the generation time for red snapper to be 13 years, thus the maximum rebuilding period for red snapper would be 19.5 years. Given the rebuilding time period started in 1990, the target date could be revised to 2009. Thus, the Council selected a 6 mp TAC based on using the revised rebuilding time period ending in 2009 and assuming that a 50% reduction in shrimp bycatch could be achieved by 1994.

The RFSAP reviewed a revised stock assessment in 1994 (Goodyear 1994). The panel recommended an acceptable biological catch (ABC) range of 4–6 mp based on several management options that could rebuild the stock by 2009, given shrimp trawl bycatch reductions were possible in either 1994, 1995, 1996, or incrementally through 1998 (RFSAP 1994). The projections also showed unless drastic reductions in bycatch were achieved, the likelihood of achieving 20% SPR was minimal. The Council maintained TAC at 6 mp for both 1994 and 1995 through a regulatory amendment.

New information on red snapper life history and shrimp trawl bycatch became available for the 1995 stock assessment (Goodyear 1995). These included an increase in red snapper longevity (53 years), a decrease in the natural mortality rate (0.2–0.1), and indications that BRDs could achieve a 50% bycatch reduction. As a result, a new generation time was estimated (19.6 years) resulting in a revised rebuilding target date of 2019 (=1990 + 1.5 × 19.6). The RFSAP provided an ABC range to the Council of 6 mp to 10 mp, but cautioned these recommendations were based on: 1) actual shrimp trawl bycatch mortalities for 1995 and 1996 are no greater than the projected estimates; 2) the recreational sector stays within its allocation; 3) the 50% bycatch reduction in the shrimp fishery is achieved in 1997; and 4) projected increases in red snapper recruitment are realized (RFSAP 1995). Additionally, the RFSAP pointed out the stock was operating at a dangerously low SPR. Given this advice, the Council implemented TAC for 1996 at 9.12 mp in a 1995 regulatory amendment. This TAC was derived from bag and size limits that suggested a five-fish bag limit and 15-in minimum size limit for the recreational fishery would result in a harvest of 4.47 mp (Holiman 1995). Given the 51/49% split between the commercial and recreational fisheries, the commercial quota was 4.65 mp. In setting this TAC, the Council assumed a minimum of a 37% reduction in shrimp trawl bycatch in 1997 and a 50% reduction in bycatch by

1998. Additional updates on the red snapper stock were provided to the RFSAP in 1996 (Goodyear 1996), but the RFSAP did not make any recommendations regarding TAC (RFSAP 1996).

The 1998 assessment assumed shrimp trawl bycatch reductions beginning in 1996 would not occur until 1998 (Goodyear 1997). Under these assumptions, the RFSAP advised that in order to achieve 20% SPR in 2019, either the TAC must be reduced to approximately 3.6 million pounds at the currently planned bycatch reduction level of 44% in 1998 with the requirement of BRDs, or bycatch mortality must reduced by approximately 66% of baseline levels to maintain the current 9.12 mp TAC (RFSAP 1997). They also advised harvest in the fishery could support a 12 mp TAC, although this would require a 77% reduction in shrimp trawl bycatch from baseline levels. The RFSAP also suggested the Council consider a constant *F* rebuilding strategy so TAC could increase as the stock size increased (RFSAP 1997). To accomplish this, TAC would need to be reduced to between 3 and 6 mp; however, the Council rejected this

Table 2. Changes in commercial red snapper quota, size limits, and season length by year.

Year	Size Limit (Inches TL)	Calendar Days Open	Quota (million pounds)	Commercial Harvest (million pounds)
1984-1989	13	365	na	na
1990[1]	13	365	3.1	2.65
1991	13	236[2]	2.04	2.21
1992	13	53+42=95[3]	2.04 + emergency	3.03
1993[4]	13	94	3.06	3.37
1994	14	77	3.06	3.22
1995	15	50+2=52[5]	3.06	2.93
1996	15	65+22=87[6]	4.65	4.31
1997	15	53+20=73[7]	4.65	4.81
1998[8]	15	42+30=72	4.65	4.68
1999	15	45+25=70[9]	4.65	4.87
2000	15	38+28=66[10]	4.65	4.84
2001	15	56+23=79	4.65	4.63
2002	15	64+27=91	4.65	4.78
2003	15	67+27=94	4.65	4.41
2004	15	70+35=105	4.65	4.67
2005	15	80+51=131	4.65	4.04

[1] Bottom longlines prohibited within 50 fathoms west of Cape San Blas, FL, and within 20 fathoms elsewhere.
[2] First year commercial red snapper fishery was closed.
[3] Season re-opened April 4–May 15 with 1,000-pound trip limit.
[4] First year of two-tiered system of trip limits; 2,000 pounds for boats with endorsements and 200 pounds for other boats with reef fish permits.
[5] Season re-opened for 36 hours Nov 1–2. Two-tiered system of trip limits.
[6] First year of planned spring (3.06 million pounds) and fall (for the remaining unfilled quota) seasons.
[7] The fall season opened for the first 15 days of each month or until the quota is filled.
[8] First year of license limitation system with trip limits of 2000 pounds for Class 1 boats and 200 pounds for Class 2 boats.
[9] The fall season opened during the first 10 days of each month or until the quota is filled.
[10] The spring and fall season opened during the first 10 days of each month or until the quota is filled.

idea. Subsequent analyses from NOAA Fisheries Service projected step-wise increases in bycatch reductions from 45% in 1998 to greater than 60% in years following 2000 had a 50% or greater probably of rebuilding the stock to 20% SPR by 2019 with a 9 mp TAC (Schirripa 1998). Therefore, the Council selected to maintain TAC at 9.12 mp in a 1998 regulatory amendment (Table 1).

The 1999 assessment used an age-structured assessment program (ASAP) rather than the virtual population analysis models used in previous assessments (Schirripa and Legault 1999). This model provided greater flexibility, provided internally consistent estimates of management parameters of interest (i.e., the *F* that can sustain maximum sustainable yield (MSY) and stock biomass capable of producing MSY (B_{MSY}), and improved evaluating uncertainty in characterizing stock status. The RFSAP evaluated ABC under several combinations of shrimp trawl bycatch reduction levels, levels of steepness of the spawner-recruit curve, and constant catch versus constant fishing mortality rate harvest strategies (RFSAP 1999). Further, they provided biomass-based status determination criteria following Restrepo et al.'s (1998) guidance on the use of precautionary approaches to National Standard 1 of the Magnuson-Stevens Act.

The assessment suggested a high degree of uncertainty about the stock (Schirripa and Legault 1999). Estimates of MSY ranged from 22 to 205 mp, and estimates of minimum stock size threshold (calculated as $(1 - M) \times B_{MSY} = 0.9 \times B_{MSY}$) would be 2.2 to 3.7 billion pounds. This resulted in a range of maximum ABC recommendations of 5.8 to 9.12 mp under the constant catch scenario. Under the constant *F* scenario, the maximum ABC recommendations were 2.0 to 3.5 mp in 2000, and 2.4 to 4.2 mp in 2001.

While the RFSAP strongly endorsed the constant *F* approach over the constant catch rebuilding scenario (RFSAP 1999), the RFSAP was concerned with the hardships associated with proposed reductions in TAC (e.g., by ~50%) necessary to achieve a constant *F* scenario in a single year as outlined in the harvest strategy. Based upon this concern, the RFSAP recommended NOAA Fisheries Service consider alternatives that could lessen the impacts of moving to a constant *F* in a single year, such as either a phased reduction in TAC over two or three years or no changes in current TAC, but capping long-term yields at historical values of 15–20 mp. However, the Council was reluctant to reduce TAC to the levels prescribed by the RFSAP. Thus, NOAA Fisheries Service staff developed a decision-tree approach to managing the stock based on levels of bycatch reduction and periodic assessments (Powers et al. 2000). To minimize the adverse effects to the directed fishery, TAC in this plan was maintained at 9.12 mp.

In May 2001, the Council submitted to NOAA Fisheries Service a regulatory amendment for the Reef Fish FMP to set a red snapper rebuilding plan time period through 2032. The plan used as its basis the rebuilding plans provided in Powers et al. (2000). However, in July 2002, NOAA Fisheries Service determined the regulatory amendment would have a reasonably foreseeable significant adverse effect on both the shrimp and (potentially) the directed red snapper fisheries. Therefore, NOAA Fisheries Service recommended the Council develop the rebuilding plan in an amendment to the Reef Fish FMP, as well as analyze current and additional rebuilding alternatives in greater detail through an environmental impact statement. The revised plan was developed by the Council in Amendment 22 and was based on projections from the 1999 assessment indicating the red snapper stock could rebuild to B_{MSY} within the longest time period recommended by NOAA Fisheries Service guidelines (31 years for red snapper; RFSAP 1999). The plan maintained TAC at 9.12 mp, projected an end to overfishing

Table 3. Changes in recreational red snapper *Lutjanus campechanus* size limits, bag limits, season length, and allocation/quota.

Year	Size Limit (Inches TL)	Daily Bag Limit (Number of Fish)	Season Length (days)	Allocation/Quota (Million Pounds)	Recreational Harvest (Million Pounds)
1984	13[1]	no bag limit[2]	365	na	3.09
1990	13	7	365	na	1.36
1991	13	7	365	1.96	2.10
1992	13	7	365	1.96	3.62
1993	13	7	365	2.94	5.57
1994	14	7	365	2.94	4.53
1995	15	5	365	2.94	3.69
1996	15	5	365	4.47	3.47
1997	15	5	330[3]	4.47	4.37
1998	15	4[4]	272[5]	4.47	4.35
1999	15[6]	4	240[7]	4.47	4.35
2000	16	4	194[8]	4.47	3.33
2001	16	4	194	4.47	3.56
2002	16	4	194	4.47	4.87
2003	16	4	194	4.47	4.60
2004	16	4	194	4.47	5.02
2005	16	4	194	4.47	4.59

[1] For-hire boats exempted until 1987.
[2] Allowed to keep 5 undersized fish per day.
[3] Fishery closed on November 27, 1997.
[4] Bag limit was 5 fish from January through April, 1998.
[5] Fishery closed on September 30, 1998.
[6] Size limit was 18 inches from June 4 through August 29, 1999.
[7] Fishery closed on August 29, 1999.
[8] Fishing season opens at 12:01 a.m. April 21 and closes at 12:00 midnight October 31.

between 2009 and 2010, and projected rebuilding the stock to B_{MSY} by 2032. However, it was dependent on large reductions in bycatch mortality through technological means such as BRDs, and reductions in effort due to an economic downturn in the shrimp trawl fishery.

The most recent stock assessment was conducted through the SEDAR process. This assessment used data through 2003 and concluded while the red snapper stock was still overfished and undergoing overfishing, the stock was showing small signs of improvement (SEDAR 2005b). However, the assessment also concluded reductions in red snapper *F*s in both the directed and shrimp trawl fisheries were warranted to maintain rebuilding. SEDAR (2005b) provided precautionary advice to the Council in selecting TAC. Because of uncertainty in the stock–recruitment relationship and the effects of shrimp trawl bycatch, the SEDAR indicated the emphasis should focus on short-term (5–10 year) goals that rebuild the stock in the desired direction rather than on specific rebuilding targets, or how to attain them. The SEDAR also indicated the Council needs to determine what limitations shrimp trawl bycatch has on the ultimate red snapper stock status (SEDAR 2005b). Thus, selecting a TAC needs to balance the tradeoff between bycatch reduction and rebuilding stock biomass to a practicable level given the extent that shrimp trawl bycatch can be reduced. The Council is using this advice in an amendment addressing both red snapper rebuilding and shrimp trawl bycatch (Strelcheck and Hood 2007, this volume).

Commercial Fishery

The directed commercial fishery in the GOM has been managed with size limits, trip limits, limited entry, season closures, and a quota. The quota, once met, causes the fishery to be closed. The first regulation placed on this fishery was a 13-in TL minimum length limit in the initial FMP (Tables 1 and 2). The purpose of this regulation was to increase the yield in the fishery by 18–25%. This measure also increased the likelihood of red snapper being able to spawn before caught. In 1994, a stepped increase in the minimum size over a 5-year period (1994–1998) from 14-in TL to 16-in TL was implemented through Amendment 5 (Table 1). This increase was projected to increase the yield per recruit and biomass yield from the fishery. By using stepped increases, adverse effects on the fishery would be minimized. Through Amendment 12 (implemented in 1997), the Council tried to hold the commercial size limit at 14-in TL because industry indicated a smaller fish was more desirable in the market and discard mortality rates (estimated to be 33%) were too low. However, the Secretary disapproved this measure continuing the current stepped-size limit increase. On the basis of a new stock assessment (Schirripa and Legault 1997), the increase in minimum size to 16-in TL was canceled through a 1997 regulatory amendment on the advice of the RFSAP (1997) who concluded potential gains by the fishery would be offset by decreases in yield per recruit, having no or a negative effect on rebuilding. In December 1998, the Council submitted a regulatory amendment to reduce the minimum size limit to 14-in (Table 1). However, NOAA Fisheries Service once again disapproved the measure because it did not provide a clear economic or biological benefit.

With the use of quotas to manage the commercial fishery, the fishery had to be closed once the quota was met. In 1991, the commercial quota was reduced from 3.1 mp to 2.04 mp. The fishery was able to meet this quota prior to the end of the fishing year, and thus was closed on August 25, 1991 (Tables 2 and 4). However, this closure of the fishery led to a shift in fishing effort such that the 1992 fishery had to close by February 22, 1992 (53 d). The short 1992 season created several problems to the fishery including depressed prices from flooding the market due to an influx of nontraditional fishermen. To alleviate the adverse economic and social affects of the early closure, the Council requested an emergency rule to open the season from April 3 to May 14, 1992, with a 1,000-lb trip limit.

To reduce the adverse economic conditions in the fishery observed in 1992, NOAA Fisheries Service, at the request of the Council, published an emergency rule establishing a 2,000- and 200-lb red snapper trip limit endorsements. The 2,000-lb daily trip limit endorsement went to fishermen able to demonstrate landings of at least 5,000 lb whole weight two of three years (1990, 1991, and 1992) while the 200-lb daily trip limit endorsement went to interested reef fish permit holders as a reasonable bycatch allowance. The red snapper endorsements were then extended by Amendment 6 in 1993 and developed into transferable Class 1 (2,000 lb trip limit) and Class 2 (200 lb trip limit) licenses in 1994 in Amendment 5. In 1992, the Council also limited the number of reef fish permits with a moratorium on the issuance of new permits through Amendment 4; this has been continued through Amendments 9, 17, and 24 (Table 1).

The Council has adjusted the commercial seasons to work with industry to improve the economic environment for the fishery. The first adjustment occurred for the 1993 fishing year when the opening of the fishery was delayed from January 1 to February 16 in order to accommodate new trip limit endorsements being put in place through emergency regulations (Table 1). This delay to a February opening for the fishery was continued for following years to ensure the commercial red snapper fishery was open during Lent, when the industry indicated they obtain higher prices, and to keep the fishery closed during January when weather conditions are worst.

In 1995, the commercial season initially closed April 15, but there was still about 220,000 lb of red snapper to be harvested (Tables 2 and 4). Rather than rolling these pounds into the 1996 commercial quota, the fishermen indicated they would rather harvest these pounds in the fall in order to have some income from red snapper fishing prior to the holidays. The Council requested, and NOAA Fisheries Service approved, a 36-h mini-season in November 1995 (Tables 1 and 4). In 1996, the commercial quota was raised from 3.06 to 4.65 mp. A 1996 regulatory amendment delayed the release of the 1.59 mp to September 15 so the commercial fishery could receive an economic benefit similar to the previous fishing year (Table 1).

With the 1992 commercial quota being filled in just 53 d, the Council recognized the effort capacity in the fishery was excessively high. In addition, this type of derby effect created other problems such as market gluts, depressed prices, and unsafe fishing conditions by forcing fishermen to fish in bad weather (Waters 2001; 2003). Thus, the Council developed an individual fishing quota (IFQ) program for the commercial fishery in Amendment 8, which was approved by NOAA Fisheries Service (Table 1). However, this amendment was never implemented because Congress put in place a moratorium on the development or implementation of new IFQ programs until October 1, 2000, with the 1996 Sustainable Fisheries Act. The rationale for the moratorium was in response to concerns about the social and economic effects of IFQs.

With the IFQ program delayed, the Council tried to ameliorate the negative economic conditions of the red snapper derby fishery by using mini-seasons. In 1997, the fall subquota was divided up into 15-d mini-seasons beginning at noon on the first day of the month and ending at noon on the 15th day of the month in September (note the September opening was reduced by one day so that the fishery did not overlap with the Labor Day holiday weekend) (Table 4). The purpose of these mini-seasons was to extend the number of months the fishery could be open. It was thought this could mitigate some of the effects of a derby fishery by reducing the amount of fish flooding the market at any one time. These mini-seasons were further reduced to 10 d beginning in September 1999 by a regulatory amendment. Economic analyses suggested shorter seasons would provide further economic benefits to the fishery (Waters and Antozzi 1997).

In 2001, the Council reinitiated the development of the IFQ program through an IFQ profile. Congress dictated before a red snapper IFQ program could be implemented, there needed to be two referenda voted on by the Class 1 license holders. The first referendum asked whether red snapper fish-

Table 4. Dates the red snapper *Lutjanus campechanus* commercial fishing season has been open from 1990 to 2005. * denotes a monthly opening begins and ends at noon rather than midnight. # denotes a monthly opening begins at midnight and ends at noon.

Year	Jan	Feb	Mar	Apr	May	June	July	Aug	Sept	Oct	Nov	Dec	Full Days	Half days	Total Cal-endar Days	Total hours
1990	1–31	1–28	1–31	1–30	1–31	1–30	1–31	1–31	1–30	1–31	1–30	1–31	365		365	8,760
1991	1–31	1–28	1–31	1–30	1–31	1–30	1–31	1–24					236		236	5,664
1992	1–31	1–22#		3–30	1–14								94	1	95	2,268
1993		16–28	1–31	1–30	1–20								94		94	2,256
1994		10–28	1–31	1–27									77		77	1,848
1995		24–28	1–31	1–14							1–2#		51	1	52	1,236
1996		1–29	1–31	1–5					15–30	1–6			87		87	2,088
1997		1–28	1–25						2–15*	1–6*			69	4	73	1,704
1998		1–15*	1–15*	1–12*					1–15*	1–15*			62	10	72	1,608
1999		1–15*	1–15*	1–15*					1–10*	1–10*	1–5*		58	12	70	1,536
2000		1–10*	1–10*	1–10*	1–8*					1–10*	1–10*	1–8*	52	14	66	1,416
2001		1–10*	1–10*	1–10*	1–10*	1–10*	1–6*			1–10*	1–10*	1–3*	61	18	79	1,680
2002		1–10*	1–10*	1–10*	1–10*	1–10*	1–7*	1–7*		1–10*	1–10*	1–7*	71	20	91	1,944
2003		1–10*	1–10*	1–10*	1–10*	1–10*	1–10*	1–7*		1–10*	1–10*	1–7*	74	20	94	2,016
2004		1–10*	1–10*	1–10*	1–10*	1–10*	1–10*	1–10*		1–10*	1–10*	1–15*	85	20	105	2,520
2005		1–10*	1–10*	1–10*	1–10*	1–10*	1–10*	1–10*	1–10*	1–10*	1–10*	1–31*	109	22	131	3,144

ermen supported further consideration of an IFQ program. The fishermen qualified to vote in this election voted overwhelmingly for the Council to proceed with the development of an IFQ program in February 2004 (Phil Steele, NOAA Fisheries Service, personal communication[2]). The Council thus began to develop the program in Amendment 26, which was approved by the qualified fishery participants in the second referendum in February 2006. This amendment was implemented in time for the 2007 fishing season.

Recreational Fishery

The directed recreational fishery in the GOM has been managed with size limits, bag limits, season closures, and quotas. The first regulation placed on this fishery in 1990 was a 13-in TL minimum length limit (Table 1). Like the commercial fishery, the purpose of this regulation was to increase the yield in the fishery by 18–25% and increase the likelihood of red snapper being able to spawn before being harvested. The first bag limit (seven fish per person per day) was put in place in 1990 through Amendment 1 to reduce the recreational harvest by 20% and assist in rebuilding the stock. In 1994, Amendment 5 created a stepped increase in the minimum size over a 5 year period (1994–1998) from 14-in TL to 16-in TL. However, because of allocation overages in the recreational fishery (Table 3), the increase from 14 to 15 in was accelerated by one year with a reduction in the bag limit to 5 fish to achieve a 43% reduction in recreational harvest. The increase in minimum sizes was also projected to increase the yield per recruit and biomass yield from the fishery, thus assisting in rebuilding the stock more quickly.

The first year the recreational fishery needed to be closed prior to the end of the fishing year was 1997 when the recreational quota was projected to be filled by November 26, 1997 (Table 3). In 1998, NOAA Fisheries Service projections indicated the fishery would meet its quota by October 1 causing the for-hire industry to discuss with the Council the need for a longer season so the industry could remain economically viable. In response, the Council requested NOAA Fisheries Service implement an emergency rule to reduce the bag limit from 5 to 4 fish (Table 1). This reduction was made permanent in a 1998 regulatory amendment. In addition, a zero-bag limit for captain and crew was implemented for the charter and headboat fisheries.

Projections for the 1999 fishing year indicated the fishery would close on August 5, 1999. Representatives from the for-hire industry were concerned this earlier closure would create economic harm to their industry and requested the Council and NOAA Fisheries Service examine ways to extend the season through August 28. Thus, a temporary 18-in minimum size limit was implemented to achieve the desired season. However, this measure was very unpopular with the angling public, and so in a 2000 regulatory amendment, the current 16-in minimum size limit was implemented (Table 1). This larger size limit, in conjunction with the reduced bag limit, was projected to provide the recreational fishery with a six to seven month season.

In evaluating the red snapper fishing season and taking into account seasons desired by anglers, the Council determined a spring, summer, and fall fishery was most desirable. It was also determined October was economically a more important month than April. Therefore, the fishing season selected by the Council was from April 15 to October 31. However, the Council decided to reinstate the captain and crew bag limit, projected to shorten the fishing season by three to ten days. Thus the season was further shortened to April 21 to October 31. Selection of this season was not without controversy. South

[2]Phil Steele, NOAA Fisheries Service, Southeast Regional Office, 263 13th Ave. S., St. Petersburg, Florida 33705.

Texas fishermen asked for a winter season in January and February, which are important months for them. However, scenarios including a winter season would eliminate several weeks from the spring-fall season. Thus, the Council rejected this idea in favor of the longer spring-fall season, which they reasoned would most benefit the fishery as a whole.

Recreational effort was further constrained in Amendment 20, which established a 3-year moratorium on the issuance of any additional reef fish charter vessel/headboat permits (Table 1). This cap on permit numbers was needed because effort since the 1980s had more than doubled and the for-hire sector of the fishery was responsible for approximately two-thirds of recreationally caught red snapper. This moratorium, set to expire in June 2006, was renewed indefinitely in Amendment 25.

Shrimp Fishery

As mentioned above, the 1988 stock assessment indicated juvenile red snapper bycatch from the shrimp fishery is a major contributor to red snapper *F* (Goodyear 1988). In 1990, the Council proposed seasonal closures for some shrimping grounds to reduce bycatch by 50% in a reef fish regulatory amendment. However, development of these measures was halted in 1990 when Congress placed a 3-year moratorium on regulations so NOAA Fisheries Service could evaluate different methods for bycatch reduction. This moratorium was extended by one year so NOAA Fisheries Service could complete the Cooperative Shrimp Bycatch Characterization Project (NOAA Fisheries Service 1995). In 1995, the Council began work on Amendment 9 to the Shrimp FMP which was implemented in 1998. This amendment established the use of BRDs west of Cape San Blas and established criteria to certify different BRD designs for use in the fishery. The requirement for BRDs in shrimp trawls was extended east of Cape San Blas in 2004 through Shrimp Amendment 10.

The Council had considered area closures, seasonal closures, and limited access programs as alternatives to BRDs to reduce bycatch. However, these measures were considered impracticable. Juvenile red snapper are on the shrimp grounds year-round and in areas of high shrimp concentrations making them difficult to avoid either temporally or spatially (Nichols 1990). The Council also considered limited access programs requiring permits, which at the time were considered difficult to implement due to the complexity of the fishery and uncertainties regarding revocations and administrative fees.

Once certified BRDs were placed on shrimp trawls, the shrimp bycatch fishing mortality rate on red snapper was estimated to be potentially reduced by an estimated 40% in the shrimp fishery (Nichols, undated). Field tests conducted by NOAA Fisheries have demonstrated BRDs may be able to reduce the fishing mortality rate for red snapper in the shrimp fishery by as much as 70% with only small reductions in shrimp catch (Watson et al. 1999). However, as reported by SEDAR, there has been a decline in BRD performance since 1998 (Foster 2005; SEDAR 2005b). This decline, particularly in the fisheye design, is likely due to changes in fishing techniques to minimize shrimp loss as the nets are hauled back aboard shrimp vessels. Actual bycatch reduction of juvenile red snapper from BRDs is currently estimated to be below 15%. Currently, the Council is evaluating new information on reductions in shrimp trawl effort due to an economic downturn in the fishery. The Council will weigh the effects of this change in determining what future actions will be required to achieve appropriate reductions in juvenile red snapper bycatch.

Summary

Management of red snapper in the GOM EEZ has entered its 25th year, yet many management challenges remain for this species. The stock has not been rebuilt even though the initiation of a rebuilding plan began in 1990. Three factors account for this lack of progress. One is the stock had been fished to a very low level (at least 1% of 20% SPR) (SEDAR 2005b). Another is the Council, as documented in this paper, generally chose the higher end of ABCs provided by stock assessments, thus delaying rebuilding. This choice was based in part on balancing the need for stock rebuilding while minimizing the adverse effects of limiting TAC on the directed fishery. The third factor is the high level of *F* placed on the red snapper stock from the shrimp trawl fishery acts to limit recruitment.

While rebuilding may not be proceeding as quickly some would like, the Council and NOAA Fisheries Service have, over time, become better able to manage the directed fishery within its quota. With the exception of 1992 (emergency season reopening) to 1993, the commercial fishery has not exceeded its quota by more than 5%, and frequently has landed less than its quota (Table 2). However, closing the fishery once the quota was met led to the development of a derby fishery, no matter how the season was manipulated. The derby fishery should disappear with the introduction in 2007 of the red snapper IFQ program.

Holding the recreational fishery to their allocation of 49% of TAC has been problematic. Prior to 1995 when this sector was given an allocation rather than held to a quota, landings in some years nearly doubled the fishery's allocation (e.g., 1993; Table 3). However, with the quota and the ability of NOAA Fisheries Service to close the fishery once the quota is projected to be filled, landings have stayed near or below the sector's quota by modifying fishing season length in conjunction with size and bag limits.

Managing red snapper bycatch in the shrimp trawl fishery has also been problematic. While BRDs have been introduced into the fishery to reduce bycatch, their performance has not met expectations. Other methods to reduce bycatch such as seasonal or area closures are thought to be impracticable because shrimp and juvenile red snapper share the same areas in high concentrations throughout the year. However, this limitation may change as the spatial-temporal concentrations of juvenile red snapper are better understood through investigations like those of Diamond and Wang (2006). Additionally, red snapper bycatch may be reduced as shrimp trawl effort declines from factors such as lower-priced imports (Haby et al. 2003), higher fuel costs, and fleet damage from hurricanes.

The most recent stock assessment (SEDAR 2005b) included new information on red snapper and the shrimp trawl fishery testing previous views of the fisheries including a greater influence of discard mortality from the directed fishery and the effectiveness of BRDs. Strelcheck and Hood (2007, this volume) discuss these challenges, as well as challenges in balancing competing interests from the various fishing sectors, environmental organizations, and mandates from within the Magnuson-Stevenson Act.

Acknowledgments

We would like to thank David Nieland and two anonymous reviewers for providing constructive input during the development of this manuscript.

References

Camber, C. I. 1955. A survey of the red snapper fishery of the Gulf of Mexico, with special reference to the Campeche Banks. Florida State Board of Conservation, Technical Series No 12:1–64.

Carpenter, J. S. 1965. A review of the Gulf of Mexico red snapper fishery. U.S. Fish and Wildlife Service Circular 208.

Diamond, S. L., and Y. Wang. 2006. Analysis of the relationship between shrimp and juvenile red snapper distributions on large temporal and spatial scales. Meeting Abstracts, 2006 Spring Meeting, Southern Division of American Fisheries Society, San Antonio, Texas.

Foster, D. 2005. Funnel BRD performance. Report for the Gulf of Mexico Fishery Management Council. Harvesting Systems and Engineering Division, National Marine Fisheries Service, Pascagoula Laboratory, Pascagoula, Mississippi.

Gulf of Mexico Fishery Management Council. 1981. Environmental impact statement and fishery management plan for the reef fish resources of the Gulf of Mexico. Gulf of Mexico Fishery Management Council, Tampa, Florida.

Goodyear, C. P. 1988. Recent trends in the red snapper fishery of the Gulf of Mexico. NOAA Fisheries Service, SEFSC, Miami Laboratory, CRD 87/88–16.

Goodyear, C. P. 1992. Red snapper in U. S. waters of the Gulf of Mexico. National Oceanic and Atmospheric Administration. NOAA Fisheries Service, SEFSC contribution MIA-91/92–70.

Goodyear, C. P. 1994. Red snapper in U. S. waters of the Gulf of Mexico. National Marine Fisheries Service, SEFSC contribution MIA-93/94–63.

Goodyear, C. P. 1995. Red snapper in U.S. waters of the Gulf of Mexico, National Marine Fisheries Service, Southeast Fisheries Science Center, Miami Laboratory, Report MIA-95/96–05. Miami, Florida.

Goodyear, C. P. 1996. An update of red snapper in U.S. waters of the Gulf of Mexico. National Marine Fisheries Service, Southeast Fisheries Science Center, Miami Laboratory Contribution MIA-95/96–60. Miami, Florida.

Goodyear, C.P. 1997. An evaluation of the minimum reduction in the 1997 red snapper shrimp bycatch mortality rate consistent with the 2019 recovery target. April 10, 1997, report prepared for the Gulf of Fisheries Management Council, Tampa, Florida.

Goodyear, C. P., and P. Phares. 1990. Status of red snapper stocks of the Gulf of Mexico—report for 1990. National Marine Fisheries Service, Southeast Fisheries Science Center CRD 89/90-05, Miami, Florida.

Haby, M. G., R. J. Miget, L. L. Falconer, and G. L Graham. 2003. A review of current conditions in the Texas shrimp industry, an examination of contributing factors, and suggestions for remaining competitive in the global shrimp market. Texas Cooperative Extension Sea Grant College Program, TAMU-SG-03–701. Texas A&M University, College Station, Texas.

Holiman, S. G. 1995. Reef fish economic assessment for the Gulf of Mexico recreational fishery. National Marine Fisheries Service, Southeast Regional Office, St. Petersburg, Florida.

Nichols, S. No date. An update on the overall effectiveness of BRDs. Report prepared for the gulf of Mexico Fisheries Management Council, Tampa, Florida.

Nichols, S. 1990. The spatial and temporal distribution of the bycatch of red snapper by the shrimp fishery in the offshore waters of the U.S. Gulf of Mexico. Mississippi Laboratories. Pascagoula, Mississippi.

NOAA Fisheries Service. 1995. Cooperative research program addressing finfish bycatch in the Gulf of Mexico and South Atlantic shrimp fisheries: a report to congress. National Marine Fisheries Service, St. Petersburg, Florida.

Powers, J. E., C. M. Legault, and R. E. Crabtree. 2000. Updated projections for Gulf of Mexico red snapper. National Marine Fisheries Service, Southeast Fisheries Science Center, St. Petersburg, Florida.

Restrepo, V. R., G. G. Thompson, P. M. Mace, W. L. Gabriel, L. L. Low, A. D. MacCall, R. D. Methot, J. E. Powers, B. L. Taylor, P. R. Wade, and J. F. Witzig. 1998. Technical guidance on the use of precautionary approaches to implementing national standard 1 of the Magnuson-Stevens Fishery Conservation and Management Act. NOAA Technical Memorandum, NMFS-F/SPO-31.

RFSAP. 1992. Final report of the Reef Fish Stock Assessment Panel. Gulf of Mexico Fishery Management Council. Tampa, Florida.

RFSAP. 1994. 1994 report of the Reef Fish Stock Assessment Panel. Gulf of Mexico Fishery Management Council. Tampa, Florida.

RFSAP. 1995. 1995 report of the Reef Fish Stock Assessment Panel. Gulf of Mexico Fishery Management Council. Tampa, Florida.

RFSAP. 1997. October 1997 report of the Reef Fish Stock Assessment Panel. Gulf of Mexico Fishery Management Council. Tampa, Florida.

RFSAP. 1999. September 1999 Report of the Reef Fish Stock Assessment Panel. Gulf of Mexico Fishery Management Council. Tampa, Florida.

Schirripa, M. J. 1998. Status of the red snapper in U.S. waters of the Gulf of Mexico: updated through 1997. NOAA/NOAA Fisheries Service Sustainable Fisheries Division Contribution, SFD-97/98–30.

Schirripa, M. J., and C. M. Legault. 1997. Status of the red snapper in U.S. waters of the Gulf of

Mexico: updated through 1996. Southeast Fisheries Science Center, Miami Laboratory, Miami MIAC97/98–05.

Schirripa, M.J., and C.M. Legault. 1999. Status of the red snapper in U.S. waters of the Gulf of Mexico: updated through 1998. National Marine Fisheries Service, Southeast Fisheries Science Center. Sustainable Fisheries Division contribution SFD-99/00–75. Miami Laboratory, Miami, Florida.

SEDAR (Southeast Data, Assessment, and Review). 2005a. Stock assessment report of SEDAR 7. Southeast Fisheries Science Center, Miami Laboratory, Miami, Florida.

SEDAR (Southeast Data, Assessment, and Review). 2005b. Gulf of Mexico red snapper consensus summary report. Southeast Fisheries Science Center, Miami Laboratory, Miami, Florida.

Strelcheck, A. J., and P. B. Hood. 2007. Rebuilding red snapper: recent actions and future management challenges. Pages 385–396 *in* W. F. Patterson, III, J. H. Cowan, Jr., G. R. Fitzhugh, and D. L. Nieland, editors. Red Snapper Ecology and Fisheries in the U.S. Gulf of Mexico. American Fisheries Society, Symposium 60, Bethesda, Maryland.

Wallace, R. K., and K. M. Fletcher. 2000. Understanding Fisheries Management: A Manual for Understanding the Federal Fisheries Management Process. Mississippi-Alabama Sea Grant Consortium Publication MASGP-00–005.

Waters, J. R. 2001. Quota management in the commercial red snapper fishery. Marine Resource Economics 16:65–78.

Waters, J. R. 2003. Review of the U.S. commercial red snapper fishery in the Gulf of Mexico. SEFSC, Social Science Research Group. Working Paper Series SEFSC-SSRG-02.

Waters, J. R., and W. Antozzi. 1997. Implications of different types of derbies in the Gulf of Mexico red snapper fishery. NMFS, SEFSC, Miami, Florida.

Watson, J., D. Foster, S. Nichols, A. Shah, E. Scott-Denton, and J. Nance. 1999. The development of bycatch reduction technology in the southeastern United States shrimp fishery. Marine Technology Society Journal 33(2):55–56.

American Fisheries Society Symposium 60:285–300, 2007

Indices of Larval Red Snapper Occurrence and Abundance for Use in Stock Assessment

David S. Hanisko[1], Joanne Lyczkowski-Shultz, and G. Walter Ingram
National Marine Fisheries Service
Southeast Fisheries Science Center
Mississippi Laboratories, 3209 Frederic St., Pascagoula, Mississippi 39567 USA

Abstract.—Red snapper *Lutjanus campechanus* larval occurrence and abundance during Southeast Area Monitoring and Assessment Program (SEAMAP) Summer Shrimp/Bottomfish (1982–2003) and Fall Plankton (1986–2003) surveys were examined to identify the time series of ichthyoplankton data that might best reflect trends in the red snapper spawning population in the U.S. Gulf of Mexico (GOM). Since bongo nets were more effective than neuston nets at capturing red snapper larvae only catches from bongo nets were used to estimate annual occurrence and abundance, i.e. the SEAMAP larval red snapper index. The summer survey was conducted during the peak of red snapper spawning in June and July, but limited and inconsistent coverage during this survey did not permit development of a reliable Gulfwide (U.S. continental shelf) index of larval abundance. In contrast, the fall survey conducted near the end of the spawning season in September yielded a 16 year time series over which to examine trends in red snapper abundance throughout the GOM. Although occurrence and abundance of red snapper larvae were lower during September than in June and July, estimates from both summer and fall surveys showed the same inter-annual patterns and were highly correlated. Larvae were eight times more abundant and occurred in five times as many samples in the western than in the eastern GOM. Separate standardized indices of relative abundance were generated for the western and eastern GOM. The standardization procedure accounted for the effects of year, time of day, depth and subregion in the western GOM, but only for subregion in the eastern GOM. Larval indices of red snapper abundance suggest an increased spawning stock in both the western and eastern GOM after 1995.

Introduction

Red snapper *Lutjanus campechanus* are found along the U.S. Atlantic Coast and throughout the Gulf of Mexico (GOM). The U.S. GOM red snapper population supports a popular and economically valuable fishery resource utilized by both recreational and commercial sectors. The fishery is managed by the Gulf of Mexico Fishery Management Council under the Reef Fish Management Plan. The most recent population assessment of red snapper in the GOM resulted in

[1] Corresponding author: david.s.hanisko@noaa.gov.

the stock being classified as overfished and undergoing over fishing (Porch 2007, this volume). Among the fishery independent indices used in this assessment was a larval index based on a 22 year time series of Southeast Area Monitoring and Assessment Program (SEAMAP) resource surveys. A description of these surveys and plankton collection methodologies is presented by Lyczkowski-Shultz and Hanisko (2007, this volume) along with a general summary of information on red snapper larvae taken during SEAMAP surveys. The objective of this companion paper is to present standardized larval indices based on the SEAMAP time series of ichthyoplankton data that might best reflect trends in the red snapper spawning population in the GOM.

Methods

Surveys and Collections:

SEAMAP resource surveys have been conducted in the Gulf of Mexico by the National Marine Fisheries Service since 1982 in cooperation with the states of Florida, Alabama, Mississippi, Louisiana and Texas. Red snapper larvae were captured primarily during two annual SEAMAP surveys that cover the spawning area (continental shelf) and season (summer to early fall) of this species (Lyczkowski-Shultz and Hanisko 2007). The Summer Shrimp/Bottomfish (*SB*) survey, 1982–present, is conducted over the U.S. continental shelf from the U.S./Mexico border to 88° West longitude from mid June through July. The SEAMAP Fall Plankton (*FP*) survey, 1986–present, is conducted over the U.S. continental shelf from the U.S/Mexico border to south Florida from mid August to early October with the majority of samples taken during the month of September. Only data from those two surveys were used to examine the potential of a SEAMAP larval red snapper index.

Plankton sampling on SEAMAP resource surveys is conducted around the clock at predetermined stations arranged in a fixed, systematic grid across the U.S. Exclusive Economic Zone of the GOM. Most systematic grid locations or SEAMAP stations (designated by a unique SEAMAP or 'B' number) are located at ~56 km or 0.5 degree intervals along this grid. Sampling at each location is conducted with paired 61-cm, 0.333 mm mesh bongo nets and/or a single, 2 × 1 m, 0.947-mm mesh neuston net following established SEAMAP collection protocols (SEAMAP 2004). Neuston nets are towed horizontally in the top 0.5 m of the water column, while bongo nets are towed in an oblique manner to within 2–5 m of the bottom or a maximum depth of 200 m. Catches of larvae are standardized to account for sampling effort and expressed as the number of larvae under 10 m^2 of sea surface (larvae/10 m^2) for bongo nets, and as the number of larvae per 10 min tow (larvae/10 min) for neuston nets.

All snapper larvae were examined and identified by ichthyoplankton specialists at the Southeast Fisheries Science Center, Mississippi Laboratories (Lyczkowski-Shultz and Hanisko 2007). Red snapper larvae were identified using descriptions in Drass et al. (2000) and Lindeman et al. (2005). Body length of larvae was measured to the nearest 0.1 mm. Only red snapper larvae greater than 3.8 and less than 6.3 mm were used in our analysis because snapper larvae smaller than 3.8 mm cannot be reliably identified to species, while snapper larvae over 6.0 mm were not effectively captured by bongo and neuston nets presumably due to avoidance (Lyczkowski-Shultz and Hanisko 2007).

Diel period designation for each SEAMAP sample was based on the start time of sample collection. Samples taken after sunrise and before sunset were assigned to the daytime period, and samples after sunset and before sunrise to the nighttime period. Sunrise and sunset for each sample date was calculated using station latitude, longitude and Julian date based on formulae in Seidelmann (1992).

Sample Selection and Data Comparisons:

Plankton data used for this analysis were limited to a single neuston and/or bongo sample from each SEAMAP station taken during the *SB* and *FP* surveys. In cases where more than one sample was taken at a grid location during a survey, the sample taken closest to the targeted location was chosen. When SEAMAP stations were sampled by more than one vessel during the survey, priority was given to samples taken by NMFS vessels as they conduct a majority of surveys each year and therefore provided the most consistent temporal and spatial coverage. Only data from *SB* surveys in 1986, 1987, 1994, 1997, and 2000–2002 that sampled the entire extent of the intended survey area were included in our analysis. Data from all years of the *FP* survey were used with the exception of 1998 when tropical storms severely curtailed sampling. Samples from the *FP* surveys were restricted to those stations sampled during at least 10 years of the survey time series to account for annual variability in spatial coverage (Figure 1).

We examined the relative efficiency of neuston and of bongo nets at capturing red snapper larvae by comparing catches in day and night samples. Only *FP* survey samples from stations where both the neuston and bongo samples were taken within the same diel period were considered. Efficiency was measured by comparing the percent occurrence, mean abundance, and the diel percentage of total abundance of red snapper larvae in day and night samples. Diel percentage of total abundance was calculated by dividing the total summed red snapper larval abundance of all day or night samples by the total summed abundance of all samples. Coefficients of variation (CV; standard error/mean) were calculated for each gear and year of the *FP* survey. Average annual CVs were used as an indicator of consistency over the time series. Chi-square tests were used to test for

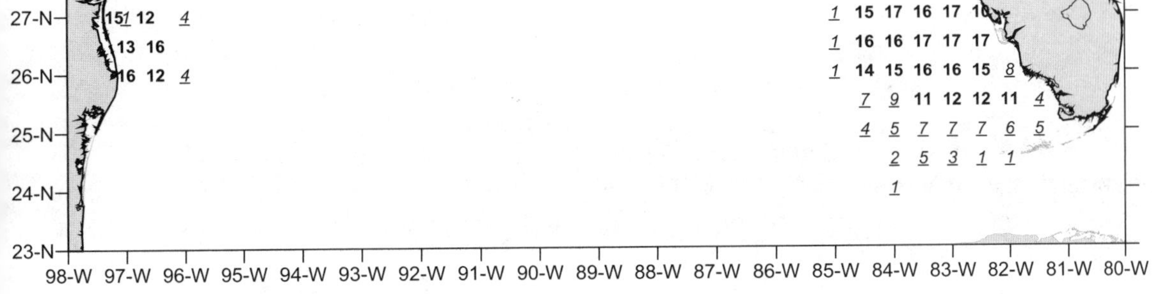

Figure 1. Number of samples taken at each SEAMAP B-number location during all years of the Fall Plankton survey. Bold numbers represent B-number locations which were sampled during at least 10 years of the survey and retained in our analysis, and the underlined italic numbers B-number locations dropped from the analysis.

equal proportion of day and night captures of larvae between the two gear types.

Trends in percent occurrence and abundance of red snapper larvae for the *SB* and *FP* surveys were compared by correlation analysis. All comparisons were carried out using only bongo samples collected west of 087.75°W longitude during the 1986 & 1987, 1994, 1997, and 2000–2002 *SB* and *FP* surveys. The selected years are those where both surveys sampled the full spatial area of the western GOM.

Regional and sub-regional differences in larval red snapper abundance and occurrence were assessed using data from the *FP* surveys. Western and eastern regions were separated at the mouth of the Mississippi River (089.17°W longitude) as delineated by the 2005 red snapper stock assessment (Porch 2007, this volume). The western region was further divide into Texas (TX) and Louisiana (LA) subregions at the TX/LA state line (~093.80°W longitude); and the eastern region into Mississippi/Alabama (MS/AL) and Florida (FL) subregions at the AL/FL state line (~087.25°W longitude). The positions used to separate the subregions are slightly shifted from the actual state lines to accommodate the systematic sampling grid of the plankton surveys. Percent occurrence, mean abundance and the regional or subregional percentage of total abundance were calculated for each region or subregion of the GOM. Regional or subregional percentage of total abundance was calculated by dividing the total summed abundance of all samples in a region or subregion by the total abundance of all samples.

Larval red snapper nominal percent occurrence, nominal mean abundance and model based estimates of standardized relative abundance with associated CVs (standard error/mean) were calculated by year for the western and eastern GOM (as defined above) utilizing the FP survey time series of observations. Standardized indices of relative red snapper abundance based on larval occurrence and abundance were estimated using a delta-lognormal model (Lo et al. 1992). Indices based on this model are a mathematical combination of yearly estimates from two distinct generalized linear models: a binomial model which describes proportion of positive catches (i.e., occurrence) and lognormal model which describes variability in only the nonzero abundance data. A backward selection approach using the GLMMIX and MIXED procedures (Patetta 2002) in SAS (Version 9.1.3 of the SAS System for Windows © 2003, SAS Institute Inc., Cary, North Carolina) was employed to provide yearly index values for both the binomial and lognormal sub-models, respectively. The effects tested for inclusion in each submodel were year, time of day (day or night), sub-region (TX and LA or MS/AL and FL) and water depth. For the binomial sub-models, a logistic-type generalized linear mixed model was employed, and model fit was evaluated using the fit statistics provided by PROC GLMMIX in SAS. Likewise, for the lognormal sub-model, a generalized linear mixed model was used to describe the nonzero abundance data, and model fit was evaluated using the fit statistics provided by PROC MIXED in SAS. The year effect is integral to the calculation of annual estimates and is forced into the standardization procedure regardless of significance when at least one other parameter is significant. Years when no red snapper larvae were collected were dropped from the analyses since an index developed using delta-lognormal methodology cannot be calculated from data containing only zero catches. Also, when the lognormal submodels did not converge or did not retain any significant effects, only the logistic model describing occurrence was used to develop the indices.

Results

Mean abundance, percent occurrence and the percentage of total abundance of red snapper larvae were higher during *FP* nighttime

Table 1. Mean abundance (Mean) and percent occurrence (%O) with number of samples (*N*), standard error (*SE*), and percentage of total abundance (% Total) of day and night caught red snapper *Lutjanus campechanus* larvae captured in neuston and bongo nets during the Fall Plankton survey.

			Abundance			Occurrence	
Gear	Diel Period	*N*	Mean	*SE*	% Total	%O	*SE*
Neuston	Day	790	0.02	0.01	5.00	1.65	0.45
	Night	729	0.50	0.08	95.00	14.68	1.31
	Day + Night	1519	0.25	0.04	100.00	7.90	0.69
Bongo	Day	790	0.25	0.04	23.00	5.06	0.78
	Night	729	0.92	0.13	77.00	11.93	1.20
	Day + Night	1519	0.57	0.07	100.00	8.36	0.71

Table 2. Percent coefficient of variation (standard error/mean) of annual abundance (*A*) and percent occurrence (%O) of red snapper *Lutjanus campechanus* larvae captured in neuston and bongo nets during the Fall Plankton survey.

		Nueston		Bongo	
Year	*N*	*A*	%O	*A*	%O
1986	107	65.74	49.29	44.94	43.87
1987	110	62.64	49.31	60.84	43.89
1988	51	100.00	100.00	100.00	100.00
1989	51	70.74	70.00	58.42	56.57
1990	62	67.22	48.75	53.05	48.75
1991	58	100.00	100.00	44.77	43.12
1992	106	31.98	25.02	38.72	36.70
1993	110	38.99	28.73	38.65	36.74
1994	118	42.04	39.94	67.81	43.95
1995	114	56.64	32.13	35.09	25.14
1996	112	44.88	27.40	35.83	32.11
1997	115	40.12	32.14	29.60	24.18
1999	108	33.10	26.13	42.97	36.72
2000	105	46.47	27.30	27.96	21.56
2001	106	47.24	43.86	43.21	32.04
2002	86	40.08	25.70	32.48	25.70

sampling for both neuston and bongo nets (Table 1). No difference between gears (α = 0.05, P = 0.1415) was observed in the occurrence of larvae during nighttime hours. However, the occurrence of red snapper larvae in bongo net samples was found to be significantly higher than nueston samples (α = 0.05, P = <0.0001) during the day. The diel percentage of total red snapper larval abundance was skewed in favor of nighttime catches for both gears. However, the percentages of diel total abundance were more equitably distributed between day and night samples for the bongo than for the neuston (Table 1). Sampling variability over the time series was less variable for the bongo than the neuston. Annual CV on mean abundance for the neuston averaged 56%, and annual CV on percent occurrence averaged 45%. While annual CV on mean abundance for the bongo averaged 47%, and annual CV on percent occurrence averaged 41% (Table 2). Overall, the bongo was more effective at catching larvae over the 24 h time period than the neuston with less year to year sampling variability. Therefore all further analyses of red snapper larvae in SEAMAP collections were solely based on bongo net samples.

Day and night occurrence, mean abundance and diel percentage of total abundance from selected years of the *SB* and *FP* surveys west of 087.75°W longitude differed between the surveys (Table 3). Mean abundance and occurrence during the *FP* survey were considerably higher at night than during the day. However, mean abundance and occurrence during the *SB* survey were similar between day and night. The number of night samples available from the *SB* survey was less than half the number of day samples. Whereas, the number of day and night samples available from the *FP* survey were about the same. The difference in the observed diel pattern between the two surveys was likely caused by disparity in the number of night and day samples collected during the *SB* survey. Therefore comparison between the two surveys was confined to daytime samples only.

Mean abundance during the *SB* survey was two times greater than during the *FP* survey. Occurrence was also higher during the *SB* survey (13%) than during the *FP* survey (9%) (Table 3). Annual ratios (*SB*/*FP*) of mean abundance ranged from 1.14 to 2.97, and annual ratios (*SB*/*FP*) of occurrence between the two surveys ranged from 0.73 to 1.92 (Table 4). Mean abundance and occurrence between the *SB* and *FP* surveys were highly correlated within years. The correlation was 80% ($n = 7$, $r = 0.795$) between *SB* and *FP* annual abundance and 75% ($n = 7$, $r = 0.747$) between *SB* and *FP* annual occurrence.

Larval red snapper occurrence and abundance during the *FP* survey were an order of magnitude higher in the western than in the eastern GOM (Figure 2). Larvae were nine times more abundant and occurred in five times as many samples in the western than in the eastern GOM. The western GOM accounted for 88% of the total GOM larval abundance from the 16 years of *FP* surveys. In contrast, the eastern GOM accounted for only 12% of the total GOM larval abundance. Larval abundance, occurrence and percentage of total GOM abundance were similar between the TX and LA sub-regions. However, the MS/AL and FL sub-regions in the eastern GOM were quite different. Larvae were four times more abundant and occurred in four times as many samples in the MS/AL sub-region than in the FL sub-region. The disproportionately smaller MS/AL subregion also accounted for nearly half of the 12% of total abundance in the eastern GOM.

Table 3. Mean abundance (Mean) and percent occurrence (%*O*) with number of samples (*N*), standard error (*SE*), and percentage of total abundance (% Total) of day and night caught red snapper larvae captured in bongo nets during the 1986, 1987, 1994, 1997, and 2000–2002 Summer Shrimp/Bottomfish (*SB*) and Fall Plankton (*FP*) surveys in the western Gulf of Mexico.

			Abundance			Occurrence	
Survey	Diel Period	*N*	Mean	*SE*	% Total	%O	*SE*
SB	Day	222	0.96	0.20	72.68	13.06	2.27
	Night	102	0.78	0.24	27.32	13.73	3.42
	All	324	0.90	0.16	100.00	13.27	1.89
FP	Day	224	0.46	0.11	21.95	8.93	1.91
	Night	193	1.88	0.38	78.05	20.73	2.93
	All	417	1.12	0.19	100.00	14.39	1.72

Table 4. Daytime abundance (A) and percent occurrence (B) of red snapper larvae captured in bongo nets during the 1986, 1987, 1994, 1997, and 2000–2002 Summer Shrimp/Bottomfish (*SB*) and Fall Plankton (*FP*) surveys in the western Gulf of Mexico with associated standard errors (*SE*) and number of samples (*N*). *SB*/*FP* ratios are the annual abundance or percent occurrence (% Occurrence) of the *SB* survey divided by the *FP* survey. Correlation is the Pearson correlation coefficient of annual abundance or percent occurrence for all years listed.

(A)

Year	*N*	*SB* Abundance	*SE*	*N*	*FP* Abundance	*SE*	*SB*/*FP* Ratio	Correlation
1986	41	0.48	0.28	30	0.42	0.24	1.14	$r = 0.80$
1987	29	0.42	0.31	32	0.14	0.14	2.97	
1994	28	0.79	0.62	34	0.00	0.00		
1997	34	1.25	0.70	35	0.65	0.28	1.93	
2000	34	1.51	0.60	28	0.96	0.47	1.57	
2001	26	1.16	0.52	30	0.59	0.41	1.97	
2002	30	1.17	0.57	35	0.51	0.28	2.30	

(B)

Year	*N*	*SB* % Occurrence	*SE*	*N*	*FP* % Occurrence	*SE*	*SB*/*FP* Ratio	Correlation
1986	41	7.32	4.12	30	10.00	5.57	0.73	$r = 0.75$
1987	29	6.90	4.79	32	3.13	3.13	2.21	
1994	28	7.14	4.96	34	0.00	0.00		
1997	34	14.71	6.17	35	14.29	6.00	1.03	
2000	34	20.59	7.04	28	14.29	6.73	1.44	
2001	26	19.23	7.88	30	10.00	5.57	1.92	
2002	30	16.67	6.92	35	11.43	5.46	1.46	

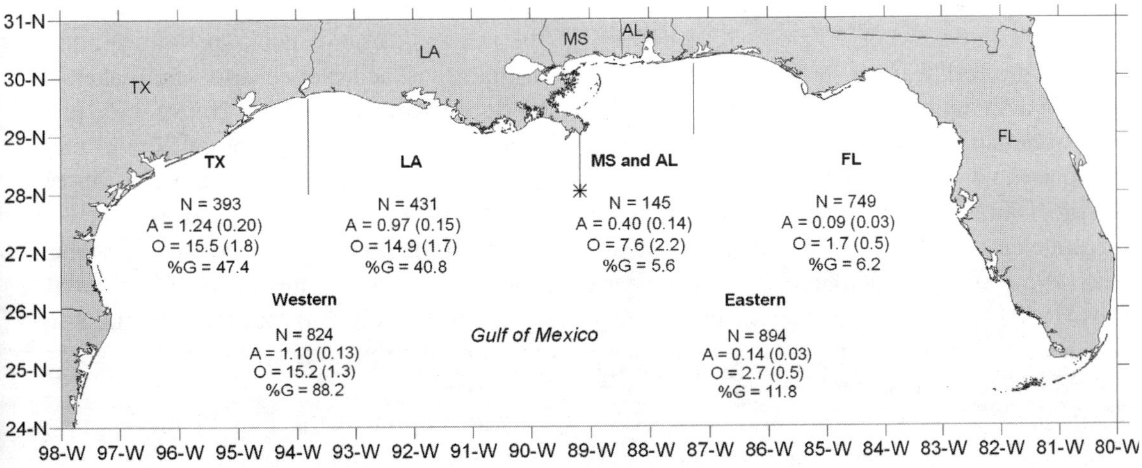

Figure 2. Mean abundance (A), percent occurrence (O) and percentage of Gulf of Mexico total abundance (%G) of red snapper *Lutjanus campechanus* larvae from the western and eastern Gulf of Mexico, and the Texas (TX), Louisiana (LA), Mississippi/Alabama (MS and AL) and Florida subregions collected during the Fall Plankton survey. Values in parenthesis are standard errors of abundance and percent occurrence. Western and eastern Gulf of Mexico regions are separated at the mouth of the Mississippi River (*) and the subregions by the plotted demarcation lines.

Model based standardized indices of relative abundance were generated using the delta-lognormal procedure for both the western and eastern GOM utilizing samples collected during the Fall Plankton survey. The modeling exercise for the western GOM index identified the following significant effects: year ($P = 0.0138$) and time of day ($P = <0.0001$) for the binomial submodel; and time of day ($P = 0.0075$), subregion ($P = 0.0159$) and water depth ($P = 0.0150$) for the lognormal submodel. The year effect ($P = 0.0935$) was not significant but was retained in the lognormal sub-model. For the eastern GOM index the modeling exercise identified subregion ($P = 0.0002$) as the only significant effect in the binomial submodel. The year effect ($P = 0.9086$) was not significant but was retained in the binomial sub-model. No significant factors were identified in the lognormal sub-model. Therefore, the eastern standardized index of relative abundance is based solely on the binomial submodel describing the occurrence of larvae.

The western standardized index indicated a substantial increase in red snapper larvae after 1994 (Tables 5 and Figure 3). In the western GOM CVs of the standardized index of annual abundance ranged from 29% to 75% and in general were below 40% after 1994. Nominal indices based on larval occurrence and abundance in the western GOM showed similar trends in abundance at similar levels of precision. A standardized index of relative abundance based on the binomial sub-model (occurrence) was calculated for the eastern GOM (Table 5 and Figure 3). Both the standardized and nominal indices for the eastern GOM indicated very low levels of red snapper abundance with an increase after 1995. CVs of the eastern abundance indices in almost all years were greater than 50%.

Discussion

Lyczkowski-Shultz and Hanisko (2007) review the early life history of red snapper in the Gulf of Mexico based on over 7,900 neuston and 7,000 bongo collections from SEAMAP plankton surveys 1982–2003. The objective of this companion study was to develop standardized larval indices of relative red snapper abundance generated from those collections that might best reflect trends in the size of the red snapper spawning populations in the western and eastern U.S. GOM. After examination of the survey data we have concluded that the most reliable and spatially consistent index is the one based on abundance and occurrence of red snapper larvae estimated from bongo net samples taken during the Fall Plankton survey of shelf waters from Brownsville, TX to south Florida.

Use of Fall Plankton survey data as a basis for an index of the spawning population size is complicated by the timing of this survey (~September to early October) which is near the end of the red snapper spawning season (Futch and Bruger 1976; Collins et al. 1996, 2001; Woods 2003; Fitzhugh et al. 2004). However, occurrence and abundance of red snapper larvae were lower in September than during the SEAMAP Summer Shrimp/Bottomfish survey in June and July (the nominal time of peak spawning) annual estimates of occurrence and abundance for the two survey types (*SB* and *FP*) were highly correlated, and the ratios of annual mean abundance and occurrence between the two were fairly consistent from year to year. Thus we conclude that larval red snapper abundance as measured during the fall plankton survey effectively approximates reproductive output of the red snapper population at least in the western GOM.

The timing of the Fall Plankton survey may have biased our estimates of the spawning population of red snapper among the various regions and subregions of the GOM. Sampling during the Fall Plankton survey typically begins in early September off south Texas and continues eastward to south Florida through the end of September and occasion-

Table 5. Nominal abundance (*NA*), nominal percent occurrence (%*O*), and standardized relative abundance (*SRA*) of red snapper *Lutjanus campechanus* larvae collected during the Fall Plankton survey with associated percent coefficient of variation (%CV, (standard error/mean) and number of samples (*N*) by year for the western (A) and eastern (B) Gulf of Mexico.

(A)

		Nominal Abundance		Nominal Occurrence		Standardized Relative Abundance	
Year	*N*	*NA*	%CV	%O	%CV	*SRA*	%CV
1986	50	0.31	49.54	8.00	48.45	0.33	51.93
1987	56	0.95	69.46	5.36	56.68	0.78	71.99
1988	28	0.00		0.00		.	
1989	29	0.59	57.55	10.34	55.63	0.74	63.20
1990	32	0.97	45.34	15.63	41.74	0.88	49.22
1991	32	0.62	49.23	12.50	47.52	0.56	52.04
1992	56	0.53	37.77	12.50	35.68	0.47	38.82
1993	56	0.55	37.65	12.50	35.68	0.48	38.50
1994	56	0.91	72.14	7.14	48.62	0.77	75.43
1995	56	1.99	35.78	21.43	25.82	1.78	36.25
1996	56	1.12	34.70	16.07	30.81	1.04	36.13
1997	55	1.65	28.71	25.45	23.29	1.71	28.94
1998						.	
1999	52	0.42	44.72	9.62	42.93	0.43	47.06
2000	55	2.01	31.97	25.45	23.29	1.80	29.80
2001	47	1.25	48.49	12.77	38.54	1.19	50.43
2002	54	1.43	32.86	22.22	25.70	1.47	32.64
2003	54	2.33	31.48	29.63	21.17	2.19	29.20

(B)

		Nominal Abundance		Nominal Occurrence		Standardized Relative Abundance	
Year	*N*	*NA*	%CV	%O	%CV	*SRA*	%CV
1986	62	0.08	100.00	1.61	100.00	0.02	100.11
1987	60	0.14	70.30	3.33	70.11	0.03	70.17
1988	27	0.21	100.00	3.70	100.00	0.04	99.04
1989	29	0.00		0.00			
1990	36	0.00		0.00			
1991	35	0.10	100.00	2.86	100.00	0.03	99.48
1992	53	0.00		0.00			
1993	59	0.00		0.00			
1994	65	0.06	100.00	1.54	100.00	0.02	100.15
1995	61	0.10	70.13	3.28	70.12	0.03	70.19
1996	62	0.00		0.00		.	
1997	63	0.03	100.00	1.59	100.00	0.02	100.13
1998						.	
1999	61	0.33	69.11	4.92	56.76	0.05	56.82
2000	58	0.51	53.19	6.90	48.67	0.07	48.69
2001	62	0.15	58.14	4.84	56.78	0.05	56.84
2002	39	0.09	100.00	2.56	100.00	0.03	99.63
2003	62	0.40	52.49	6.45	48.75	0.06	48.81

ally into mid October. Lyczkowski-Shultz and Hanisko (2007) report a sharp decline in the occurrence and abundance of red snapper larvae from September to October. The decline may be indicative of the abrupt termination of spawning in this species as was suggested by Woods (2003) based on the low incidence during the spawning season of red snapper ovaries exhibiting over 50% atresia. Declining reproductive output from September to mid October and the west to east progression of the Fall Plankton survey may lead to an underestimation of larval occurrence and abundance in the eastern region of the survey area.

Given the likelihood of underestimating larval occurrence and abundance in the eastern GOM the larval data still seems to reflect the !relative regional and subregional differences in GOM red snapper population. Based on larval occurrence and abundance the relative red snapper population was four to eight times greater in the western than the eastern GOM. The current stock assessment for red snapper estimated the unfished abundance of the western population to be three times greater than the eastern population (Porch 2007, this volume). At the subregional level, larval occurrence and abundance indicated a decreasing trend in the red snapper abundance from Texas to Florida. The percentage of the total number of age-0 and age-1 red snapper caught off TX (69%), LA (23%) and MS/AL (7%) during 1988 to 2006 SEAMAP Fall Groundfish trawl surveys show a similar pattern (Nichols, NMFS, personal communication). Gold and Saillant (2007, this volume) and Saillant and Gold (2004) estimated the population of red snapper off TX, LA and MS/AL based on genetic variance effective size. Population estimates for TX and MS/AL were similar, but the estimate for LA was at least an order of magnitude higher than for TX and MS/AL. Larval occurrence and abundance also suggest a higher abundance of red snapper off LA than MS/AL, but in contrast to the variance effective size estimates suggested that red snapper abundance off TX and LA were similar. The only estimate of relative population size between subregions in the eastern GOM was the larval data. Although the MS/AL and FL subregions each contributed about the same percent (5.6 and 6.2) to total Gulfwide abundance of larval snapper the mean abundance of larvae off MS/AL was four times greater than off FL. The greater concentration of larvae off MS/AL may be attributed to production from the adult spawning stock associated with the high concentration of artificial reefs in the area (Szedlmayer and Shipp 1994; Minton and Heath 1998; Shipp 1999; Patterson and Cowen 2003).

The standardized larval indices of abundance presented here for the western and eastern GOM differed from the larval indices used in the most recent red snapper stock assessment (SEDAR7 2005; Porch 2007, this volume). The initial indices were based on the size adjusted abundance of 3.8–8.3 mm larvae taken in bongo nets during both the SEAMAP Summer Shrimp/Bottomfish and Fall Plankton surveys, and only the year effect was accounted for in the delta-lognormal model (Hanisko et al. 2004; Lyczkowski-Shultz et al. 2004).. In contrast, the current indices were based on the abundance of 3.8–6.3 mm larvae in bongo nets taken during the Fall Plankton survey, and were unadjusted for size. The model generating these indices attempted to account for the effects of year, time of day and subregion. The different formulations of the larval indices revealed similar trends in relative red snapper abundance over time. Both versions suggested an increased adult spawning stock in the eastern GOM, as did the 2005 assessment (Porch 2007). However, the eastern GOM indices were of limited value in resolving annual changes in population size due to their low precision. The 2005 stock assessment in the western GOM indicated little or no increase

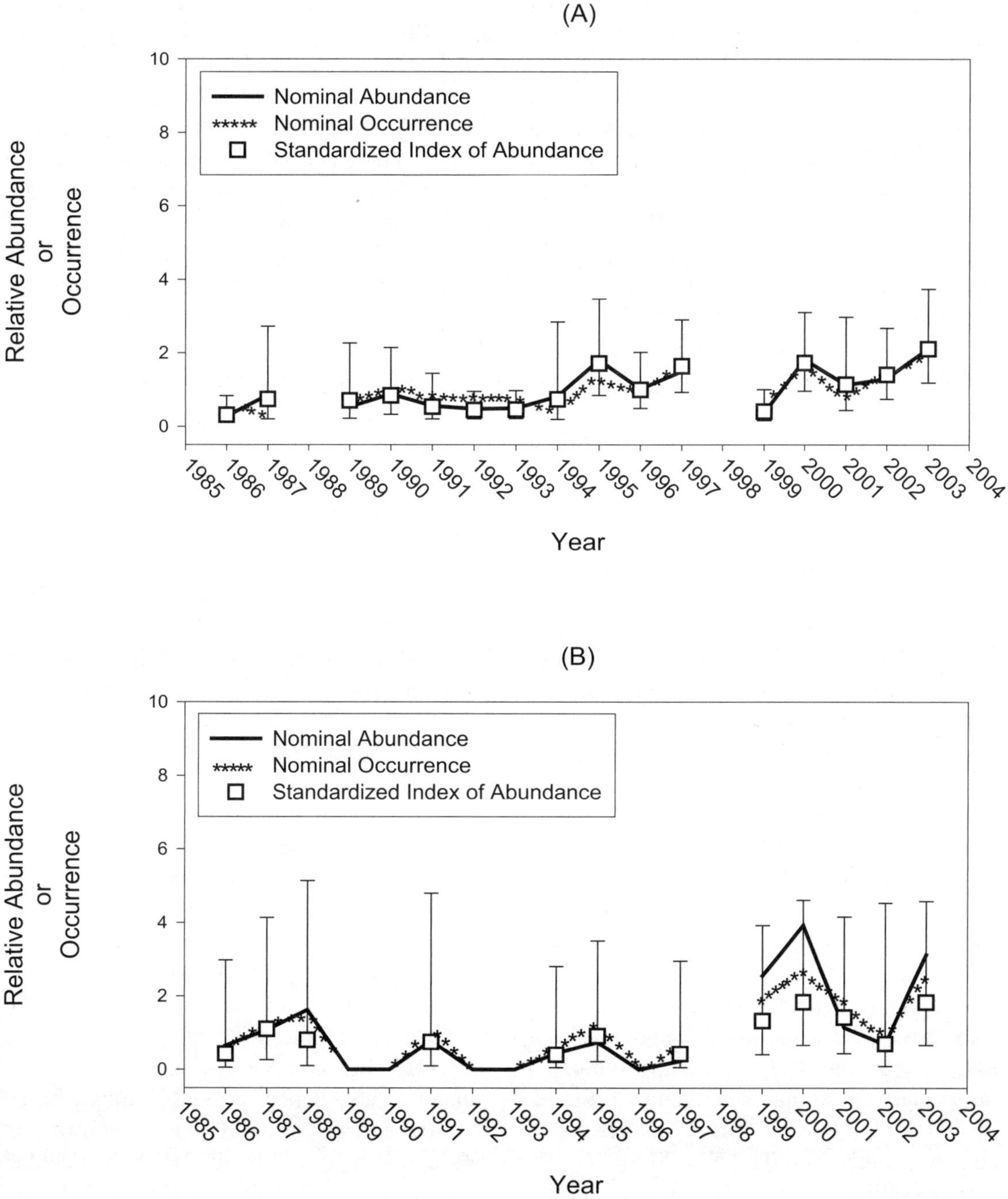

Figure 3. Nominal abundance, nominal percent occurrence, and standardized relative abundance of red snapper *Lutjanus campechanus* larvae collected during the Fall Plankton survey by year for the western (A) and eastern (B) Gulf of Mexico. Error bars associated with the standardized index are asymmetrical 95% confidence intervals. Annual values of nominal abundance, nominal percent occurrence, and standardized relative abundance are scaled by their respective means.

in the population of red snapper, while both (current and initial) indicated an increased spawning stock.

In addition to the initial larval indices, three other indices were used by the recent stock assessment to identify trends in the adult spawning stock: a SEAMAP reef fish video survey index (VIDEO; Gledhill and Ingram 2004), an index of commercial hand line catches (CHL; McCarthy and Cass-Calay 2004) and an index of recreational catch reported from Marine Recreational Fisheries Statistics Survey Data (MRFSS; Cass-Calay 2004). The trends indicated by the larval, VIDEO, CHL and MRFSS indices were in general agreement for the eastern GOM. In the western GOM, the 2005 assessment was not able to reconcile the increasing trend of the initial larval index with the flat or declining trends indicated by the other indices of adult abundance (Porch 2007, this volume). The current larval index indicated a less dramatic increase over the time series with little or no increase from 1995 to 2003, but still indicated higher abundances of red snapper after 1995 than the other indices. Confidence intervals estimated for the current larval index and the VIDEO, MRFSS and CHL indices suggested that the difference among the trends in the indices may not be statistically significant (Cass-Calay 2004; Gledhill and Ingram 2004; McCarthy and Cass-Calay 2004).

Potential explanations for the discrepancy among the trends suggested by the VIDEO, MRFSS and CHL and larval indices in the western GOM may be linked to differences in sampled habitat, subregional coverage and age selectivity. The SEAMAP VIDEO survey provides data on the adult population from natural reef and hard bottom habitats but does not index the spawning stock from artificial reefs in the western GOM. Artificial reefs, predominantly offshore oil and gas platforms, harbor large numbers of red snapper and are a major destination of commercial and recreational fisherman targeting the species (Witzig 1986; Reggio 1987; Stanley and Wilson 2000, 2003; Nieland and Wilson 2003). Data in the MRFSS index was limited to catches reported solely from the state of LA, as data available from the Texas Department of Parks and Wildlife was incompatible with MRFSS and was not included (Cass-Calay 2004). The MRFSS and CHL indices select predominantly for age 2 to age 4 and age 3 to age 5 fish respectively, and may not adequately represent older fish. Allman et al. (this volume) compared age compositions between the recreational, commercial handline and longline sectors and found that age 2 to age 4 fish accounted for 90% of the recreational sector with less with 0.3% of all fish greater than age 10, where as age 3 to age 5 fish dominated CHL catches with 1% of fish greater than age 10. The age distribution from commercial longline catches underscores the age selectivity of the MRFSS and CHL indices. Red snapper by age 5 were fully recruited to the commercial long-line fishery with over 22% of fish greater than age 10 (Allman et al. 2007, this volume). In contrast to the VIDEO, MRFSS and CHL indices, the larval index references the majority of the spawning area in the western GOM, and the reproductive output of the adult spawning stock regardless of habitat and age.

In general, the eastern and western larval indices indicated two distinct periods of larval abundance and occurrence: 1986–1994 when larval occurrence and abundance was extremely low, and 1995–2003 when occurrence and abundance were two times greater than the earlier period. The inception of the *FP* survey in 1986 coincided with the decline of the red snapper fishery in the mid to late 1980s. During this time, the total catch of red snapper fell from 4.7 million kg (10.3 million pounds) in 1982 to 1.8 million kg (4.0 million pounds) in 1990, and the fishery was supported primarily by age 1 to age 3 fish (Hood and Steele 2004). The depletion of the adult spawning stock during this period may be reflected in the low levels of larval red snapper occurrence and abundance indicated by the larval indices from 1986 to 1994. Increased occurrence and abundance after 1995 may reflect higher recruitment from above average year classes during the period from 1989 to 1991 which coincided with the lower total catches in the early 1990s, and recruitment from strong year classes during the mid 1990s (Allman et al. this volume; Nichols 2004; Turner and Porch 2004; SEDAR7 2005). Most female red snapper mature by age 2 and nearly all (95%) by age 5 (Fitzhugh et al. 2004). Therefore, fish recruited to the population in the late 1980s and early 1990s would begin contributing to larval production between 1991 and 1996, and the 1994 to 1996 recruits between 1996 and 2001. This corresponds well with increased larval occurrence and abundance during SEAMAP Fall Plankton surveys after 1995.

The red snapper spawning stock in the GOM was estimated to be much lower than it had been historically, but estimated recruitment for both the western and eastern components of the stock have been above the long term average. Porch (2007) indicates that the recruitment estimates are well above those for an unfished population despite the indicated decrease in the spawning potential of the current stock, and suggests as a possible interpretation that red snapper stocks may have become more productive over the last two decades. Under this hypothesis, the increase in red snapper larval occurrence and abundance may reflect an increase in the reproductive output of red snapper and not an increase in the size of the spawning population.

The idea that the size or biomass of a fish stock can be estimated from egg or larva abundance data as measured during field surveys has been around since the end of the 19th century (Heath 1992). Use of ichthyoplankton data to estimate the biomass of fish populations over time, either in absolute or relative terms, is based on the assumption that population parameters such as fecundity, spawning frequency, hatching success, development, growth and mortality are unvarying from spawning season to spawning season (i.e., year to year) or even within a spawning season (Heath 1992; Hunter and Lo 1993). More often than not this assumption is difficult, if not impossible, to verify due to the lack of specific information on early life stage vital rates or the cost of obtaining such information. Despite the shortcomings, ichthyoplankton abundance and presence/absence data continues to be used in contemporary, age structured stock assessment models both in the U.S. and worldwide. These models are enhanced by inputs of fishery-independent indices of relative stock abundance which are considered to be without bias (as opposed to fishery-dependent data sources and indices) because they are based on statistical sampling design. Lack of fishery-independent data are considered to a great impediment to fishery assessments (NMFS 2001). As a result of this ichthyoplankton surveys continue to become an increasingly important source of fishery-independent data for fish stocks such as Pacific sardine (Lo and Macewicz 2006), bocaccio rockfish (Ralston and Ianelli 1998), cowcod rockfish (Butler et al. 2002), and Atlantic bluefin tuna (Scott et al. 1993; Scott and Turner 2002; Ingram et al. 2006).

The value of larval data in red snapper assessments remains problematic as our results have shown. Inclusion of data on the abundance of smaller (<4 mm) red snapper larvae in SEAMAP collections that now can be identified only to family or genus (*Lutjanus* sp.) may improve not only the precision of the larval index but its value as a practical gauge of spawning biomass as well. Use of molecular genetic techniques to identify the smallest field collected snapper larvae to species holds great promise. There are plans at SEFSC/NMFS Mississippi Laboratories to begin identifying small snapper larvae in SEAMAP samples using genetic techniques in the near future. Modeling the effects of environmental factors influencing larval occurrence and abundance may also improve the precision of the larval indices and provide a better understanding of their distribution. Environmental effects were not investigated for the current indices as considerable work still needs to be completed regarding the identification of corresponding environmental data within the SEAMAP ichthyoplankton database. More consistent plankton sampling during SEAMAP summer trawl surveys in coming years, i.e. during peak months of red snapper spawning, may also enhance the contribution of the red snapper larval index in future population assessments.

Acknowledgments

The following individuals are gratefully acknowledged for their significant contributions to this work: Pam Bond, Connie Cowan, Denice Drass, Gelenn Zapfe, Lanora Lange and Rosanne Brasher, NOAA/NMFS Mississippi Laboratories, Pascagoula, MS; Malgorzata Konieczna, Hanna Skolska and the Ichthyoplankton Group, Sea Fisheries Institute, Plankton Sorting and Identifications Center, Szczecin and Gdynia, Poland; Kim Williams at the SEAMAP Archiving Center, Fish and Wildlife Resaearch Institute, St. Petersburg, FL; and Cindy Gavins, Gulf Coast Research laboratory/USM, Ocean Springs, MS. We would also like to recognize the enduring efforts of the crews of the NOAA Ships Chapman, Oregon II and Gordon Gunter; and the dedication of the biologists of the NMFS, Gulf States resource agencies, Johnson Controls International and IAP World Services that participate on SEAMAP cruises making this historical data series possible. I also thank Will Patterson and two anonymous reviewers for their helpful comments on the manuscript.

References

Allman, R. J., and G. R. Fitzhugh. 2007. Temporal age progressions and relative year-class strength of Gulf of Mexico red snapper. Pages 311–328 *in* W. F. Patterson, III, J. H. Cowan, Jr., G. R. Fitzhugh, and D. L. Nieland, editors. Red snapper ecology and fisheries in the U.S. Gulf of Mexico. American Fisheries Society, Symposium 60, Bethesda, Maryland.

Butler, J. L., L. D. Jacobson, J. T. Barnes, and H. G. Moser. 2002. Biology and population dynamics of the cowcod (*Sebastes levis*) in the southern California Bight. Fishery Bulletin 101:260–280.

Cass-Calay, S. L. 2004. Revised catch rate indices of red snapper (*Lutjanus campechanus*) landed during 1981–2003 by the U.S. Gulf of Mexico recreational fishery. Southeast Data, Assessment and Review 7: Gulf of Mexico red snapper. Available: http://www.sefsc.noaa.gov/sedar/download/SEDAR7-AW-04v2.pdf?id=DOCUMENT. (March 2007).

Collins, L. A., G. R. Fitzhugh, L. Mourand, L. A. Lombardi, W. T. Walling, Jr., W. A. Fable, M. R. Burnett, and R. J. Allman. 2001. Preliminary results from a continuing study of spawning and fecundity in the red snapper (Lutjanidae: *Lutjanus campechanus*) from the Gulf of Mexico, 1998–1999. In L. R. Creswell, editor. Proceedings of the 52nd Annual Gulf and Caribbean Fisheries Institute 52:34–47.

Collins, L. A., A. G. Johnson, and C. P. Keim. 1996. Spawning and annual fecundity of the red snapper (*Lutjanus campechanus*) from the northeastern Gulf of Mexico. Pages 174–188 *in* F. Arreguin-Sanchez, J. L. Munro, M. C. Balgos, and D. Pauly editors, Biology, fisheries and culture of tropical groupers and snappers. ICLARM Conference Proceedings 48.

Drass, D. M., K. L. Bootes, J. Lyczkowski-Shultz, B. H. Comyns, G. J. Holt, C. M. Riley, and R. P. Phelps. 2000. Larval development of red snapper, *Lutjanus campechanus*, with comparisons to co-occurring snapper species. Fishery Bulletin. 98:507–527.

Fitzhugh, G. R., M. S. Duncan, L. A. Collins, W. T. Walling and D. W. Oliver. 2004. Characterization of red snapper (*Lutjanus campechanus*) reproduction: for the 2004 Gulf of Mexico SEDAR. Southeast Data, Assessment and Review 7: Gulf of Mexico red snapper. Available: http://www.sefsc.noaa.gov/seda/download/SEDAR7_DW35.pdf?id=DOCUMENT. (March 2007).

Futch, R. B., and G. E. Bruger. 1976. Age, growth, and

reproduction of red snapper in Florida waters. Pages 165–183 *in* H. R. Bullis and A. C. Jones editors. Proceedings: Colloquium on snapper-grouper fishery resources for the western central Atlantic Ocean. Florida Seagrant College Report 17:165–185.

Gledhill, C. T., and W. Ingram. 2004. SEAMAP reef fish survey of offshore banks. Southeast Data, Assessment and Review 7: Gulf of Mexico red snapper. Available: http://www.sefsc.noaa.gov/sedar/download/SEDAR7_DW15.pdf?id=DOCUMENT. (March 2007).

Gold, J. R., and E. Saillant. 2007. Population structure of red snapper in the northern Gulf of Mexico. Pages 201–216 *in* W. F. Patterson, III, J. H. Cowan, Jr., G. R. Fitzhugh, and D. L. Nieland, editors. Red snapper ecology and fisheries in the U.S. Gulf of Mexico. American Fisheries Society, Symposium 60, Bethesda, Maryland.

Hanisko, D. S., J. Lyczkowski-Shultz, and G. W. Ingram, Jr. 2004. Alternate fishery independent larval indices of abundance for red snapper. Southeast Data, Assessment and Review 7: Gulf of Mexico red snapper. Available: http://www.sefsc.noaa.gov/sedar/download/SEDAR7-RW%2007.pdf?id=DOCUMENT. (March 2007).

Heath, M. R. 1992. Field investigations of the early life stages of marine fish. Advances in Marine Biology 28:1–174.

Hood, P., and P. Steele. 2004. History of red snapper management in federal waters of the U.S. Gulf of Mexico, 1982 to 2004. Southeast Data, Assessment and Review 7: Gulf of Mexico red snapper. Available: http://www.sefsc.noaa.gov/sedar/download/SEDAR7_DW40.pdf?id=DOCUMENT. (March 2007).

Hunter, J. R., and N. C.-H. Lo. 1993. Ichthyoplankton methods for estimating fish biomass introduction and terminology. Bulletin of Marine Science 53(2): 723–727.

Ingram, G. W., Jr., W. J. Richards, G. P. Scott, and S. C. Turner. 2006. Development of indices of bluefin tuna (*Thunnus thynnus*) spawning biomass in the Gulf of Mexico using delta-lognormal models. ICCAT Working Document. SCRS/06/082.

Lindeman, K. C., W. J. Richards, J. Lyczkowski-Shultz, D. M. Drass, C. B. Paris, J. M. Leis, M. Lara, and B. H. Comyns. 2005. Lutjanidae: snappers. Pages 1549–1595 *in* W. J. Richards, editor. Early stages of Atlantic fishes: an identification guide for the western central North Atlantic. CRC Press, Boca Raton, Florida.

Lo, N. C. -H., L. D. Jacobson, and J. L. Squire. 1992. Indices of relative abundance from fish spotter data based on delta-lognormal models. Canadian Journal of Fisheries and Aquatic Sciences 49:2515–1526.

Lo, N. C. -H., and B. J. Macewicz. 2006. Spawning biomass of Pacific sardine (*Sardinops sagax*) off California in 2005. NOAA Technical Memorandum NMFS. NOAA-TM-NMFS-SWFSC-387

Lyczkowski-Shultz, J., and D. S. Hanisko. 2007. A time series of observations on red snapper larvae from SEAMAP surveys, 1982–2003: seasonal occurrence, distribution, abundance and size. Pages 3–23 *in* W. F. Patterson, III, J. H. Cowan, Jr., G. R. Fitzhugh, and D. L. Nieland, editors. Red snapper ecology and fisheries in the U.S. Gulf of Mexico. American Fisheries Society, Symposium 60, Bethesda, Maryland.

Lyczkowski-Shultz, J., D. S. Hanisko, and G. W. Ingram, Jr. 2004. The potential for incorporating a larval index of abundance for stock assessment of red snapper, *Lutjanus campechanus*. Southeast Data, Assessment and Review 7: Gulf of Mexico red snapper. Available: http://www.sefsc.noaa.gov/sedar/download/SEDAR7_DW14.pdf?id=DOCUMENT. (March 2007).

McCarthy K. J., and S. L. Cass-Calay. 2004. Standardized catch rates of red snapper (*Lutjanus campechanus*) from the United States commercial handline fishery in the Gulf of Mexico during the 1996–2003: additional indices. Southeast Data, Assessment and Review 7: Gulf of Mexico red snapper. Available: http://www.sefsc.noaa.gov/sedar/download/SEDAR7-AW-09.pdf?id=DOCUMENT. (March 2007).

Minton, V., and S. R. Heath. 1998. Alabama's artificial reef program: building oases in the desert. Gulf of Mexico Science 16:105–106.

Nichols. S. 2004. Calibration among the separate trawl survey programs to extend the time series for juvenile snapper indexes. Southeast Data, Assessment and Review 7: Gulf of Mexico red snapper. Available: http://www.sefsc.noaa.gov/sedar/download/SEDAR7_DW2.pdf?id=DOCUMENT. (March 2007).

Nieland, D. L., and C. A. Wilson. 2003. Red snapper recruitment to and disappearance from oil and gas platforms in the northern Gulf of Mexico. Pages 73–81 *in* D. R. Stanley and A. Scarborough-Bull, editors. Fisheries, reefs, and offshore development. American Fisheries Society, Symposium 36, Bethesda, Maryland.

NMFS (National Marine Fisheries Service). 2001. Marine Fisheries Stock Assessment Improvement Plan. Report of the National Marine Fisheries Service National Task Force for Improving Fish

Stock Assessments. U.S. Department of Commerce, NOAA Technical Memorandum. NMFS-F/SPO-56.

Patetta, M. 2002. Longitudinal Data Analysis with Discrete and Continuous Responses Course Notes. SAS Institute Inc., Cary, North Carolina.

Patterson, W. F., III, and J. H. Cowan, Jr. 2003. Site fidelity and dispersion of red snapper associated with artificial reefs in the northern Gulf of Mexico. Pages 181–193 *in* D. R. Stanley and A. Scarborough-Bull, editors. Fisheries, reefs, and offshore development. American Fisheries Society, Symposium 36, Bethesda, Maryland.

Porch, C. E. 2007. An assessment of the red snapper fishery in the U.S. Gulf of Mexico using a spatially-explicit age-structured model. Pages 355–384 *in* W. F. Patterson, III, J. H. Cowan, Jr., G. R. Fitzhugh, and D. L. Nieland, editors. Red snapper ecology and fisheries in the U.S. Gulf of Mexico. American Fisheries Society, Symposium 60, Bethesda, Maryland.

Ralston, S., and J. N. Ianelli. 1998. When lengths are better than ages: the complex case of bocaccio. Pages 451–468 in F. Funk, T. J. Quinn II, J. Heifetz, J. N. Ianelli, J. E. Powers, J. F. Schweigert, P. J. Sullivan, and C.-I. Zhang, editors. Fishery stock assessment models. Alaska Sea Grant College Program Report No. AK-SG-98–01, University of Alaska Fairbanks, Alaska.

Reggio, V. C., Jr. 1987. Rigs-to-reefs: the use of obsolete petroleum structures as artificial reefs. U.S. Department of the Interior, Minerals Management Service, Gulf of Mexico OCS Region, OCS Report MMS87–0015, New Orleans, Louisiana.

Saillant, E., and J. R. Gold. 2004. Population structure and variance effective size of red snapper (*Lutjanus campechanus*) in the northern Gulf of Mexico. Fishery Bulletin 104:136–148.

Scott, G. P., and S. C. Turner. 2002. Updated index of bluefin tuna (*Thunnus thynnus*) spawning biomass from Gulf of Mexico ichthyoplankton surveys. International Commission for the Conservation of Atlantic Tunas, Working Document SCRS/02/91.

Scott, G. P., S. C. Turner, C. B. Grimes, W. J. Richards, and E. B. Brothers. 1993. Indices of larval bluefin tuna, *Thunnus thynnus*, abundance in the Gulf of Mexico: modeling variability in growth, mortality and gear selectivity. Bulletin of Marine Science 53(2):912–929.

SEAMAP (Southeast Area Monitoring and Assessment Program). 2004. Environmental and biological atlas of the Gulf of Mexico 2001, Number 118. Gulf States Marine Fisheries Commission, Ocean Springs, Mississippi.

SEDAR (Southeast Data, Assessment, and Review). 2005. Stock assessment report of SEDAR 7–10 Gulf of Mexico Red Snapper. SEDAR Assessment Report 1.

Seidelmann, P. Kenneth editor. 1992. Explanatory Supplement to the Astronomical Almanac. University Science Books, Sausalito, California.

Shipp, R. L. 1999. The artificial reef debate: are we asking the right questions? Gulf of Mexico Science 17:51–55.

Stanley, D. R., and C. A. Wilson. 2000. Variation in the density and species composition of fishes associated with three petroleum platforms using dual beam hydroacoustics. Fisheries Research 47:161–172.

Stanley, D. R., and C. A. Wilson. 2003. Seasonal and spatial variation in the biomass and size frequency distribution of fish associated with oil and gas platforms in the northern Gulf of Mexico. Pages 123–153 *in* D.R. Stanley and A. Scarborough-Bull, editors. Fisheries, reefs, and offshore development. American Fisheries Society, Symposium 36, Bethesda, Maryland.

Szedlmayer, S. T., and R. L. Shipp. 1994. Movement and growth of red snapper, *Lutjanus campechanus*, from an artificial reef area in the northeastern Gulf of Mexico. Bulletin of Marine Science 55:887–896.

Turner, S. C., and C. E. Porch. 2004. Alternate indices of abundance of juvenile red snapper from the Gulf of Mexico from SEAMAP surveys 1972–2003. Southeast Data, Assessment and Review 7: Gulf of Mexico red snapper. Available: http://www.sefsc.noaa.gov/sedar/download/SEDAR7-AW-25.pdf?id=DOCUMENT. (March 2007).

Witzig, J. 1986. Rig fishing in the Gulf of Mexico 1984: marine recreational fishing results. Pages 103–105 *in* Proceedings, Sixth Annual Gulf of Mexico Information Transfer Meeting. U.S. Department of the Interior, Minerals Management Service, Gulf of Mexico OCS Region, OCS Study/MMS86–0073. New Orleans, Louisiana.

Woods, M. K. 2003. Demographic differences in reproductive biology of female red snapper (*Lutjanus campechanus*) in the northern Gulf of Mexico. Master's thesis, University of South Alabama, Mobile, Alabama.

American Fisheries Society Symposium 60:301–310, 2007

Red Snapper in the Northern Gulf of Mexico: Age and Size Composition of the Commercial Harvest and Mortality of Regulatory Discards

David L. Nieland[1] and Andrew J. Fischer
Department of Oceanography and Coastal Sciences
School of the Coast and Environment
Louisiana State University, Baton Rouge, Louisiana 70803 USA

M. Scott Baker, Jr.
North Carolina Sea Grant College Program
University of North Carolina–Wilmington, Wilmington, North Carolina 28409 USA

Charles A. Wilson, III
Louisiana Sea Grant College Program
Louisiana State University, Baton Rouge, Louisiana 70803 USA

Abstract.—Red snapper *Lutjanus campechanus* were sampled from commercial landings from the Gulf of Mexico (GOM) off Louisiana from October 2001 to May 2004. Fork length (FL), eviscerated weight, otoliths (both sagittae), and sex determinations were taken from 2,900 specimens; FL was subsequently converted to total length (TL) with the equation TL = 1.073 (FL) + 3.56. Red snapper ages (N = 2,867) estimated from counts of opaque annuli in otoliths ranged from 1 to 14 years; however, the vast majority (97.7%) of these were ages 2 to 6 years and the modal age was 3 years. Total lengths among 2,897 specimens ranged from 278 to 940 mm, modal TL was 400 mm, and 98% of all specimens were less than 600 mm TL. We also investigated the fate of red snapper regulatory discards (individuals <381 mm [15 in] TL) during 16 trips on working commercial vessels; over two-thirds of 4,839 red snapper assigned among four discard fate categories (ranging from alive and vigorous to dead) were returned to the water either in moribund or dead condition. Among 399 potential discards retained for age and length analyses, 86% were between 12 and 15 in (305–381 mm) TL and 85% were 2 years of age. The minimum size regulation appears to do little to protect juvenile red snapper from commercial fishing mortality. Heavy red snapper mortality, which begins as bycatch mortality in shrimp trawls, continues as discard mortality at sub-legal lengths when they first recruit to the offshore fishing grounds, and persists as harvest mortality among the youngest legal year- and size-classes. If the minimum size limit is intended to provide a respite from such mortality, a reconsideration of the utility of the minimum length regulation in the commercial harvest of red snapper may be warranted.

[1] Corresponding author: dniela@lsu.edu.
Current address: Louisiana Sea Grant College Program, Louisiana State University, Baton Rouge, Louisiana 70803 USA

Introduction

Red snapper *Lutjanus campechanus* in the Gulf of Mexico (GOM) have been intensively managed by the Gulf of Mexico Fisheries Management Council (GMFMC) since 1991. Both the commercial and recreational fisheries are currently constrained by size limits, trip or creel limits, seasonal closures, and quotas in an attempt to achieve a spawning potential ratio (SPR) of 30% and to allow populations to recover; however, the species has remained overfished in the GOM (Schirripa and Legault 1997,1999; SEDAR 2005). Accurate information on the age and size composition of the red snapper commercial harvest is necessary to monitor year-class strength, conduct stock assessments, and document population recovery.

The longevity of red snapper can no longer be disputed. Recent studies estimating red snapper ages from otolith annuli have incrementally increased estimates of the potential life span of the species to over 50 years (Szedlmayer and Shipp 1994; Manooch and Potts 1997; Patterson 1999; Baker and Wilson 2001; Patterson et al. 2001a; Wilson and Nieland 2001). However, few of these age classes are represented among the red snapper commercial harvest from the northern GOM off Louisiana during any given year. Our previous studies (Wilson et al. 1994; Wilson and Nieland 1998; Wilson and Nieland 2000; Wilson and Nieland 2001) have been overwhelmingly populated with specimens at the youngest end of the red snapper age spectrum and older specimens have become increasingly rare in the harvest.

Minimum size regulations have been applied to both the recreational and commercial red snapper fisheries with the purpose of both providing a respite from heavy fishing mortality and allowing opportunity to spawn. The GMFMC successively applied minimum size limits for commercially harvested red snapper of 13 in (330 mm) total length (TL) in 1984, 14 in (356 mm) TL in 1994, and 15 in (381 mm) TL in 1995. The current 15 in minimum size likely has resulted in a drastic increase in the numbers of commercial regulatory discards; anecdotes of large numbers of red snapper discards drifting behind commercial boats are common along the Louisiana coast. A 1998 fishery independent "snapshot" of red snapper randomly sampled from among the mortalities resulting from the explosive removal of an offshore petroleum platform showed that the majority (55%) of the red snapper population associated with the platform was less than the legal minimum size (Nieland and Wilson 2003). Similar platforms in the northern GOM are very popular red snapper fishing locations among both recreational and commercial fishers.

The greatest importance of discard mortality estimates is in their application to assessments of populations; however, to date most discard mortalities applied to the red snapper commercial fishery have been based on little direct observation. A mid-1990s observer program aboard handline (commercial) red snapper vessels documented that only 1.6% of total catch was discarded dead (Goodyear 1995); however, due to protrusion of the eyes and stomachs among the discards that swam down, it was suggested that many of these small red snapper suffered delayed mortality. Indeed, Goodyear (1995) variously applied discard mortalities of 10–33% in his analyses of red snapper yield per recruit at several different minimum size limits. Schirripa and Legault (1997, 1999) applied discard mortalities of 33% and 20% to the commercial and recreational fisheries, respectively, in their assessments of red snapper populations in the GOM. In addition to overt mortality, delayed mortality (as above) due to barotrauma (Rummer and Bennett 2005) and subsequent predation of struggling discards by dolphins, barracudas, jacks, sharks, and even brown pelicans (personal observations) may also significantly increase mortality among red snapper regulatory discards in both fisheries.

The primary objectives of this research were to describe the distributions of ages and sizes and to estimate release mortality and age structure of regulatory discards in the red snapper commercial fishery in the northern GOM. Our specific goals were to: 1) Randomly sample the commercial harvest of red snapper from the northern GOM during 2001–2004 seasons and use counts of otolith annuli to estimate the ages of same; 2) determine the distributions of ages and lengths within these catches; 3) compare age and length distributions to previous studies in 1995–1997 and 1997–2000; and 4) investigate both the catch-and-release mortality and the age composition of red snapper regulatory discards.

Methods

Sample collections spanned four years and six red snapper commercial fishing seasons commencing in October 2001 and ending in May 2004. All of our sampling efforts focused on the red snapper commercial landings in Cameron, Louisiana, where a substantial portion of the Louisiana red snapper fishing fleet is based. Also, several boats from Panama City, Florida, routinely offload their catches in Cameron. The majority of these snapper were caught within NOAA Fisheries statistical grids 16, 17, and 18 which, not coincidentally, historically have been the leading contributors to the total commercial harvest of red snapper in the GOM (Schirripa and Legault 1999). Our sample population was drawn from those catches that were available on the sampling days; catch location and depth of capture were unavailable in all cases. Randomization of specimens sampled was attempted by simply selecting the next available individual from a moving conveyor belt. Fork length (FL) in mm, eviscerated weight (EW) in kg, and sex (when it could be determined with certainty) were recorded and sagittal otoliths were removed from all specimens. Red snapper otolith sections were prepared and ages were estimated as described in Cowan et al. (1995) and Wilson and Nieland (2001). Total length was estimated from FL with the equation TL = 1.073 (FL) + 3.56 (Wilson and Nieland 2001). Data from similar red snapper sampling efforts undertaken in 1995–1997 (Wilson and Nieland 1998; sampled in Leeville, LA) and 1997–2000 (Wilson and Nieland 2000; sampled in Cameron, LA) are used for comparative purposes.

The mortality of red snapper regulatory discards was assessed aboard appropriately permitted commercial fishing vessels operating out of Port Fourchon, LA. We (one person per vessel per trip) assessed the condition of undersize red snapper returned to the water during the course of normal fishing operations; no discards were assessed either during severely inclement weather or at night. We used four condition states taken from Patterson et al. (2001c):

1) Fish oriented toward the bottom and swam down vigorously,
2) Fish appeared disoriented upon entering the water, but soon oriented toward the bottom and swam down slowly,
3) Fish appeared very disoriented upon entering the water and remained at the surface, and
4) Fish was either unresponsive or dead upon entering the water.

At each fishing opportunity, water depth and fishing depth were recorded. The relationship between capture depth and percent of likely mortalities (discards in categories 3 and 4) was examined with linear regression of data from all fishing opportunities in which five or more discards were assessed. Further, each trip a maximum of 25 obvious red snapper mortalities from among the potential dis-

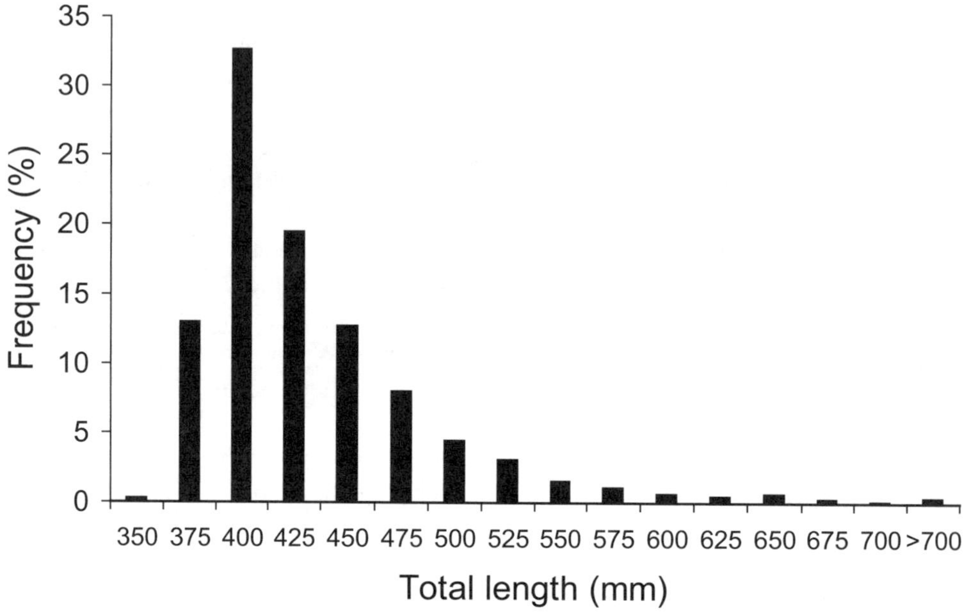

Figure 1. Total length frequency histogram for red snapper *Lutjanus campechanus* from the commercial harvest of the northern Gulf of Mexico, 2001–2004. Sample size = 2,897.

cards were sampled by our personnel for age and length analysis as outlined above.

Results

Red snapper (N = 2,900) were sampled from the commercial harvest of the species during the project period: 593 in 2001, 734 in 2002, 887 in 2003, and 686 in 2004; 42% were sampled from October to December and 58% were sampled from February to May. Among all specimens 884 were males, 981 were females, and 1,035 were of unresolved gender (the fish are landed in eviscerated condition). Total lengths, EW, and ages are available for 2,897 and 2,619 and 2,867 specimens, respectively.

Red snapper in our sample population from the commercial harvest ranged from 278 to 940 mm TL; however, due to the 15 in TL minimum size applied to the commercial red snapper fishery, only four specimens under 370 mm TL were sampled. The distribution of TL binned in 25 mm increments is distinctly unimodal with that mode seen at 400 mm (Figure 1). Fully 98% of all specimens were under 600 mm TL. Figure 2 shows the distributions of TL for red snapper similarly sampled from the commercial harvest during 1995–1997, 1997–2000, and 2001–2004. Length distributions during the 1995–1997 and 1997–2000 sampling efforts were quite similar and showed only minor variations; however, the 2001–2004 sample population was considerably enriched at TL < 450 mm and substantially depauperate at TL ≥ 600 mm. Modal lengths were in the vicinity of 400–425 mm and the preponderances of individuals were less than 600 mm TL in all three cases.

During the 2001–2004 sampling effort, red snapper of ages 1–14 years were encountered; however, the vast majority (97.7%) of these were ages 2–6 years, the modal age was 3 years, and only six specimens were age 10 years or older (Figure 3). The sampling efforts of 1995–1997 and of 1997–2000, during which the maximum ages observed were 48

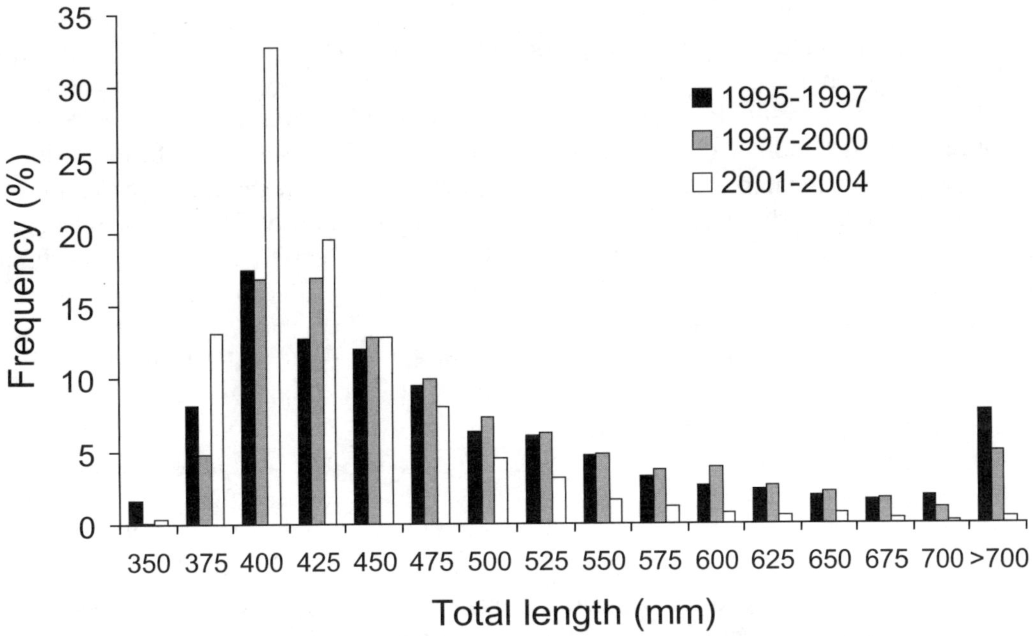

Figure 2. Total length frequency histogram for red snapper *Lutjanus campechanus* from the commercial harvest of the northern Gulf of Mexico, 1995–1997, 1997–2000, and 2001–2004. Samples sizes = 2,091, 2,947, and 2,897, respectively.

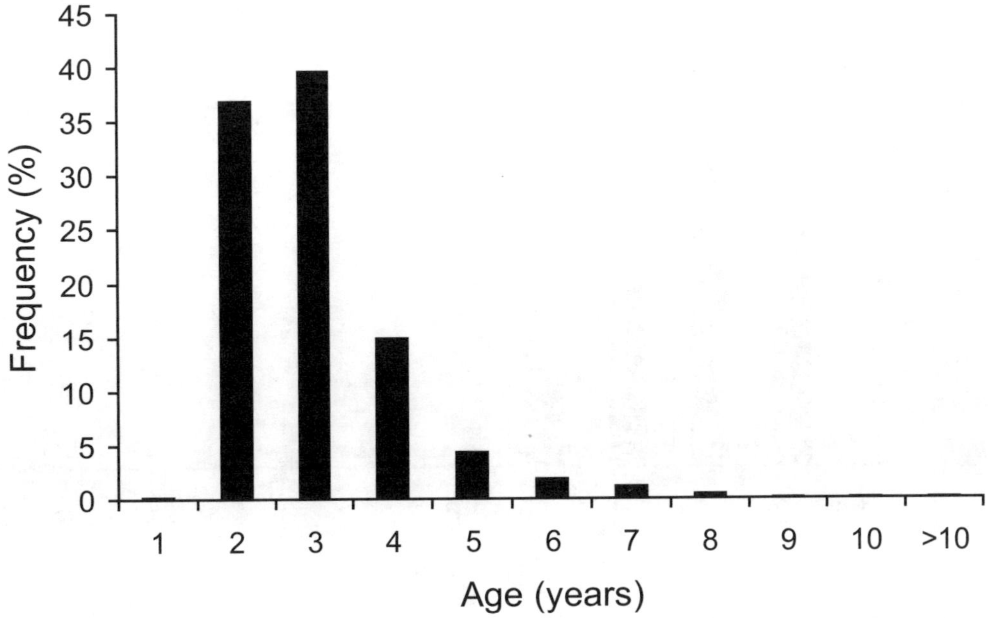

Figure 3. Age frequency histogram for red snapper *Lutjanus campechanus* from the commerical harvest of the northern Gulf of Mexico, 2001–2004. Sample size = 2,867.

years and 39 years, respectively, also showed that the bulk of the commercial catch during these years was individuals of ages 2–6 years (Figure 4). The modal age of 3 years has remained unchanged over the last decade, but, fluctuation notwithstanding, the numbers of age 2 red snapper in the harvest now is nearly equal to that of age 3 individuals. There have been concomitant decreases in the proportions of individuals at all ages older than 3 years. Specimens over age 10 years, a group that in our experience has never been abundantly represented in the harvest, reached their lowest numbers during the most recent sampling effort (Figure 4).

Mortality among red snapper regulatory discards in this study was both very high and related to depth of capture. A total of 4,839 discards from 273 fishing opportunities during 16 fishing trips was observed and characterized for release condition; all were caught with multi-hook (#10 circle hook) gear deployed from "bandit" reels at depths of 9–85 m (mean = 46 m). Among the discards 778 (16%) swam down vigorously, 714 (15%) swam down slowly or erratically, 1,765 (36%) were alive but could not swim down, and 1,582 (33%) were dead. There was a significant positive relationship between capture depth and percent mortality (proportion of discards in categories 3 and 4) (Figure 5) for all fishing opportunities where five or more discards were assessed:

$$\% \text{ mortality} = 0.70 \times \text{depth (m)} + 35.88 \; (F = 35.65, P < 0.0001, r^2 = 0.16).$$

Among 399 potential red snapper regulatory discards retained for age and size analyses, TL ranged from 248 to 380 mm (10–15 in) with a mean and modal TL of 335 mm (13 in); 86% of all specimens were ≥ 300 mm (12 in). Ages among these ranged from 1 year to 4 years with 2 year old individuals (86%) clearly dominating the sample population.

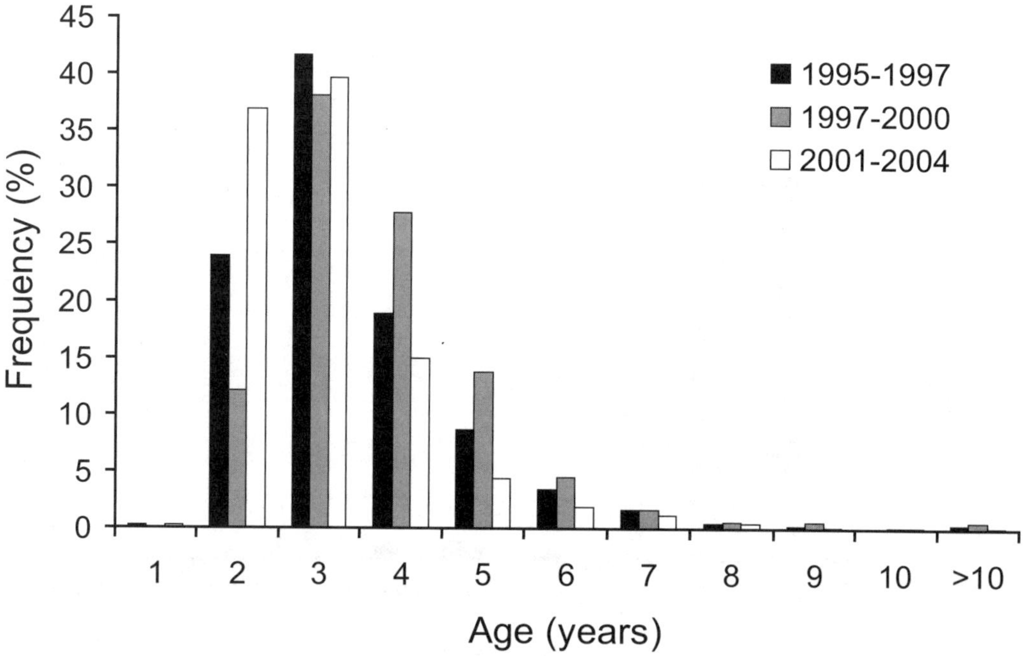

Figure 4. Age frequency histogram for red snapper *Lutjanus campechanus* from the commerical harvest of the northern Gulf of Mexico, 1995–1997, 1997–2000, and 2001–2004. Sample sizes = 2,083, 2,908, and 2,867, respectively.

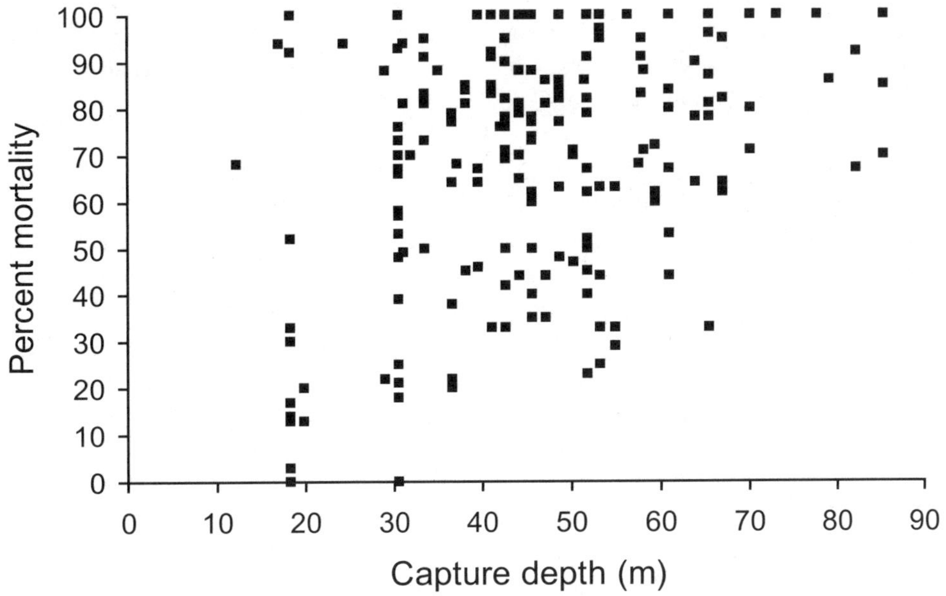

Figure 5. Percent mortality (categories 3 and 4) at capture depth for red snapper *Lutjanus campechanus* regulatory discards from the commercial fishery in the northern Gulf of Mexico, 2001–2003. Only fishing opportunities in which five or more discards were assessed are plotted.

Discussion

Heavy red snapper fishing mortality, which begins as bycatch mortality in shrimp trawls (Schirripa and Legault 1997,1999; SEDAR 2005), continues as discard mortality at sub-legal lengths when juveniles first recruit to the offshore fishing grounds, and persists as harvest mortality among the youngest legal year- and size-classes. Red snapper commercial fishermen are currently allowed a Gulf-wide quota of 2,114 metric tons (mt) (4.65 million lb), two-thirds of which is allocated to a winter season beginning in February and one-third to an autumn season which has begun in either September or October. Additionally, both the winter and autumn seasons are currently open to red snapper harvest only for the first 10 d of each successive month until the quota is achieved. The trip limit of 0.91 mt (2000 lb) for holders of federally issued reef fish permits and the modest number of available fishing days has resulted in a derby fishery that necessitates maximum catch in a minimum of time.

To compete in this derby fishery, many red snapper fishermen have concentrated their efforts at the numerous offshore oil and gas platforms of the northern GOM, particularly those closest to port. These easily located structures can hold large numbers of red snapper (Continental Shelf Associates 1982; Putt 1982; Stanley and Wilson 1990; Stanley 1994). The usual routine (called rig hopping) involves test fishing at successive platforms until a large and readily caught population of red snapper is found. Under the appropriate conditions and with some good fortune, a single trip may last less than one day. Thus, red snapper are harvested as close to port as is possible (usually in relatively shallow waters) and as soon as they achieve legal size as fast-growing 2 year olds, as 3 year olds, or shortly thereafter; they disappear from the fishery, due either to mortality or to emigration to alternative, more remote habitats that are largely unfished, within a few years. The heavy harvest sustained by the younger age-classes of red snapper in the northern GOM appears to have produced populations showing symptoms (decreasing size at age (Nieland et al. 2007, this volume), decreasing size at maturity (Woods 2003)) of overfishing and concomitant juvenescence.

Efforts to estimate discard mortality in the red snapper recreational fishery have produced numbers comparable to those reported above for the commercial fishery. Two studies simulating techniques used in the red snapper recreational fishery have estimated mortality of regulatory discards to range between 1% and 44% and increasing with depth of capture (Gitschlag and Renaud 1994; Render and Wilson 1994). Patterson (2001b) calculated a red snapper discard mortality of 13% from analyses of release condition and recapture rates for red snapper caught with recreational gear in relatively shallow waters off Alabama. Among red snapper less than 18 in (450 mm) TL released from headboats in Texas waters, 15.2% floated off and 1.4% were discarded dead (Dorf 2003).

Given the gamut of life-threatening circumstances that a red snapper regulatory discard must face, it may not be unreasonable to expect a near 100% mortality of discards in the commercial fishery. Based on qualitative characteristics of the discard release conditions we observed, fully 69% of specimens returned to the water were either near death (category 3) as evidenced by their failure to resubmerge or dead (category 4). Additional mortality due to either piscine or mammalian predators may occur on specimens that are trying, perhaps struggling, to return to depth. Should an individual survive the catch and release experience and should it avoid various predators as it swims down, there is also the possibility of long-term mortality due to internal injuries (Rummer and Bennett 2005). Additional studies are needed to determine the level of, as well as spatial and temporal patterns in, both short-term and long-term mortality of discarded red snapper.

Depth of capture had a significant effect on the proportions of discards in the most severely distressed categories: As capture depth increased, the percentage of all discards recorded in categories 3 and 4 increases (Figure 5); however, neither is the magnitude of the effect very strong (slope = 0.70) nor is the relationship very tight (r^2 = 0.16). Whereas discards in Categories 1 and 2 are more likely to be observed at the shallowest capture depths, high percentages of moribund and dead discards are also seen at depths as shallow as 20 m. According to Boyle's law, as pressure decreases 50%, the volume of gas will increase by a factor of two. Thus a red snapper brought to the surface from 80 m will experience a twofold increase in volume of the air bladder at 40 m, an additional twofold increase at 20 m, another at 10 m, and another when at the surface. This exponential increase in gas volume in the air bladder at decreasing depths suggests that the greatest potential for internal injury/trauma may actually occur as the fish approaches the surface from even the shallowest of depths.

No commercial fisher wants to catch undersized fish; it is a waste of time and energy, it increases overhead, and it is potentially detrimental to the population from which one is harvesting. Indeed, fishers generally use hooks of a size that will exclude smaller fish from becoming hooked. However, even with precautions in place, undersized fishes are going to be caught. The larger diameter, manually, electrically or hydraulically powered reels (often called "bandit" reels) used almost exclusively in the commercial fishery assuredly bring hooked fish to the surface faster than could normal recreational gears. Thus the prospect of injury to hooked fish due to hydrostatically-induced barotrauma (eyes bulging, intestine protruding from anus, air bladder distended and stomach protruding from mouth, etc.) is enhanced.

Discard mortality observed in this study (69%) is considerably higher than the current estimate (33%) used historically in red snapper stock assessment models. These findings certainly warrant investigating the sensitivity of these models to high estimates of discard mortality. Even if this increased level of mortality has little effect on the calculations of allowable catch and quotas, it may still warrant reconsideration of the utility of the minimum length regulation on the commercial harvest of red snapper. However, both the intense fishing mortality and the heavy discard mortality of young red snapper will continue to negatively impact populations of the species in the northern GOM.

Postscript: The most recent red snapper stock assessment (SEDAR 2005), which was prepared concurrently with this manuscript, recognizes the high release mortality experienced by regulatory discards in both the commercial and recreational fisheries. As a result, regulatory discard mortalities in the commercial fishery ranging from 71% to 82% were applied to analyses of the red snapper population in the GOM. Based on the SEDAR findings, the GMFMC recently has decreased the total allowable catch (TAC) for the combined fisheries in the GOM from 4,145 mt (9.12 million lb) to 2,954 mt (6.5 million lb) with the prospect of additional decreases in TAC in the near future. The minimum size for harvest of red snapper in the commercial fishery was also decreased from 15 in (381 mm) to 13 in (330 mm). The commercial fishery for red snapper is now also managed under an Individual Fishing Quota System (IFQ) that should operate to either diminish or completely end the derby fishing seen in previous years.

Acknowledgments

The authors would like to thank the captains and crews of the commercial fishing vessels for allowing us to observe aboard their vessels, USA Fish and Trosclair Canning Company for giving us access to their commercial red snapper catch, and Bret Blackmon for assistance in otolith processing. Funding for this research was provided by the U.S. Department of Commerce Marine Fisheries Initiative (MARFIN) Program (grant numbers NA57FF0287, NA77FF0544, and NA17FF2007).

References

Baker, M. S., Jr., and C. A. Wilson. 2001. Use of bomb radiocarbon to validate otolith section ages of red snapper *Lutjanus campechanus* from the northern Gulf of Mexico. Limnology and Oceanography 46:1819–1824.

Continental Shelf Associates. 1982. Study of the effect of oil and gas activities on reef fish populations in Gulf of Mexico OCS area. OCS Report MMS 82–10. U.S. Department of Interior, Minerals Management Service, Gulf of Mexico Outer Continental Shelf Region, New Orleans, Louisiana.

Cowan, J. H., Jr., R. L. Shipp, H. K. Bailey, IV, and D. W. Haywick. 1995. Procedure for rapid processing of large otoliths. Transactions of the American Fisheries Society 124:280–282.

Dorf, B. A. 2003. Red snapper discards in Texas waters—a fishery dependent onboard study of recreational headboat discards and landings. Pages 155–166 *in* D. R. Stanley and A. Scarborough-Bull, editors. Fisheries, reefs, and offshore development. American Fisheries Society, Symposium 36, Bethesda, Maryland.

Gitschlag, G. R., and M. L. Renaud. 1994. Field experiments on survival rates of caged and released red snapper. North American Journal of Fisheries Management 14:131–136.

Goodyear, C. P. 1995. Red snapper in U.S. waters of the Gulf of Mexico. Contribution MIA-95/96–05. National Marine Fisheres Service, Southeast Fisheries Science Center, Miami Laboratory, Miami, Florida.

Manooch, C. S., III, and J. C. Potts. 1997. Age and growth of red snapper, *Lutjanus campechanus*, Lutjanidae, collected along the southeastern United States from North Carolina through the east coast of Florida. Journal of the Elisha Mitchell Scientific Society 113:111–122.

Nieland, D. L., and C. A. Wilson. 2003. Red snapper recruitment to and disappearance from oil and gas platforms in the northern Gulf of Mexico. Pages 73–81 *in* D. R. Stanley and A. Scarborough-Bull, editors. Fisheries, reefs, and offshore development. American Fisheries Society, Symposium 36, Bethesda, Maryland.

Nieland, D. L., C. A. Wilson, III, and A. J. Fischer. 2007. Declining size at age among red snapper in the northern Gulf of Mexico off Louisiana, USA: Recovery or collapse? Pages 329–335 *in* W. F. Patterson, III, J. H. Cowan, Jr., G. R. Fitzhugh, and D. L. Nieland, editors. Red Snapper Ecology and Fisheries in the U.S. Gulf of Mexico. American Fisheries Society, Symposium 60, Bethesda, Maryland.

Patterson, W. F., III. 1999. Aspects of the population ecology of red snapper *Lutjanus campechanus* in an artificial reef area off Alabama. Doctoral dissertation. University of South Alabama, Mobile, Alabama

Patterson, W. F., III, J. H. Cowan, Jr., C. A. Wilson, and R. L. Shipp. 2001a. Age and growth of red snapper, *Lutjanus campechanus*, from an artificial reef area off Alabama in the northern Gulf of Mexico. Fishery Bulletin 99:617–627.

Patterson, W. F., III, G. W. Ingram, Jr., R. L. Shipp, and J. H. Cowan, Jr. 2001b. Indirect estimation of red snapper (*Lutjanus campechanus*) and gray triggerfish (*Balistes capriscus*) release mortality. Proceedings of the Gulf and Caribbean Fisheries Institute 53:526–536.

Patterson, W. F., III, J. C. Watterson, R. L. Shipp, and J. H. Cowan, Jr. 2001c. Movement of tagged red snapper in the northern Gulf of Mexico. Transactions of the American Fisheries Society 130:533–545.

Putt, R. E., Jr. 1982. A quantitative study of fish populations associated with a platform within Buccaneer oil field, northwestern Gulf of Mexico. Master's thesis. Texas A&M University, College Station, Texas.

Render, J. H., and C. A. Wilson. 1994. Hook-and-line mortality of caught and released red snapper around oil and gas platform structural habitat. Bulletin of Marine Science 55:1106–1111.

Rummer, J. L., and W. A. Bennett. 2005. Physiological effects of swim bladder overexpansion and catastrophic decompression on red snapper. Transactions of the American Fisheries Society 134:1457–1470.

Schirripa, M. J., and C. M. Legault. 1997. Status of the red snapper in the U. S. waters of the Gulf of Mexico: updated through 1996. Contribution MIA-97/98–05, National Marine Fisheries Service, Southeast Fisheries Science Center, Miami Laboratory, Miami, Florida.

Schirripa, M. J., and C. M. Legault. 1999. Status of the red snapper in the U. S. waters of the Gulf of Mexico: updated through 1998. Contribution MIA99/00–75, National Marine Fisheries Service, Southeast Fisheries Science Center, Miami Laboratory, Miami, Florida.

SEDAR (Southeast Data, Assessment, and Review). 2005. Stock assessment report of SEDAR 7. Gulf of Mexico red snapper. Available: www.gulfcouncil.org (October 2007).

Stanley, D. R. 1994. Seasonal and spatial abundance and size distribution of fishes associated with a petroleum platform in the northern Gulf of Mexico. Doctoral dissertation. Louisiana State University, Baton Rouge, Louisiana.

Stanley, D. R., and C. A. Wilson. 1990. A fishery dependent based study of fish species composition and associated catch rates around petroleum platforms off Louisiana. Fishery Bulletin 88:719–730.

Szedlmayer, S. T., and R. L. Shipp. 1994. Movement and growth of red snapper, *Lutjanus campechanus,* from an artificial reef area in the northeastern Gulf of Mexico. Bulletin of Marine Science 55(2–3):887–896.

Wilson, C. A., and D. L. Nieland. 1998. Age and size distribution of commercially harvested red snapper *Lutjanus campechanus* in the northern Gulf of Mexico. Final report to U. S. Department of Commerce, National Marine Fisheries Service, Marine Fisheries Initiative, Cooperative Agreement NA57FF0287.

Wilson, C. A., and D. L. Nieland. 2000. Age and size distribution of commercially harvested red snapper *Lutjanus campechanus* in the northern Gulf of Mexico. Final report to U. S. Department of Commerce, National Marine Fisheries Service, Marine Fisheries Initiative, Cooperative Agreement NA77FF0544.

Wilson, C. A., and D. L. Nieland. 2001. Age and growth of red snapper *Lutjanus campechanus* from the northern Gulf of Mexico off Louisiana. Fishery Bulletin 99:653–664.

Wilson, C. A., J. H. Render, and D. L. Neiland (*sic*). 1994. Life history gaps in red snapper, *Lutjanus campechanus*, swordfish, *Xiphias gladius*, and red drum, *Sciaenops ocellatus*, in the northern Gulf of Mexico; age distribution, growth, and some reproductive biology. Final report to U.S. Department of Commerce, National Marine Fisheries Service, Marine Fisheries Initiative, Cooperative Agreement NA17FF0383–02.

Woods, M. K. 2003. Demographic differences in reproductive biology of female red snapper (*Lutjanus campechanus*) in the northern Gulf of Mexico. Master's thesis. University of South Alabama, Mobile, Alabama.

American Fisheries Society Symposium 60:311–328, 2007

Temporal Age Progressions and Relative Year-Class Strength of Gulf of Mexico Red Snapper

Robert J. Allman[1] and Gary R. Fitzhugh
National Marine Fisheries Service
Southeast Fisheries Science Center
3500 Delwood Beach Road, Panama City, Florida 32408 USA

Abstract.—Red snapper *Lutjanus campechanus* sagittal otoliths were sampled from U.S. Gulf of Mexico commercial vertical hook and line, longline and recreational landings over a twelve year period (1991–2002). Our objectives were to examine the empirical age structure of red snapper through space and time, to gauge the relative year-class strength over time, and to compare the impact of strong year-classes upon annual age structure by fishing sector. The recreational fishery selected the youngest fish with a mode at 3 years and a mean age of 3.2 years. The commercial vertical hook and line fishery selected for slightly older fish with a mode of 3 years and a mean age of 4.1 years. The commercial longline fishery selected the oldest individuals with fish first fully recruited to the fishery by age 5 and, a mean age of 7.8 years. Only the commercial longline fishery age distributions were significantly different between the eastern and western Gulf of Mexico. Based on age progressions, strong 1989 and 1995 year-classes were dominant in the landings of the recreational and commercial vertical hook and line fisheries and the 1995 year-class was dominant in the commercial longline landings. A relative year-class index further highlighted these results, and we noted a significant correlation in year-class strength between recreational and commercial vertical hook and line sectors. The year-class index for combined sectors was also significantly correlated between eastern and western Gulf of Mexico with 1989 and 1995 year-classes similarly dominating both regions. An empirical age progression year-class index could be valuable in correlation with early life abundance indices of red snapper and serve to provide inference about the relative error of recruitment data.

Introduction

The economic and management importance of Gulf of Mexico red snapper *Lutjanus campechanus* has driven many life history studies including a recent focus on growth, demographics, and age structure (Patterson 1999; Nieland and Wilson 2000; Wilson and Nieland 2001; Allman et al. 2002; Fischer et al. 2004; Mitchell et al. 2004). Much of this work has highlighted spatial differences in growth and demographics in eastern versus western Gulf of Mexico. Examining spatial trends is particularly important for reef fish which can show affinities for patchy habitats and often exhibit high degrees of site fidelity during some phase of their life history. It is

[1] Corresponding author: robert.allman@noaa.gov.

increasingly being realized that matching the relevant spatial scale to population attributes is an important prerequisite for meeting fishery management objectives (Sale 1998; Gust 2004).

While an improved understanding of spatial differences in life history and demographics is much needed, it is no less important to understand the degree to which life history and demographic attributes can change on temporal scales. Differences in recruitment patterns (i.e., year-class strength) can have a dramatic effect on the demographics of reef fish populations (Sissenwine 1984). Fluctuations in stock size have been attributed to changes in the age structure through the recruitment of strong year- classes (Laevastu and Favorite 1988). To date, studies of red snapper growth and age structure have been conducted within one year or using data sets aggregated over a few years. The degree that spatial and temporal differences in age structure may occur has not been fully examined.

Because of the importance of red snapper and controversies over its management, sampling and aging of the catch has increased since the early 1990s and this has allowed us to develop a time series of age structure from the landed catch. Our objectives were to examine the age structure of red snapper through space and time, to compare empirical age structure patterns by fishing sector and gauge the relative year-class strength over a 12 year time series. To do so, we employed the use of a relative year-class index (YCI) to estimate and compare year-class strength for the principal fishing sectors and for the east and west regions of the U.S. Gulf of Mexico.

Methods

Red snapper were randomly sampled by port agents from Texas to the west coast of Florida from February 1991 through December 2002 mainly through the trip interview program (www.sefsc.noaa.gov/tip.jsp). Fish were sampled from recreational and commercial landings. All fish were measured to total length (TL) or were converted to TL from fork length (FL) using the equation: TL (mm) = 1.061x FL (mm) + 2.601, r^2 = 0.99 (Allman et al. 2002). Sagittal otoliths (hereafter referred to as otoliths) were collected with corresponding fishery data. The left otolith was weighed to the nearest 0.0001 g. If the left otolith was not whole, the right otolith was weighed. A paired t-test found no significant difference between right and left otolith weights ($P = 0.20$).

Otolith processing and aging

Otoliths were processed with a high-speed thin sectioning machine utilizing the methods of Cowan et al. (1995). Two transverse cuts were made through the otolith core to a thickness of 0.5 mm. Due to recent advances in otolith preparation techniques, red snapper otoliths which were sectioned and aged during the early 1990s were ground, polished and re-aged. Otolith sections were assigned an age based on the count of annuli (opaque zones observed with reflected light at 40x counted along the dorsal side of the sucal groove in the transverse plane), including any partially completed opaque zones on the otolith margin) and the degree of marginal edge completion. For example, otoliths collected after 1 January were advanced a year in age if their edge type was a nearly complete translucent zone. Typically, marine fishes in the southeastern U.S. complete annulus formation by late spring to early summer (Patterson et al. 2001; Wilson and Nieland 2001; Garcia et al. 2003; Allman et al. 2005a). Therefore an otolith with two completed annuli and a large translucent zone would be classified as age 3 if the fish was caught during spring, in expectation that a third annulus would have soon formed. For otoliths collected after 30 June all fish were assigned an age equal to the annulus count since opaque zone formation is typically complete. By this traditional method, an annual age-cohort is based on a calendar year rather than time since spawning (Jearld 1983; Vanderkooy and Guindon-Tisdel 2003). Red snapper otolith based ages have been validated (Baker and Wilson 2001) and the timing of annulus formation determined (Allman et al. 2005b).

Three otolith readers within our laboratory aged red snapper otoliths. To establish an estimate of reader precision (i.e., repeatability of age estimates), we prepared a reference collection of 300 red snapper otolith sections representing the ranges of age classes, seasons, sexes, years, preparation quality and collection locations (Campana 2001). Average percent error (APE; Beamish and Fournier 1981) was used to compare age estimates among readers for the reference collection. We considered an APE $\leq$ 5% acceptable for moderately long-lived species with relatively difficult to read otoliths (Morison et al. 1998; Campana 2001). The two most common sources of reader variation in red snapper otoliths were found to be interpretation of the first annulus and edge type. Criteria for identification and interpretation of both first annulus and edge type in red snapper were established in a previous study (Allman et al. 2005b). We examined the linear relationship between otolith weight and estimated age to identify possible systematic bias in age determinations (Morison et al. 1998). We added or subtracted one year from the estimated ages. If estimated ages were biased in either direction, the x-intercept would be closer to the origin by adding or subtracting one year. We restricted the regression to ages 1 through 10 where an overall, significant linear relationship between otolith weight and age was apparent ($F_{1,3128} = 6254$; $P < 0.001$).

Data analysis

To examine potential regional differences within fishing sector, red snapper which were sampled randomly from Florida, Alabama and Mississippi were classified as eastern Gulf of Mexico (hereafter eastern gulf) and those collected from Louisiana and Texas as western gulf. Age distributions were compared regionally with the Kolmogorov-Smirnov two-sample test (KS). A time series of age data was used to calculate a relative year-class index. Only the most common age classes that were well represented in the catches (2–6 years) with at least three consecutive years of data were used to construct the index. Year-class strength was estimated stepwise beginning with the calculation of the percentage age distributions in the annual age samples. Thereafter the mean percentage age distribution for the whole period was established. In the next step the different year-classes in different years were expressed as percentages of this distribution (Böhling et al. 1991). The index assumes total mortality does not affect any one year-class differently from another. Year-class indices were compared by fishing sector and region with Pearson correlation (Minitab, Inc. 1997).

Results

A total of 29,312 red snapper otoliths were randomly selected and aged for the period 1991 to 2002. Collections were fairly evenly divided between the commercial sector (54%) and the recreational sector (46%). Likewise, collections were almost evenly distributed between the eastern gulf (51%) and western gulf (49%). Due to increased funding for sampling of red snapper, the majority of otoliths were collected after 1997 (Table 1). Ages were successfully assigned to 96.5% (28,302) of otoliths read.

The red snapper reference collection APE for the three otolith readers was below the 5% benchmark (APE = 4.8%), so we assumed reader age interpretations were consistent. In addition, aging results from a red snapper otolith reference collection exchange indicated good agreement between

Table 1. Numbers of U.S. Gulf of Mexico red snapper *Lutjanus campechanus* aged by fishing sector within region. CHL = commercial vertical hook and line, CLL = commercial longline and REC = recreational.

	East			West		
Year	CHL	CLL	REC	CHL	CLL	REC
1991	210	13	273	25		631
1992	141	15	474	214		514
1993	169	31	725	355	31	1,180
1994	199	9	754	505		429
1995	157	21	380	48		10
1996	10	6	221			
1997	33	16	156			
1998	239	27	1,670	1,099	358	932
1999	770	111	1,692	2,061	76	457
2000	1,041	135	663	1,171	316	260
2001	1,185	92	638	1,133	191	74
2002	1,217	186	1,239	1,841	360	125

Gulf of Mexico laboratories (Allman et al. 2005b). The regression of otolith weight on age was estimated separately with age, age + 1 and age – 1 to test whether the first increment was correctly assigned. There was no indication that otoliths were consistently biased. The intercept for the original ages was closer to the origin (0.006) than the age + 1 data (–0.146) or the age – 1 data (0.158) (Figure 1). In addition, the 95% confidence intervals for the original age data included zero (–0.012 to 0.024), while the age + 1 and age – 1 data did not (–0.168 to –0.124 and 0.143 to 0.173, respectively), however otolith weight and age were not strongly related ($r^2 = 0.67$).

Red snapper ranged in age from 1 to 57 years. A comparison of age distributions indicated differences by fishing sector, region and sampling year. The commercial longline fishery selected the oldest individuals with fish fully recruited to the fishery by age 5, mean age was 7.8 years, and 22% of individuals were greater than or equal to 10 years (Figure 2A). The commercial vertical hook and line fishery selected for younger fish with a mode of 3 years, mean age was 4.1 years, and 1% of fish greater than or equal to 10 years (Figure 2B). The recreational fishery selected the youngest fish with fish fully recruited to the fishery at 3 years, with 90% of individuals 2, 3 and 4 (mean age = 3.2 years) and only 0.3% of fish were greater than or equal to 10 years (Figure 2C). Only the commercial long-line fishery age distributions were significantly different between eastern and western gulf (KS = 0.33, $P < 0.001$) (Figure 3A). Recruitment to the commercial long-line fishery was by age 4 in the east and age 5 in the west with a mean age

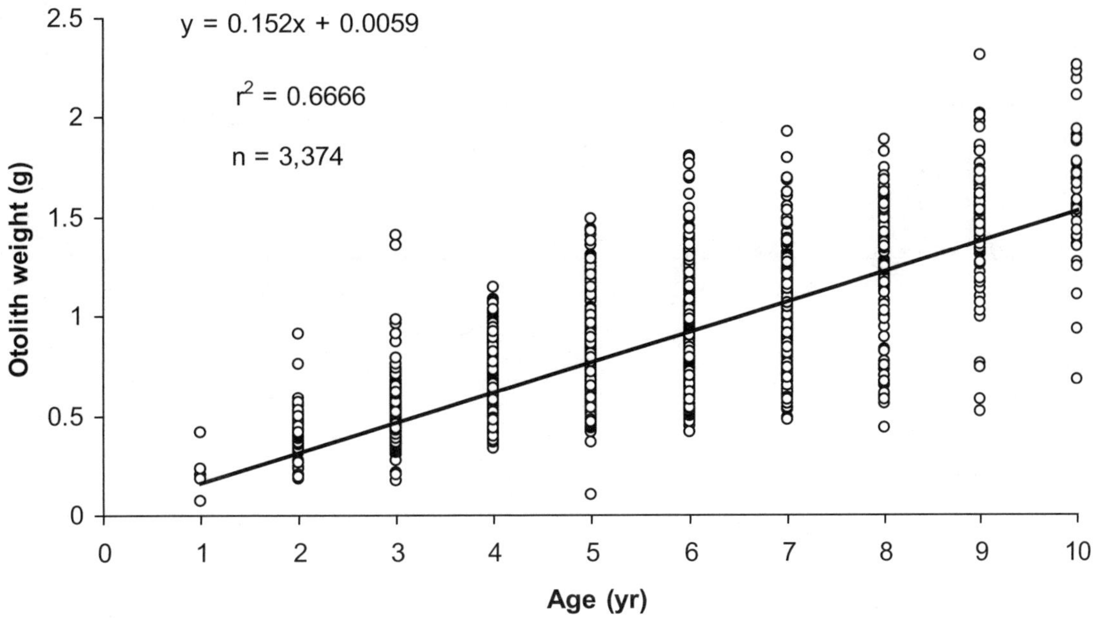

Figure 1. Red snapper *Lutjanus campechanus* otolith weight by estimated age with the fitted regression equation.

of 7 years for the east and 8.2 years for the west. The age distribution from the commercial vertical hook and line fishery was similar between east and west with recruitment at age 3 for both regions and mean ages of 4.1 and 4.2 years, respectively (Figure 3B). Similarly, the recreational fishery suggested little difference between east and west with recruitment at age 3 and mean ages of 3.1 and 3.4 years, respectively (Figure 3C).

Age frequency distribution by sampling year revealed potential changes in the age at recruitment through time. The annual recruitment pattern of red snapper from the recreational fishery indicated recruitment alternated between age 2 or 3 prior to 1998 but only age 3 was noted during and after 1998 (Figure 4). Recruitment to the commercial vertical hook and line fishery alternated between age 3 or 4 all years (Figure 5). Recruitment to the commercial long-line fishery for the few years with large sample size was from age 4 to 6 (Figure 6).

There was evidence for strong 1989 and 1995 year-classes in the age structure of the recreational and commercial vertical hook and line fisheries and for 1995 in the commercial longline fishery. Generally these strong year-classes could be followed for 2–3 consecutive years (Figures 4–6). Strong year-classes followed a progression through the different fishing sectors, first appearing in the recreational fishery, then a year later appearing in the commercial vertical hook and line and then the commercial longline fishery the following year. For example, evidence for the dominance of the 1995 year-class first emerged with a large proportion of 2 year olds (>70%) in the recreational fishery in 1997, a year later this year-class was dominant in the commercial vertical hook and line fishery as 3 year olds, and in 1999 in the commercial longline fishery as 4 year olds (Figure 7 A–C). The 1989 year-class first was dominant in the recreational fishery as two year olds in 1991 and then in the commercial vertical hook and line fishery as three year olds in 1992 (Figure 7 D&E). We

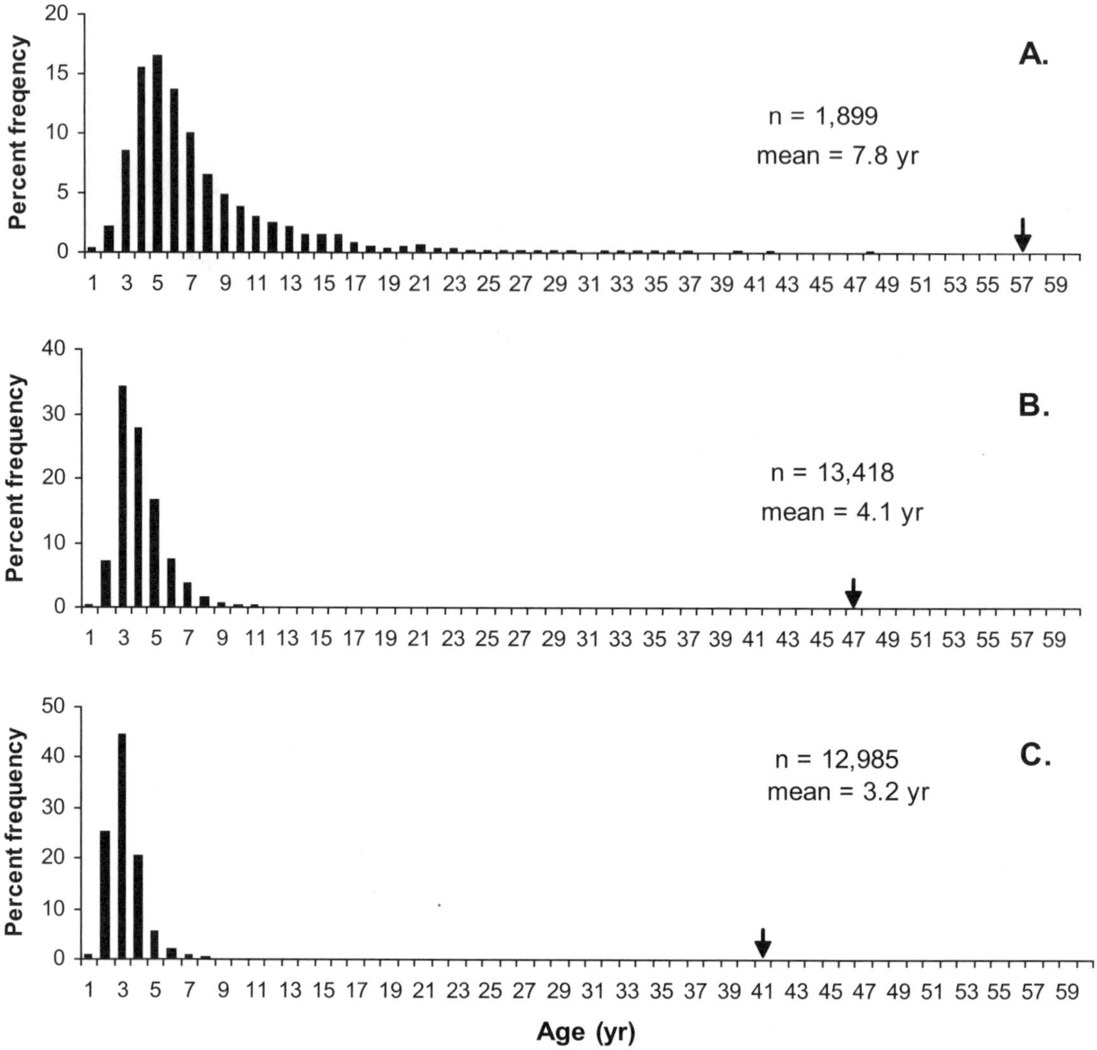

Figure 2. Age frequency distributions for red snapper *Lutjanus campechanus* (1991–2002): (A) commercial longline, (B) commercial vertical hook and line and (C) recreational. Arrows indicate maximum age.

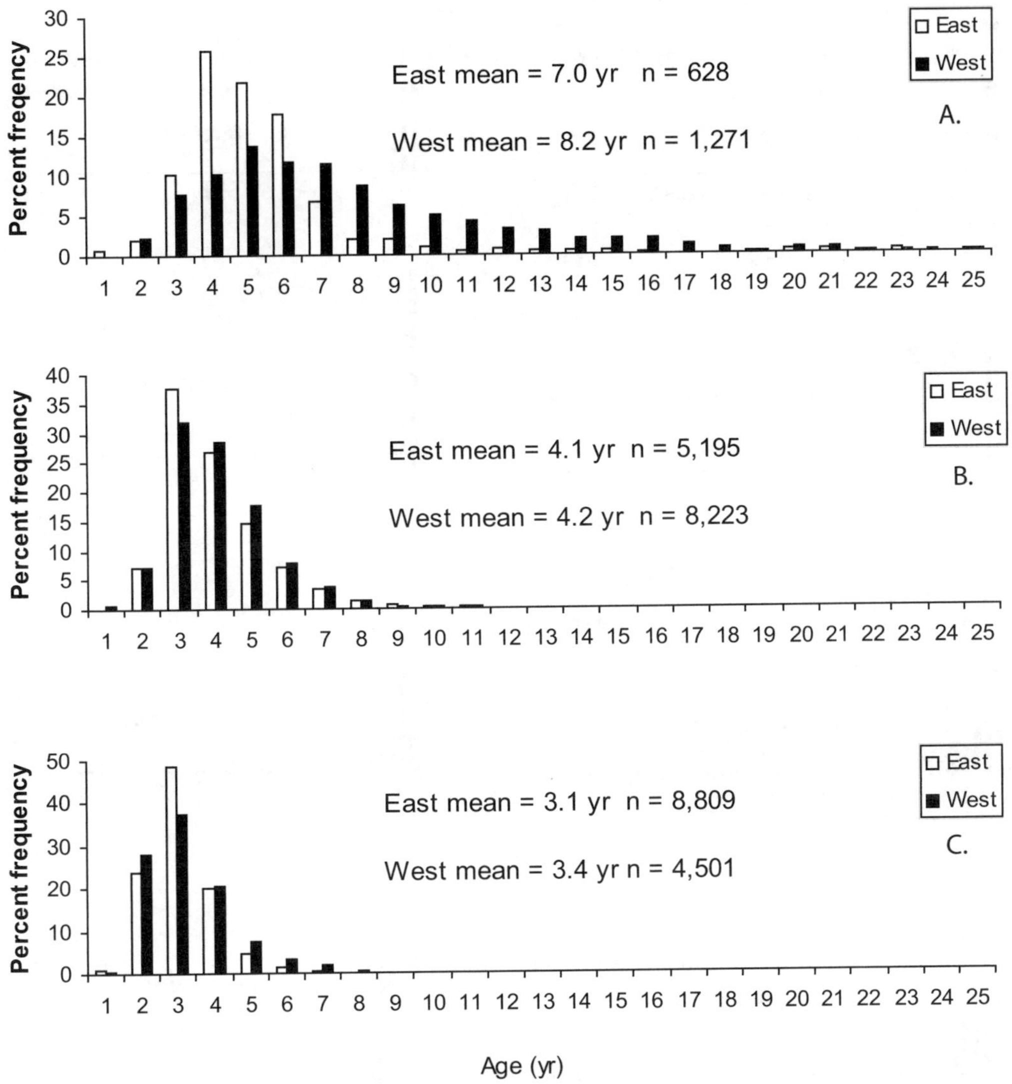

Figure 3. Age frequency of red snapper *Lutjanus campechanus* by sampling region (1991–2002): (A) commercial longline, (B) commercial vertical hook and line, and (C) recreational.

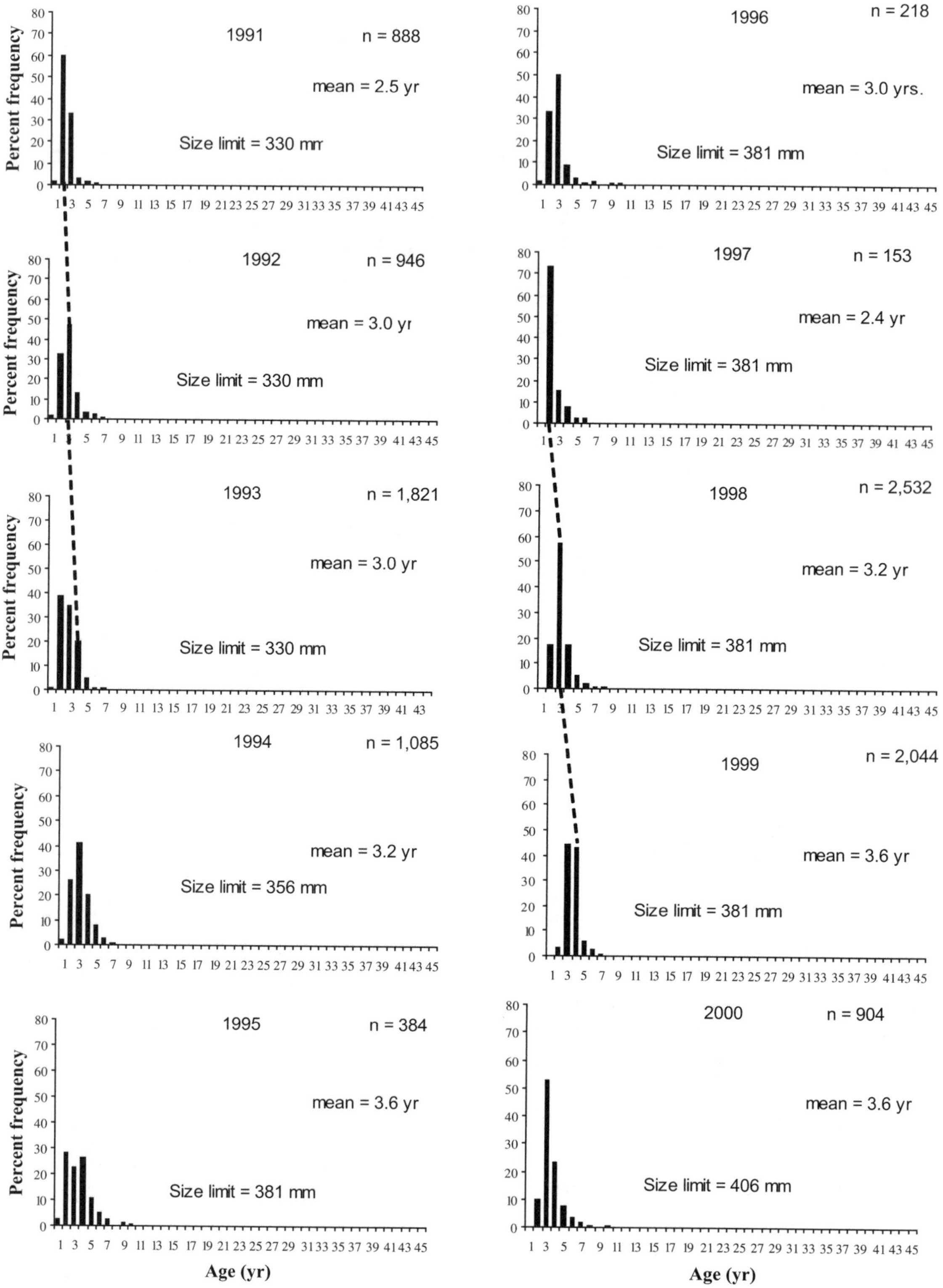

Figure 4. Age distribution of the recreational fishery by year. Dashed lines indicate year-class progression.

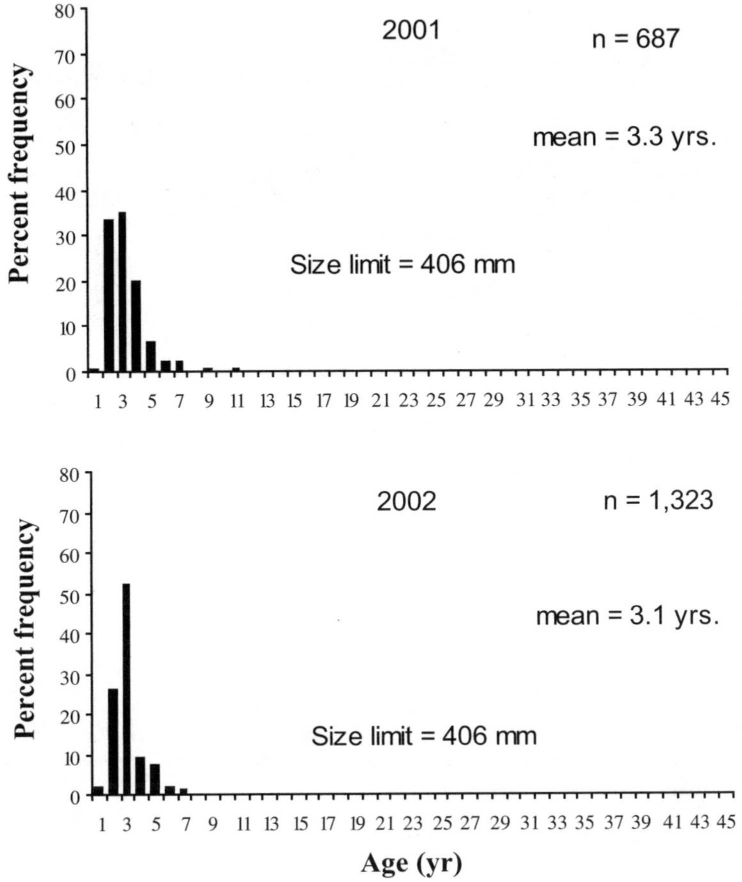

Figure 4. (Continued)

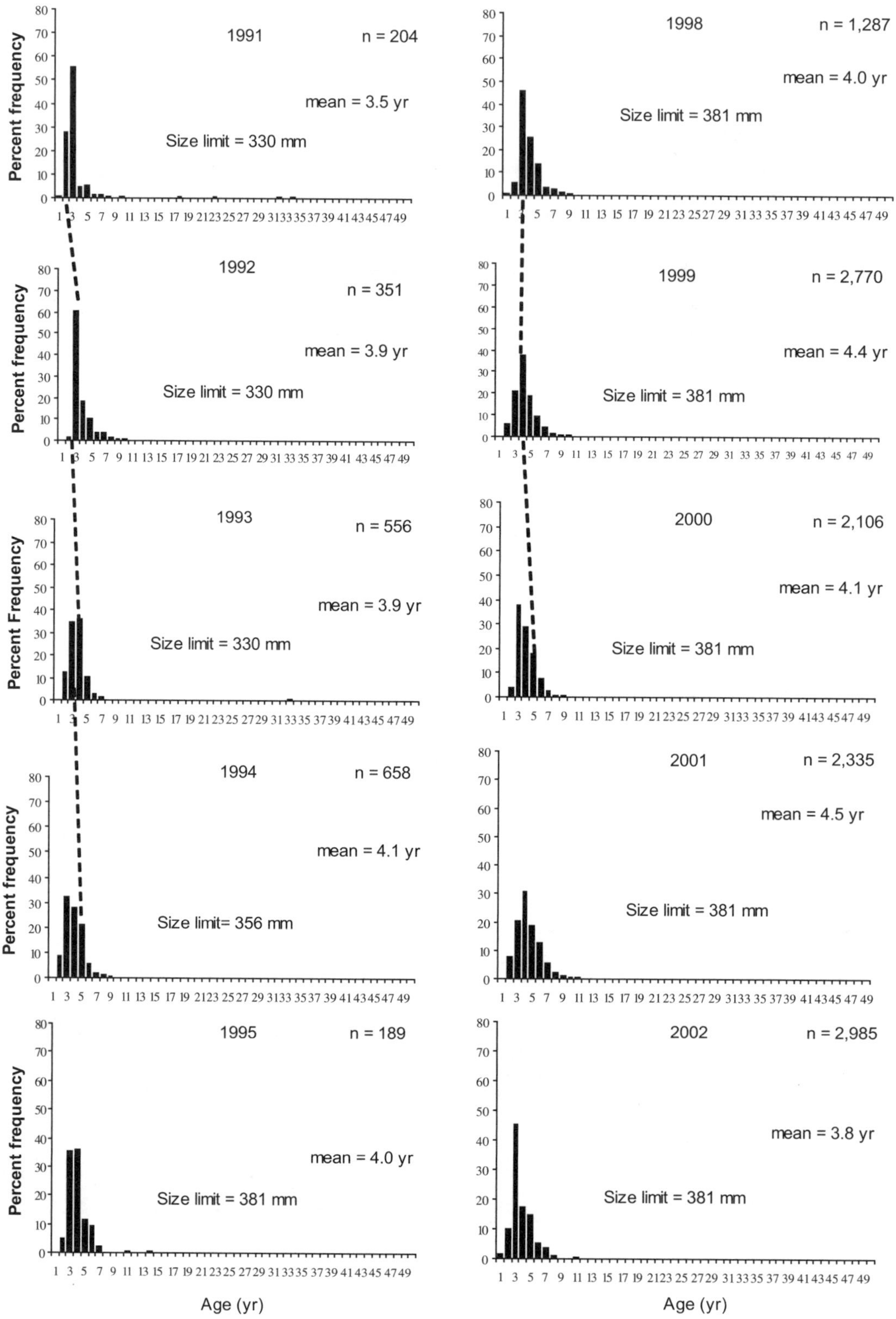

Figure 5. Age distribution of the commercial vertical hook and line fishery by year. Dashed lines indicate year-class progression.

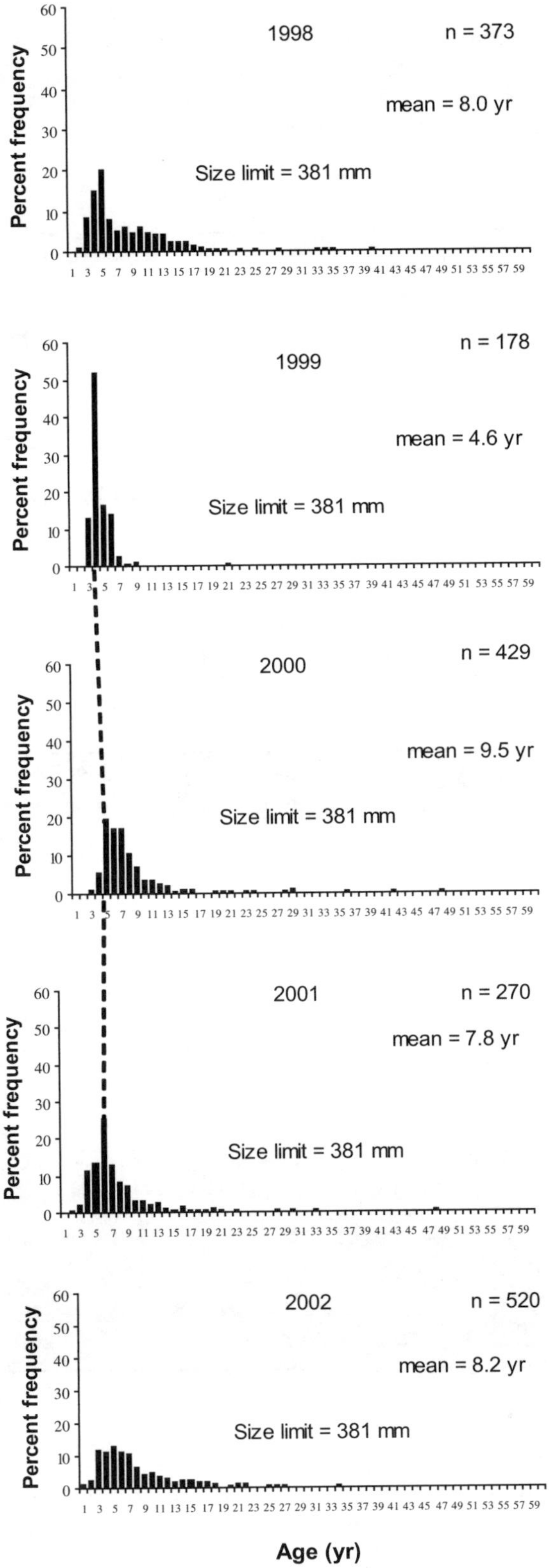

Figure 6. Age distribution for the commercial longline fishery by year. Dashed lines indicate year-class progression.

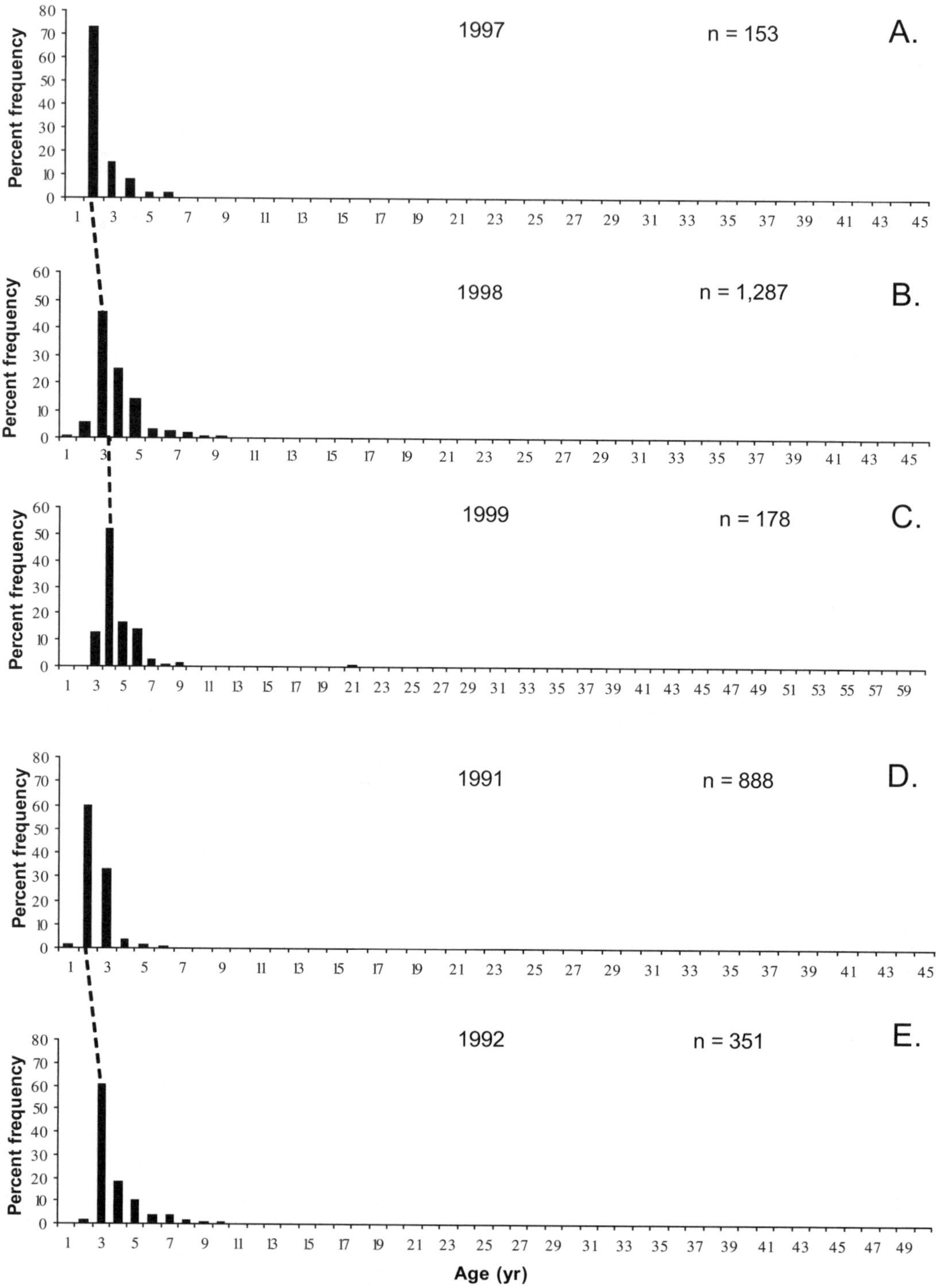

Figure 7. Progression of 1995 year-class (dashed line) through recreational (A), commercial vertical hook and line (B) and commercial longline sectors (C) and 1989 year-class through the recreational (D) and commercial vertical hook and line sectors (E).

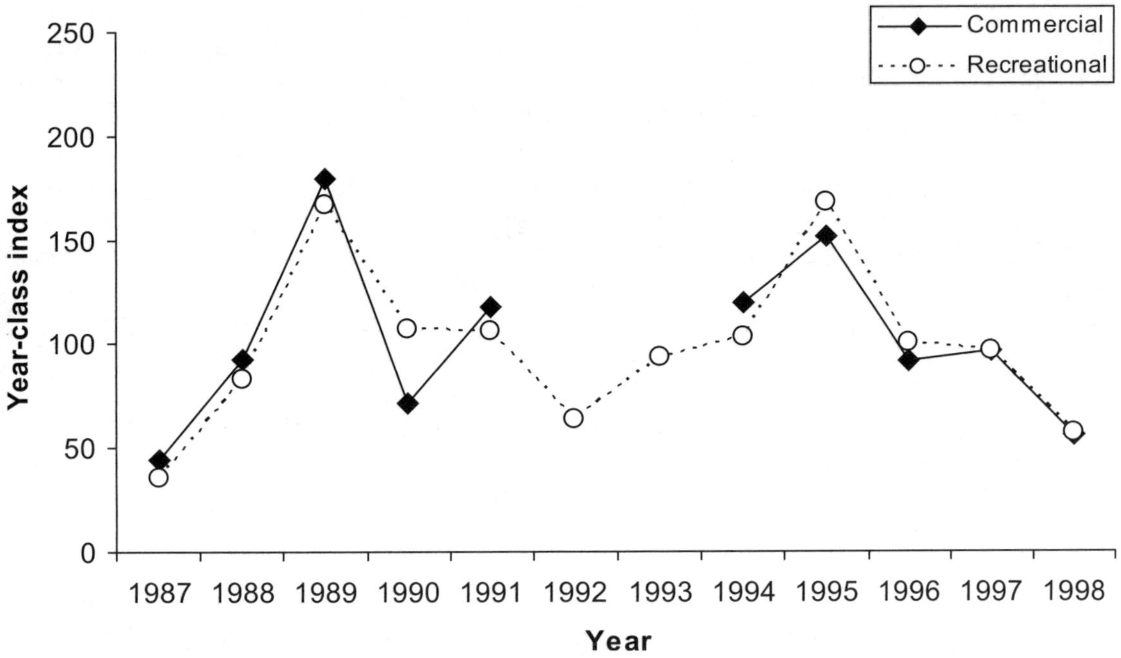

Figure 8. Red snapper *Lutjanus campechanus* year-class index calculated from Gulf of Mexico vertical hook and line sectors (eastern and western gulf combined).

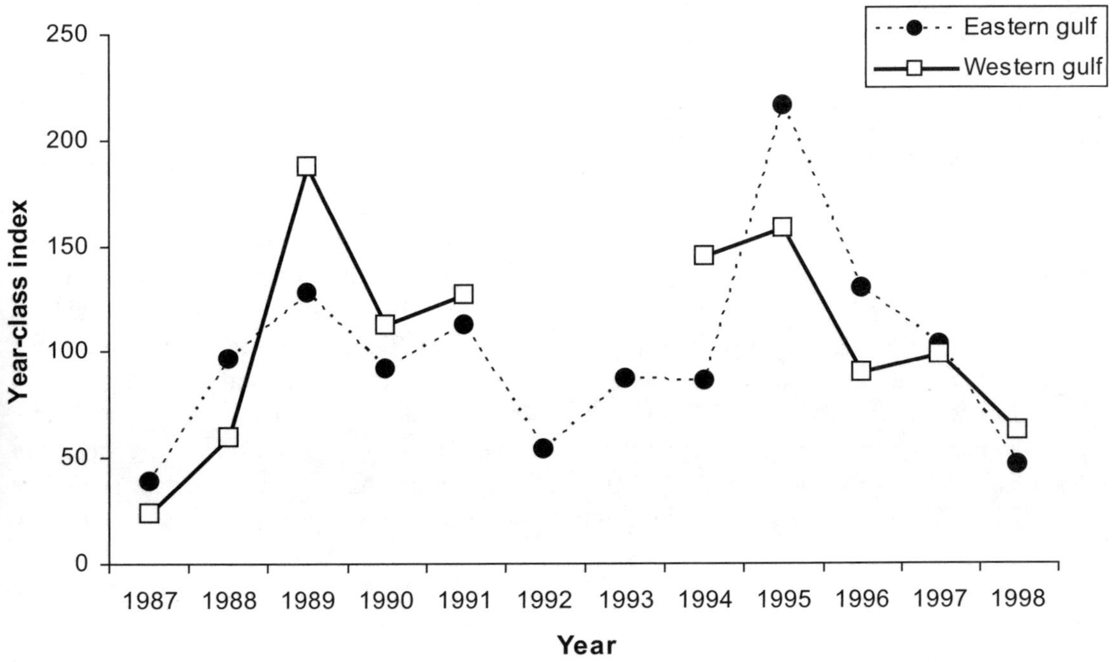

Figure 9. Red snapper *Lutjanus campechanus* year-class index calculated for the eastern and western Gulf of Mexico (commercial and recreational vertical hook and line sectors combined).

did not have commercial longline samples in large enough numbers prior to 1998 to resolve the 1989 year-class.

A relative year-class index (YCI) more clearly revealed the two dominant year-classes, with peaks in index values for the recreational and commercial vertical hook and line fisheries during 1989 and 1995 (Figure 8). Additionally, there was a significant correlation in year-class strength between these two fishing sectors (Pearson correlation = 0.924, $P < 0.001$). The YCI for combined sectors was also significantly correlated between regions (Pearson correlation = 0.674, $P = 0.032$) with 1989 and 1995 year-classes similarly dominating in both regions (Figure 9).

Discussion

An annual aging program based on port sampling of red snapper otoliths enabled us to track multiple year age progressions among the principal gulf red snapper fishing sectors. It was apparent that consistent patterns in year-class strength were detectable over a period of several years. During 1991 to 2002, examination of annual age structure revealed that 1989 and 1995 year-classes were evident as relatively strong year-classes for recreational and commercial vertical hook and line as was 1995 for the longline sector. We noted differences in the age progression patterns among the three fishing sectors which suggest some differences in age selectivity. In the recreational and commercial vertical hook and line fisheries, 1989 and 1995 year-class dominance was evident in the progression of age 2 to age 4 red snapper in 1991 to 1993 and 1997 to 1999. In the longline fishery, the dominance of the 1995 year-class was observed in the progression of age 4 to age 6 in 1999–2001.

This shift in age tracking of younger fish in the vertical hook and line fisheries versus older fish in the long-line sector has been noted before and may reflect differences in gear, locations, depths fished, fish behavior or habitat selection (Allman et al. 2002; Mitchell et al. 2004). Our observations reveal increased complexity in making gear selectivity comparisons because age structure is not static from year to year and dominant ages caught by different gears can overlap. For example, fish as old as age 4 dominated in the vertical hook and line fisheries (1993, 1999, and 2001) while fish as young as age 4 dominated in the longline sector (1999 only; a shorter record to compare). But the overall pattern appeared to be an age progression over time with a year-class moving through a "gauntlet" of the respective fisheries. These empirical observations support the modeled assessment results showing vulnerability peaking at younger ages in the vertical hook and line sector compared to vulnerability of older fish taken in the longline sector (Porch 2007, this volume).

Size limits increased over the period of study for recreational and commercial vertical hook and line sectors (Hood and Steele 2004). We thus expected a shift to an older age structure over time, but changes were not readily apparent. The recreational size limit increased from 330 mm TL (13 in) in 1991 to 356 mm TL (14 in) in 1994, 381 mm TL (15 in) in 1995, and 406 mm TL (16 in) in 2000. Commercial vertical hook and line size limits increased similarly, except that the commercial size limit remained at 381 mm TL (15 in) after 1995. During the period of record for longline samples (1998–2002), the size limit also remained at 381 mm TL (15 in). These size limit changes basically split the middle of the size range for age 2 red snapper (compare raw size-at-age data in Nieland and Wilson 2000; Wilson and Nieland 2001; Allman et al. 2002). For example, age 2 dominated the recreational age structure in 1991 when the size limit was 330 mm TL, dominated again when the size limit increased to 381 mm TL (1997), and even exceeded 30% of the age structure when the size limit increased to 406 mm TL (2001). So the effect of size limit changes to age structure in the recreational sector seemed minimal. However, 2 year olds were not as frequent in the commercial vertical hook and line fishery and were virtually absent from the longline sector. They became even less frequent over the period of our age record, and after the size limit increased to 356 mm TL and then 381 mm TL, 2 year olds were never more than about 10% of the commercial vertical hook and line landings. This influence of size limits upon the age structure in the commercial fishery, as opposed to the recreational sector, suggests greater selection for larger and faster growing 2 year olds from a species exhibiting broad variation in size-at-age. The contrast between recreational and commercial vertical hook and line age structure also indicated differences in selectivity, albeit, more subtle than differences exhibited by the longline sector.

Our observations about age differences among gears are consistent with other studies showing that red snapper undergo a general ontogenetic shift in habitat and depth. Nieland and Wilson (2000) reported that age 1 and younger fish dominated benthic habitats sampled by survey trawls, whereas age 2 and age 3 fish dominated the samples obtained from explosive removal of an oil platform. Nieland and Wilson (2000) further indicated that commercial vertical hook and line collections of red snapper in Louisiana waters revealed few age 2 fish, and in a follow-up paper, they speculated that fish older than about age 6 become rare in the vertical hook and line catches due to mortality and emigration away from the oil platforms (Wilson and Nieland 2001). Mitchell et al. (2004) also suggested older red snapper may be less reef obligate, based on research longline catches.

The progression of strong year-classes in aging data has been used to corroborate the method of aging (Morison et al. 1998; Campana 2001). Our finding of age progressions independently observed among fishery sectors, as well as good precision among readers, provides strong support that our method of aging is consistent for the age range commonly observed (ages 2–6). To address concerns about bias in age assignment possibly due to first annulus interpretation (Allman et al. 2005b), we examined the otolith weight versus age relationship. By adding or subtracting one year to our original ages and re-estimating regression parameters, we mimicked the effect of being consistently biased in our assignment of the first annulus. As the regression of otolith weight with our original ages was closest to a zero intercept and was the only regression to have zero occur within the 95% confidence interval, the analysis provides evidence that our method of aging was not biased.

Our conclusions concerning age frequencies and year-class dominance in the catch are also based upon the assumptions that otolith samples collected by the various fishery-dependent programs are of sufficient sample size and represent simple random samples of the catch. Regarding sample sizes, Thompson (1987) showed that about 510 fish ages is a conservative and sufficient sample size such that all age-class proportions (viewed as multinomial proportions) should be within 5% of the statistical population proportions with 95% confidence. Otolith sample sizes for the principal stratification levels in our study, for example, fishing sector within a given year, sometimes fell below 510 samples. Regarding simple random sampling of age structures, it is known that the approach in theory may be the simplest (Quinn and Deriso 1999) but the practice can be difficult to impossible to achieve (Pope 1988; Aanes and Pennington 2003), thus even a quasi-random approach can be useful (Pope 1988). Therefore, the assumptions of sufficient sample size and simple random sampling were likely not always met in our estimates of age frequencies. However, consistent age progression observed in different fishery sectors suggests that the year-class signal determined from empirical age estimates may be robust to degrees of nonrandom sampling and low sample size. This does not eliminate concern over other possible nonrandom sampling effects, and thus every effort should be made to develop strategies for random sampling and to establish adequate stratification levels.

We generated an index of year-class strength by tracking progressions of the most common ages; thus a relative year-class index (YCI) based upon empirical age frequencies (Böhling et al. 1991). The 1989 and 1995 year-classes stood out over the 12 year record examined. Empirical tracking of dominant age classes in some fish populations has enabled inferences about recruitment processes in earlier life history stages (Böhling et al. 1991, McFarlane and Beamish 1992; Doherty and Fowler 1994; Russ et al. 1996). Our temporal comparison of year-class progressions and development of the YCI for separate gulf regions contributes to the understanding of spatial differences in stock dynamics and recruitment. Red snapper in the gulf traditionally have been managed as a single unit stock but development of a two stock, two region model was undertaken recently (Southeast data, assessment, and review 2005). This new management approach, viewing gulf red snapper as either one or two stocks, underscores how our understanding of stock structure and spatial dynamics is evolving. There were several lines of investigation that led to a two stock management approach. A recent study found variation in red snapper growth rates from east to west (Fischer et al. 2004). We detected differences in the age structure for the commercial longline fishery with older fish occurring more frequently in the western gulf, as did a research longline survey (Mitchell et al. 2004). However, our observed age progressions and the YCI indicated that year-class patterns were consistent between the eastern and western Gulf of Mexico. While we cannot discount that there may be differences in the stock structure between the eastern and western gulf, our results support a hypothesis that recruitment processes influencing year-class strength operate at large spatial scales.

Inference of recruitment trends from age structure has been common in fisheries but the approach requires simplifying assumptions. Hjort and Lea (1914) were the first to infer recruitment pulses from year-class trends. Subsequent development of indices have been based on samples measured in one year (common in freshwater systems assuming equal vulnerabilities across ages and years, e.g., Maceina 1997; Isermann et al. 2002; Cowx and Frear 2004) and from age progressions over time (assuming total mortalities experienced among year-classes are similar; e.g., Böhling et al. 1991, McFarlane and Beamish 1992; King et al. 2000; McGlennon et al. 2000). Age structured assessment models in theory can separate recruitment effects from cumulative mortality, vulnerability by age and yearly exploitation; but in practice, clear distinctions are difficult to make (Walters and Martell 2004). An age-progression approach may have utility for inference contrasts with other approaches if it is a reasonable assumption that exploitation rates have been relatively constant over the period of record (McFarlane and Beamish 1992). For example, comparison of multiple year-class indices, determined from different ontogenetic stages has commonly been conducted to determine the period when year-class strength is established (Helle et al. 2000). But, these correlations often reveal more about the relative error inherent in the various indices (Mukhina et al. 2003). Our empirical YCI is integrated over five years (the common ages), rather than measured in one year, and has traits of consistency and minimal age assignment bias in its estimation. We pose that our YCI could be valuable as a measurement standard for other indices of red snapper year-class strength and could specifically serve as the oldest stage year-class estimate. Currently three other red snapper indices are being developed, including a larval index based upon a plankton survey and age 0 and age 1 indices based upon length categories of red snapper captured in a trawl survey. Soon all red snapper indices will have a common record exceeding 15 years, and it is our hope that a correlation analysis can be conducted. We would expect closest correlation of the empirical YCI with the age 1 index, followed by the age-0 index, and then by the larval index. Deviations from this pattern, observed from a matrix of the index correlations, could then be useful to gauge the relative error inherent in any particular index.

Acknowledgments

We thank the many port agents who sampled red snapper through the years, especially Debbie Fable and June Weeks for their large contributions to this effort. We also want to acknowledge Nancy Evou, Chris Gardner, K. J. Starzinger, Bob Farsky and Laura Goetz for all their work preparing red snapper otoliths. Bill Fable helped with otolith reading. Linda Lombardi-Carlson helped organize and proof the database. Pete Sheridan provided helpful suggestions.

References

Aanes, S., and M. Pennington. 2003. On estimating the age composition of the commercial catch of Northeast Arctic cod from a sample of clusters. ICES Journal of Marine Science 60:297–303.

Allman, R. J., L. R. Barbieri, and C. T. Bartels. 2005a. Regional and fishery-specific patterns of age and growth of yellowtail snapper, *Ocyurus chrysurus*. Gulf of Mexico Science 2:211–223.

Allman, R. J., G. R. Fitzhugh, K. J, Starzinger and R. A. Farsky. 2005b. Precision of age estimation in red snapper (*Lutjanus campechanus*). Fisheries Research 73:123–133.

Allman, R. J., L. A. Lombardi-Carlson, G. R. Fitzhugh, and W. A. Fable. 2002. Age structure of red snapper (*Lutjanus campechanus*) in the Gulf of Mexico by fishing mode and region. Gulf and Caribbean Fisheries Institute 53:482–495.

Baker, M. S., Jr., and C.A. Wilson. 2001. Use of bomb radiocarbon to validate otolith section ages of red snapper, *Lutjanus campechanus*, from the northern Gulf of Mexico. Limnology and Oceanography 46:1819–1824.

Beamish, R. J., and D. A. Fournier. 1981. A method for comparing the precision of a set of age determinations. Canadian Journal of Fisheries and Aquatic Sciences 38:982–983.

Böhling, P., R. Hudd , H. Lehtonen, P. Karås, E. Neuman, and G. Thoresson. 1991. Variations in year-class strength of different perch (*Perca fluviatilis*) populations in the Baltic Sea with special reference to temperature and pollution. Canadian Journal of Fisheries and Aquatic Sciences 48:1181–1187.

Campana, S. E. 2001. Accuracy, precision and quality control in age determination, including a review of the use and abuse of age validation methods. Journal of Fish Biology 59:197–242.

Cowan, J. H., Jr., R. L. Shipp, H. K. Bailey, IV, and D. W. Haywick. 1995. Procedure for rapid processing of large otoliths. Transactions of the American Fisheries Society 124:280–282.

Cowx, I. G., and P. A. Frear. 2004. Assessment of year class strength in freshwater recreational fish populations. Fisheries Management and Ecology 2004:117–123.

Doherty, P. J., and A. J. Fowler. 1994. An empirical test of recruitment limitation in a coral reef fish. Science 263:935–939.

Fischer, A. J., M. S. Baker and C. A. Wilson. 2004. Red snapper (*Lutjanus campechanus*) demographic structure in the northern Gulf of Mexico based on spatial patterns in growth rates and morphometrics. Fishery Bulletin 102:593–603.

Garcia, E. R., J. C. Potts, R. A. Rulifson, and C. S. Manooch III. 2003. Age and growth of yellowtail snapper, *Ocyurus chrysurus*, from the southeastern United States. Bulletin of Marine Science 72:909–921.

Gust, N. 2004. Variation in the population biology of protogynous coral reef fishes over tens of kilometers. Canadian Journal of Fisheries and Aquatic Sciences 61:205–218.

Helle, K., B. Bogstad, C. T. Marshall, K. Michalsen, G. Ottersen, and M. Pennington. 2000. An evaluation of recruitment indices for Arcto-Norwegian cod (*Gadus morhua* L.). Fisheries Research 48:55–67.

Hjort, J., and E. Lea. 1914. The age of herring. Nature (London) 94:255–256.

Hood, P., and P. Steele. 2004. History of red snapper management in federal waters of the U.S. Gulf of Mexico 1984–2004. SEDAR7 Data Workshop Document 40.

Isermann, D. A., W. L. McKibbin, and D. W. Willis. 2002. An analysis of methods for quantifying crappie recruitment variability. North American Journal of Fisheries Management 22:1124–1135.

Jearld., A., Jr. 1983. Age determination. Pages 301–324 *in* L. A. Nielsen and D. L. Johnson, editors. Fisheries Techniques. American Fisheries Society, Bethesda, Maryland.

King, J., G. Mcfarlane, and R. Beamish. 2000. Decadal-scale patterns in the relative year class success of sablefish (*Anoplopoma fimbria*). Fisheries Oceanography 9(1):62–70.

Laevastu, T., and F. Favorite. 1988. Fishing and stock fluctuations. Fishing News Books Ltd., Farnham, Surrey, UK.

Maceina, M. J. 1997. Simple application of using residuals from catch-curve regressions to assess year-class strength in fish. Fisheries Research 32:115–121.

McFarlane, G. A., and R. J. Beamish. 1992. Climatic influence linking copepod production with strong year-classes in sablefish, *Anoplopoma fimbria*. Canadian Journal of Fisheries and Aquatic Sciences 19:743–753.

McGlennon, D., G. K. Jones, J. Baker, W. B. Jackson, and M. A. Kinloch. 2000. Ageing, catch-at-age and relative year-class strength for snapper (*Pagrus auratus*) in northern Spencer Gulf, South Australia. Marine and Freshwater Research 51:669–677.

Minitab, Inc. 1997. Minitab user's guide: Data analysis and quality tools, release 12. Minitab, Inc., State College, Pennsylvania.

Mitchell, K. M., T. Henwood, G. R. Fitzhugh, and R. J. Allman. 2004. Distribution, abundance, and age structure of red snapper (*Lutjanus campechanus*) caught on research long-lines in the U.S. Gulf of Mexico. Gulf of Mexico Science 22:164–172.

Morison, A. K., P. C. Coutin, and S. G. Robertson. 1998. Age determination of back bream, *Acanthopagrus butcheri* (Sparidae), from the Gippsland Lakes of south-eastern Australia indicates slow growth and episodic recruitment. Marine and Freshwater Research 49:491–8.

Mukhina, N. V., C. T. Marshall, and N. A. Yaragina. 2003. Tracking the signal in year-class strength of Northeast Arctic cod through multiple survey estimates of egg, larval and juvenile abundance. Journal of Sea Research 50:57–75.

Nieland, D. L., and C. A. Wilson. 2000. Red snapper in the northern Gulf of Mexico: fishery dependent and fishery independent characterization of age and length. Gulf and Caribbean Fisheries Institute 51:129–139.

Patterson, W. F., III. 1999. Aspects of the population ecology of red snapper *Lutjanus campechanus* in an artificial reef area off Alabama. Doctoral dissertation. University of South Alabama. Mobile, Alabama.

Patterson, W. F., III., J. H. Cowan Jr., C. A. Wilson, and R. L. Shipp. 2001. Age and growth of red snapper, *Lutjanus campechanus*, from an artificial reef area off Alabama in the northern Gulf of Mexico. Fishery Bulletin 99:617–627.

Pope, J. G. 1988. Collecting fisheries assessment data. Pages 63–82 *in* J. A. Gulland, editor. Fish Population Dynamics. Wiley, Chichester, UK.

Porch, C. E. 2007. An assessment of the red snapper fishery in the U.S. Gulf of Mexico using a spatially-explicit age-structured model. Pages 355–384 *in* W. F. Patterson, III, J. H. Cowan, Jr., G. R. Fitzhugh, and D. L. Nieland, editors. Red snapper ecology and fisheries in the U.S. Gulf of Mexico. American Fisheries Society, Symposium 60, Bethesda, Maryland.

Quinn, T. J., and R. B. Deriso. 1999. Quantitative fish dynamics. Oxford University Press, New York.

Russ, G. R., D. C. Lou, and B. P. Ferreira. 1996. Temporal tracking of a strong cohort in the population of a coral reef fish, the coral trout, *Plectropomus leopardus* (Serranidae: Epinephelinae), in the central Great Barrier Reef, Australia. Canadian Journal of Fisheries Aquatic Science 53:2745–2751.

Sale, P. F. 1998. Appropriate spatial scales for studies of reef-fish ecology. Australian Journal of Ecology 23:202–208.

Southeast Data, Assessment, and Review. 2005. Stock assessment report of SEDAR 7—Gulf of Mexico Red Snapper. SEDAR Assessment Report 1. Charlestown, South Carolina.

Sissenwine, M. P. 1984. Why do fish populations vary? Pages 59-94 *in* R. M. May, editor. Exploitation of marine communities: report of the Dahlem Workshop on exploitation of marine communities, Berlin, Germany, April 1–6, 1984. Life Science Research Report No. 32. Springer-Verlag, Berlin.

Thompson, S. K. 1987. Sample size for estimating multinomial proportions. The American Statistician 41:42–46.

Vanderkooy, S. and K. Guindon-Tisdel. 2003. A practical handbook for determining the ages of Gulf of Mexico fishes. Gulf States Marine Fisheries Commission, Publication Number 111.

Walters, C. J., and S. J. D. Martell. 2004. Fisheries Ecology and Management, Princeton University Press, Princeton, New Jersey.

Wilson, C. A., and D. L Nieland. 2001. Age and growth of red snapper, *Lutjanus campechanus*, from the northern Gulf of Mexico off Louisiana. Fishery Bulletin 99:653–664.

American Fisheries Society Symposium 60:329–335, 2007

Declining Size at Age Among Red Snapper in the Northern Gulf of Mexico off Louisiana, USA: Recovery or Collapse?

David L. Nieland[1] and Charles A. Wilson, III
Louisiana Sea Grant College Program
Louisiana State University
Baton Rouge, Louisiana 70803 USA

Andrew J. Fischer
Department of Oceanography and Coastal Sciences
School of the Coast and Environment
Louisiana State University
Baton Rouge, Louisiana 70803 USA

Abstract.—The red snapper, *Lutjanus campechanus*, is currently both overfished and undergoing overfishing in the Gulf of Mexico (GOM) waters adjacent to the southeastern United States. From October 1998 through May 2004, we sampled 6,159 red snapper landed at a commercial dock in Cameron, LA, for morphometric data and otoliths for age estimation. Despite the species' potential lifespan of more than 50 years, the harvest is almost totally dominated by individuals of ages 2–6 years. Over the course of our sampling we have observed striking, statistically significant decreases in mean total length (TL) at age for red snapper of ages 2–6 years. Density dependence theory tells us that, within a population of fishes that is increasing in numbers, a decrease in resources (food, habitat, etc.) per individual might be manifested in a compensatory decrease in growth rate. Thus, the declines in red snapper mean TL at age may be an expression of recovery of an overfished population. Conversely, the heavy commercial and recreational harvest of young red snapper in the northern GOM, many of them at the very minimum TL required for retention, may have resulted in an inadvertent selection for the survival of slow-growing individuals.

Introduction

The red snapper, *Lutjanus campechanus*, is currently both overfished (low biomass, B) and undergoing overfishing (excess fishing mortality, F) in the Gulf of Mexico (GOM) waters adjacent to the southeastern United States (SEDAR7 2005). As a consequence, the recreational and commercial fisheries for red snapper in these waters are among the most rigorously managed fisheries in the GOM. Both fisheries are variously regulated by the enforcement of size limits, trip or creel

[1] Corresponding author:dniela@lsu.edu.

limits, seasonal closures, and quotas with the expectation of achieving a spawning potential ratio (SPR) of 30% and of allowing populations to recover.

Although the red snapper in the GOM has been recognized as a single, panmictic stock for assessment purposes, several studies have shown that local populations may be behaving differentially to the selective regimes, including extent of overfishing, they are experiencing. Indeed, demographic variations in population genetics (Camper et al. 1993; Gold et al. 1997, 2001; Gold and Saillant 2007, this volume; Saillant et al. 2003; Saillant and Gold 2004), growth rates (Fischer et al. 2004), and reproduction (Jackson et al. 2007, this volume) may have resulted, at least to a degree, from excessive reductions in the numbers of individuals in local populations of red snapper.

The efficacy of the management of red snapper might be perceived in changes in density dependent processes such as growth, survival, and reproduction over the span of several years. Such processes are compensatory if they vary in response to change in population density: population growth rate slows at high densities and population growth rate increases at low densities (Rose et al. 2001). Within a population of red snapper that is decreasing in numbers, an increase in resources (food, habitat, etc.) per individual might be manifested in an increase in growth rate. Conversely, intraspecific competition for limited resources within a red snapper population that is increasing in numbers might curtail growth of individuals. Any changes in growth rates would most likely be demonstrated during the period of fastest growth, up to age 10 years in red snapper (Wilson and Nieland 2001).

The commercial harvest of red snapper produces significant dockside revenue (>$11 million in 2005) (National Marine Fisheries Service 2007) and is among the most highly regulated (381 mm total length (TL) minimum size, 909 kg trip limit, 2.11 million kg annual quota) (Gulf of Mexico Fishery Management Council 2006) fisheries in the GOM. From 1998 to 2004 we sampled red snapper from the commercial harvest in the waters off the Louisiana coast. The primary objective of this research was to describe the age and size composition of the red snapper commercial harvest in the northern GOM (Nieland et al. 2007, this volume). However, these 6.5 years of data were subsequently applied to an investigation of possible changes in TL at age that may have resulted from the management strategies applied to the red snapper fisheries.

Methods

Red snapper was sampled opportunistically during commercial harvest seasons from October 1998 to May 2004, a period when between 34% and 64% of the total GOM harvest of red snapper was landed in Louisiana (National Marine Fisheries Service 2007). To assure broad seasonal coverage, we attempted to extend our sampling efforts to include as many monthly openings as was practicable. All of our sampling efforts were focused at a commercial dock in Cameron, Louisiana, where an important share of the total red snapper commercial harvest is landed. Our sample population was drawn from those catches that were available on the sampling days; randomization of specimens was attempted by simply selecting the next available individual from a moving conveyor belt.

Fork length (FL) in mm, eviscerated weight (EW) in kg, and gender (when apparent) were recorded for each specimen. The sagittal otoliths from each specimen were removed and placed in labeled envelopes; all undamaged, intact otoliths were subsequently weighed to the nearer 0.1 mg. Red snapper total length (TL) was estimated from FL with the equation TL = 1.073 (FL) + 3.56 (Wilson and Nieland 2001). As there were few significant changes to the commercial harvest regulations applied to red snapper during our sampling period, we assume that fishing tactics and fisherman behavior were reasonably constant throughout this period.

The left sagitta (in those few instances where the left sagitta was damaged or unavailable, the right sagitta was substituted) of each specimen was sectioned following the protocols described either in Cowan et al. (1995) or in Wilson and Nieland (2001). Opaque annuli were counted, and ages, plus year of birth or cohort, were estimated as described in Wilson and Nieland (2001). Mean TL at age was graphed and compared for specimens of ages 2–6 years by sample year. A sample year began in September and extended through May of the following year; no red snapper were sampled during June, July, and August, the presumed months of their most vigorous growth. This arrangement both ensures that all specimens included together in a sample year have experienced the previous summer's growth maximum and assumes minimal growth outside the summer months.

Mean TL at ages 2–6 years were compared both with a one factor analysis of variance (ANOVA) of log_{10} transformed TL by sample year and with a Tukey's Studentized Range (HSD) Test (SAS Institute, Inc. 2001). An alpha level of 0.05 was applied in all instances.

Results

Our sample population of 6,159 red snapper was drawn from the commercial harvest of the species off Louisiana during the 6.5 year period from October 1998 to May 2004. Among these 2,018 were males, 2,223 were females, and 1,918 were of unresolved gender (the fish are landed in eviscerated condition). Total lengths (N = 6,159) ranged from 278 to 953 mm; however, due to the 15 in (381 mm) TL minimum size applied to the commercial fishery, only 237 specimens less than the regulatory minimum were sampled. The distribution of TL binned in 25 mm increments is distinctly unimodal with that mode seen at 400 mm (Figure 1). Fully 97.5% of all specimens were less than 700 mm TL. Red snapper ages (N = 6,077) ranged from 0 to 36 years, but the preponderance (97.3%) of these were ages 2–6 years, the modal age was 3 years, and only 20 specimens were 10 years or older (Figure 2).

Mean TL at age by sample year for red snapper ages 2–6 years are shown in Figure 3. Mean TL for 2 year old individuals, ranging from 407 to 430 mm, demonstrated little variation during the sampling period. Similarly, age 3 specimens, while perhaps showing slightly greater variation in mean TL, have remained in the 425–475 mm range. The variations and declines in mean TL evidenced in the 4, 5, and 6 year olds are more striking. Mean TL among red snapper at age 4 years has shown a consistent decline from about 525 mm to about 445 mm. Red snapper at age 5 showed mean TL decreasing abruptly from 590 mm in 1999–475 mm in 2002 and subsequently increasing to 495 mm the following two years. The most striking changes in mean TL, from 692 mm in 2000 down to 507 mm in 2002 and back up to 565 mm in 2004, are seen among red snapper of age 6 years. Although the decreasing trends in red snapper mean TL at age are obvious, they are also statistically significant decreases. Both the ANOVA and the Tukey's (Table 1) demonstrated the differences among mean TL at age among sample years.

Discussion

As stated above, we could reasonably expect that compensatory processes espoused in density dependence theory would produce individuals of smaller size-at-age in an expanding population of red snapper. Thus

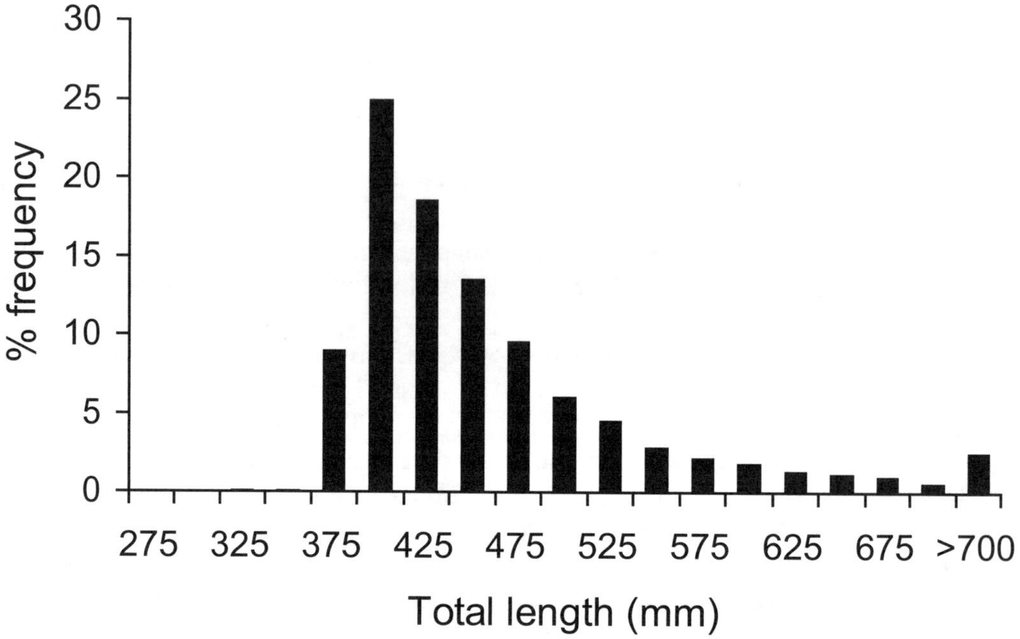

Figure 1. Total length frequency histogram for red snapper *Lutjanus campechanus* sampled from the commercial harvest of the northern Gulf of Mexico off Louisiana, 1998–2004. Total sample size = 6,077 specimens.

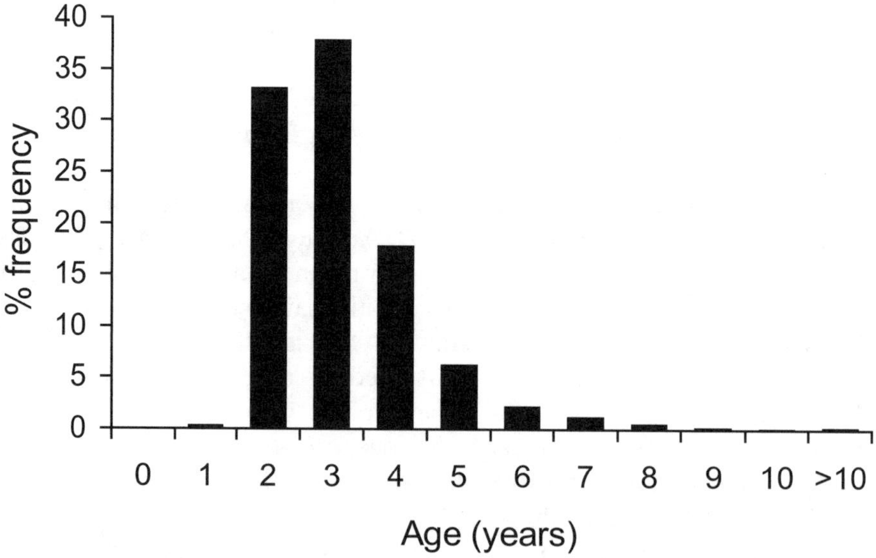

Figure 2. Age frequency histogram for red snapper *Lutjanus campechanus* sampled from the commercial harvest of the northern Gulf of Mexico off Louisiana, 1998–2004. Total sample size = 6,152 specimens.

Table 1. Analyses of variance and Tukey's Studentized Rage (HSD) Tests on red snapper *Lutjanus campechanus* mean total length at age by sample year. Within each age, similar letters indicate no difference in mean total length (α = 0.05).

	ANOVA		Tukey's (HSD) comparisons of mean TL at age by sample year					
Age (years)	*F*	*P*	1999	2000	2001	2002	2003	2004
2	28.83	<0.0001	B	A	C	D	C	D
3	51.5	<0.0001	B	A	B	C	C	D
4	38.23	<0.0001	A	A	B	B	C	C
5	18.63	<0.0001	A	A	B	C	BC	BC
6	16.2	<0.0001	A	A	B	C	C	BC

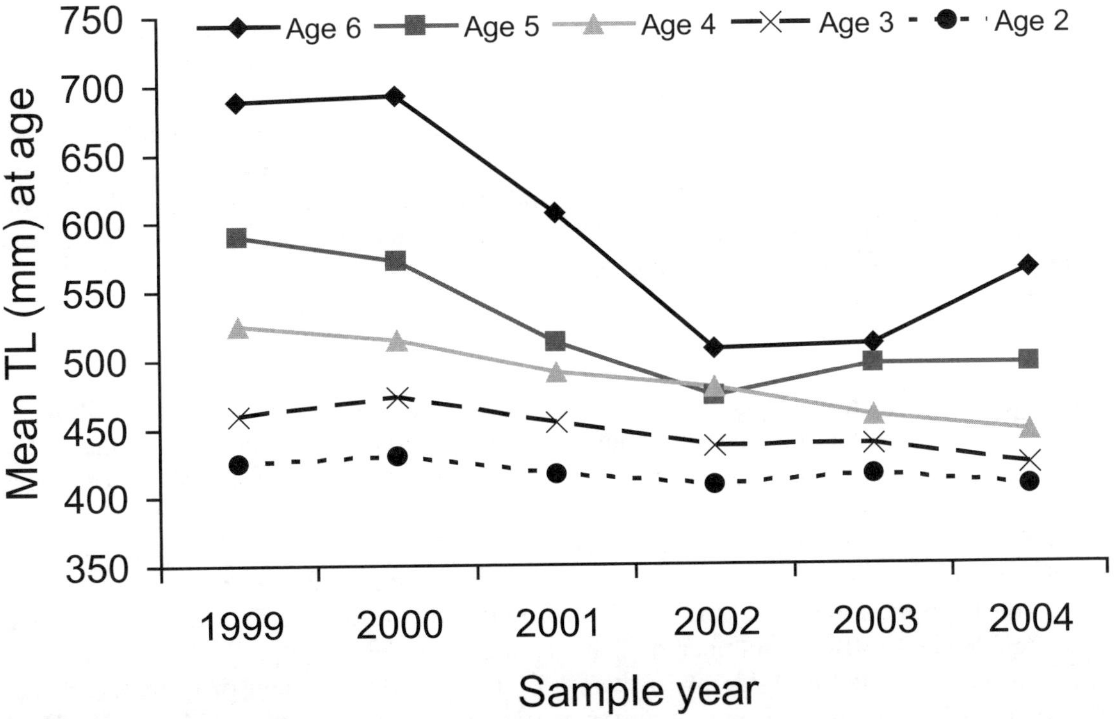

Figure 3. Mean total length at age for red snapper *Lutjanus campechanus* from the commercial harvest of the northern Gulf of Mexico off Louisiana by sample year (1999–2004). Sample year sample size ranges: 165–554 (mean = 338) at age 2 years, 309–445 (mean = 383) at age 3 years, 74–248 (mean = 177) at age 4 years, 35–109 (mean = 61) at age 5 years, and 8–32 (mean = 23) at age 6 years.

the declines in red snapper TL seen in red snapper sampled in Cameron, LA may be an expression of recovery of overfished populations. Conversely, the heavy commercial (Figure 2) and recreational harvest of young red snapper, the vast majority of them at the very minimum TL required for harvest, may have resulted in both a selective removal of individuals predisposed to rapid growth and an inadvertent selection for the survival of slow-growing individuals. Such selective harvest may also result in declines in fecundity, egg volume, larval size at hatch, larval viability, larval growth rates, food consumption rate and conversion efficiency, vertebral number, and willingness to forage (Walsh et al. 2006) and reduce the capacity for population recovery. Severe overexploitation previously has been invoked as a major contributory factor in the decreases in size at age experienced by red porgy *Pagrus pagrus* in both the western north Atlantic Ocean and the eastern Gulf of Mexico (Harris and McGovern 1997; Hood and Johnson 2000; Vaughan and Prager 2002) and by the vermilion snapper *Rhomboplites aurorubens* in the South Atlantic Bight (Zhao et al. 1997).

However, the age and size structure of the commercial catch in Louisiana may not be representative of the red snapper population Gulf-wide. The regulations applied to the fishery and the very nature of the fishery itself may be more important determinants of the composition of the commercial harvest. The 15 in minimum size, the 2,000 lb trip limit, and a restricted number of fishing days have resulted in a "derby" fishery that dictates maximum catch in a minimum of time. Thus, red snapper are harvested as soon as they achieve legal size as fast-growing 2 year olds, as 3 year olds, or shortly thereafter (Figure 1); they disappear from the fishery, due either to mortality or to emigration to alternative habitats, within a few years (Figure 2). The heavy harvest sustained by the younger age-classes of red snapper in the northern GOM appears to have produced populations showing symptoms (decreasing size at age (Fischer et al. 2004; this study), decreasing size at maturity (Woods 2003)) of overfishing and concomitant juvenescence.

Total length (Figure 1) and age (Figure 2) histograms indicate that the commercial harvest of red snapper in the northern GOM is dominated by relatively small (375–625 mm TL) and relatively young (2–6 years) individuals; larger specimens over age 10 years are becoming less frequently observed in the commercial harvest (Nieland et al. 2007). The concentrated commercial and recreational harvest of young red snapper may have brought about the selective elimination of individuals predisposed to fast growth and an unintentional selection for the survival of slow-growing individuals. This and the similarly dramatic decreases in numbers of older, larger individuals (Nieland et al. 2007) may be the portent of detrimental changes in the red snapper populations of the northern GOM.

Acknowledgments

USA Fish, Cameron, LA, and many commercial fishermen graciously allowed us to sample their red snapper; we thank the employees of Trosclair Canning Company, Cameron, LA, for their assistance and patience. The comments of Brigitte Nieland and two anonymous reviewers contributed greatly to the manuscript. Support for our red snapper research program was provided by the U.S. Department of Commerce Marine Fisheries Initiative (MARFIN) Program (grant numbers NA90AA-H-MF762, NA57FF0287, and NA77FF0544) and the Gulf and South Atlantic Fisheries Development Foundation, Inc. (project number 70–05–29840).

References

Camper, J. D., R. C. Barber, L. R. Richardson, and J. R. Gold. 1993. Mitochondrial DNA variation among red snapper (*Lutjanus campechanus*) from the Gulf of Mexico. Molecular Marine Biology and Biotechnology 2:154–161.

Cowan, J. H., Jr., R. L. Shipp, H. K. Bailey, IV, and D. W. Haywick. 1995. Procedure for rapid processing of large otoliths. Transactions of the American Fisheries Society 124:280–282.

Fischer, A. J., M. S. Baker, Jr., and C. A. Wilson. 2004. Red snapper (*Lutjanus campechanus*) demographic structure in the northern Gulf of Mexico based on spatial patterns in growth rates and morphometrics. Fishery Bulletin 102:593–603.

Gold, J. R., E. Pak, and L. R. Richardson. 2001. Microsatellite variation among red snapper (*Lutjanus campechanus*) from the Gulf of Mexico. Marine Biotechnology 3:293–304.

Gold, J. R., and E. Saillant. 2007. Population structure of red snapper from the northern Gulf of Mexico. Pages 201–216 *in* W. F. Patterson, III, J. H. Cowan, Jr., G. R. Fitzhugh, and D. L. Nieland, editors. Red snapper ecology and fisheries in the U.S. Gulf of Mexico. American Fisheries Society Symposium 60, Bethesda, Maryland.

Gold, J. R., F. Sun, and L. R. Richardson. 1997. Population structure of red snapper from the Gulf of Mexico as inferred from analysis if mitochondrial DNA. Transactions of the American Fisheries Society 126:386–396.

Gulf of Mexico Fishery Management Council. 2006. Commercial fishing regulations pamphlet for regulations as of November 2005. Gulf of Mexico Fishery Management Council, Fishing Regulations. Available: www.gulfcouncil.org. (May 2006).

Harris, P. J., and J. C. McGovern. 1997. Changes in the life history of red porgy, *Pagrus pagrus*, from the southeastern United States. Fishery Bulletin 95:732–747.

Hood, P. B., and A. K. Johnson. 2000. Age, growth, mortality, and reproduction of red porgy, *Pagrus pagrus*, from the eastern Gulf of Mexico. Fishery Bulletin 98:723–735.

Jackson, M. W., J. H. Cowan, Jr., and D. L. Nieland. 2007. Demographic differences in northern Gulf of Mexico red snapper reproductive maturation: implications for the unit stock hypothesis. Pages 217–227 *in* W. F. Patterson, III, J. H. Cowan, Jr., G. R. Fitzhugh, and D. L. Nieland, editors. Red snapper ecology and fisheries in the U.S. Gulf of Mexico. American Fisheries Society Symposium 60, Bethesda, Maryland.

National Marine Fisheries Service. 2007. Annual commercial landings statistics survey. National Marine Fisheries Service, Fisheries Statistics Division. Available: www.st.nmfs.gov/st1/commercial/landings/annual_landings.html. (May 2007).

Nieland, D. L., A. J. Fischer, M. S. Baker, Jr., and C. A. Wilson. 2007. Red snapper in the northern Gulf of Mexico: Age and size composition of the commercial harvest and mortality of regulatory discards. Pages 301–310 *in* W. F. Patterson, III, J. H. Cowan, Jr., G. R. Fitzhugh, and D. L. Nieland, editors. Red snapper ecology and fisheries in the U.S. Gulf of Mexico. American Fisheries Society Symposium 60, Bethesda, Maryland.

Rose, K. A., J. H. Cowan, Jr., K. O. Winemiller, R. A. Myers, and R. Hilborn. 2001. Compensatory density dependence in fish populations: importance, controversy, understanding and prognosis. Fish and Fisheries 2:293–327.

Saillant, E., and J. R. Gold. 2004. Population structure and variance effective size of red snapper (*Lutjanus campechanus*) in the northern Gulf of Mexico. Fishery Bulletin 104:136–148.

Saillant, E., T. A. Mousseau, and J. R. Gold. 2003. Genetic variation and relatedness of juvenile red snapper sampled from shrimp trawls in the northern Gulf of Mexico. Transactions of the American Fisheries Society 132:1229–1235.

SEDAR7. 2005. Stock Assessment Report of SEDAR7 - Gulf of Mexico Red Snapper. SEDAR7 Assessment I, Section III. Stock Assessment Workshop Report. Available: www.gulfcouncil.org. (April 2006).

Vaughan, D. S., and M. H. Prager. 2002. Severe decline in abundance of the red porgy (*Pagrus pagrus*) population off the southeastern United States. Fishery Bulletin 100:351–375.

Walsh, M. R., S. B. Munch, S. Chiba, and D. O. Conover. 2006. Maladaptive changes in multiple traits caused by fishing: impediments to population recovery. Ecology Letters 9:142–148.

Wilson, C. A., and D. L. Nieland. 2001. Age and growth of red snapper *Lutjanus campechanus* from the northern Gulf of Mexico off Louisiana. Fishery Bulletin 99:653–664.

Woods, M. K. 2003. Demographic differences in reproductive biology of female red snapper (*Lutjanus campechanus*) in the northern Gulf of Mexico. Master's thesis. University of South Alabama, Mobile, Alabama.

Zhao, B., J. C. McGovern, and P. J. Harris. 1997. Age, growth, and temporal change in size at age of the vermilion snapper from the South Atlantic Bight. Transactions of the American Fisheries Society 126:181–193.

American Fisheries Society Symposium 60:337–353, 2007

Reconstructing the Commercial Landings of Red Snapper in the Gulf of Mexico from 1872 to 1963

Clay E. Porch[1] and Stephen C. Turner
Southeast Fisheries Science Center
Sustainable Fisheries Division
75 Virginia Beach Drive, Miami, Florida 33149 USA

Michael J. Schirripa
Northwest Fisheries Science Center
2032 Southeast OSU Drive, Newport, Oregon 97365 USA

Abstract.—Statistics on the commercial landings of red snapper *Lutjanus campechanus* in the United States have been recorded by the National Marine Fisheries Service (NMFS) and its predecessors as far back as 1880. However, there are many gaps in the historical record. Censuses were conducted sporadically prior to 1949 and the water body where the catch was taken was not consistently recorded until 1963. Past assessments of the status of red snapper in the U.S. Gulf of Mexico have avoided these problems by using only data collected after the early 1960s. Unfortunately, there are indications the stock may have already been depressed by that time and the resulting lack of contrast in the data has made it difficult to estimate important benchmarks such as the maximum sustainable yield. For this reason, the participants of several SEDAR (Southeast Data and Assessment Review) workshops strongly recommended reconstructing the catches as far back as possible using whatever auxiliary information might be available. This paper uses statistics from several state and private agencies to supplement NMFS landings statistics and then uses historical accounts dating back to the 1840s to help fill the remaining gaps. The data suggest that a substantial red snapper fishery existed as early as 1872 and that by the turn of the century the landings were comparable to those of recent times, albeit heavily dependent on snapper grounds located in Mexico. The trends of the reconstructed landings are qualitatively consistent with major events in the history of the fishery, but considerable uncertainty remains over the many uncanvassed years between 1890 and the 1920s.

Introduction

The fishery for red snapper *Lutjanus campechanus* began in the early 1840s with the accidental discovery of a large aggregation by Captain James Keeny, a fisherman from Connecticut who used to come south each winter to fish the Gulf. Collins (1887) reports Captain Keeny's account of the adventure as follows: "On one occasion when I was on my way to New Orleans with a cargo

[1] Corresponding author: clay.porch@noaa.gov

of beach fish (pompano, sheepshead, redfish, etc.), I got becalmed when several miles offshore. We had just finished eating, and the cook came on deck and threw over some refuse from the table. The vessel lay motionless, and very soon many strange looking red fish were seen in the water alongside, eagerly feeding on the material the cook had thrown overboard. We quickly baited some lines and threw them out, and the fish bit as fast as we could haul them in. Nearly two hundred snappers were caught, which we took to New Orleans, where they sold like hot cakes."

The Gulf of Mexico was poorly charted in those days, and fishermen knew little of the offshore grounds. It is said that Captain Keeny and his crew did not even realize they were "on soundings," i.e., in less than 100 fathoms, when they made their discovery (Collins 1887). Nevertheless, within a few years several red snapper banks were located and the potential of the resource was quickly recognized. The first serious attempts to develop a commercial market began in the late 1840s, but the trade was mostly local owing to difficulties in transporting and preserving the catch (Warren 1898). The commercial fishery is said to have begun in earnest in 1872 when Mr. S. C. Cobb brought four New England live-well "smacks" down to Pensacola, Florida and built the first fish house dedicated to processing and shipping red snapper (Jarvis 1935; Bortone et al. 1997). Pensacola proved to be an ideal port for landing red snapper because, in addition to having a railroad terminal, the city wharves were in saltwater, making it possible to land the fish alive and fresh (Stearns 1887). Initially the primary markets were in New Orleans and Mobile, but by the late 1870s the market had expanded to include cities throughout the Mississippi valley and as far north as New York (Goode and Gill 1903).

The snapper grounds exploited during the early days of the fishery were located primarily along the edge of the continental shelf between Mobile Bay, Alabama, and Cape St. George, Florida. Those grounds were soon depleted however, and the fleet moved progressively further south towards the Florida Middle Grounds (a 1,000 km^2 bank about 140 km south of Apalachicola and 120 km northwest of Tarpon Springs). By 1883 the Middle Grounds were also becoming unproductive, prompting Stearns (1883) to remark "Most of the old fishing grounds are barren, and smacks have to go farther each year to find new ones." In 1885 the research vessel *Albatross* found productive snapper grounds even further to the south off Tampa and the Dry Tortugas (Collins 1885). At about the same time new snapper grounds were also found off Texas (Camber 1955). These discoveries, in concert with the strong market demand, revitalized the fishery and new fish houses were established throughout the Gulf of Mexico. In the early 1890s a few vessels began making trips to productive snapper grounds found on the Campeche Banks off Mexico, but preserving the catch proved difficult until about the turn of the century when cheap manufactured ice became widely available. After that the landings soared to a record high of almost 14 million pounds in 1902 (Figure 1). By 1910 most snapper vessels were spending at least part of their time fishing off the Campeche Banks or elsewhere in Mexico, usually during the winter when the threat of hurricanes was over (Camber 1955). This pattern continued until the early 1980s when the government of Mexico excluded the U.S. fleet.

Statistics on the historical landings of red snapper in the United States have been collected by the National Marine Fisheries Service (NMFS) and its predecessors since 1880, but there are several gaps in the historical record (Figure 1). Moreover, prior to 1963, the statistical summaries that are typically available include only the port of landing and not the water body where the

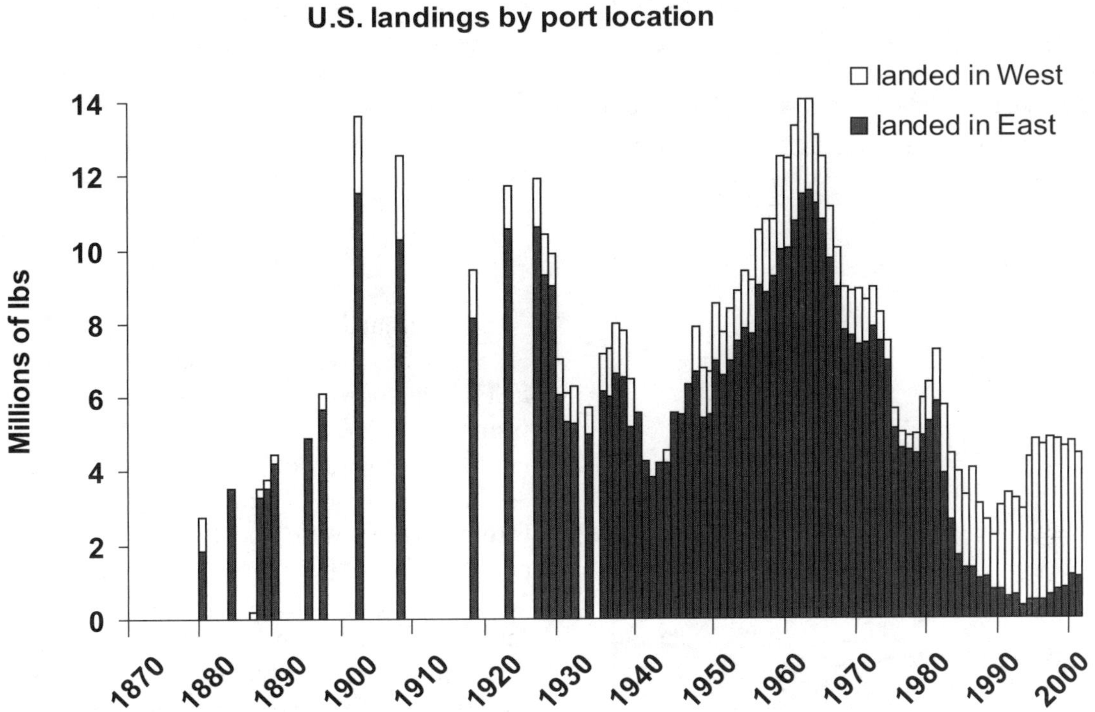

Figure 1. Census estimates of red snapper *Lutjanus campechanus* landings at U.S. ports located east or west of the Mississippi River (without regard to fishing location).

catch was taken. These omissions are of particular concern inasmuch as a large fraction of the total U.S. landings is known to have been from outside of U.S. waters. For this reason, past assessments of the status of red snapper in the Gulf of Mexico have relied on the more complete data collected since the 1960s. Unfortunately, there are indications the abundance of the stock may have already been depressed by this time and the corresponding data do not exhibit any strong temporal trends. As a result it has been difficult to reliably estimate important benchmarks such as the maximum sustainable yield. To remedy this, the participants of several Southeast Data and Assessment Review workshops recommended reconstructing the catches as far back as possible using whatever auxiliary information might be available (SEDAR 2004). This paper details a first attempt at such a reconstruction using clues from several historical references.

Methods

As mentioned previously, the Federal records prior to 1963 are incomplete. Censuses were not always conducted annually in every region (owing to budget limitations, of course). Moreover, the older statistics were published in greater or lesser detail under the auspices of a succession of Federal agencies including the U.S. Fish Commission, Bureau of Fisheries, Fish and Wildlife Service, Bureau of Commercial Fisheries, and Bureau of the Census. The NMFS Fishery Statistics Division has summarized many of these reports (Anonymous 1990), but they did not include information from studies that were focused particularly on the red snapper fishery. Camber (1955) provided more complete statistics by including data collected during Federally-sponsored studies of the red snapper fishery as well as statistics published since 1938 by the Florida State Board of Conservation.

Table 1. Number of vessels fishing for red snapper *Lutjanus campechanus* out of Pensacola (Camber 1955) and the ratio of total landings (in thousands of pounds) to number of vessels for years when landings surveys were conducted. Gray areas denote linear interpolations for years with no vessel information.

Year	Number of vessels	Landings per vessel
1872	4	
1873	6	
1874	9	
1875	11	
1876	13	
1877	11	
1878	10	
1879	11	
1880	14	105.9
1881	21	
1882	26	
1883	24	
1884	25	142. 1
1885	27	
1886	33	
1887	34	
1888	34	
1889	35	99.1
1890	34	122.7
1891	36	
1892	37	
1893	39	
1894	40	
1895	42	116.3
1896	36	
1897	36	147.6

Even so, substantial gaps remain in the historical record and neither set of summaries records where the fish were caught. The following sections detail how these gaps were filled based on auxiliary information gleaned from the literature.

Estimating landings during years that were not surveyed

The landings statistics are especially spotty prior to 1927, being limited to 1880, 1884, 1889, 1890, 1895, 1902, 1908, 1918, and 1923. However, Townsend (1900) gives annual statistics on the number of red snapper vessels operating out of the primary port of Pensacola from 1872 to 1897. This information suggests that, while the number of vessels increased rapidly during the late 1800s, the average landings per vessel was relatively stable at around 120,000 lbs (Table 1). Thus, it seems reasonable to estimate the landings in Florida for years that were not surveyed from the number of vessels and the average landings per vessel observed during the nearest surveyed years (as shown in Table 2). Unfortunately, we were unable to locate similar annual indices of effort for time periods after 1897 or for ports other than Pensacola. While Jarvis (1935) and Camber (1955) do provide some information in this regard, their vessel counts come from the same censuses as the landings data and therefore have the same gaps. Moreover, the average efficiency of the vessels increased markedly with the advent of steam and diesel powered motors during the 1920s and again after World War II with the increased availability of diesel engines, depth recorders and other gear innovations. Nevertheless, the technology and number of vessels does not appear to have changed too much from one census to the next, suggesting that the landings during the intervening periods may be reasonably approximated by linear interpolation (Table 2).

Partitioning landings by fishing location

This section describes how the U.S. red snapper landings were partitioned according to the water body where the fish were caught. Three water bodies are considered: waters outside of U.S. jurisdiction (primarily Mexico) and waters within U.S. jurisdiction lying east or west of the Mississippi River. The division at the Mississippi river accommodated requests by SEDAR 7 pan-

Table 2. Total landings by state from 1872 to 1963 (the inception of the commercial fishery to the point where the U.S. fishery statistics routinely included the water body from which the catch was taken). The shaded regions refer to gaps that were filled as described in the text.

	State where catch was landed					
year	TX	LA	MS	LA	FL	Gulfwide
1872	0	0	0	0	423795	423795
1873	0	0	0	0	671009	671009
1874	0	0	0	0	918223	918223
1875	0	0	0	0	1165438	1165438
1876	0	0	0	0	1377335	1377335
1877	0	0	0	0	1165438	1165438
1878	0	0	0	0	1059489	1059489
1879	0	0	0	0	1165438	1165438
1880	0	900000	0	360000	1483284	2743284
1881	0	810000	0	323000	2710916	3843916
1882	0	719000	0	287000	3356373	4362373
1883	11667	629000	0	250000	3098190	3988857
1884	23333	539000	0	213000	3551275	4326608
1885	35000	448000	0	176000	3159290	3818290
1886	46667	358000	0	140000	3861354	4406021
1887	75000	131000	0	103000	3939362	4248362
1888	65000	150000	0	86000	3224000	3525000
1889	22000	250000	0	51000	3469369	3792369
1890	5000	240000	0	62000	4172942	4479942
1891	93000	183000	0	105000	4243585	4624585
1892	155000	152000	0	143000	4434308	4884308
1893	217000	122000	0	181000	4625030	5145030
1894	279000	91000	0	220000	4815753	5405753
1895	341000	61000	0	258000	4886396	5546396
1896	403000	30000	0	297000	4708100	5438100
1897	465000	0	0	335000	5314487	6114487
1898	786000	0	0	961000	5695000	7442000
1899	1106000	0	0	1587000	6290000	8983000
1900	1427000	0	0	2214000	6885000	10526000
1901	1747000	0	0	2840000	7479000	12066000
1902	2068000	0	0	3466000	8074066	13608066
1903	2099000	0	0	3328000	8005000	13432000
1904	2129000	0	0	3189000	7936000	13254000
1905	2160000	0	0	3051000	7867000	13078000
1906	2191000	0	0	2912000	7797000	12900000
1907	2221000	0	0	2774000	7728000	12723000
1908	2252000	0	0	2635000	7659000	12546000
1909	2151000	6000	10000	2451000	7616000	12234000
1910	2050000	12000	20000	2268000	7573000	11923000
1911	1949000	18000	29000	2084000	7530000	11610000
1912	1848000	24000	39000	1900000	7487000	11298000
1913	1748000	30000	49000	1717000	7445000	10989000
1914	1647000	36000	59000	1533000	7402000	10677000
1915	1546000	42000	69000	1349000	7359000	10365000
1916	1445000	48000	78000	1165000	7316000	10052000
1917	1344000	54000	88000	982000	7273000	9741000
1918	1243000	60000	98000	798000	7230000	9429000
1919	1196000	83000	99000	832000	7678000	9888000
1920	1149000	106000	100000	867000	8126000	10348000
1921	1103000	129000	102000	901000	8575000	10810000
1922	1056000	152000	103000	936000	9023000	11270000
1923	1009000	175000	104000	970000	9471000	11729000
1924	1015000	148000	112000	1026000	9179000	11480000

Table 2. Continued.

	State where catch was landed					
year	TX	LA	MS	LA	FL	Gulfwide
1925	1021000	121000	120000	1083000	8886000	11231000
1926	1026000	94000	128000	1139000	8594000	10981000
1927	1237306	72000	218706	1058650	9312667	11899329
1928	1055162	48000	97328	1300522	7891835	10392847
1929	804140	80000	90864	1227601	7700143	9902748
1930	929578	76400	188725	847918	5001672	7044293
1931	690664	78835	68236	863201	4392992	6093928
1932	985291	66884	36812	681573	4539465	6310025
1933	810346	72942	80006	816037	4227833	6007163
1934	635400	79000	123200	950500	3916200	5704300
1935	771000	98000	224050	989000	4360150	6442200
1936	906600	117000	324900	1027500	4804100	7180100
1937	1141200	148100	303800	1168200	4550500	7311800
1938	1279000	85000	173900	1193100	5260800	7991800
1939	1156300	90900	40400	1020000	5493700	7801300
1940	1233100	104400	26200	1255400	3891300	6510400
1941	1223000	93000	80000	1492263	4058484	6946747
1942	849000	66000	53000	1275062	2988486	5231548
1943	662000	53000	39000	1001293	2824923	4580216
1944	475000	39000	26000	1092384	3108629	4741013
1945	288200	25500	12000	1360500	2845800	4532000
1946	500000	46000	26000	1703822	3866392	6142214
1947	712000	66000	40000	1514391	4029488	6361879
1948	923000	87000	54000	1851800	4508038	7423838
1949	1054400	169900	135900	1343200	5184500	7887900
1950	1233200	142000	65600	993900	4353600	6788300
1951	1116800	9200	2400	1228800	4313200	6670400
1952	1523400	64600	0	1458800	5499900	8546700
1953	1100700	44100	28900	1418000	5135600	7727300
1954	1344700	45200	68000	1403600	5524000	8385500
1955	1261700	70900	147000	1173200	6209700	8862500
1956	1533500	43700	271400	1065100	6499716	9413416
1957	1443000	28200	549600	932900	6201459	9155159
1958	1399000	87800	1109900	1417900	6487173	10501773
1959	1665100	313000	1021700	1819100	5993778	10812678
1960	1152600	426000	1468500	1720100	6046170	10813370
1961	1828900	677200	2151700	1784200	6044949	12486949
1962	1742300	693900	2175900	1893400	5966250	12471750
1963	2168700	387900	1885800	2314900	6568758	13326058

elists that were based on recent evidence in support of possibly distinct populations in the eastern and western Gulf.

Florida and Alabama

1872–1933.—The commercial fishery essentially began in 1872 with four vessels fishing locally off Pensacola. By 1890, the Pensacola fleet had grown to 34 vessels and smaller fleets were established in several other Florida ports as well as in Mobile, Alabama. All of these fleets fished almost exclusively off Florida until late in 1890 when a few vessels began fishing on the Campeche Banks off Mexico (Stearns 1883, 1885, 1887; Smith 1894). The fraction of U.S. landings coming from the Campeche banks appears to have been rather low until about 1895 when ice became more readily available, making it easier to preserve the catch during the long return trip from Mexico. By 1910 most of the larger vessels (over 5 tons) from Pensacola and Mobile were fishing primarily on the Campeche Banks (Camber 1955). Jarvis (1935) and Carpenter (1965) estimated that 75% of all landings by vessels (crafts over 5 tons) were from Mexico until about 1933, when the proportion dropped to about 50%. For lack of more detailed information, we assume that the proportion of the vessel landings that came from U.S. waters decreased linearly from 100% in 1894 to 25% in 1910 and then remained constant at about 25% until 1933, when it increased to 50%. The remaining vessel landings are assumed to have come from the Dry Tortugas grounds (cf. Jarvis 1935).

The landings in Florida and Alabama by boats (defined as crafts under 5 tons) were small early in the fishery, but increased to more substantive levels during the 1920s (Table 3). Owing to their small size, most boats fished fairly close to the ports where the landings were made rather than in distant foreign waters. Hence the proportion of the total landings for boats and vessels combined that came from U.S. waters may be calculated from the proportion of the total landings made by vessels, $p(\text{vsl})$, and the proportion of the vessel landings from U.S. waters, $p(\text{US}|\text{vsl})$:

$$(1)\ p(\text{US}) = 1 - p(\text{vsl}) + p(\text{US}|\text{vsl}) \times p(\text{vsl})$$

Values for $p(\text{vsl})$ were linearly interpolated between census years until after 1932, when they were fixed at the 1932 level through 1951 (the statistical bulletins published after 1932 no longer reported landings by vessel class). The results are summarized in Table 4.

1934 to 1951.—Camber (1955) provided annual estimates of the proportion of vessel landings that came from Mexico between 1934 and 1951 (based on records kept by the major fish houses in Florida and Alabama, which he considered accurate). He published his statistics under the headings "Pensacola" and "west Florida," but stated that the data included vessels based in Mobile, which operated much like the Pensacola fleet. He also stated that the number of trips to areas west of the Mississippi River (e.g., the "Galveston Lumps" off Texas) was small until depth recorders became available in the 1950s and the area

Table 3. Estimates of the catch of vessels over 5 tons and of smaller boats from Collins and Smith (1891), Radcliffe (1920) , Sette (1925), and Jarvis (1935).

	Catch by state and operating units									
	W. Florida		Alabama		Mississippi		Louisiana		Texas	
Year	Vessels	Boats	Vessels	Boats	Vessels	Boats	Vessels	Boats	Vessels	Boats
1890	3092620	376750	51000	0	0	0	250400	0	0	22000
1891	3904342	268600	62375	0	0	0	240500	0	0	4800
1918	6011297	1218871	798400	0	85752	12480	0	60000	1243002	0
1923	7964907	1506360	970000			103618		175000	1008960	
1927	7609494	1703173	1037722	20928		218706		72000	1234026	3280
1928	6492700	1398503	1131245	169277		97328		48000	1019452	35710
1929	5797225	1902918	1184488	43113		90864		80000	732337	71803
1930	3983215	1018457	820441	27477		188725		76400	835136	94442
1931	3132965	1260027	849451	13750		68236		78835	650827	39837
1932	3096876	1442589	612938	68635		36812		66884	895637	89654

Table 4. Fraction of the total landings made in U.S. waters, *p(US)*, as deduced from the fraction of the total landings made by vessels, *p(vsl)*, and the fraction of vessel landings that came from U.S. waters *p(US|vsl)*. Landings from boats (under 5 tons) are assumed to have come from U.S. waters. Shaded regions represent interpolations or assumptions that are not based on published accounts.

	West Florida			Alabama			Texas		
Year	*p(vsl)*	*p(US|vsl)*	*p(US)*	*p(vsl)*	*p(US|vsl)*	*p(US)*	*p(vsl)*	*p(US|vsl)*	*p(US)*
1889	0.89	1.00	1.00	1.00	1.00	1.00	0.00	1.00	1.00
1890	0.94	1.00	1.00	1.00	1.00	1.00	0.00	1.00	1.00
1891	0.94	1.00	1.00	1.00	1.00	1.00	0.08	1.00	1.00
1892	0.94	1.00	1.00	1.00	1.00	1.00	0.17	1.00	1.00
1893	0.95	1.00	1.00	1.00	1.00	1.00	0.25	1.00	1.00
1894	0.95	1.00	1.00	1.00	1.00	1.00	0.33	1.00	1.00
1895	0.96	0.95	0.95	1.00	0.95	0.95	0.42	0.95	0.98
1896	0.96	0.91	0.91	1.00	0.91	0.91	0.50	0.91	0.96
1897	0.97	0.86	0.86	1.00	0.86	0.86	0.58	0.86	0.92
1898	0.97	0.81	0.82	1.00	0.81	0.81	0.67	0.81	0.87
1899	0.97	0.77	0.78	1.00	0.77	0.77	0.75	0.77	0.83
1900	0.98	0.72	0.73	1.00	0.72	0.72	0.83	0.72	0.77
1901	0.98	0.67	0.68	1.00	0.67	0.67	0.92	0.67	0.70
1902	0.99	0.63	0.63	1.00	0.63	0.63	1.00	0.63	0.63
1903	0.98	0.58	0.59	1.00	0.58	0.58	1.00	0.58	0.58
1904	0.97	0.53	0.54	1.00	0.53	0.53	1.00	0.53	0.53
1905	0.96	0.48	0.50	1.00	0.48	0.48	1.00	0.48	0.48
1906	0.95	0.44	0.47	1.00	0.44	0.44	1.00	0.44	0.44
1907	0.94	0.39	0.43	1.00	0.39	0.39	1.00	0.39	0.39
1908	0.93	0.34	0.39	1.00	0.34	0.34	1.00	0.34	0.34
1909	0.92	0.30	0.36	1.00	0.30	0.30	1.00	0.30	0.30
1910	0.91	0.25	0.32	1.00	0.25	0.25	1.00	0.25	0.25
1911	0.90	0.25	0.33	1.00	0.25	0.25	1.00	0.25	0.25
1912	0.89	0.25	0.33	1.00	0.25	0.25	1.00	0.25	0.25
1913	0.88	0.25	0.34	1.00	0.25	0.25	1.00	0.25	0.25
1914	0.87	0.25	0.35	1.00	0.25	0.25	1.00	0.25	0.25
1915	0.86	0.25	0.36	1.00	0.25	0.25	1.00	0.25	0.25
1916	0.85	0.25	0.36	1.00	0.25	0.25	1.00	0.25	0.25
1917	0.84	0.25	0.37	1.00	0.25	0.25	1.00	0.25	0.25
1918	0.83	0.25	0.38	1.00	0.25	0.25	1.00	0.25	0.25
1919	0.83	0.25	0.38	1.00	0.25	0.25	1.00	0.25	0.25
1920	0.84	0.25	0.37	1.00	0.25	0.25	1.00	0.25	0.25
1921	0.84	0.25	0.37	1.00	0.25	0.25	1.00	0.25	0.25
1922	0.84	0.25	0.37	1.00	0.25	0.25	1.00	0.25	0.25
1923	0.84	0.25	0.37	1.00	0.25	0.25	1.00	0.25	0.25
1924	0.83	0.25	0.38	1.00	0.25	0.25	1.00	0.25	0.25
1925	0.83	0.25	0.38	0.99	0.25	0.26	1.00	0.25	0.25
1926	0.82	0.25	0.39	0.99	0.25	0.26	1.00	0.25	0.25
1927	0.82	0.25	0.39	0.98	0.25	0.26	1.00	0.25	0.25
1928	0.82	0.25	0.38	0.87	0.25	0.35	0.97	0.25	0.28
1929	0.75	0.25	0.44	0.96	0.25	0.28	0.91	0.25	0.32
1930	0.80	0.25	0.40	0.97	0.25	0.27	0.90	0.25	0.33
1931	0.71	0.25	0.47	0.98	0.25	0.26	0.94	0.25	0.29
1932	0.68	0.25	0.49	0.90	0.25	0.33	0.91	0.25	0.32
1933	0.68	0.50	0.66	0.90	0.21	0.29	0.91	0.50	0.55
1934	0.68	0.43	0.61	0.90	0.17	0.25	0.91	0.43	0.48
1935	0.68	0.45	0.62	0.90	0.18	0.26	0.91	0.45	0.50
1936	0.68	0.48	0.65	0.90	0.19	0.27	0.91	0.48	0.53
1937	0.68	0.50	0.66	0.90	0.20	0.28	0.91	0.50	0.55

Table 4. (Continued)

	West Florida			Alabama			Texas		
Year	*p(vsl)*	*p(US|vsl)*	*p(US)*	*p(vsl)*	*p(US|vsl)*	*p(US)*	*p(vsl)*	*p(US|vsl)*	*p(US)*
1938	0.68	0.58	0.71	0.90	0.20	0.28	0.91	0.58	0.62
1939	0.68	0.65	0.76	0.90	0.19	0.27	0.91	0.65	0.68
1940	0.68	0.58	0.71	0.90	0.18	0.26	0.91	0.58	0.62
1941	0.68	0.49	0.65	0.90	0.18	0.26	0.91	0.49	0.54
1942	0.68	0.52	0.67	0.90	0.21	0.29	0.91	0.52	0.56
1943	0.68	0.44	0.62	0.90	0.18	0.26	0.91	0.44	0.49
1944	0.68	0.47	0.64	0.90	0.17	0.25	0.91	0.47	0.52
1945	0.68	0.42	0.60	0.90	0.15	0.23	0.91	0.42	0.47
1946	0.68	0.53	0.68	0.90	0.26	0.34	0.91	0.53	0.57
1947	0.68	0.55	0.69	0.90	0.24	0.32	0.91	0.55	0.59
1948	0.68	0.52	0.67	0.90	0.26	0.34	0.91	0.52	0.56
1949	0.68	0.57	0.71	0.90	0.25	0.32	0.91	0.57	0.61
1950	0.68	0.52	0.67	0.90	0.20	0.28	0.91	0.52	0.56
1951	0.68	0.63	0.75	0.90	0.31	0.38	0.91	0.63	0.66

was better charted. Hence, it appears the bulk of the landings from domestic waters during this period continued to come from east of the Mississippi River. Therefore, equation (1) was applied to Florida and Alabama using Camber's statistics for west Florida and Pensacola, respectively.

1952 to 1963.—Carpenter (1965) states that complete statistics of red snapper landings from Mexico were collected by the U.S. Fish and Wildlife service during 1954 to 1963, however he provided only Gulf-wide summaries of the proportion taken from foreign waters. Similar Gulf-wide summaries are available after 1958 in the general section of the annual volumes of Fishery Statistics for the United States. We have been unable to locate any documents with the corresponding state-specific statistics; however we make the following observations:

(a) The proportions of the landings in Florida and Texas that came from foreign waters were relatively constant from 1963 to 1966 and are similar to the proportions in given by Camber (1955). The landings for Alabama were not classified by water body in 1963, but the proportion from foreign waters can be calculated from the corresponding proportions for the other states and the Gulf-wide summary. The resulting value, 66%, is also similar to the proportions given by Camber (1955).

(b) The landings in Mississippi were low and exclusively from U.S. waters until 1957, when they began to increase rapidly. An average of only 26% of the catch was taken in U.S. waters during 1963 (see below)

(c) The landings in Louisiana were small and there are no historical accounts of any foreign catch during this time (see below)

Based on these observations, linear transitions are assumed for the U.S. proportions (*F*) for Alabama (from 38% in 1951 to 34% in 1963), Mississippi (from 100% in 1956 to 26% in 1963) and Louisiana (100% for all years). The U.S. proportions for Texas and Florida (F_{Tx} and F_{Fl}) were determined from the Gulf-wide statistics using the landings by state (*C*), the interpolated fractions for Louisiana, Mississippi and Alabama described above, and assuming the ratio of the fractions

Table 5. Fraction of the landings in each state that came from U.S. waters and for all states combined (Total). Shaded regions represent values based on interpolations or assumptions that are not based on reliable published accounts. The values in bold (TX and FL) were inferred using equation (2).

year	TX	LA	MS	AL	FL	Total
1951	0.66	1.00	1.00	0.38	0.75	0.67
1952	**0.56**	1.00	1.00	0.38	**0.63**	0.58
1953	**0.46**	1.00	1.00	0.38	**0.52**	0.49
1954	**0.36**	1.00	1.00	0.37	**0.41**	0.40
1955	**0.34**	1.00	1.00	0.37	**0.38**	0.39
1956	**0.38**	1.00	1.00	0.37	**0.43**	0.44
1957	**0.42**	1.00	0.89	0.36	**0.48**	0.48
1958	**0.67**	1.00	0.79	0.36	**0.76**	0.70
1959	**0.64**	1.00	0.68	0.36	**0.73**	0.65
1960	**0.73**	1.00	0.58	0.35	**0.82**	0.70
1961	**0.66**	1.00	0.47	0.35	**0.75**	0.64
1962	**0.70**	1.00	0.37	0.35	**0.79**	0.65
1963	0.74	1.00	0.26	0.34	0.66	0.57

for Texas and Florida r was constant at the 1963–1966 average (0.89):

$$(2)\quad \begin{aligned} F_{Fl,y} &= \frac{F_{total,y}C_{total,y} - F_{Al,y}C_{Al,y} - F_{Ms,y}C_{Ms,y} - C_{La,y}}{C_{Fl,y} + rC_{Tx,y}} \\ F_{Tx,y} &= rF_{Fl,y} \end{aligned}$$

The results of these calculations are summarized in Table 5.

As mentioned previously, the statistics published between 1932 and 1962 do not include estimates of the fraction of the landings in Florida and Alabama that came from east or west of the Mississippi River. Camber (1955) states that in recent years (presumably the early 1950s) appreciable quantities of red snapper were caught on the "western grounds," located near the 100 fathom lime (200 m) off Texas and Louisiana. He also mentions that vessels from Pensacola and Mobile fished in this area (as well as vessels from Galveston). In the absence of any other information, it is assumed that the fraction of the catch taken by eastern-based fleets from western waters since 1951 increased linearly to the mean fraction from the years 1963–1966 (30% for Florida and 9% for Alabama, see Table 6).

Mississippi

Radcliffe (1920) and Jarvis (1935) indicate that the snapper boats operating out of Mississippi were relatively small and fished exclusively in U.S. waters. This appears to have remained the case through the early 1950s; Camber (1955, p. 14) indicates that the main Mississippi fleet (out of Pascagoula) did not fish on the major foreign fishing grounds. However, during the late 1950s the landings of red snapper in Mississippi began to increase rapidly, apparently due to an increase in the number of vessels fishing in foreign waters (the U.S. fishery statistics for 1963–1966 indicated that only 27% of the landings in Mississippi were taken from U.S. waters). For lack of more detailed information, we assume that the proportion of Mississippi landings from U.S. waters decreased linearly from 100% in 1956 to 26% by 1963.

Vessels operating out of Pascagoula are cited by Camber (1955) as fishing on both sides of the river, but no percentages are given. The 1963–1966 U.S. fishery statistics indicate that 95% of the Mississippi landings from U.S. waters were taken west of the river. It is unclear that an equally large fraction applies to the earlier years of the fishery, but the Mississippi landings are small until the mid 1950s and any errors introduced by applying this fraction should have a negligible impact on the final catch series.

Louisiana

The landings of red snapper in Louisiana, like those in Mississippi, were relatively low and of local origins until the late 1950s, after which time they increased rapidly. However, unlike Mississippi, the increase in landings does not appear to have been the result of increased fishing in foreign waters. The fleet appears to have expanded offshore into so-called "western grounds" near the 100 fathom line. The U.S. fishery statistics for 1963–1966 indicate that all of the landings in Louisiana were caught in U.S. waters. Hence, it appears safe to assign all Louisiana landings to U.S. waters west of the Mississippi River.

Texas

No statistics are available on the proportion of the landings in Texas that came from foreign waters prior to 1963. Smith (1894, 1895) reported that three vessels from Galveston went fishing on the Campeche Banks in 1892, but the venture was plagued by an inadequate supply of ice and difficulties in securing labor. Camber (1955) indicated that, at the turn of the century, most of the landings in Texas were still being made by smaller vessels operating in local waters. Camber (1955, p. 39) does provide statistics for the number of vessels fishing on the Campeche Banks in 1951, which when combined with data on the total number of vessels in the 1951 U.S. fishery statistics bulletin, suggests that the proportion of vessels that fished regularly off the Campeche Banks was about the same for Texas (5 or 6 of 18) as for Florida (24 of 75). The proportions of the landings from foreign waters during 1963–1966 are also quite similar between the two states (about 38% for Texas and 30% for Florida). Based on these observations, we apply the proportions obtained for Florida (above) to Texas through 1951. For 1952 to 1962 we make the somewhat less restrictive assumption that the ratio of the proportions for Texas and Florida are constant (at the 1963–1966 level) and apply the method discussed in connection with equation (2) above. We recognize that neither assumption is especially well supported, however the landings in Texas during this period were typically on the order of 15% of the total, so we do not expect the error in this regard to be large.

Camber (1955) indicates that waters east of the Mississippi River were fished exclusively by boats operating out of ports from that region (e.g., Pascagoula, Mobile, Pensacola, Panama City, Tampa) and that the waters west of the Mississippi River were

Table 6. Fraction of landings in each state that came from the adjacent region (i.e., from the same side of the Mississippi River as the port was located). Here the state of Mississippi is considered to lie in the eastern region.

Year	TX	LA	MS	AL	wFL
1872-1891	1.00	1.00	0.05	1.00	1.00
1892-1949	1.00	1.00	0.05	1.00	1.00
1950-1962	1.00	1.00	0.05	0.91	0.70

fished by vessels operating out of Galveston, Freeport, Brownsville and Pascagoula. This is consistent with the 1963–1966 U.S. fishery statistics, which indicate that 100% of the landings in Texas and Louisiana from U.S. waters were taken west of the Mississippi.

Other Adjustments

Two other adjustments were made to the landings data, one to account for cases where the landings were recorded in dressed (gutted) weight rather than whole weight and another to account for the fraction of the red snapper landings that were actually *L. campechanus*.

Conversion from dressed weights

Camber (1955) reported that red snapper were landed whole prior to 1934 and dressed after that. However, the descriptions in the various U.S. fishery statistics bulletins (e.g., Anderson and Peterson 1953; p. 461) generally indicate that the dressed landings were converted to whole weight assuming a whole weight to dressed weight ratio of 1.11. The only exception appears to be Florida, in which case the landings from 1956 to 1985 were never converted from dressed to whole weight (E. Snell, SEFSC, personal communication).

Correction for species composition

Dealers frequently marketed other species as "red snapper," including silk snapper (yellow eye snapper, *L. vivanus*), gray snapper *L. Griseus*, lane snapper *L. synagris*, mutton snapper *L. analis,* and vermillion snapper *Rhomboplites aurorubens*. However, Camber (1955) noted that typically 95% or more of the landings were in fact *L. campechanus* (then called *L. aya*). He adjusted the landings statistics of Florida producers from 1880 to 1951 based on his review of fish house records and samples of the catch. His results are summarized in Table 7. Here it assumed with no further information that these fractions apply equally to the landings in other areas. The value of 96% is assumed to continue through 1962 inasmuch as port agents apparently did not make a concerted effort to distinguish the red snapper catch by species until 1986 (G. Davenport, SEFSC, personal communication.).

Results and Discussion

Estimates of the annual landings of red snapper caught in the U.S. Gulf of Mexico (east and west of the Mississippi River) were obtained by multiplying four quantities: the total landings in whole weight (Table 2), the proportion from U.S. waters (Tables 4 and 5), the proportion of landings from U.S. waters that came from east or west of the Mississippi River (Table 6), and the fraction of the landings sold as red snapper that actually were *L. campechanus*. The results are shown in Table 7 and Figure 2.

The landings from waters west of the Mississippi River appear to have been relatively small, generally under 1 million pounds, until the late 1940s. The landings from waters east of the Mississippi River, on the other hand, show a strong increasing trend that reflects the initial expansion of the fishery from a few vessels fishing locally off Pensacola in 1872 to dozens of vessels fishing as far as the Dry Tortugas by 1902. After 1902 the landings from the east appear to have diminished rapidly, primarily in response to a shift in effort to the Campeche Banks (Figure 1). The landings from the east increased somewhat during the early 1920s owing to an increase in the number of smaller boats (<5 tons) fishing locally and the widespread use of steam engines. The landings dropped again during the 1930s, which is consistent with anecdotal accounts of a decline in market demand during the Great Depression (Camber 1955). The landings began to increase during the mid 1930s as the economy

Table 7. A reconstructed history of red snapper *Lutjanus campechanus* landings (in lbs) from U.S. waters.

Year	Red Snapper and others		Proportion *L. campechanus* (Gulf-wide)	Red Snapper (*L. campechanus*)	
	East	West		East	West
1872	423796	0	0.99	419574	0
1873	671009	0	0.99	664325	0
1874	918223	0	0.99	909076	0
1875	1165438	0	0.99	1153828	0
1876	1377335	0	0.99	1363615	0
1877	1165438	0	0.99	1153828	0
1878	1059489	0	0.99	1048934	0
1879	1165438	0	0.99	1153828	0
1880	1843284	900000	0.99	1824922	891034
1881	3033916	810000	0.99	3003739	801943
1882	3643373	719000	0.99	3607188	711859
1883	3348190	640667	0.99	3314988	634313
1884	3764275	562333	0.99	3727003	556765
1885	3335290	483000	0.99	3302316	478225
1886	4001354	404667	0.99	3961855	400672
1887	4042362	206000	0.99	4002518	203970
1888	3310000	215000	0.99	3277425	212884
1889	3520369	272000	0.99	3485777	269327
1890	4234942	245000	0.99	4192270	242531
1891	4348585	276000	0.99	4304834	273223
1892	4577308	307000	0.99	4531325	303916
1893	4806030	339000	0.99	4757824	335600
1894	5035753	370000	0.99	4985319	366294
1895	4896949	394839	0.99	4847980	390891
1896	4571590	414865	0.99	4525762	410706
1897	4880880	427242	0.99	4831831	422949
1898	5423822	685942	0.99	5369317	679049
1899	6108691	915215	0.99	6047304	906018
1900	6589836	1095365	0.99	6523614	1084358
1901	6963091	1216611	0.99	6893118	1204385
1902	7308770	1302840	0.99	7235733	1289821
1903	6640382	1217420	0.99	6574083	1205265
1904	6008148	1128370	0.99	5948213	1117114
1905	5404274	1036800	0.99	5350410	1026466
1906	4930276	964040	0.99	4881180	954440
1907	4378625	866190	0.99	4335060	857572
1908	3853806	765680	0.99	3815496	758069
1909	3447084	660812	0.99	3412777	654236
1910	2972403	543525	0.99	2942785	538109
1911	2969664	532836	0.99	2940039	527520
1912	2966329	523098	0.99	2936702	517874
1913	2962939	513611	0.99	2933312	508475
1914	2958322	503873	0.99	2928706	498829
1915	2953059	494136	0.99	2923461	489183
1916	2947103	483447	0.99	2917530	478596
1917	2940801	473709	0.99	2911256	468950
1918	2925910	463972	0.99	2896481	459305
1919	3111272	476173	0.99	3079966	471382

Table 7. (Continued)

Year	Red Snapper and others		Proportion *L. campechanus* (Gulf-wide)	Red Snapper (*L. campechanus*)	
	East	West		East	West
1920	3228246	488374	0.99	3195750	483458
1921	3402973	501777	0.99	3368705	496724
1922	3577532	513978	0.99	3541492	508800
1923	3745059	526179	0.99	3707316	520876
1924	3727033	508289	0.99	3689539	503176
1925	3639188	490399	0.99	3602644	485474
1926	3608224	472259	0.99	3572054	467525
1927	3896569	591829	0.99	3857579	585907
1928	3478754	431155	0.99	3444187	426871
1929	3695890	421321	0.99	3658800	417093
1930	2256050	559149	0.99	2233495	553559
1931	2272708	346288	0.99	2249781	342794
1932	2440472	415464	0.99	2416037	411305
1933	3025882	591087	0.99	2995666	585185
1934	2637816	502370	0.99	2611546	497367
1935	2994761	696661	0.97	2902548	675210
1936	3394723	904123	0.95	3219475	857449
1937	3341912	1059607	0.94	3139239	995347
1938	4096645	1041120	0.97	3973613	1009853
1939	4460916	917750	0.97	4326942	890188
1940	3107244	891646	0.97	3001832	861397
1941	3038129	825124	0.96	2917622	792396
1942	2386241	594977	0.98	2329540	580839
1943	2006007	415111	0.97	1953460	404237
1944	2263894	309890	0.97	2206368	302015
1945	2039656	173169	0.97	1985188	168545
1946	3204031	357116	0.96	3086345	343998
1947	3276483	524804	0.96	3161005	506307
1948	3655589	658640	0.96	3520166	634241
1949	4103870	941437	0.97	3965333	909656
1950	2316791	1792748	0.96	2230589	1726045
1951	2695334	1750439	0.96	2586570	1679804
1952	2955945	2001141	0.96	2836665	1920390
1953	2369639	1416738	0.96	2274018	1359569
1954	2070236	1315562	0.96	1986697	1262476
1955	2262983	1461874	0.96	2171666	1402884
1956	2574421	1854330	0.96	2470537	1779504
1957	2648857	2112996	0.96	2541969	2027731
1958	4357859	3532154	0.96	4182008	3389623
1959	4029521	3535484	0.96	3866920	3392818
1960	4471641	3765292	0.96	4291199	3613353
1961	4145286	4391495	0.96	3978014	4214287
1962	3945536	4124420	0.96	3786324	3957989
1963	3906025	3709840	0.96	3748408	3560139

recovered, but dipped once again during World War II when many fishermen enlisted and the remaining crews were forced to fish in shallow waters to avoid the threats of marauding U-boats (Camber 1955). The landings from both sides of the River, but especially the west, increased rapidly after the war owing in part to several technological innovations such as the fathometer, reels, and wire line, which opened up new fishing grounds in deeper water (Siebenaler and Brady 1952). The landings continued to increase through the late 1950s and early 1960s, presumably owing to what Carpenter (1965) described as a tremendous increase in the size of the commercial fleet.

The expansiveness of the shaded regions in Tables 4 through 7 attests to the substantial gaps in the historical record of Gulf of Mexico red snapper landings. As discussed above, a number of interpolations or other assumptions had to be made which were not directly grounded in the U.S. census data or other reliable published accounts. In many cases even severe violations of these assumptions would have little impact on the overall result because the landings from the affected strata are negligible (e.g., Louisiana and Mississippi prior to the late 1950s). In other cases, the fisheries involved may not have changed their mode of operation for an extended time period, so the interpolated values are relatively constant and errors are likely to be relatively small. Nevertheless, there are several gaps which are of particular concern.

Perhaps the greatest uncertainties lie in the magnitude of the landings during the many uncanvassed years between 1890 and the 1920s, and in the fraction of those landings that came from the Campeche Banks. Here the total landings were linearly interpolated between years with census estimates, which seems consistent with anecdotal accounts of a rapidly growing fishery during the early years. After 1902, however, the landings appear to have declined

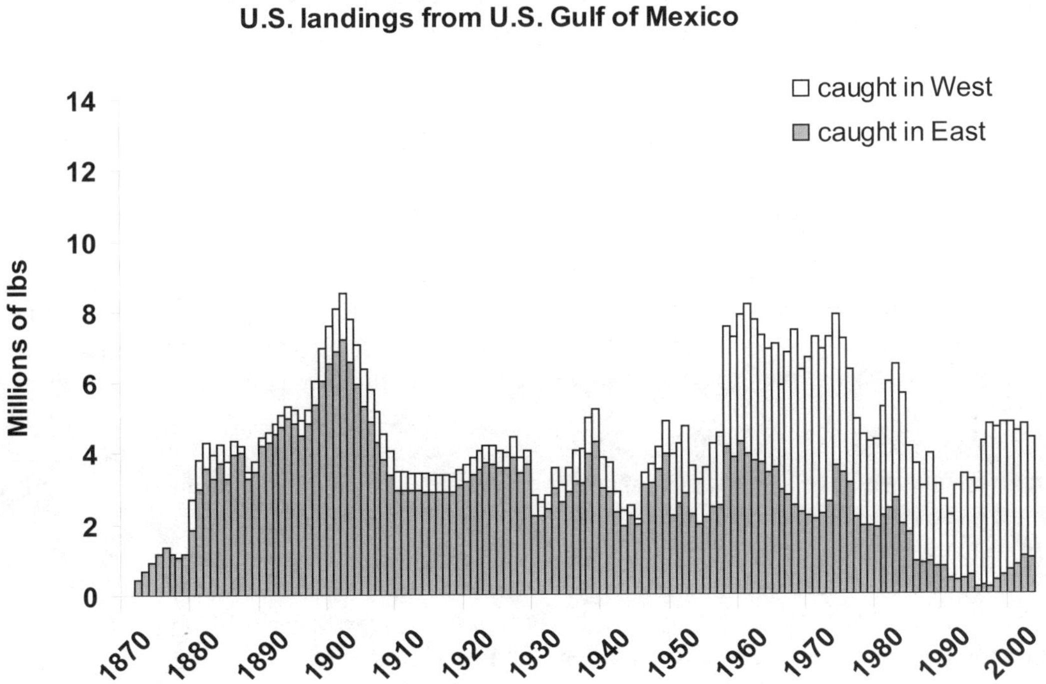

Figure 2. Reconstructed landings of red snapper *Lutjanus campechanus* caught in the U.S. Gulf of Mexico (east and west of the Mississippi River).

somewhat. Jarvis (1935) states that severe hurricanes in 1906, 1916, 1917, and 1926 damaged the fleet and resulted in reduced catches; one firm alone apparently losing 14 of its vessels in the 1906 storm. In light of this, it is probably fair to say that the interpolation approach correctly reflects the overall decline between 1902 and the 1920s, but overestimates the landings during the years affected by hurricanes (to a degree we cannot reasonably surmise). Of equally uncertain status is the nature of the transition from negligible foreign catches prior to 1900 to about 75% foreign in 1910 (and constant at that level thereafter). The anecdotal accounts we have found seem to confirm that the move to Campeche was neither abrupt nor complete owing to the expense of ice, inexperience of the captains in uncharted waters and disputes over the rights of American vessels to fish in waters claimed by Mexico (e.g., Bortone et al. 1997). Nevertheless, there is no particular reason to assume the transition was precisely linear and the question remains of some importance given that the landings at that time are among the highest on record.

Also of particular concern are the landings in Texas after about 1900. The references we have found so far do not give much indication of the fraction of the landings that came from foreign waters. It is known that at least three vessels from Galveston, Texas were fishing off the Campeche Banks as early as 1892 (Smith 1894) and that by 1897 vessels from Galveston were fishing off Mexico on a regular basis (Carpenter 1965). This, plus the observation that the trends in Florida and Texas were similar during the 1950s and 1960s, is essentially all the evidence we have to back our assumption that percentage of the Texas landings coming from foreign waters was roughly the same as for Florida. It is of interest to note, however, that the total landings in Texas were generally under 2 million pounds for most of its recorded history, far less than recent levels. Hence, it seems unlikely that errors in this regard would have an especially great effect on one's perception of the fishery or the results of the SEDAR stock assessment (Porch 2007, this volume).

We recommend that subsequent research focus on locating additional observations relating to the fraction of the landings coming from foreign waters. In this regard we note that the landings statistics for a given year are often summarized to various degrees in multiple documents. For example, the water body where the catches were made appears in the NMFS electronic database, but not in 1963 Fishery Statistics summaries (Lyles 1965). In addition, we sometimes have the fraction of the landings that came from foreign waters on a Gulf-wide basis, in which case state-specific estimates must also have been made. This suggests to us that more complete statistics were generally collected, but simply not published in the various digests we have available to us. Perhaps the more detailed data still exists somewhere in State and Federal archives.

Finally, as mentioned earlier, past stock assessments of Gulf red snapper have generally focused on the years after 1962, when detailed landings are available. However, it was found that the lack of contrast in those data made it difficult to reliably estimate measures of stock productivity such as MSY. Comparisons made during SEDAR 7 showed that the estimates of stock abundance obtained with the extended series (1872–2003) were rather similar to those obtained using only the more recent data (1963–2003). However, the estimates of MSY (conditioned on recent fishing patterns) obtained with the shorter time series were highly uncertain and more than three times larger than the corresponding estimates from the extended time series (over 20 mp compared with about 6 mp). The high MSY value obtained for the shorter time series was viewed as unlikely because the stock was estimated to be overfished even though the historical landings have probably never approached 20 mp. For these and other reasons, the participants in SEDAR 7 elected to base their advice on the models that employed the extended time series developed here (for more details see Cordue 2005; SEDAR 2005; Porch 2007).

Acknowledgments

We thank two anonymous reviewers for their helpful comments. We also thank the members of the SEDAR 7 assessment workshop for their comments on an earlier draft.

References

Anderson, A. W., and C. E. Peterson. 1953. Fishery statistics of the United States 1950. United States Department of the Interior. Fish and Wildlife Service Statistical Digest 27. Washington, D.C.

Anonymous. 1990. Historical Catch Statistics Atlantic and Gulf States 1879–1989. Current Fishery Statistics 9010. United States Department of Commerce. National Marine Fisheries Service, Fisheries Statistics Division. Historical Series 5-9 (revised). Silver Spring, Maryland.

Bortone, S. A., R. N. Besser, and C. R. McNeil. 1997. The importance of Silas Stearns and the red snapper fishery to North American ichthyology. Pages 459–467 *in* T. W. Pietsch and W. D. Anderson, Jr., editors. Collection building in ichthyology and herpetology. American Society of Ichthyologists and Herpetologists, Special Publication No. 3. Allen Press, Lawrence, Kansas.

Camber, C. I. 1955. A survey of the red snapper fishery of the Gulf of Mexico, with special reference to the Campeche Banks. State of Florida Board of Conservation Technical Series 12. Marine Laboratory, St. Petersburg, Florida.

Carpenter, J. S. 1965. A review of the Gulf of Mexico red snapper fishery. United States Department of the Interior. Fish and Wildlife Service Circular 208:1–25. Washington, D.C.

Collins, J. W. 1885. The red snapper grounds in the Gulf of Mexico. Bulletin of the United States Fish Commission 5:145–146.

Collins, J. W. 1887. Notes on the red snapper fishery. Bulletin of the United States Fish Commission 6:299–300.

Collins, J. W., and H. M. Smith. 1891. A statistical report on the fisheries of the Gulf states. Bulletin of the United States Fish Commission 11:93–184.

Cordue, P. L., editor. 2005. Gulf of Mexico Red Snapper Consensus Summary Report. South Atlantic Fishery Management Council, New Orleans, Louisiana.

Goode, G. B., and T. N. Gill. 1903. American fishes: a popular treatise upon the game and food fishes of North America with especial reference to habits and methods of capture. L. C. Page and Co., Boston, Massachusetts.

Jarvis, N. D. 1935. Fishery for red snappers and groupers in the Gulf of Mexico. Investigational Reports of the United States Bureau of Fisheries 26:1–29.

Lyles, C. 1965. Fishery statistics of the United States. 1963. Bureau of Commercial Fisheries. Fish and Wildlife Service, Statistical Digest 57.

Porch, C. E. 2007. An assessment of the red snapper fishery in the U.S. Gulf of Mexico using a spatially-explicit age-structured model. Pages 355–384 *in* W. F. Patterson, III, J. H. Cowan, Jr., G. R. Fitzhugh, and D. L. Nieland, editors. Red snapper ecology and fisheries in the U.S. Gulf of Mexico. American Fisheries Society, Symposium 60, Bethesda, Maryland.

Radcliffe, L. 1920. Fishery industries of the United States. Bureau of Fisheries, Document 892:129–191. Washington, D.C.

SEDAR 7. 2004. Southeast Data Assessment and Review. Stock assessment report of SEDAR 7: Gulf of Mexico red snapper. South Atlantic fishery Management Council, Charleston, South Carolina.

Sette, O. E. 1925. Fishery Industries of the United States. Bureau of Fisheries, Document 1010:263–261. Washington, D.C.

Siebenaler, J. B., and W. Brady. 1952. A high speed manual commercial fishing reel. State of Florida Board of Conservation Technical Series 4. Marine Laboratory, St. Petersburg, Florida.

Smith, H. M. 1894. Report of the division of statistics and method of the fishing. Report of the United States Commissioner of Fisheries 18:183–184.

Smith, H. M. 1895. Report of the division of statistics and method of the fishing. Report of the United States Commissioner of Fisheries 19:68–70.

Stearns, S. 1883. Fluctuations in the fisheries of the Gulf of Mexico and the proposed investigation of them. Bulletin of the United States Fish Commission 3:467–468.

Stearns, S. 1885. The fisheries of Pensacola, Florida. Bulletin of the United States Fish Commission 5:245–247.

Stearns, S. 1887. Some of the fisheries of western Florida. Bulletin of the United States Fish Commission 6:465–467.

Townsend, C. H. 1900. Statistics of the fisheries of the Gulf states. Report of the United States Fish Commission 25:105–169.

Warren, A. F. 1898. The red snapper fisheries; their past, present, and future. Bulletin of the United States Fish Commission 17:331–335.

American Fisheries Society Symposium 60:355–384, 2007

An Assessment of the Red Snapper Fishery in the U.S. Gulf of Mexico Using a Spatially-Explicit Age-Structured Model

Clay E. Porch[1]
National Marine Fisheries Service
Southeast Fisheries Science Center
Sustainable Fisheries Division
75 Virginia Beach Drive, Miami, Florida 33149 USA

Abstract.—Red snapper *Lutjanus campechanus* have been fished in the Gulf of Mexico since before the Civil War. The size and efficiency of the commercial fleet increased greatly during the 1960s, but without a corresponding increase in catch, suggesting that red snapper populations throughout the Gulf of Mexico were by that time fully-exploited and perhaps even overfished. Nevertheless, most assessments of red snapper in the U.S. Gulf of Mexico have been based on data collected since the 1980s owing to a combination of gaps in the catch data and limitations of the models employed. The lack of contrast in the more recent data makes it difficult to develop meaningful estimates of stock status, particularly in relation to abundance-based reference points such as the equilibrium spawning biomass at maximum sustainable yield. This paper presents a flexible age-structured model that includes information dating back to the inception of the fishery. The results suggest that the populations of red snapper in the U.S. Gulf of Mexico are well below the levels corresponding to a spawning potential ratio of 30%. They also suggest the stock will not to recover to that level in the foreseeable future without substantial reductions in both the catch of adults by the directed fleets and the bycatch of juveniles by the offshore shrimp fishery.

Introduction

Red snapper *Lutjanus campechanus* have long been esteemed as one of the finest food-fishes in the Gulf of Mexico. The first fisheries for red snapper developed during the late 1840s to supply local markets in New Orleans (Louisiana) and Mobile (Alabama), but the trade was interrupted by the Civil War (Bortone et al. 1997). The fishery was reestablished after the war by several persistent New Englanders who found that packing red snapper on ice would keep them fresh long enough to ship to lucrative northern markets. Production on a commercial scale appears to have begun in 1872 with the formation of the Pensacola Fish Company (Warren 1898). From there the fishery expanded rapidly, fueled by the growing demand as more and more markets were reached through the improving railroad system. By the turn of the century record landings were being made by dozens of vessels

[1] Corresponding author: clay.porch@noaa.gov

operating throughout the Gulf of Mexico. As one dealer put it, "No man who is willing to buy a red snapper has lacked the opportunity" (Jordan and Evermann 1923).

Troubling signs that all was not well with the red snapper fishery were evident as early as the 1880s. Stearns (1883), for example, remarked that "most of the old fishing grounds are barren, and smacks have to go farther each year to find new ones." Collins (1885) went so far as to say "...it is probable that this species is being more or less rapidly depleted, and the day may not be far distant when the services of the Commission will be needed to keep up the supply, if not to prevent the practical destruction of the important fishery now prosecuted for red snapper." Concerns about depletion and overfishing have echoed through the Gulf ever since.

Several technological innovations introduced after World War II (diesel engines, depth recorders, wire line, and reels) allowed vessels to fish farther away and in deeper waters than ever before. Nevertheless, by the early 1950s most of the snapper banks off west Florida were considered impoverished and catch per unit effort was observed to be declining on the Campeche Banks (Camber 1955). Continued declines in catch per unit effort and in the size of the fish caught raised concerns among fishermen, who attributed the declines to an increase in the number of snapper vessels and an increase in the bycatch of juvenile snapper by shrimp trawlers (Moe 1963; Lyles 1965, Bradley and Bryan 1975). A further blow was dealt during the 1980s when the U.S. fleet was excluded from the Campeche Banks by the Mexican government and had to redirect its effort towards domestic waters.

The first quantitative analysis of the red snapper population in the Gulf of Mexico was conducted by Nelson and Manooch (1982). They estimated fishing mortality rates on the order of 0.6 $year^{-1}$ off Louisiana and 0.23 $year^{-1}$ off West Florida (based on catch-curves applied to age samples from the commercial landings between 1978 and 1979). The estimates for Louisiana in particular were several times greater than commonly assumed levels of adult natural mortality (0.1 to 0.2 $year^{-1}$) and could have been interpreted as evidence of overfishing. Goodyear (1988), using data collected between 1984 and 1986, showed that fishing mortality was not constant with age and probably declined from about 0.75 at age 2 to about 0.3 at age 8 or 9. He also estimated spawning biomass to be less than 5% of unfished levels. At about the same time, Nichols et al. (1987) estimated that upwards of 12 million juvenile (age 0 and age 1) red snapper were killed as a bycatch of offshore shrimp trawlers and Powers et al. (1987) showed that eliminating this bycatch could increase the long term yield to the red snapper fishery by as much as 90%. As a result of these and subsequent studies, the fishery was declared overfished and a series of regulatory amendments were introduced to help rebuild the stock (see Hood et al. 2007, this volume).

Several stock assessments were conducted by the National Marine Fisheries Service during the 1990s, the lastest of these being Schirripa and Legault (1999). Each indicated that the stock remained in an overfished condition and that both the directed harvest and trawl bycatch needed to be reduced substantially. Nevertheless, the results were generally regarded as highly uncertain and managers often opted for total allowable catch limits (TACs) that were at or above the maximum recommended levels (Hood et al. 2007). Much of the uncertainty can be attributed to the fact that the assessments were based only on recent data collected since the 1980s, when red snapper populations are likely to have already been depressed. The lack of contrast in such data, coupled with the lack of historical perspective, makes it difficult to develop meaningful estimates of stock status, particularly in relation to abundance-based reference points such as maximum sustainable yield.

The restriction to data collected since the 1980s is a consequence of both gaps in the available historical record and limitations of the methodology employed. The former has partly been addressed by Porch et al. (2007, this volume), who have supplemented the extant landings database with observations gleaned from historical documents. The latter can be addressed by the use of modern age-structured statistical algorithms that can accommodate missing data and incorporate auxiliary data sources that cannot be tapped by simpler methodologies. The purpose of this paper is to present the age-structured statistical algorithm CATCHEM_AD; its mathematical underpinnings and two particular applications of it that provided the basis for much of the management advice generated during the latest assessment (SEDAR 2004; Cordue 2005). As such, the reader should appreciate that the paper covers only a portion of the work that was presented during the lengthy SEDAR assessment process and is not intended to supersede that assessment. In addition, certain features that were not used in the original assessment, but which may become useful in the future, are discussed in anticipation of forthcoming information on red snapper movement and recruitment patterns.

Model Structure

The basic units in the model are stock, cohort, fleet, habitat, era, year, season and age-class. As used here, a "stock" designates any population of animals with similar life history characteristics for which a distinct accounting is desired. For example, a stock may be identified with a growth-morph, sex, or species. The term "cohort" is used to refer to members of a given stock that were born during the same year and season. A "fleet" refers to any entity that harvests or samples any of the stocks (e.g., a fishery, scientific survey or predator species). The term "habitat" represents any form of spatial domain where the concentration of the stock or fleet may vary in important ways from the overall mean density. The time unit "era" is used to distinguish three periods of exploitation: a "prehistoric" era, during which no data are available; a "data" era, when presumably there are data useful for estimating abundance and mortality; and a "future" era, when mortality rates are assumed (input). The duration of the prehistoric era is set equal to the number of seasonal age-classes so as to generate a complete age-structure by the beginning of the first year of the data era. The time units "year" and "season" are used in the conventional sense, except that the number of seasons per year may be defined by the user. The calculations are done on a seasonal basis to accommodate seasonal movement and fishing patterns and to mimic the effect of temporally protracted spawning by allowing for multiple cohorts per year. The model tracks the abundance of each cohort throughout its life span as shown in Table 1.

The age-classes range from 1 to A, where r is the age (in seasons) associated with age-class 1 and subsequent age-classes are incremented forward by one season. The last age-class, A, is not cumulative, i.e., fish are assumed to have a maximum life span of $A + r - 1$ seasons. The calendar year y and season s are inferred from the cohort c and age-class a as

$$\text{(1)}\quad y\{c,a\} = y_0 + \text{int}\left\{\frac{c+a-2}{n\{s\}}\right\}$$

$$s\{c,a\} = c+a-1-\text{int}\left\{\frac{c+a-2}{n\{s\}}\right\}n\{s\}$$

where y_0 is the first year of the prehistoric period, $\text{int}\{x\}$ is the integer portion of quantity x and $n\{s\}$ is the number of seasons in a year. Hereafter, the notation $\{c,a\}$ will be omitted for compactness, with the implicit under-

Table 1. Method of accounting used in the stock assessment algorithm. The entries represent a cohort, with cohort 1 being born in season 1 of year 1, cohort 2 being born in season 2 of year 1, and so on. In this example there are three years of data with each year having two seasons, and therefore there are six seasonal age-classes. Thus, in order to have a complete age-structure by the first season of the data period (season 1 of year 4), it is necessary to track the first five cohorts recruited immediately prior to the data era.

		Prehistoric era						Data era						Future era	
Year		1	1	2	2	3	3	4	4	5	5	6	6	7	8...
Season		1	2	1	2	1	2	1	2	1	2	1	2	1	2...
Age	1	1	2	3	4	5	6	7	8	9	10	11	12	13	14...
	2		1	2	3	4	5	6	7	8	9	10	11	12	13...
	3			1	2	3	4	5	6	7	8	9	10	11	12...
	4				1	2	3	4	5	6	7	8	9	10	11...
	5					1	2	3	4	5	6	7	8	9	10...
	6						1	2	3	4	5	6	7	8	9...

standing that s and y are derived quantities. Otherwise, curly braces are used throughout to distinguish function arguments from calculation precedence.

Population Dynamics Model

The progression from one age-class to the next is modeled as

$$(2) \qquad N_{cah} = \sum_{j} \tilde{N}_{caj}\, T(h|j,a,s)$$

$$(3) \qquad \tilde{N}_{c,a+1,h} = \begin{cases} R_c \tau_h & a = 0 \\ N_{cah} e^{-Z_{asyh}} & 1 \le a < A \\ 0 & a \ge A \end{cases}$$

where the subscript h (or j) identifies habitat. The subscript indexing stock has been omitted for convenience of notation, but the equations should be understood to depend on stock as well. The variable R_c denotes the initial recruitment to cohort c, τ_h denotes the probability that a new recruit will start out in habitat h, and Z denotes the instantaneous total mortality rate. The variables $\tilde{N}_{cah}$ and N_{cah} denote the number of survivors in habitat h before and after a movement event, which is assumed to occur instantaneously at the beginning of the season (T denotes the transfer probability described below). Note that the movements of red snapper between habitats were not modeled in this paper owing to insufficient data. However, a brief description of the underlying model is provided here in anticipation of future applications with more informative data.

Movement.— The conditional probability T that a fish will transfer to habitat h given its current location j is modeled as a diffusive process where the net pull towards a given habitat is a function of the difference between the intrinsic attraction of a habitat (β_1) and the difficulty in getting to it (β_2):

$$(4) \quad T(h|j) \propto e^{-(\beta_2 - \beta_1)}$$

Specifically, β_1 is expressed as a categorical variable that varies by habitat, age-class and season and β_2 is expressed as the effective distance between habitats (x_{hk}) divided by the diffusion velocity of each age-class in distance units per season (u_a, which may or may not be proportional to swimming speed):

$$(5)\quad T(h|j,a,s) = \frac{\tau_{ash} e^{-x_{hj}/u_a}}{\sum_i \tau_{asi} e^{-x_{ij}/u_a}}$$

The summation occurs over all possible habitats, ensuring that the all fish are accounted for, i.e., $\Sigma_h T(h|j,a,s) = 1$ for all j,a,s. Essentially, this is a discrete version of the Joseph and Sendner (1958) diffusion equation immersed in an inhomogeneous advection field. The τ parameters can be thought of as the relative distribution of the cohort among habitats that would be achieved with an infinite diffusion velocity u. Purely diffusive motion is achieved when the τ parameters are identical and the matrix of distance parameters x_{hj} is symmetric ($x_{hj} = x_{jh}$).

Mortality.—The instantaneous mortality rate Z is modeled as the sum of coefficients reflecting natural (M) and fishing-related (F) causes:

$$(6)\quad Z_{asyh} = M_a + \sum_i F_{iasyh}$$

where i indexes a particular source of fishing mortality, hereafter referred to as a fleet. The fishing mortality rate parameters are further decomposed into separable age-dependent and time-dependent effects:

$$(7)\quad F_{iasyh} = q_{iy} v_{ia} f_{iy} \xi_{iay} \delta_{ish} / n\{s\}$$

where q represents the "catchability" of the most vulnerable age-class, v represents the relative vulnerability of the remaining age-classes, f is the total effort exerted by the fleet over all seasons, and ξ is the probability that a fish will die once it is caught (i.e., the probability it was either landed or suffered post-release mortality). The quantity δ_{ish} equals 1 or 0 depending on whether the fleet does or does not operate in season s or habitat h. Essentially this model assumes that the fishing effort of any given fleet may vary from year to year, but is otherwise spread evenly over the seasons and habitats the fleet is operating. The vulnerability parameters implicitly include the effects of factors such as gear selectivity and the fraction of the stock exposed to the fishery.

Inter-annual variations in f and q are modeled as first-order, lognormal auto-regressive processes, e.g.,

$$(8)\quad \begin{aligned} f_{iy} &= \mu\{f_{iy}\} e^{\varepsilon_{iy}} \\ \varepsilon_{iy} &= \rho\{f_{iy}\} \varepsilon_{i,y-1} + \eta_{iy} \end{aligned}$$

where μ and ρ represent the median and correlation coefficient of the f_{iy}, respectively, and the η_{iy} are normal distributed random variables with mean zero and standard deviation $\sigma\{f_{iy}\}$. In the present application, the $\mu\{f_{iy}\}$ are model inputs, hopefully based on some index of the relative effort expended by each fishery. For σ sufficiently large, the f_{iy} essentially become free parameters and the values of $\mu\{f_{iy}\}$ become arbitrary. However, the absence of data during the 'prehistoric period' generally precludes the estimation of unconstrained changes in effort. Accordingly, for the prehistoric period σ is set to 0, such that $f_{iy} = \mu\{f_{iy}\}$.

Recruitment and the definition of spawning potential.—The recruitment to the first age-class of each cohort (R) is modeled as a first-order, lognormal auto-regressive process,

$$(9)\quad \begin{aligned} R_c &= \mu\{R_c\} e^{\varepsilon_y} \\ \varepsilon_y &= \rho\{R_c\} \varepsilon_{y-1} + \eta \end{aligned}$$

where μ is the median recruitment, ρ is the correlation coefficient and η is a normally distributed random variate having mean 0 and standard deviation $\sigma\{R_c\}$ (ostensibly representing the effect on recruitment of fluctuations in the environment). The recruitment standard deviation and correlation coefficient can be difficult to estimate without a good index of recruitment and may have to be fixed to some moderate values (say $\sigma_R = 0.4$ and $\rho = 0.5$.)

The median recruitment can be a constant or specified as truncated Ricker or Beverton–Holt functions that have been recast in terms of the maximum lifetime reproductive rate (α), virgin recruitment during peak season (R_0) and spawning potential relative to virgin levels during peak season (ϕ):

$$(10)\quad \mu\{R_c\} = \begin{cases} R_0\phi_c\alpha^{\,1-\phi_c} & \text{Ricker} \\ R_0\dfrac{\alpha\phi_c}{1+(\alpha-1)\phi_c} & \text{Beverton and Holt} \end{cases}$$

(see appendix 1). Note that α is related to the so called "steepness" parameter (ψ, the recruitment when spawning potential is 20% of the unfished level relative to R_0):

$$(11)\quad \psi = \begin{cases} 0.2\alpha^{4/5} & \text{Ricker } (0.2 \le \psi) \\ \dfrac{\alpha}{\alpha+4} & \text{Beverton and Holt } (0.2 \le \psi < 1) \end{cases}$$

The parameters R_0 and α are generally estimated during the model fitting (see below), whereas the spawning potential ϕ is derived from other modeled quantities. In the case of a single unit stock and a single habitat, the calculation of ϕ is straightforward,

$$(12)\quad \begin{aligned} \phi_c &= S_c / S_0 \\ S_c &= \sum_a E_{as} N_{\hat{c}a} \qquad (\hat{c} = c - r - a + 1) \\ S_0 &= R_0 \sum_a E_{a\hat{s}} \exp\left(-\sum_{j=1}^{a-1} M_j\right) \end{aligned}$$

Here S_c represents the spawning potential that contributed to cohort c and S_0 is the spawning potential of the unexploited state during the peak spawning season (indexed by the subscript $\hat{s}$). The variable E_{as} represents a measure of the per-capita number of eggs produced by each age-class. The subscript $\hat{c}$ indexes the cohort that was age a at the time of spawning (r seasons prior to the recruitment of cohort c).

When there are multiple stocks and multiple habitats a number of alternatives present themselves. One extreme is to assume that all members of stock k contribute to the net spawning potential of that stock regardless of their current location, as might occur if the adults generally migrate back to the spawning habitat or the larvae are spatially well-mixed. In that case

$$(13)\quad \begin{aligned} \phi_{ck} &= S_{ck} / S_{0k} \\ S_{ck} &= \sum_a E_{ask} \sum_h N_{\hat{c}ahk} \\ S_{0k} &= R_{0k} \sum_a E_{ask} \exp\left(-\sum_{j=1}^{a-1} M_{jk}\right) \end{aligned}$$

Alternatively, one might wish to model a situation where the members of the various stocks spawn more opportunistically. For example, if all adult fish located in a spawning habitat during the spawning season contribute to the spawning potential of the habitat, regardless of their original stock affiliation, then one obtains:

$$(14)\quad S_{ck} = \sum_a \sum_j E_{asj} N_{\hat{c}a\hat{h}\{k\}j}$$

where the subscripts j and k are used to index stocks and $\hat{h}\{k\}$ identifies the spawning habitat associated with stock k. The expression for S_{0k} is tedious to write in this scenario, but can be obtained from equations (2) and (3) with recruitment fixed to the stock specific values of R_{0k} and zero fishing mortality. Various scenarios in between (13) and (14) may be admitted by choosing (14) and altering the movement coefficients such that some fraction of the stock migrates into the assigned spawning habitats.

Data Models

The basic data structure in the model is the "fleet," which is defined here as an entity with relatively constant selection characteristics (i.e., vulnerability coefficients). In this sense a fleet can include a collection of individuals with different selection habits as long as the aggregate selection pattern does not vary much through time. Fishery-independent surveys may be regarded as fleets with negligible catch. Predators other than humans may also be treated as a "fleet" if there are some data relating to their consumption of the stocks in question.

The basic catch equation for each fleet is

$$(15) \qquad C_{iasy} = \sum_h \frac{F_{iasyh}}{\xi_{iay} Z_{asyh}} \tilde{N}_{cah} (1 - e^{-Z_{asyh}})$$

The term catch here refers to the total number of fish caught, including both the number landed and the number released. The fishing mortality rate coefficient F, on the other hand, refers to the fraction that are killed and explicitly includes ξ (equation 7); therefore the ξ terms cancel out in equation (15). In the present application there are four basic types of data associated with the seasonal catches of each fleet– total catch C_{isy}, an index of abundance I_{isy}, age composition p_{iasy} and length composition p_{ilsy}:

$$(16) \quad C_{isy} = \sum_a C_{iasy}$$

$$(17) \quad I_{isy} = Q_i C_{isy} / f_{isy}$$

$$(18) \quad p_{iasy} = C_{iasy} / C_{isy}$$

$$(19) \quad p_{ilsy} = \sum_a p_{iasy} g\{l|a\}_{isy}$$

where Q is a scaling coefficient used when the units of the index are not consistent with the units used for C or f, and g is a function of the growth parameters that expresses the probability that a fish from age-class a is length l. Note that the calculations are made over the entire life span of each cohort, but provision is made for cases where the data are censored (e.g., when the last category in the data are cumulative for fish older than a certain age or larger than a certain size).

One issue of concern is how best to deal with the situation where some fraction of the catch is discarded (released) and subsequently dies, i.e., how to parameterize ξ. Under the presumption that discarded fish are mostly below the size limit L,

$$(20) \quad \xi_{iay} = 1 - (1 - d_{ias}) G_{L|a}$$

where d is the fraction of released fish that die and $G_{L|a}$ is the probability that a captured fish will be smaller than the size limit given that it is age a. Estimates of commercial landings (harvest H) are often available, but seldom the number caught (C) or discarded (D). Assuming discarded fish are mostly below the size limit, one obtains

$$(21a) \quad H_{iasy} = (1 - G_{L|a}) C_{iasy}$$

$$(21b) \quad D_{iasy} = G_{L|a} C_{iasy}$$

The total number killed (K) is simply the sum of the harvest and the number of discarded fish that die ($d_{ias} D_{iasy}$).

Reference Points

Current law requires federal fishery management plans to include an "MSY control rule" that is comprised of two limit reference points known as the maximum fishing mortality threshold (MFMT) and the minimum stock size threshold (MSST). When the abundance of the stock drops below the MSST, provisions must be made to rebuild the stock to the level that would support the maximum sustainable yield (MSY), or some proxy thereof, within a time frame that is as short as possible commensurate with the productivity of the stock and the needs of the fishing community. A common practice of the Gulf of Mexico Fishery Management Council has been to set MFMT equal to the fishing mortality rate that will produce MSY (F_{MSY}) and to set MSST equal to a fraction (1-M) of the spawning potential required to sustain MSY (S_{MSY}).

The computation of reference points such as S_{MSY} is complicated by the existence of multiple fleets operating in multiple habitats on multiple stocks. For example, the MSY obtained by maximizing over all stocks simultaneously will generally be lower than the sum of values obtained when each stock is treated as though it were harvested independent of the others. Similarly, maximizing over all fleets can lead to a situation where fleets that are less efficient in terms of yield are allocated negligible effort. The approach taken here is to fix the relative allocation of effort among the 'directed fleets' and absolute allocation of effort among 'bycatch fleets' to the values desired by managers (say some particular fraction of the estimates for recent years). This tact admits the possibility that the management policies relating to the bycatch fleets may be based on considerations other than their take from the stocks in question. The MSY is then computed by rescaling the combined effort of the directed fleets so as to maximize their landings over the selected stocks and habitats. Strictly speaking, such a procedure does not produce the MSY because it is conditioned on a potentially suboptimal allocation strategy (Goodyear 1996; Powers 2005). A better term would perhaps be the maximum sustainable marginal yield (MSMY).

A difficulty with employing reference points based on MSY or MSMY concepts is that they can lead to a situation where less productive stocks are extirpated as a consequence of optimizing the exploitation of more productive stocks. A less risk-prone policy would be to adopt a strategy based on maintaining the spawning potential ratio for each stock above some predetermined minimum. The spawning potential ratio (SPR) is defined as the lifetime spawning potential of each recruit expected with a given level of fishing divided by the corresponding expectation with no fishing (Goodyear 1993). In the simplest case of one fishery, one stock, one season and one habitat it is computed as

$$(22) \quad SPR = \frac{\sum_{a=1}^{A} E_a e^{-\sum_{j=0}^{a-1} Fv_j + M_j}}{\sum_{a=1}^{A} E_a e^{-\sum_{j=0}^{a-1} M_j}}$$

More generally, SPR is equivalent to ϕ when recruitment is constant. Thus, it may be calculated for various combinations of fishing effort by initializing the recursion implied by equation (14) with the same arbitrary recruitment values for all scenarios (including no fishing).

Parameter Estimation

A Bayesian approach to estimation is adopted wherein one seeks to develop a 'posterior' probability density for the vector of parameters Θ that is conditioned on the data, $P(\Theta \mid data)$. By application of Bayes rule it is easy to show that

$$P(\Theta \mid data) \propto P(data \mid \Theta)\, P(\Theta)$$

where $P(data \mid \Theta)$ is the sampling density (likelihood function) and $P(\Theta)$ is the prior density (in this case the analyst's best guess of the probability density for Θ). Estimates for Θ may be obtained by integrating the posterior (classical Bayes moment estimator)

$$(23) \quad \bar{\theta}_i = \int \theta_i\, P(data \mid \Theta)\, P(\Theta) d\theta_i \quad , \quad \theta_i \in \Theta$$

or, as done here, by minimizing its negative logarithm (highest posterior density estimator, Bard 1974)

$$(24) \quad \min_{\Theta} \left\{ -\log_e P(data \mid \Theta) - \log_e P(\Theta) \right\}$$

Sampling densities.—Sampling densities, also known as likelihood functions, measure the disparity between the model predictions and observed data. Catch, index and effort data are assumed to be normal or lognormal distributed, e.g.,

$$(25)\quad -\log_e P(\boldsymbol{C}|\Theta) = \begin{cases} 0.5\sum_i\sum_y\sum_s\left(\dfrac{C_{isy}^{obs}-C_{isy}}{\sigma\{C_{isy}\}}\right)^2 - \log_e\left(\sigma^2\{C_{isy}\}\right) & normal \\ 0.5\sum_i\sum_y\sum_s\left(\dfrac{\log_e(C_{isy}^{obs}/C_{isy})}{\sigma\{\log_e C_{isy}\}}\right)^2 - \log_e\left(\sigma^2\{\log_e C_{isy}\}\right) & lognormal \end{cases}$$

where the superscript *obs* distinguishes the observed data from the value predicted by the model. The variable $\sigma\{\}$ is the standard deviation of the enclosed quantity.

Data describing the age and length composition of a sample ought to be multinomially distributed provided measurement error is low and the samples were truly taken at random. In that case, the appropriate log-likelihood functions for the age and length composition of the catch are

$$(26)\quad \begin{aligned} -\log_e P(\boldsymbol{p}_a|\Theta) &= \sum_i\sum_y\sum_s n_{isy}\sum_a p_{iasy}^{obs}\log_e p_{iasy} \\ -\log_e P(\boldsymbol{p}_l|\Theta) &= \sum_i\sum_y\sum_s n_{isy}\sum_l p_{ilsy}^{obs}\log_e p_{ilsy} \end{aligned}$$

where again the superscript *obs* distinguishes the observed data from the value predicted by the model and n indicates the effective sample size input by the analyst.

Prior densities.—Prior densities are similar to sampling densities in that they measure the disparity between the model predictions of a parameter and other information known about it. The difference is that sampling densities express the probability of observing some information (data) given the model estimates, whereas prior densities express the probability of observing the model estimates given some information (prior knowledge). When possible prior densities should be based on data, otherwise one may choose to adopt functional forms that are relatively uninformative over the plausible range of parameter values. For example, the logarithm of the natural mortality rate might be treated as uniformly distributed between –5 and 2. The primary advantage of using uninformative priors is that the potential for introducing biases is minimized. On the other hand, if the data relating to a particular parameter are too sparse, the solution may be so uncertain as to be rendered meaningless. This observation has led some to develop prior densities based on expert opinion (e.g., Wolfson et al. 1996; Punt and Walker 1998) or analyses of other species (e.g., Liermann and Hilborn 1997; Maunder and Deriso 2003).

One parameter of special concern in the analysis of Gulf of Mexico red snapper is the steepness of the stock–recruitment relationship. Previous analyses have estimated this parameter to be close to the mathematical limit of 1.0 (Anonymous 1999), suggesting it may not be well determined. A possible alternative is to develop a prior based on a subset of the values collected by Myers et al. (1999) that corresponds to larger, highly fecund fishes with long life spans (the 'periodic' strategists of Rose et al. 2001). There is, of course, the potential for introducing bias when one or more of the priors are based on expert opinion or otherwise subjective information. However, the same sorts of bias can be introduced by conducting sensitivity analyses where the unknown parameters are fixed to various values selected by the analysts. It might be best to incorporate this uncertainty in a more rigorous fashion.

Covariance parameters.—It is not generally possible to obtain consistent estimates for all of the elements of the covariance matrix associated with the objective function, i.e., the correlation coefficients and variances. As is typically done in most stock assessments, each individual data record is assumed to be independent of the other data, implying the correlation coefficients are zero. In the case of the fishery (survey) data, the variances associated with sampling variability are often estimated extraneous to the population model (e.g., during the standardization procedure). However, there may be additional variance owing to fluctuations in the distribution of the stock relative to the survey habitat (IWC 1994). To accommodate such possibilities, the variance parameters for the catches and indices of abundance may be modeled as

$$\begin{aligned} \sigma^2\{C_{iy}\} &= \chi^2\{C_{iy}\} + \lambda\{C_i\}\sigma^2 \\ \sigma^2\{I_{iy}\} &= \chi^2\{I_{iy}\} + \lambda\{I_i\}\sigma^2 \end{aligned} \tag{27}$$

where the χ^2 are the annual observation variances associated with each type of data (estimated outside the model), σ^2 reflects some overall process variance (estimated within the model), and the λ are constant multipliers (usually fixed by the analyst based on a careful consideration of the inherent variability of the underlying processes). The variances corresponding to the age and length composition data are implicit functions of sample size, which is controlled on input.

The model has been implemented using the nonlinear optimization package AD model Builder (Otter Research Ltd.[2]), which provides facilities for estimating the mode and shape of the posterior distribution.

Application to Red Snapper

Model Structure

The model described above was applied to information on red snapper populations in the U.S. Gulf of Mexico during the years from 1872 to 2003. Five fisheries were designated for each of two regions east and west of the Mississippi River: handline, longline (after 1979), recreational (after 1945), closed-season bycatch (after 1990) and offshore shrimp trawl bycatch (after 1947). Three four-month seasons were modeled, starting in January. Spawning was assumed to occur during the second season. Each cohort was modeled for 30 years, the contribution of very old animals being assumed negligible. The initial age structure of the populations east and west of the Mississippi River were set to virgin levels as the fishery that existed prior to 1872 was by all accounts negligible. Subsequent recruitment of age 0 or age 1 fish was modeled as a Beverton–Holt function of the relative number of eggs produced (with a one-year lag in the case of age-1). Recruitment was allowed to deviate from the Beverton–Holt expectations according to equation (9) with $\rho = 0$ and CV = 0.4 (no deviations were estimated prior to 1930 owing to insufficient data). Separate recruitment parameters were estimated for the populations east and west of the Mississippi River. Note that the choice of starting with age 1 implies that density-dependent natural processes constitute the dominant source of mortality during the first year of life, whereas starting with age 0 implies that density-independent mortality sources (such as shrimp bycatch) dominate after settlement.

Parameter specifications.—A total of 861 parameters were estimated; 4 for the two spawner-recruit curves, 210 recruitment deviations, an overall variance parameter, and 646 parameters pertaining to the ten "fleets." The vulnerability and catchability coefficients for each specific fleet were assumed to be relatively unchanged through time, but allowed to vary with age and among fleets (totaling 76 estimable parameters). The effect of minimum size limits was modeled explicitly as described above. While SEDAR 7 participants recognized that improvements in fishing gear since the 1870s may have resulted

[2] Otter Research Ltd. 2001. An introduction to AD MODEL BUILDER Version 4.5. Box 2040, Sidney B.C. V8L 3S3, Canada.

in changes in the vulnerability schedule, it was felt that these changes were relatively small and that the major effect would be an increase in the efficiency (effective effort) of each fleet. Accordingly, the effective effort of each fleet was allowed to vary by year essentially as a free parameter (556 estimable parameters). The scaling coefficients Q for each index of abundance were assumed to be constant through time (12 parameters). The vulnerability coefficients for the fishery independent surveys were fixed to 0 for age 1 and 1.0 for ages 2 and older. Natural mortality was fixed to 0.98 $year^{-1}$ for age-0, 0.6 $year^{-1}$ for age-1 and 0.1 $year^{-1}$ thereafter (SEDAR 2004). The degree of intermixing between the populations of red snapper east and west of the Mississippi was assumed to be negligible.

The fecundity at age (including maturity) was set to the vector derived from the age-conditioned model described by Porch (2004), normalized to have a maximum value of 1 at age 30. Thus, the spawning potential (S) of the adult population is not expressed in terms of the number of eggs produced, but in terms of the effective number of fully-productive spawners. In other words, S is the number of age 30 animals required to produce the same number of eggs as the estimated adult population.

Likelihoods and priors.—The catch, effort, and relative abundance indices were assumed to be approximately lognormal distributed. Age composition was assumed to be multinomial distributed. A lognormal prior (Nowlis 2004a) was imposed on α with a median value of 13.3 and log-scale variance of 1.28 (equivalent to a mean steepness of about 0.86). The remaining parameters were treated as free parameters constrained to lie with bounds that encompassed the range of plausible values (essentially the same as specifying uninformative priors over the feasible range).

Data employed

Landings data.—The commercial landings from 1963 to 2003 are discussed by Turner et al. (2004) and the landings prior to 1963 are discussed by Porch et al. (2004) and Porch et al. (2007). The annual recreational harvest since 1981 is based on the NMFS Marine Recreational Fisheries Statistics Survey[3] (MRFSS), Texas Parks and Wildlife Survey and NMFS headboat survey as described by Turner et al. (2004). The recreational harvest statistics used for earlier years (1946–1980) were reconstructions based on U.S. census data (Scott 2004). It is assumed that prior to 1946 the recreational take was negligible in comparison to the commercial take owing to the relative inaccessibility of the fishing grounds (powered vessels were few and expensive, making offshore trips mostly a past time for the wealthy). The bycatch of juveniles from the offshore shrimp fishery is based on the series produced by Nichols (2004), which extends back to 1972. A time series of offshore shrimping effort, which extends back to the advent of the offshore shrimp fishery in 1948, was also used to tune the model (see Porch and Turner 2004). The catch during the closed season was derived by Turner et al. (2004).

The discards from the recreational and commercial fleets during the open season were assumed to occur predominantly due to the regulations on minimum size. They were computed on a seasonal rather than annual basis to better accommodate the rapid growth exhibited by younger red snapper. The population growth curve and coefficient of variation of length about age were fixed to the values estimated by Diaz et al. (2004).

The CVs used to weigh the landings data were fixed at 0.1 (arbitrary low value) for the commercial fleets inasmuch as they rep-

[3] Marine Recreational Fisheries Statistics Survey. National Marine Fisheries Service. Fisheries Statistics and Economics Division, 1315 East-West Highway, Silver Spring, MD 20910.

resent a census. The exceptions are for years when no census was taken, in which case the effective CVs were computed from the census estimates immediately before and after the year in question (absolute difference divided by the mean); the reasoning being that the true value likely lies somewhere between those values. The CVs for the recreational catches after 1981 came from the variance estimates produced by the MRFSS (G. Diaz, personal communication); the CVs for the catch inputs prior to 1982 were assigned arbitrary high CVs (1.0) inasmuch as they were not actually observed. The CVs for the shrimp bycatch are based on the CVs of the overall index (ages 0–2), but modified by the proportion that are not age zero (see Porch and Turner 2004). An additional process variance term was not included for the catch (cf. Equation (27); instead it is assumed that process variations in catch are adequately modeled by inter-annual deviations in recruitment and fishing mortality rates.

Indices of abundance.—Ten indices of abundance were used, 5 for each region (east or west). These include the handline CPUE series based on log books (McCarthy and Cass-Calay 2004), the MRFSS recreational indices (Cass-Calay 2004), SEAMAP larval indices (Lyczkowski-Shultz et al. 2004), SEAMAP trawl survey (Nichols et al. 2004; Turner and Porch 2004) and video surveys (Gledhill and Ingram 2004). The handline logbook indices were modeled in this case as landings per unit effort (in pounds) rather than catch per unit effort, thereby taking into account the potential discards owing to the minimum size limit and removing the major objection to their use by the SEDAR 7 Data Workshop participants. The SEAMAP larval indices were assumed to index spawning potential and the video surveys were assumed to index the combined abundance of ages 2 and older.

The CVs for the indices of abundance are based on the year-specific estimates that come from the GLM-based procedures used to standardize them (see the references cited above). These are regarded as representing observation variance. To this the model adds an internally-estimated process variance term, which is intended to represent random discrepancies between the trends in the indices and the trends in the actual population it purports to track (see equation (27).

Index values were rescaled to approach the magnitude expected for the population to facilitate the estimation of the catchability scaling parameters (e.g., it allows the initial guesses for q to be set to 1 and makes setting the upper and lower limits more intuitive).

Age composition.—The age composition data (and effective sample sizes) used for the commercial and recreational fisheries are described by Nowlis (2004b). Inasmuch as the model makes seasonal calculations with spawning occurring during the second season (mid-year), the data for each year were aggregated by the actual integer age in years (it is not necessary to shift the ages by 0.5 to track cohorts as VPA and ASAP must do). The age composition for the shrimp bycatch was based on model output from Nichols (2004). The age composition used for the closed season is described by Turner et al. (2004).

Results

Model fits to data

The age 0 and age 1 models both matched the total catch data quite well with the exception of the 1983 peak in the eastern recreational catch series and the high shrimp bycatch during some of the early years (Figure 1). The fit to these values appears poor because confidence in the data are low (i.e., the CVs associated with the data are high) and the model explicitly discounts the importance placed on them. The model also fit most of the indices of

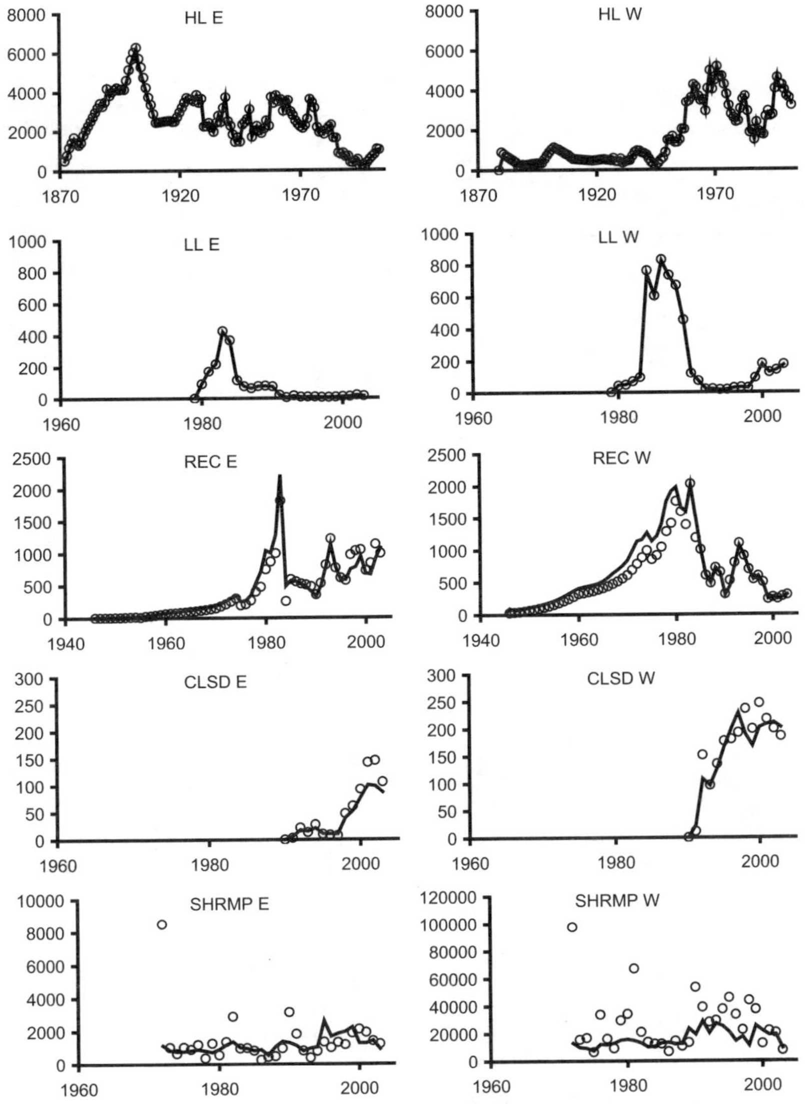

Figure 1. Age 0 model fits (lines) to the total landings (points) for each of the five "fleets" in the regions east (E) or west (W) of the Mississippi River. Figures for the handline (HL) and longline (LL) fleets are in thousands of pounds landed. Figures for the recreational fleet (REC) are in thousands of fish landed. Figures for the closed season (CLSD) and shrimp bycatch are in thousands killed. The fits obtained with the age-1 model were very similar and therefore not shown.

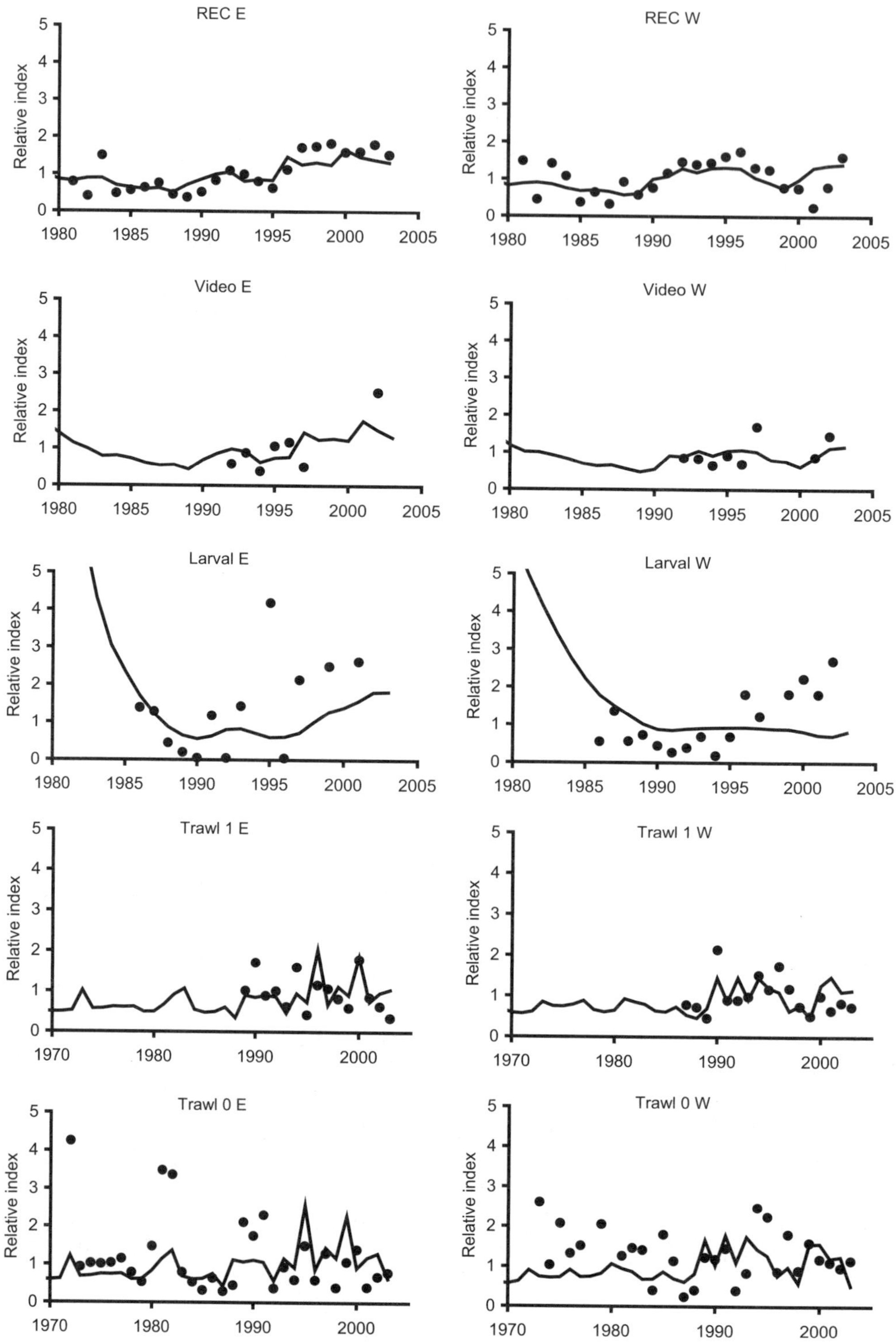

Figure 2. Age 0 model fits to indices of abundance (rescaled by the mean of the predicted values).

abundance reasonably well (Figure 2), but could not reconcile the increasing trend in the western larval index (representing spawners) with the flat or declining trends indicated by the other western indices. The model fits to the SEAMAP trawl series show a strong residual pattern where the predictions for the early years are considerably lower than the trawl values, but the predictions for the later years are considerably higher. The mismatch for the early years can be attributed to the low confidence (high CVs) associated with those data. The mismatch in more recent years reflects the influence of the bycatch data, which, in the context of relatively constant or declining effort, suggests recruitment generally has increased in recent years. The shrimp effort series were well fit (Figure 3) owing to the low CVs assigned to those data (10%).

The fits to the age composition data, aggregated over all years, appear to be quite good (Figure 4). It should be kept in mind, however, that the fits to individual years are noisier, particularly where the sample size was small.

Parameter estimates

A list of the time-invariant parameters with their highest posterior density estimates and asymptotic standard errors (computed from the inverse of the Hessian matrix) is included in Table 2. The estimates for most of these parameters appear to be reasonably precise, generally having coefficients of variation (standard error divided by the point estimate) of less than 30%. The estimates for the numerous age or time-varying parameters and associated derived quantities are shown graphically rather than tabulated (see below). The corresponding coefficients of variation were mostly less than 30%.

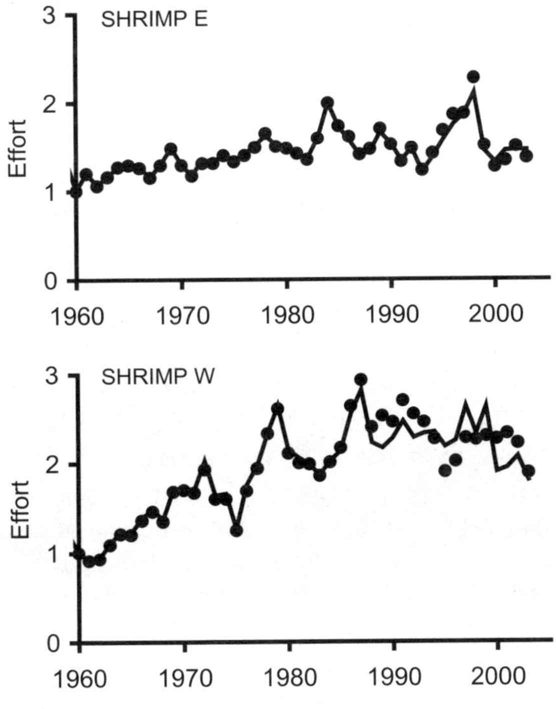

Figure 3. Age 0 model fits to the shrimp trawl effort series.

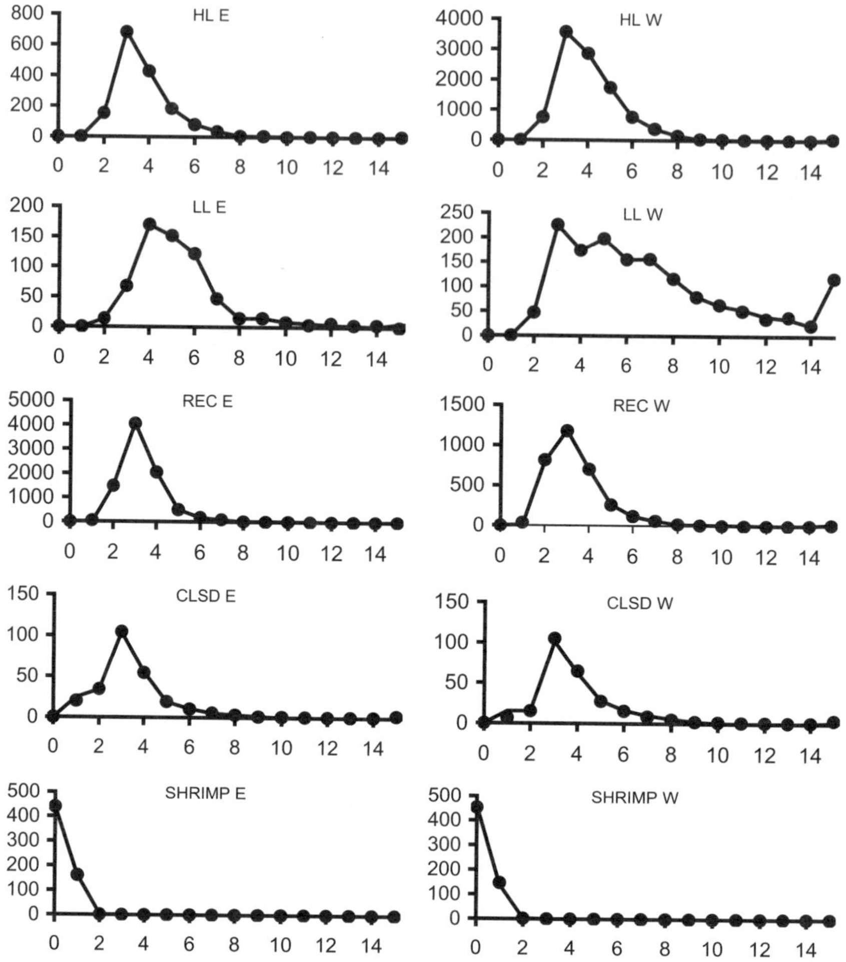

Figure 4. Age 0 model fits to the age composition data (aggregated across years).

The estimated vulnerability and apical (fully-selected) fishing mortality rates F for the age-0 and age-1 base models are shown in Figures 5 and 6, respectively. (Recall that the vulnerability coefficients reflect the probability of being caught rather than the probability of being landed, which also depends on the probability of being greater than the size limit.) The vulnerability of red snapper to the recreational and commercial hand line fleets follows a dome-shaped pattern with a peak at age 1 or 2 for the former and at age 5 for the latter. The vulnerability of red snapper to the commercial long line fleet follows a logistic pattern with older animals (10+) being the most vulnerable. The vulnerability patterns for the closed season "fleets" were between the hand line and longline. As expected, age 0 and age 1 fish were much more vulnerable to shrimp trawls than age 2 or older.

The estimated trends in F indicate persistent increase for all fleets. Although the recreational fishing rate in the east appears to have declined markedly in recent years, it remains at rather high levels. The highest rates were associated with the western shrimp bycatch followed by the eastern recreational and western commercial handline fisheries. Note,

Table 2. Time-invariant parameters estimated in the models with corresponding highest posterior density (HPD) estimates, standard errors and coefficients of variation (CV). The time or age-varying parameters (vulnerability, recruitment and effort) are shown graphically to save space. The letters E and W represent the red snapper *Lutjanus campechanus* fisheries east and west of the Mississippi, respectively. The abbreviations HL, LL, REC, CLSD, SHMP represent the commercial hand line, long line, recreational, cloased season and shrimp (bycatch) fleets, respectively. Trawl 1 and trawl 0 refer to SEAMAP trawl indices for age 1 and age 0 red snapper. The estimates for α were constrained by the imposed upper boundary condition of 150; therefore the CVs are near zero.

Parameter	HPD estimate	Std. error	CV (%)
R_0 E	6.58E+06	5.23E+05	8
α E	1.50E+02	2.27E-01	0
R_0 W	2.84E+07	2.36E+06	8
α W	1.50E+02	4.38E-01	0
q HL E	7.98E-02	1.70E-02	21
Q HL E	5.25E-02	1.97E-02	38
q HL W	9.07E-03	1.56E-03	17
Q HL W	5.27E-02	1.08E-02	20
q LL E	9.50E-03	3.29E-03	35
q LL W	7.60E-03	2.35E-03	31
q REC E	1.54E-02	3.78E-03	25
Q REC E	1.07E-01	1.56E-02	15
q REC E	1.33E-02	2.75E-03	21
Q REC W	3.26E-02	5.27E-03	16
q CLSD E	7.82E-03	3.34E-03	43
q CLSD W	9.34E-03	3.84E-03	41
q SHMP E	9.18E-02	1.12E-02	12
q SHMP W	3.17E-01	2.54E-02	8
Q Video E	1.20E-02	3.30E-03	27
Q Video W	5.22E-02	7.90E-03	15
Q Larval E	3.69E+00	8.88E-01	24
Q Larval W	2.77E+01	4.50E+00	16
Q Trawl 1 E	1.81E+00	2.24E-01	12
Q Trawl 1 W	9.62E-01	9.52E-02	10
Q Trawl 0 E	1.53E+00	1.33E-01	9
Q Trawl 0 W	4.66E-01	4.45E-02	10
σ^2	3.24E-01*	2.61E-02	8

*expressed as a coefficient of variation

however, that the high shrimp bycatch rates apply only to ages 0 and 1, whereas the lower apical *F*s estimated for the handline and recreational fleets apply to multiple age classes. Hence, the bycatch and directed fisheries have comparable cumulative impacts.

There does not appear to be a strong relationship between the number of recruits and spawning potential (*S*) in the previous years (see Figure 7). In both runs the estimates of the maximum potential spawn per recruit (α) were near the limit of 151 imposed by the model, which translates to a steepness of 0.974.

Estimated population trends

Estimates of historical trends in spawning potential and recruitment are shown in Figure 7. Prior to exploitation, the western population of red snapper is estimated to have been about three times as large and three

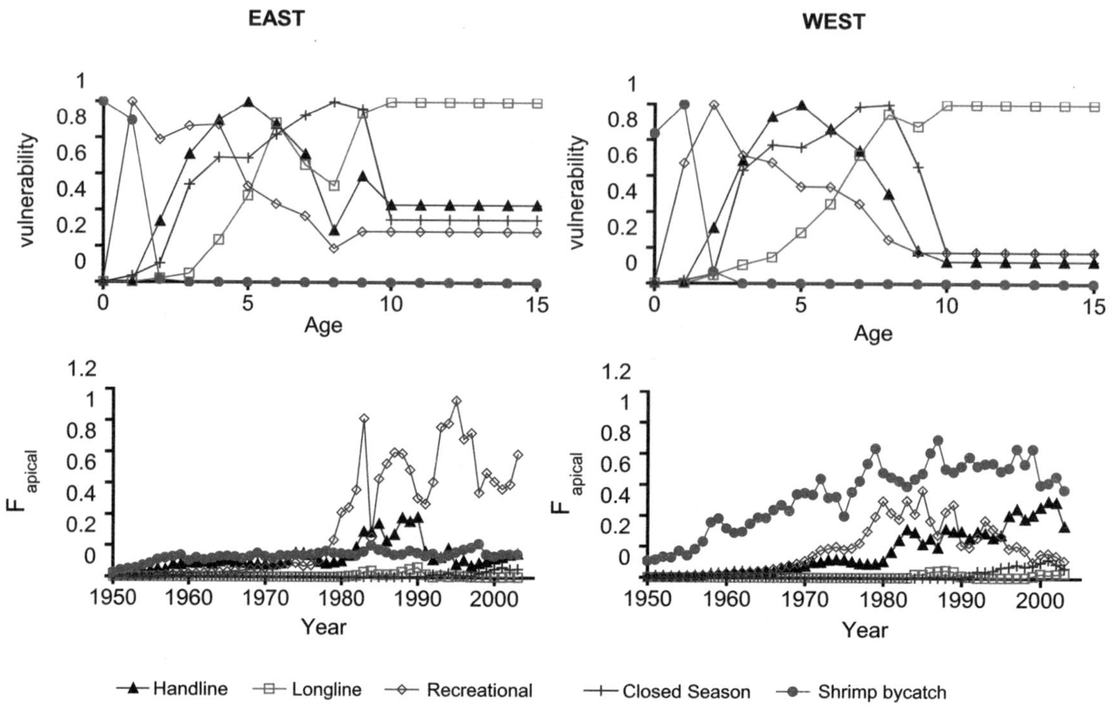

Figure 5. Age-0 model estimates of vulnerability and apical fishing mortality rate for each fleet.

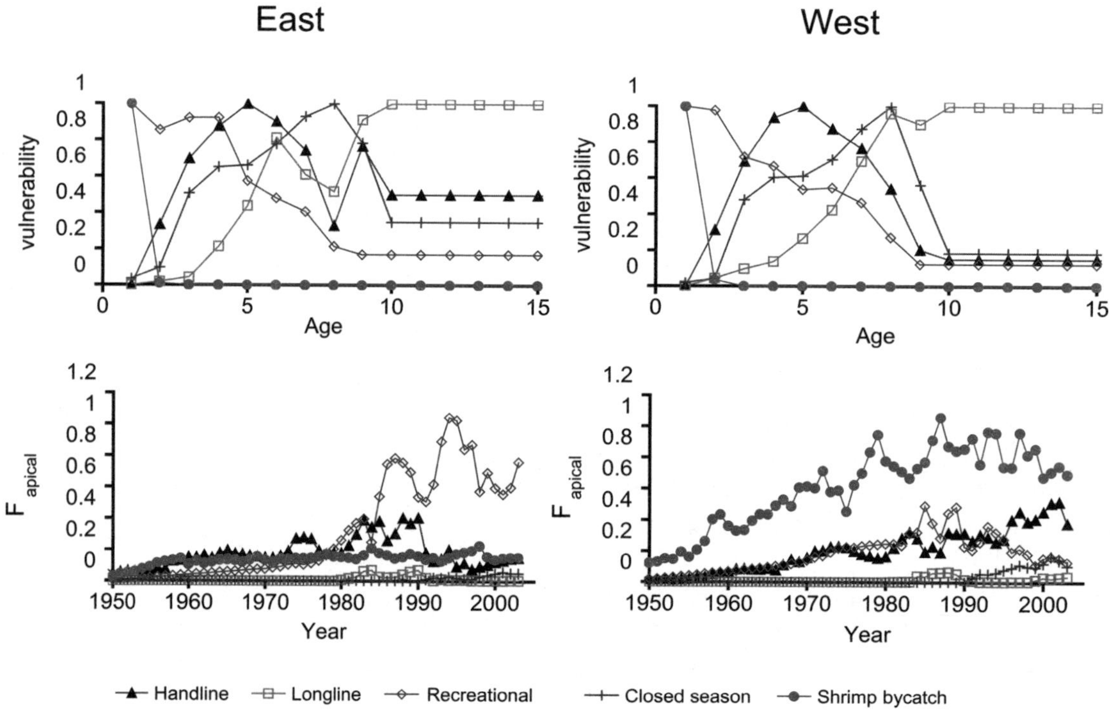

Figure 6. Age 1 model estimates of vulnerability and apical fishing mortality rate for each fleet.

times as productive as the eastern population. The eastern population began to decline very early in the history of the fishery, whereas the western population did not decline until the late 1940s. Both populations are estimated to have been reduced to less than 10% of their unfished levels by the 1980s. The declines continued into the 1990s, when both stocks were at all time lows of less than 3% of unfished levels. In recent years the western population appears to have leveled off and the eastern population appears to have increased to a little better than 4% of pre-exploitation levels.

Discussion

The results from the age 0 and age 1 assessment models are consistent in that they both suggest the populations of red snapper in the U.S. Gulf of Mexico are severely depleted. The estimated historical trends may be somewhat biased prior to the 1950s owing to ambiguities in the fraction of the total landings that came from foreign waters (see Porch et al. 2007), but nevertheless agree well with published anecdotes. The decline in the east is estimated to have begun shortly after the fishery began and continued into the early 1900s, which is consistent with the observation that some of the largest catches on record occurred during that time. The population appears to have rebounded somewhat between the 1920s and early 1940s owing to the shift to the Campeche Banks and reduced effort during events such as the Great Depression and World War II. Shortly after the war, however, the population began to plummet, reminiscent of Camber's (1955) claim that most of the snapper banks off Florida were considered impoverished. Such a decline is not surprising given the widespread adoption of postwar innovations such as diesel engines, depth finders, reels, and wire lines.

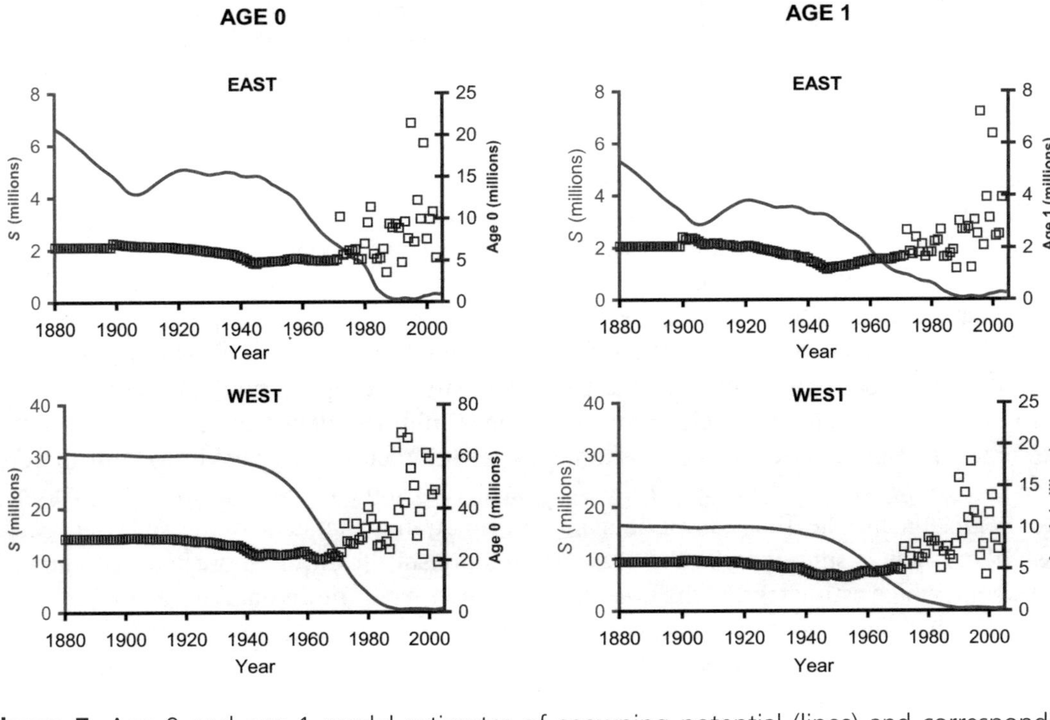

Figure 7. Age 0 and age 1 model estimates of spawning potential (lines) and corresponding number of age 0 or age 1 recruits (squares).

The continued decline through more recent years is also not unexpected inasmuch as the snapper fleet nearly doubled in size during the boat building boom of the late 1950s and early 1960s (Moe 1963; Carpenter 1965). Vessel fishing power has also continued to improve owing to various gear modifications, electronic fish-finders and navigational aids such as loran-c and GPS. Even so, there is some evidence of an increase in recent years, possibly due to regulations implemented during the 1990s to rebuild the stock (see Hood et al. 2007).

The decline in the west is estimated to have begun during the late 1940s and to have continued through the 1990s. Although productive fishing grounds had been discovered off Texas as early as the 1880s (e.g., the "Galveston Lumps"), most of the region was poorly charted and believed to be relatively unproductive insofar as red snapper were concerned. The turning point came during the 1940s with innovations such as diesel engines, depth recorders, reels and wire lines, which made it much easier to find and exploit snapper grounds in deeper waters. One of the most productive new areas was known as the "Western" grounds, a huge area off Texas and Louisiana spanning 10 mi on either side of the 100 fathom (180 m) line. Its discovery apparently led to a shift from the prevailing paradigm that the center of abundance of red snapper in U.S. waters was east of the Mississippi (c.f. Jordan and Evermann 1923) to a view that the center of abundance lay west of the Mississippi. As Camber (1955) remarks, "the largest populations seem to concentrate in the Gulf of Mexico, especially off the Yucatan Peninsula and the Texas and Louisiana coast." This view is supported by the present assessment, which estimates that the potential (unfished) abundance of the western population is about three times that of the eastern population.

Despite the apparent enormity of the western red snapper resource, the increased effort associated with the boat building boom and technological advances discussed earlier seems to have taken its toll. In fact, the models indicate that the western stock may have been depleted even more rapidly than the smaller eastern stock. It should be pointed out that none of the indices of abundance used to tune the model extend prior to the 1970s, therefore the slope of the declines is probably not very well determined. On the other hand, Bradley and Bryan (1975) reported that Texas handliners landed about 1,000 lbs (454 kg) per day prior to 1965, but felt fortunate to catch 500 lbs (227 kg) per day during the early 1970s. The implication then is that the available stock had decreased by somewhat more than 50%, which is very close to the model estimates of a 52% decline from 1960–1964 to 1970–1973.

One of the more remarkable features of the present assessment is the increase in both the magnitude and inter-annual variability of the recruitment estimates over the last several decades. The increase in variability is commensurate with the beginning of the SEAMAP trawl survey, which provides information on variations in year-class strength that allows the model to admit substantial departures from the static spawner-recruit relationship. What is more interesting is the fact that the model estimates of recruitment since 1980 are mostly well above the estimates for an unfished population despite the estimated decrease in spawning potential. The structure of the model does not require the recruitment deviations from the estimated spawner-recruit relationship to sum to zero, which to some extent decouples the "prehistoric" trends dictated by the spawner-recruit relationship from the more recent trends estimated when more data are available. Nevertheless, the estimated steepness is near 1.0 despite the much smaller value imposed in the prior, suggesting that the spawner-recruit relationship is reasonably well-determined. Thus, a possible interpretation of the peculiar pattern in recruitment is that the red snapper stocks in the Gulf of Mexico have become more productive over the last two decades.

A number of possible mechanisms have been postulated to account for the elevated recruitment estimates for recent years. They include improved oceanographic conditions for larvae, larval input from the Campeche Banks, postsettlement density-dependent mortality effects, reduced predator abundance (possibly due to shrimp trawling), increased habitat for juveniles (oil rigs and artificial reefs) and undetected increases in spawning potential (cryptic adult biomass). At present, the data available are insufficient to discriminate between these hypotheses. Regardless, it is likely that the recruitment estimates for the 1980–2003 period are better-determined than the earlier values as this is the period when most of the data are available. Accordingly, it is probably more reasonable to assume that future recruitment will follow recent levels than to assume they will follow the estimated spawner-recruit relationship (which presumably reflects the earlier history of the fishery). For this reason the participants of the SEDAR 7 review workshop recommended that forecasts and reference points be based on an alternative spawner-recruit relationship with the R_0 values for each stock set equal to the average of the recruitment estimates for 1980 to 1903 (Cordue 2005).

The determination as to whether or not a stock is over-fished depends of course on the reference point the stock is measured against. Several different reference points are routinely computed for the Gulf of Mexico red snapper fishery, but none of these have been formalized in the Gulf of Mexico Fishery Management Council's fishery management plan. As a result, the implications of changes in the stock assessment methodology or data are often confused with the implications of changing the reference points. To date, the reference points that have received the most attention are based on the MSMY or SPR achieved under two types of effort allocation schedules, referred to hereafter as the "equal proportion" and "percent shrimp reduction" schedules. The equal proportion scenario assumes the effective effort of all fleets, both directed and undirected, can be scaled down by the same proportion. This has sometimes been referred to as the "linked" or "policy neutral" scenario because all of the fleets that catch red snapper are affected to the same degree. The percent shrimp reduction schedule recognizes that the management of offshore shrimp fisheries is not completely driven by concerns about red snapper bycatch and therefore that any prescribed reduction in offshore shrimp effort will be somewhat independent of the reductions in the directed fisheries. It also assumes that the fraction of the red snapper population killed during the closed season will be similar to recent levels (here the 2001–2003 average) when the directed fisheries are scaled down. Otherwise, the relative allocation of effort among the directed fleets has been assumed to remain at recent levels.

The impact of different reference points on the perception of stock status is illustrated for the age 0 model in Figure 8. Note that the metric of comparison, $\phi = S/S_0$, is equivalent to SPR for steepness values near 1.0, as is presently the case. The spawning potential of both the East and West stocks is estimated to have been well below each of the reference points considered, therefore one must conclude that the two stocks are currently overfished. The degree of imperilment and ability of the stock to recover, however, depends on which reference point is considered most reasonable. The MSMY policies predicated on only a limited reduction in offshore shrimp bycatch are clearly more risky than a policy based on 30% SPR because they permit the stock to be fished down to within only a few percent of the unfished levels. When shrimping and closed season bycatch rates are assumed to continue at the 2001–2003 aver-

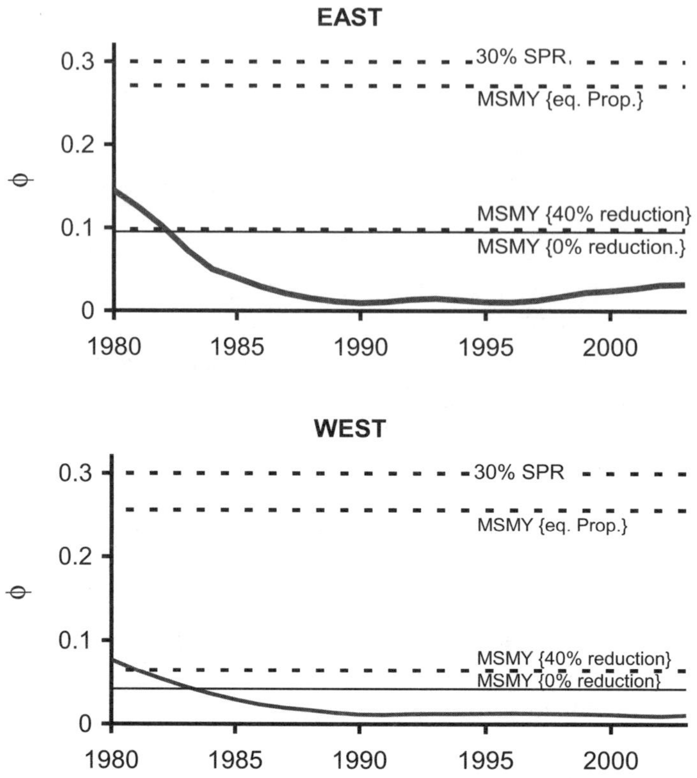

Figure 8. Age-0 model estimates of relative spawning potential ϕ (heavy solid line) compared to four reference points: ϕ_{MSMY} given current shrimp bycatch (fine horizontal line), ϕ_{MSMY} given a 40% reduction in shrimp bycatch rates, ϕ_{MSMY} when all bycatch and directed fleets reduced by the same proportion, and $\phi_{30\%SPR}$.

age, for example, ϕ_{MSMY} amounts to 4.2% and 9.5% for the west and east, respectively. When shrimping is assumed to be reduced by 40%, the ϕ_{MSMY} values are only a little higher at 6.4% and 9.8% for the west and east. In both cases the ϕ values are well below the generic recommendation of 30% adopted by the Gulf of Mexico Fishery Management Council. The values of ϕ_{MSMY} were much closer to 30% under the equal proportion reduction scenario, amounting to 26% and 27% of unfished levels for the west and east, respectively. In that case the MSMY was achieved when the *F* of all fleets was reduced by 74%, including shrimp bycatch and closed-season discards. (Recall, the percent reduction scenarios assume that closed-season discarding will continue at the 2001–2003 rate.)

One disturbing feature of previous assessments of Gulf of Mexico red snapper has been the tendency to estimate MSMY values that are much greater than the historical landings and yet predict the stock to be overfished. This paradoxical behavior is not uncommon where the time series of data are short because the model has little basis for determining the historical development of the fishery. Indeed, it was partly for this reason that SEDAR 7 participants recommended extending the historical catch series back to the inception of the fishery in 1872. The models shown here, which use the extended time series, do in fact produce more plausible MSMY estimates (Table 3). Even so, the estimates of

Table 3. Estimates of the sustainable annual landings (millions of pounds) under four harvest strategies

Strategy	Model	East	West	Total
MSMY 0% reduction	Age 0	4.5	6.8	11.3
MSMY 0% reduction	Age 1	4.4	6.3	10.7
MSMY 40% reduction	Age 0	5.1	11.5	16.5
MSMY 40% reduction	Age 1	4.7	9.3	14.1
MSMY equal proportion	Age 0	6.4	19.0	25.4
MSMY equal proportion	Age 1	5.2	12.9	18.1
SPR30% equal proportion	Age 0	6.4	18.9	25.3
SPR30% equal proportion	Age 1	5.1	12.8	18.0

MSMY for the equal proportion scenario are substantially greater than even the highest recorded landings. This is primarily a result of the use of the higher recruitment estimates for recent years (1980–2003) as indicative of the present potential of the stock, which of course implies that MSMY and S_{MSMY} levels are greater now than they were in the past. In other words, the models suggest that (1) the stocks were less productive in the past and therefore more easily overfished; and (2) the stocks are more productive now, but their potential has not been fully realized because they had already been depleted.

One possible consequence of defining a reference point based on recent increases in productivity is that the stock can appear overfished without overfishing having ever occurred. Consider, for example, a fully-exploited stock ($F = F_{MSMY}$ and $S = S_{MSMY}$) that has experienced a 50% increase in recruitment in recent years owing to a change in the environment. This would translate roughly into a 50% increase in the value of S_{MSMY}, but it would take some time for the additional recruits to contribute to the spawning potential of the stock. Accordingly, if the new higher level of S_{MSMY} were adopted as the reference point, then the stock would initially be reclassified as overfished regardless of the actions of the fishery. In practice the change from one productivity level to another may be rather gradual and difficult to establish. Nevertheless, the issue raises the prospect of developing reference points that distinguish the effect of overfishing from the transitional effect of changes in productivity.

In the case of red snapper, the transition to a higher productivity regime appears to have begun prior to 1980. If past catch levels had been sustainable before the new regime, then similar levels of catch over the last two decades should have resulted in a substantial increase in spawning potential. Instead, the eastern population has exhibited only modest gains and the western population appears not to have increased at all. This implies that, despite their increased reproductive potential, the two stocks had been so depleted in the past that they will remain unable to recover without further restrictions on fishing. The question then arises as to what level of catch and bycatch mortality might permit the recovery of the stock to its potential under the new higher recruitment scenario within a reasonable time frame.

Figure 9 presents isopleths of S/S_{MSMY} and S/S_0 (ϕ) generated from short-term projections of the western and eastern stocks to the year 2010 under various levels of TAC (Gulf-wide total allowed landings, assuming the relative levels of effort in the east and west remain

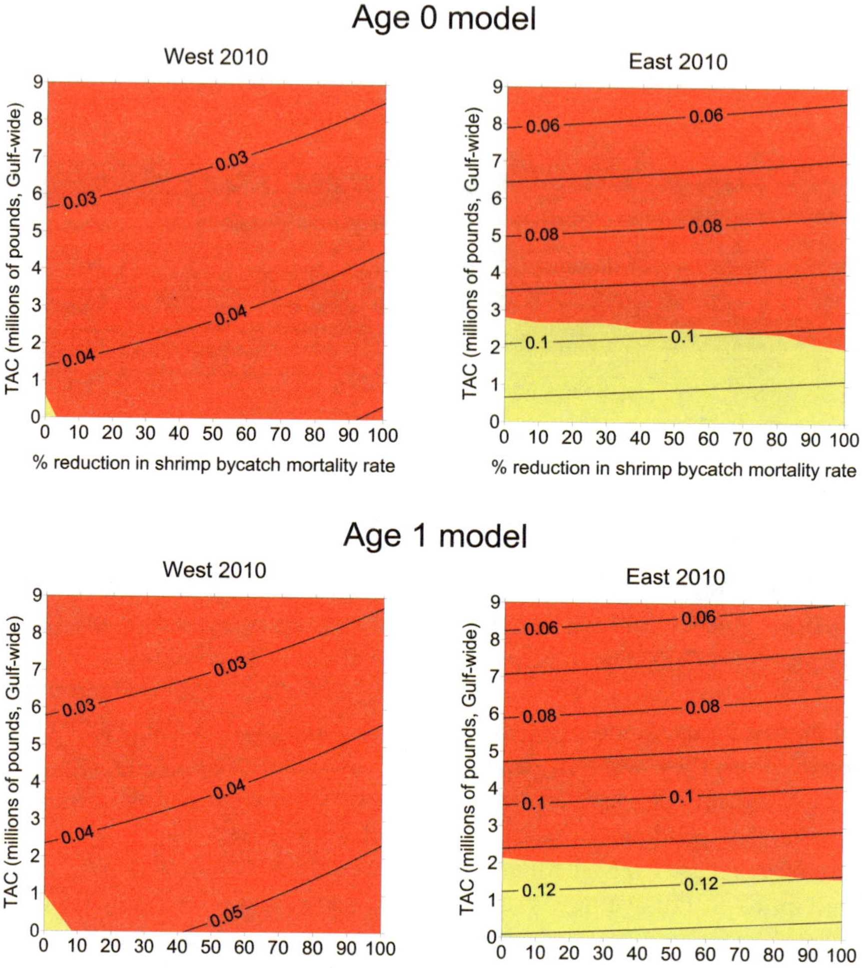

Figure 9. Isopleths of relative spawning potential in the year 2010, $\phi_{2010} = S_{2010}/S_0$, for the age 0 (top) and age 1 (bottom) models. The horizontal axis refers to the projected percent reduction in shrimp bycatch mortality rate and the vertical axis refers to the projected Gulf-wide TAC. The shading represents different levels of spawning potential relative to MSMY levels (S_{2010}/S_{MSMY}), where MSMY is conditioned on the indicated reduction in shrimp bycatch mortality. Red represents $S_{2010}/S_{MSMY} < 1$ and yellow represents $1 < S_{2010}/S_{MSMY} < 4$.

constant) and percent reductions in shrimp bycatch mortality rate (F_{shrimp}). Current (2001–2003) conditions are assumed to prevail until 2007, when the indicated TACs and bycatch reductions take effect. The value of S_{MSMY} is conditioned on the reduction in F_{shrimp} indicated on the horizontal axis. Hence, the graph should be interpreted as an indication of what might happen if managers based the MSMY definition on the actual reduction in shrimp bycatch rates (implying that managers can either control or accurately forecast future shrimp bycatch). The projections suggest the current TAC of 9 million lb (4,100 mt) is sustainable, but unlikely to foster a recovery from the current low levels (ϕ_{2003} = 2% in the west and 4% in the east). Substantial short-term gains in spawning potential might be achieved with TACs under 4 million pounds, particularly in the east. Reductions in F_{shrimp} are projected to

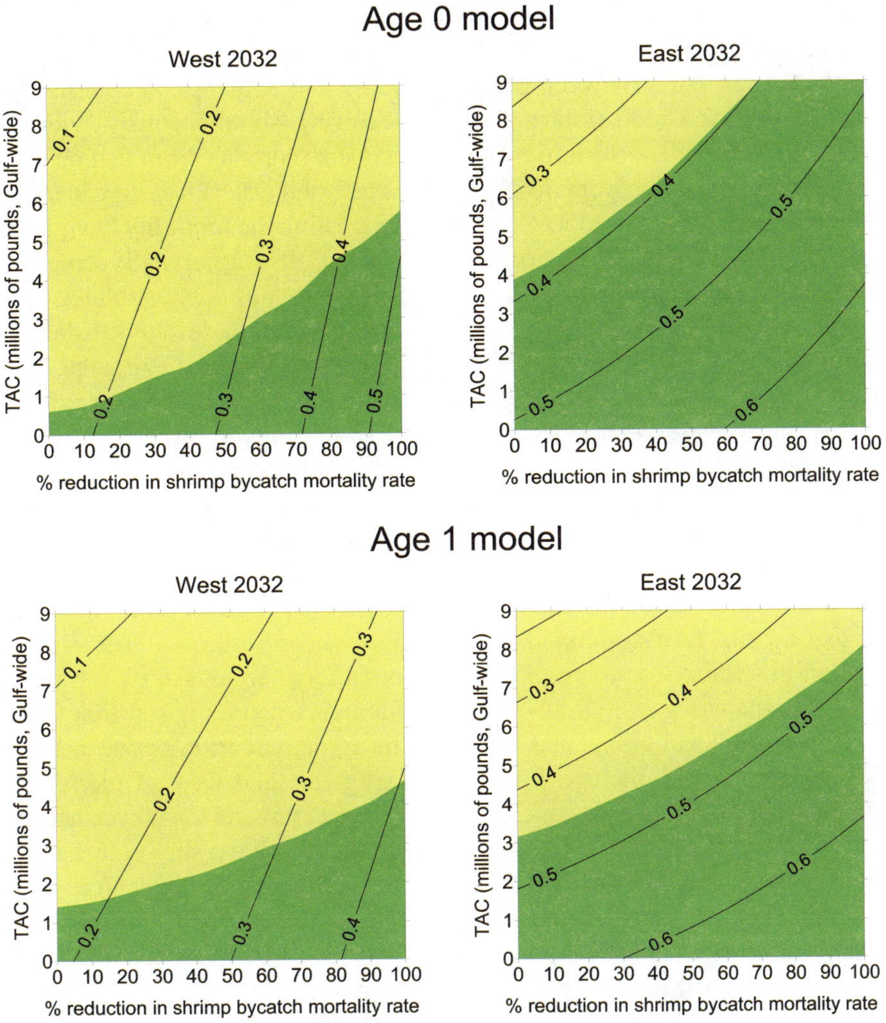

Figure 10. Isopleths of relative spawning potential in the year 2032, $\phi_{2032} = S_{2032}/S_0$, for the age 0 (top) and age 1 (bottom) models. The horizontal axis refers to the projected percent reduction in shrimp bycatch mortality rate and the vertical axis refers to the projected Gulf-wide TAC. The shading represents different levels of spawning potential relative to MSMY levels (S_{2032}/S_{MSMY}), where MSMY is conditioned on the indicated reduction in shrimp bycatch mortality. Yellow represents $1 < S_{2032}/S_{MSMY} < 4$ and green represents $S_{2032}/S_{MSMY} > 4$.

have little impact because the short term recovery in spawning potential is driven largely by the above average recruitments estimated to have occurred during the late 1990s.

The situation is more optimistic with a longer recovery time (Figure 10). The spawning potential of both stocks is projected to exceed S_{MSMY} by 2032 even at current levels of TAC and F_{shrimp}. However, it should be kept in mind that the SPR and ϕ values associated with MSMY are rather low unless the percent reduction of F_{shrimp} is large. If a more conservative reference point such as 30% SPRwere to be adopted, then either the TAC or F_{shrimp} must be reduced considerably. In the case of the east, the estimates of F_{shrimp} are substantially less than the mortality rates for the directed fisheries, therefore a recovery to $S_{30\%SPR}$ (ϕ > 30%) is predicted for TACs between 6 and 9 mp and less than a 40% reduction in F_{shrimp}. In the case of the west, where shrimp bycatch rates are much higher, a recovery to $S_{30\%SPR}$ appears unlikely with less than a 50% reduction in F_{shrimp}, even in the absence of a directed fishery. Conversely, a recovery to $S_{30\%SPR}$ is possible with the current TAC if F_{shrimp} were reduced by at least 75%.

The age 0 model assigns greater importance to the shrimp bycatch fishery than does the age 1 model (which ignores the bycatch of age 0 animals). Hence, for a given TAC, a smaller percent reduction in F_{shrimp} is estimated to be required to bring about recovery. For example, projections based on the age 0 model suggest that a 7 mp TAC combined with a 70% reduction in shrimp bycatch rate would likely allow the spawning potential of both stocks to recover the level associated with a 30% SPR. Projections with the age 1 model, however, indicate that the shrimp bycatch must be reduced by more than 80%. Not surprisingly, there has been some debate over which of the two model formulations presented is most plausible. Both formulations relate the number of recruits to past spawning potential by use of the Beverton–Holt function, which implies a belief that the survival from the egg phase to the age of recruitment is dictated primarily by density-dependent processes such as competition for resources. Thus the crux of the debate is the matter of timing. The age 0 model implies that any density-dependent effects on survival occur primarily during the planktonic phase or shortly after settlement. The age 1 model, on the other hand, implies that density-dependent effects dominate throughout the first year of life and that density-independent sources of mortality such as shrimp bycatch can be ignored.

The SEDAR 7 review workshop participants (Cordue 2005) selected the age-0 model as the most plausible of the formulations presented, stating that it "was not aware of any other assessment where the possibility that density-dependent compensatory processes occurring simultaneously with density-independent mortality from fishing (either discards or retained catch) was considered justification for treating the mortality from fishing as insignificant." It is important to remember, however, that the recruitment estimates of both models were allowed to deviate substantially from the predictions of the estimated Beverton–Holt functions. As a result, the results from the two models are not as disparate as one might imagine. Nevertheless, there is some evidence that density-dependent processes may be important during the first several years of life and it would be useful to model postrecruitment density dependence more directly.

Recent studies of red snapper life history characteristics, CPUE trends and otolith microchemistry suggest that there is a rather strong demarcation between the populations living east and west of the Mississippi river (Cowan et al. 2002). Nevertheless, tag-recapture studies have shown that red snapper occasionally move substantial distances (e.g., Patterson et al. 2001) and otolith microconstituent analyses indicate that some fish move from one side of the river to the other (Cowan et al. 2002). Hence, it may be prudent to extend the analyses presented here to allow for some degree of intermixing. The parameters for the movement model described above could be estimated from age-composition samples identified to stock by use of equations (18) and (19) with additional subscripts to reference stock, but a correction factor would be needed to account for the effect of misclassification errors.

Acknowledgments

I thank two anonymous reviewers for their many helpful comments on the current manuscript and the members of the SEDAR 7 assessment and review workshops who reviewed several earlier manuscripts as part of the SEDAR process (in particular, P. Cordue, J. Scott, and C. Walters). The use of color-shades and contour lines to represent stock status relative to the two reference points (MSMY and SPR) originated with G. Scott.

References

Anonymous. 1999. September 1999 report of the reef fish stock assessment panel. October 20, 1999. Gulf of Mexico Fishery Management Council. Tampa, Florida.

Bard, Y. 1974. Nonlinear parameter estimation. Academic Press, New York.

Bortone, S. A., R. N. Besser, and C. R. McNeil. 1997. The importance of Silas Stearns and the red snapper fishery to North American ichthyology. Pages 459–467 *in* T. W. Pietsch and W. D. Anderson, Jr., editors. Collection building in ichthyology and herpetology. American Society of Ichthyologists and Herpetologists, Special Publication No. 3.

Bradley, E., and C. E. Bryan. 1975. Life history and fishery of the red snapper (*Lutjanus campechanus*) in the northwestern Gulf of Mexico: 1970–1974. Proceedings of the Gulf and Caribbean Fisheries Institute 27:77–106.

Camber, C. I. 1955. A survey of the red snapper fishery of the Gulf of Mexico, with special reference to the Campeche Banks. State of Florida Board of Conservation Technical Series 12:1–64.

Carpenter, J. S. 1965. A review of the Gulf of Mexico red snapper fishery. United Stated Department of the Interior, Fish and Wildlife Service Circular 208:1–35.

Cass-Calay, S. L. 2004. Revised catch rate indices for red snapper (*Lutjanus campechanus*) landed during 1981–2003 by the U.S. Gulf of Mexico recreational fishery—Revised. South Atlantic Fishery Management Council, Southeast Data, Assessment and Review Working Paper SEDAR7-AW 4, Charleston, South Carolina. Available: www.sefsc.noaa.gov/sedar/ (March 2007)

Collins, J. W. 1885. The red snapper grounds in the Gulf of Mexico. Bulletin of the United States Fish Commission 5:145–146.

Cordue, P. L., editor. 2005. Gulf of Mexico red snapper consensus summary report. South Atlantic Fishery Management Council, Charleston, South Carolina. Available: www.sefsc.noaa.gov/sedar/ (March 2007).

Cowan, J. H., Jr., M. Woods, W. Patterson, III, and D. Nieland. 2002. Otolith microchemistry (and reproductive biology) portion—stock structure of red snapper in the northern Gulf of Mexico: Is their management as a single stock justified based on spatial and temporal patterns of genetic variation, otolith microchemistry, and growth rates. National Marine Fisheries Service, Marine Fisheries Initiative (MARFIN) Grant NA87FF0425.

Diaz, G. A., C. E. Porch, and M. Ortiz. 2004. Growth models for red snapper in U.S. Gulf of Mexico waters estimated from landings with minimum size limit restrictions. National Marine Fisheries Service, Southeast Fisheries Science Center, Sustainable Fisheries Division Contribution SFD-2004–038 (SEDAR7-AW 1), Miami, Florida.

Gledhill, C., and W. Ingram. 2004. SEAMAP Reef Fish Survey of Offshore Banks. South Atlantic Fishery Management Council, Southeast Data, Assessment and Review Working Paper SEDAR7-DW-17, Charleston, South Carolina. Available: www.sefsc.noaa.gov/sedar/ (March 2007).

Goodyear, C. P. 1988. Recent trends in the red snapper fishery of the Gulf of Mexico. National Marine Fisheries Service, Southeast Fisheries Science Center, Miami Laboratory Report CRD-87/88–16, Miami.

Goodyear, C. P. 1993. Spawning stock biomass per recruit in fisheries management: foundation and current use. Canadian Special Publication in Fisheries and Aquatic Sciences 120:67–81.

Goodyear, C. P. 1996. Variability of fishing mortality by age: consequence for maximum sustainable yield. North American Journal of Fisheries Management 16:8–13.

International Whaling Commission (IWC). 1994. Report of the scientific committee, Annex D. Report of the sub-committee on management procedures. Report of the International Whaling Commission 44:74–92.

Jordan, D. S., and B. W. Evermann. 1923. American food and game fishes: a popular account of all the species found in America north of the equator, with keys for ready identification, life histories, and methods of capture. Reprinted 1969, Dover Publications, New York.

Joseph, J., and H. Sendner. 1958. The horizontal diffusion in the ocean. Deutsche Hydrographische Zeitschrift 11:49–77.

Liermann, M., and R. Hilborn. 1997. Depensation in fish stocks: a hierarchic Bayesian meta-analysis. Canadian Journal of Fisheries and Aquatic Sciences 54:1976–84.

Lyczkowski-Shultz, J., D. Hanisko, and W. Ingram. 2004. The Potential for incorporating a larval index of abundance for stock assessment of red snapper, *Lutjanus campechanus*. South Atlantic Fishery Management Council, Southeast Data, Assessment and Review Working Paper SEDAR7-DW-14, Charleston, South Carolina. Available: www.sefsc.noaa.gov/sedar/ (March 2007).

Lyles, C. 1965. Fishery statistics of the United States. 1963. Bureau of Commercial Fisheries Statistical Digest 57.

Maunder, M. N., and R. B. Deriso. 2003. Estimation of recruitment in catch-at-age models. Canadian Journal of Fisheries and Aquatic Sciences 60:1204–1216.

McCarthy, K. J., and S. L. Cass-Calay. 2004. Standardized catch rates of red snapper (*Lutjanus campechanus*) from the United States commercial handline fishery in the Gulf of Mexico during 1996–2003: additional indices. South Atlantic Fishery Management Council, Southeast Data, Assessment and Review Working Paper SEDAR7-AW 9, Charleston, South Carolina. Available: www.sefsc.noaa.gov/sedar/ (March 2007).

Moe, M. A. 1963. A survey of offshore fishing in Florida. Florida State Board of Conservation, Marine Laboratory, Professional Papers Series 4:5–117.

Myers, R. A., K. G. Bowen, and N. J. Barrowman. 1999. Maximum reproductive rate of fish at low population sizes. Canadian Journal of Fisheries and Aquatic Sciences 56:2404–2419.

Nelson, R. S., and C. S. Manooch III. 1982. Growth and Mortality of red snappers in the west-central Atlantic ocean and northern Gulf of Mexico. Transactions of the American Fisheries Society 111:465–475.

Nichols, S. 2004. Estimating catch at age for red snapper in the shrimp fleet bycatch. South Atlantic Fishery Management Council, Southeast Data, Assessment and Review Working Paper SEDAR7-AW 20, Charleston, South Carolina. Available: www.sefsc.noaa.gov/sedar/ (March 2007).

Nichols, S., G. Pellegrin, Jr., and W. G. Ingram. 2004. Estimation of juvenile M for red snapper based on SEAMAP survey data. South Atlantic Fishery Management Council, Southeast Data, Assessment and Review Working Paper SEDAR7-AW 15, Charleston, South Carolina. Available: www.sefsc.noaa.gov/sedar/ (March 2007).

Nichols, S., A. Shah, G. Pellegrin, Jr., and K. Mullin. 1987. Estimates of annual shrimp fleet bycatch for thirteen finfish species in the offshore waters of the Gulf of Mexico. National Marine Fisheries Service, Southeast Fisheries Science Center, Mississippi Laboratories, Pascagoula Facility, Pascagoula, Mississippi.

Nowlis, J. 2004a. *A priori* estimates of natural mortality rates and stock-recruitment curve steepness for Gulf of Mexico red snapper. South Atlantic Fishery Management Council, Southeast Data, Assessment and Review Working Paper SEDAR7-DW 49, Charleston, South Carolina. Available: www.sefsc.noaa.gov/sedar/ (March 2007).

Nowlis, J. 2004b. Gulf of Mexico red snapper observed catch at age. South Atlantic Fishery Management Council, Southeast Data, Assessment and Review Working Paper SEDAR7-AW 19, Charleston, South Carolina. Available: www.sefsc.noaa.gov/sedar/ (March 2007).

Patterson, W. F. III, J. C. Watterson, R. L. Shipp, and J. H. Cowan, Jr. 2001. Movement of tagged red snapper in the Northern Gulf of Mexico. Transactions of the American Fisheries Society 130:533–545.

Porch, C. E. 2004. Batch-fecundity and maturity estimates for the 2004 assessment of red snapper in the Gulf of Mexico. South Atlantic Fishery Management Council, Southeast Data, Assessment and Review Working Paper SEDAR7-AW 5, Charleston, South Carolina. Available: www.sefsc.noaa.gov/sedar/ (March 2007).

Porch, C. E., G. R. Fitzhugh, M. S. Duncan, L. A. Collins, and M. W. Jackson. 2007. Modeling the dependence of batch fecundity on size and age for use in stock assessments of red snapper in U.S. Gulf of Mexico waters. Pages 229–243 in W. F. Patterson, III, J. H. Cowan, Jr., G. R. Fitzhugh, and D. L. Nieland, editors. Red snapper ecology and fisheries in the U.S. Gulf of Mexico. American Fisheries Society, Symposium 60, Bethesda, Maryland.

Porch, C. E., and S. C. Turner. 2004. Reconstructed time series of shrimp trawl effort in the Gulf of Mexico and the associated bycatch of red snapper from 1948 to 1972. South Atlantic Fishery Management Council, Southeast Data, Assessment and Review Working Paper SEDAR7-AW 23, Charleston, South Carolina. Available: www.sefsc.noaa.gov/sedar/ (March 2007).

Porch, C. E., S. C. Turner, and M. J. Schirripa. 2004. On the commercial landings of red snapper in the Gulf of Mexico from 1872 to 1962. National Marine Fisheries Service. Sustainable Fisheries Division Contribution SFD-2004–054 (SEDAR7-AW 22), Miami, Florida.

Powers, J. E. 2005. Maximum sustainable yield and bycatch minimization "to the extent practicable." North American Journal of Fisheries Management 25:785–790.

Powers, J. E., C. P. Goodyear and J. P. Scott. 1987. The potential effect of shrimp fleet bycatch on fisheries production of selected fish stocks in the Gulf of Mexico. National Marine Fisheries Service, Southeast Fisheries Science Center, Miami Laboratory CRD-87/88–06. Miami, Florida.

Punt, A. E., and T. I. Walker. 1998. Stock assessment and risk analysis for the school shark (*Galeorhinus galeus*) off southern Australia. Journal of Marine and Freshwater Research 49:719–31.

Rose, K. A., J. H. Cowan, Jr., K. O. Winemiller, R. A. Myers, and R. Hilborn. 2001. Compensatory density dependence in fish populations: importance, controversy, understanding and prognosis. Fish and Fisheries 2:293–327.

Schirripa, M. J., and C. M. Legault. 1999. Status of the red snapper in U.S. waters of the Gulf of Mexico updated through 1998. National Marine Fisheries Service, Southeast Fisheries Science Center, Sustainable Fisheries Division Contribution SFD-99/00–75. Miami, Florida.

Scott, G. P. 2004. Estimates of historical red snapper recreational catch levels using U.S. census data and recreational survey information. South Atlantic Fishery Management Council, Southeast Data, Assessment and Review Working Paper SEDAR7-AW 16, Charleston, South Carolina. Available: www.sefsc.noaa.gov/sedar/ (March 2007).

SEDAR (Southeast Data, Assessment and Review) 2004. Stock assessment report of SEDAR 7: Gulf of Mexico red snapper. Section 3. Assessment workshop. South Atlantic fishery Management Council, Charleston South Carolina. Available: www.sefsc.noaa.gov/sedar/ (March 2007).

Stearns, S. 1883. Fluctuations in the fisheries of the Gulf of Mexico, and the proposed investigation of them. Bulletin of the United States Fish Commission 3:467–468.

Turner, S. C., E. N. Brooks, G. P. Scott, and G. Diaz. 2004. Modeled age composition of Gulf of Mexico red snapper 1984–2003. South Atlantic Fishery Management Council, Southeast Data, Assessment and Review Working Paper SEDAR7-AW 18 (Revised), Charleston, South Carolina. Available: www.sefsc.noaa.gov/sedar/ (March 2007).

Turner, S. C., and Porch, C. E. 2004. Alternative indices of abundance of juvenile red snapper from the Gulf of Mexico from SEAMAP surveys 1972–2003. South Atlantic Fishery Management Council, Southeast Data, Assessment and Review Working Paper SEDAR7-AW 25, Charleston, South Carolina. Available: www.sefsc.noaa.gov/sedar/ (March 2007).

Warren, A. F. 1898. The red snapper fisheries: their past, present and future. Bulletin of the United States Fish Commission 17:331–335.

Wolfson, L. J., J. B. Kadane, and M. J. Small. 1996. Bayesian environmental policy decisions: two case studies. Ecological Applications 6:1056–1066.

Appendix

Re-parameterized spawner-recruit relationships

The number of young fish recruiting to a population (R) is often related to spawning potential (S) using one of two functional forms:

$$\text{(A. 1)}\ R = \begin{cases} aSe^{-bS} & \text{Ricker} \\ \dfrac{abS}{b+S} & \text{Beverton and Holt} \end{cases}$$

The parameter a is the slope of the curve at the origin and the parameter b controls the degree of density dependence. Notice that the domain of both functions extends from zero to infinity, whereas in practice there must be some limitation on S and R even in the absence of fishing owing to environmental constraints (call them S_0 and R_0, respectively). This being so, we obtain

$$\text{(A. 2)}\ a\frac{S_0}{R_0} = \begin{cases} e^{bS_0} & \text{Ricker} \\ 1 + S_0/b & \text{Beverton and Holt} \end{cases}$$

The ratio S_0/R_0 represents the maximum expected lifetime fecundity of each recruit and a represents the survival of recruits in the absence of density dependence. Accordingly, the product $\alpha = aS_0/R_0$ may be interpreted as maximum possible number of recruits produced by each spawner over its lifetime (Myers et al. 1999).

The dimensionless character of α makes it useful for interspecies comparisons, or for borrowing values from species with similar life history strategies. Solving for b in terms of α one obtains

$$\text{(A. 3)}\ b = \begin{cases} \log_e \alpha / S_0 & \text{Ricker} \\ S_0/(\alpha - 1) & \text{Beverton and Holt} \end{cases}$$

Substituting (A. 3) into (A. 1) gives

$$\text{(A. 4)}\ R = \begin{cases} aS\alpha^{-S/S_0} & \text{Ricker} \\ \dfrac{aS_0}{1+(\alpha-1)S/S_0} & \text{Beverton and Holt} \end{cases}$$

and, since $a = \alpha R_0/S_0$,

$$\text{(A. 5)}\ R = \begin{cases} R_0 \dfrac{S}{S_0}\alpha^{1-S/S_0} & \text{Ricker} \\ R_0 \dfrac{\alpha S/S_0}{1+(\alpha-1)S/S_0} & \text{Beverton and Holt} \end{cases}$$

Defining $\phi = S/S_0$ gives equation (9).

Note that when spawning extends over multiple seasons in the model, but the same spawner-recruit function is used for each season, then R_0 and S_0 should be interpreted as the virgin levels associated with a particular reference season. In that case, R_0 and S_0 will not necessarily be greater that the virgin values associated with other seasons.

American Fisheries Society Symposium 60:385–396, 2007

Rebuilding Red Snapper: Recent Management Activities and Future Management Challenges

ANDREW J. STRELCHECK[1] AND PETER B. HOOD
NOAA Fisheries Service
Southeast Regional Office
263 13th Avenue South, St. Petersburg, Florida 33701 USA

Abstract.—The Gulf of Mexico red snapper *Lutjanus campechanus* population was first declared overfished in the late 1980s. Subsequent stock assessments have concluded the population remains overfished and is undergoing overfishing. The first rebuilding plan for red snapper was established in 1989, setting a target of rebuilding the red snapper population by 2000. Changes in scientific advice and new information on red snapper biology resulted in several revisions to this plan. The most recent revision was approved in 2004 and set a goal of rebuilding red snapper to maximum sustainable yield by 2032. Despite the recent implementation of this plan, a 2005 red snapper population assessment indicates additional management restrictions are necessary for this plan to be successful. Managers are now confronted with several challenges to successfully recover red snapper, including addressing bycatch in the shrimp fishery and reducing discard and fishing mortality in the directed fishery. Social and economic considerations, competing conservation mandates, regulatory consistency, scientific uncertainty, political intentions, and the length of time allowed for recovery further complicate management and stock rebuilding. Managers will need to continue to focus on short-term directions and periodically adjust management strategies to ensure adequate rebuilding progress is made. This will allow managers to incorporate new information and unanticipated developments when making adjustments to red snapper and shrimp regulations, should either insufficient progress or unexpected events occur.

Introduction

Red snapper *Lutjanus campechanus* is a large, long-lived reef fish that supports economically valuable commercial and recreational fisheries in the Gulf of Mexico (GOM). They have been exploited in the GOM for over a century (Collins 1887); however, federal management of red snapper is relatively recent. Red snapper is managed in federal waters of the GOM under the Gulf of Mexico Fishery Management Council's (Council) Reef Fish Fishery Management Plan (GMFMC 1981). The first federal regulations for red snapper were implemented in 1984. Since that time, the Council has imposed many additional regulations in an effort to increase the population's abundance and total biomass (see Hood et al. 2007, this

[1] Corresponding author: andy.strelcheck@noaa.gov.

volume). However, the importance of the recreational and commercial fisheries for red snapper, shrimp trawl and directed fishery bycatch of red snapper, competing conservation mandates, political intentions, and scientific uncertainty have challenged managers to balance competing goals and interests. As a result, the population has remained overfished, although its status has slightly improved in recent years (SEDAR 2005).

The Council approved the first rebuilding plan for red snapper in 1989 (GMFMC 1989). Changes in scientific advice and new information on red snapper biology have resulted in several revisions to this rebuilding plan, the last occurring in 2004. Despite the recent revision to the plan, additional management restrictions are necessary to end overfishing of red snapper and rebuild the population to maximum sustainable yield (MSY) by 2032. The 2005 red snapper stock assessment indicates fishing mortality rates are too high for both the directed fishery and shrimp trawl fishery, and reductions in fishing mortality for both sectors are necessary at this time (SEDAR 2005; NOAA Fisheries 2006).

The Magnuson-Stevens Fishery Conservation and Management Act (M-SFCMA) requires the Council to implement conservation and management measures to rebuild overfished populations managed under a fishery management plan (FMP). Any FMP, plan amendment, or proposed regulations prepared to rebuild overfished populations must identify a timeframe for rebuilding. The timeframe must take into account the status and biology of an overfished population, as well as the needs of fishing communities and interactions of the population with the marine ecosystem. No rebuilding plan shall exceed ten years unless either biological or environmental conditions dictate otherwise. If, in the absence of fishing mortality, it will take greater than ten years to rebuild a population, then the maximum time recommended by NOAA Fisheries Service to rebuild a population is the rebuilding time calculated in the absence of fishing mortality plus one mean generation time. In the case of red snapper, the maximum time recommended for rebuilding is 31.6 years (12 years to rebuild in the absence of fishing plus a mean generation time of 19.6 years; Schirripa and Legault 1999).

The objective of this manuscript is to provide a brief overview of ongoing efforts by the Council and NOAA Fisheries Service to rebuild the red snapper population in the GOM and to discuss future challenges confronting fishery managers as the population rebuilds. We begin by discussing the current status of the population. We then provide a short summary of management actions either implemented or approved during 2006 and 2007 to rebuild red snapper. We end by discussing future challenges fishery managers will need to overcome if the red snapper population is to be successfully rebuilt.

Recent Management Activities

The following summary is intended to describe ongoing management activities by the Council and NOAA Fisheries Service to end overfishing of red snapper and to rebuild the population. For a detailed history of management prior to 2005 see Hood et al. (this volume).

The Council approved the most recent revision of the red snapper rebuilding plan in 2004 with Amendment 22 to the Reef Fish FMP (GMFMC 2004). This amendment redefined red snapper management reference points (i.e., optimum yield, maximum fishing mortality threshold, and minimum stock size threshold), established a plan to end overfishing and to rebuild the red snapper population consistent with the newly defined management benchmarks, and established a standardized methodology for collecting bycatch information in the directed fishery. The revised rebuilding plan was based on a 1999 stock assessment (Schirripa and Legault 1999) that indicated red snapper were both overfished and undergoing overfishing. The revised rebuilding plan was projected to end overfishing by 2009 or 2010 and to rebuild the red snapper population by 2032. The plan called for large reductions in bycatch mortality from the shrimp fishery to be achieved through either technological means, such as improved BRD designs, or reductions in shrimp fishing effort. The selected rebuilding plan recognized the need for periodic reviews of the stock status to ensure the rebuilding plan was adequately progressing toward the rebuilding goal. Review of the plan was designed to incorporate new information and to address unanticipated developments in the red snapper and shrimp fisheries and to make appropriate adjustments in red snapper regulations should either insufficient or unexpectedly rapid rebuilding progress occur. At the time, population biomass was estimated to be 7% relative to the biomass produced at MSY and spawning potential ratio (SPR) was estimated to range between 1.3% and 5.8% depending on the amount of recruitment assumed (Schirripa and Legault 1999; GMFMC 2004).

In 2005 a new stock assessment for red snapper was conducted. The assessment was independently reviewed through the Southeast Data, Assessment, and Review process (SEDAR 2005). An age-structured statistical model (CATCHEM), which allows for intermixing populations fished by multiple fleets, was used to determine the current status of red snapper (see Porch 2007, this volume, for more detailed information on the assessment model formulation and results). This methodology was a generalization of the previous assessment approach (Schirripa and Legault 1999) and incorporated several key changes resulting in significant differences in population estimates, such as the yield at MSY. Despite changes in methodology, the results of the assessment were consistent with those of previous assessments: the GOM red snapper population both remains overfished and is undergoing overfishing. While the directed fishery contributes a greater portion of fishing mortality than estimated by previous assessments (because of higher age 0 and age 1 natural mortality estimates and higher directed fishery release mortality rates of regulatory discards), shrimp trawl bycatch of red snapper remains a significant source of mortality in the western GOM. Recovery of the red snapper population in the western GOM is more sensitive to reductions in shrimp trawl fishing mortality and commercial fishing mortality, whereas recovery of red snapper in the eastern GOM is more sensitive to reductions in recreational fishing mortality and bycatch (NOAA Fisheries 2006). Fishing mortality (F) in both the directed fishery and shrimp trawl fishery are too high and reductions in F for both sectors are necessary to maintain the current red snapper rebuilding path (GMFMC 2004). Total F in 2003 was 3.8 times greater than the F associated with producing MSY and total red snapper biomass in 2003 was 5.7% of the level that produces MSY (SEDAR 2005). The SPR in 2003 relative to the SPR under virgin population condition was 1.5% for the entire GOM, 3.2% for the eastern GOM, and 1.1% for the western GOM (SEDAR 2005). These SPR levels are all well below the level associated with MSY (26% SPR). A 74% reduction in overall F is needed by 2009 or 2010 to end overfishing in accordance with the approved rebuilding plan.

In late 2005 the Council began developing an amendment to address the conclusions of the 2005 stock assessment (GMFMC 2007). After more than a year of deliberations, during which Hurricanes Katrina and Rita impacted fishing communities from east Texas to Alabama, the Council voted in August 2006 to delay consideration of regulatory actions needed to address overfishing until January 2007 when additional data and information was available. This decision effectively prevented NOAA Fisheries Service from implementing any permanent regulations proposed by the Council in time to address overfishing during 2007. Therefore, NOAA Fisheries Service began developing interim measures in Fall 2006 to address overfishing of red snapper during the 2007 fishing year.

The Council also approved a regulatory amendment to the Shrimp FMP in late 2005 (GMFMC 2006). The purpose of the amendment was to further reduce bycatch by establishing flexible and consistent performance standards for the certification of bycatch reduction devices (BRD) for the shrimp fishery. By modifying current performance standards, BRD performance could be improved and more BRD could be certified. Although BRD were already required in the shrimp fishery, scientific information suggested some certified BRD were not meeting the necessary performance requirements (Foster 2004). NOAA Fisheries Service is currently in the process of implementing these revisions to the BRD certification criteria, with final regulations implemented by late 2007 or 2008.

In March 2007 a District Court opinion on lawsuits filed by several environmental organizations and a recreational fishing organization concluded NOAA Fisheries Service violated the M-SFCMA, Administrative Procedures Act, and National Environmental Policy Act when it approved and implemented the red snapper rebuilding plan in Amendment 22 to the Reef Fish FMP. The Court concluded the rebuilding plan was based on flawed assumptions, did not consider the practicability of additional bycatch reduction in the shrimp trawl fishery, and did not demonstrate a 50% probability of rebuilding the red snapper by 2032. As a result, the Court ordered the Secretary of Commerce/NOAA Fisheries Service to approve a revised red snapper rebuilding plan by December 12, 2007.

Interim measures implemented by NOAA Fisheries Service in April and May 2007 were supported by the recent Court opinion. NOAA Fisheries Service implemented temporary regulations for the 2007 fishing season to address overfishing (GMFMC 2006) until the Council could adopt more permanent measures to end overfishing. These temporary regulations reduced the directed catch (TAC) from 9.12 to 6.5 million pounds (mp) (3.315 mp commercial quota and 3.185 mp recreational quota), reduced the recreational bag limit from four to two fish per angler per day, prohibited captain and crew from retaining bag limits of red snapper, reduced the commercial minimum size limit from 15 to 13 in total length, and established a goal to reduce red snapper bycatch mortality in the shrimp fishery by 50% of the bycatch mortality that occurred during 2001–2003. Temporary regulations also assumed a 10% reduction in landings owing to the impacts of hurricanes in 2005, which resulted in some reductions in fishing effort.

In June 2007, the Council approved Amendment 27/14 to the Reef Fish and Shrimp FMP; If approved and implemented by NOAA Fisheries Service, this amendment would revise the red snapper rebuilding plan, further reduce TAC from 6.5 mp to 5 mp during 2008–2010, modify recreational and commercial directed fishery regulatory measures (i.e., bag limit, season length, size limits, gear restrictions), set a goal of reducing red snapper bycatch mortality in the shrimp fishery, and establish seasonal closures if shrimp bycatch mortality does not meet the approved goal. In combination, these measures are estimated to end overfishing of red snapper by 2010. Final regulations for this action are expected to be implemented by late 2007 or early 2008.

Future Management Challenges

Despite the recent actions to the red snapper rebuilding plan discussed above, several management challenges still confront the Council and NOAA Fisheries Service if the red snapper population in U.S. waters of the GOM is to be successfully rebuilt. These challenges include issues confronting managers during the decision-making process (competing mandates, political interests, ways to reduce bycatch, scientific uncertainty), and issues confronting managers once regulations are implemented (effectiveness of regulations, compatible regulations, technological innovations). Most of these challenges have confronted managers throughout the course of red snapper rebuilding; however, understanding of these challenges has increased and lessons have been learned from past management failures allowing managers to better address them in the future.

Competing Mandates

The regulatory process is often initiated based on new scientific information that indicates a change in fishery policy is necessary. However, policymakers usually dictate what provisions the policy should contain (Watson-Wright 2005). Scientists have long argued short-term fishery impacts will be more than compensated for by long-term improvements in fishery yield (Rice, in press, cited in Watson-Wright 2005). Unfortunately, this oversimplifies both the management process and the competing interests that must be considered when setting policy. Not only must managers implement regulations to achieve biological goals, but they must also consider social and economic factors during the decision making process.

The M-SFCMA requires fishery managers to address ten national standards when developing a FMP. Although all of these national standards are relevant to red snapper, three national standards are the most influential when selecting red snapper regulatory actions: National standards 1, 8, and 9. These three standards require managers to 1) prevent overfishing, while achieving on a continuing basis optimum yield (OY), 2) consider the needs of fishing communities when setting conservation and management measures, and 3) minimize fishery bycatch and bycatch mortality to the extent practicable. In considering each of these mandates, management measures must take into account social and economic impacts while not compromising conservation goals.

The competing mandates of each of these standards make it difficult for managers to achieve the goal of each standard simultaneously when selecting management regulations. For instance, managers must balance the objectives of achieving OY in not one, but two fisheries (shrimp and red snapper) that are inextricably interlinked. Bycatch of red snapper in the shrimp trawl fishery reduces the MSY that can be obtained from the directed red snapper fishery (SEDAR 2005). Similarly, the higher directed fishery TAC is set, the more shrimp trawl effort must be reduced to end overfishing and rebuild red snapper, thereby potentially affecting the shrimp fisheries' ability to achieve OY. On top of these considerations, managers must also evaluate management actions that best balance achieving national standards 1 and 9, while minimizing social and economic consequences that inevitably result in the greatest source of controversy during the regulatory process.

Socio-economic considerations have greatly influenced red snapper regulatory actions because red snapper represents an economically important commercial and recreational fishery in the northern and western GOM and there are few supplemental offshore reef fish species to target. In the past managers have selected directed fishery red snapper TAC at or near the upper range recommended by scientific advisors in an effort to reduce social and economic impacts. These higher TAC have often been predicated on large reductions in shrimp trawl bycatch (>50%) that have never been fully realized.

Similarly, managers selected BRD as the preferred management measure for regulating shrimp trawl bycatch because at the time BRD were believed to achieve the necessary reductions in harvest, while having the fewest social and economic impacts. Undoubtedly, social and economic considerations will continue to play an important role in setting future directions for red snapper rebuilding, especially considering the impacts of recent hurricanes and declines in shrimp effort resulting from low-priced shrimp imports and high fuel prices.

Bycatch

Between 2001 and 2003, the directed red snapper fishery discarded dead approximately 1.67 million red snapper per year (includes in-season and closed season dead discards) and the shrimp trawl fleet discarded dead approximately 18.3 million juvenile (age-0, -1, and -2) red snapper per year (Figure 1; SEDAR 2005). The immense amount of bycatch in both fisheries results both in forgone yield and in lower TAC levels in the directed fishery. To successfully rebuild red snapper to desirable levels, significant reductions in both shrimp trawl and directed fishery bycatch will

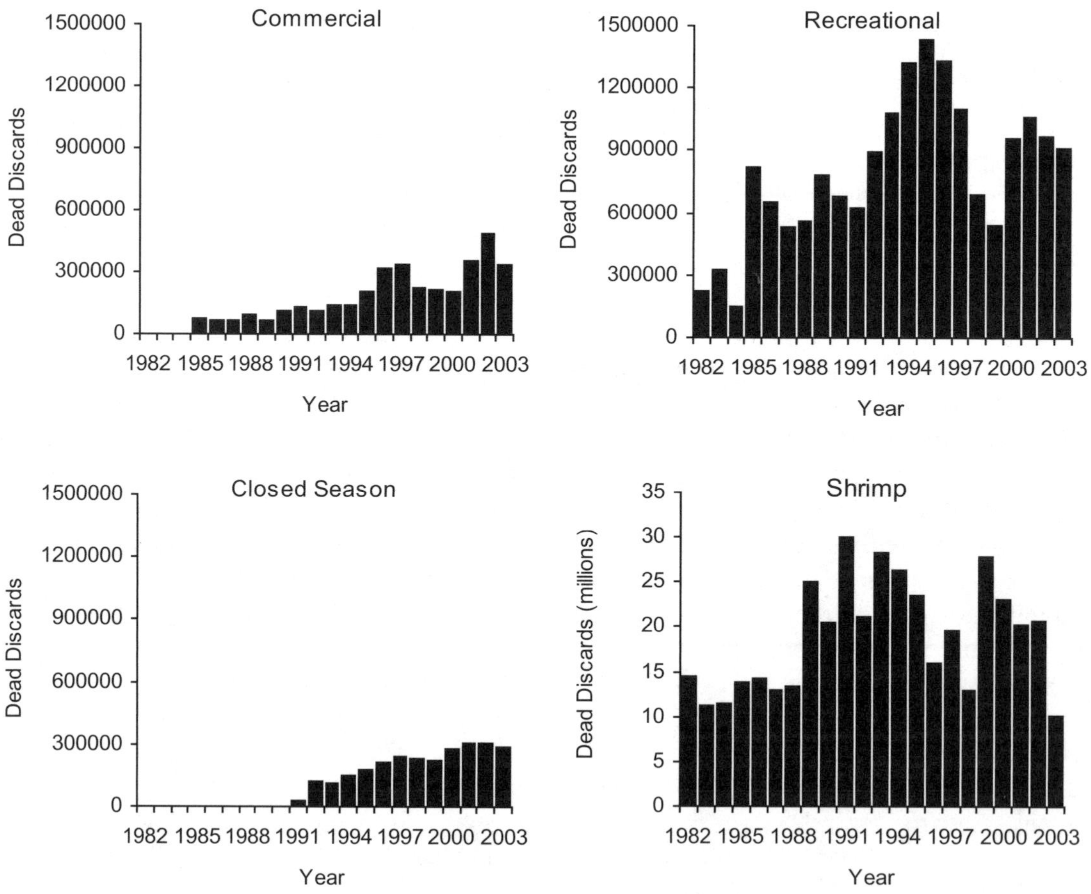

Figure 1. Commercial, recreational, closed season, and shrimp trawl dead discards of red snapper *Lutjanus campechanus*. Directed fishery and closed season discards are in thousands of fish. Shrimp trawl discards are in millions of fish. Data from C. E. Porch, NMFS, personal communication.

be needed. Historically, BRD were assumed to be the primary answer in resolving bycatch problems in the shrimp fishery. When originally certified, BRD were estimated to reduce red snapper fishing mortality in shrimp trawls by 50% or more (Watson et al. 1999; GMFMC 2004). However, more recent research indicates red snapper fishing mortality reduction from shrimp trawl BRD is now less than 12% (Foster 2004). Reasons for the large differences in BRD performance include technological changes to the overall construction of shrimp trawl gear (e.g., new turtle excluder devices, longer nets) and changes in fishing practices, such as faster towing speeds and modified retrieval procedures (Foster 2004). These actions and modifications increase shrimp retention, without concurrently maintaining fish reductions, therefore diminishing the effectiveness of BRD.

New data and information incorporated in the most recent red snapper stock assessment (SEDAR 2005) suggests directed fishery release mortality rates and discards are much greater than once thought (Wilson et al. 2004; SEDAR 2005), while shrimp trawl bycatch fishing mortality rates are less because more juvenile red snapper are assumed to die from natural causes if not killed by shrimp trawls (SEDAR 2005; NOAA Fisheries 2006). It was previously thought that reducing bycatch in the directed fishery would not measurably affect the status of the red snapper population (GMFMC 2004); however, this is no longer the case. For example, TAC in the directed fishery during 2008–2010 must be set 2.0 mp less than the maximum allowable biological catch because current management tools cannot sufficiently reduce both in-season and closed season directed fishery bycatch (GMFMC 2007).

When selecting management measures, managers will be confronted with the challenge of restricting directed harvest, which often increases bycatch and discard mortality, while simultaneously implementing regulatory actions to reduce directed fishery bycatch. Additionally, managers will need to determine new ways to address shrimp trawl bycatch mortality, such as controlling effort and implementing more efficient and effective BRD.

Political Interests and Controversy

Political interests and agendas often confront managers during the decision-making process. In some cases, political interests have prevented managers from moving forward with major policy decisions. For example, the 1990 amendment to the M-SFCMA (DOC 1990) prohibited Councils from implementing management measures to reduce shrimp trawl bycatch for a period of three years, and the 1996 reauthorization of the M-SFCMA prohibited IFQ programs from being implemented for a period of four years during the late 1990s. These actions ultimately slowed implementation of important red snapper management measures intended to assist in rebuilding the population. Although these are extreme examples of political intervention, there are also instances when political interests manifest themselves in other ways. For example, political interests and controversy may dictate the level of risk or precaution when selecting regulatory actions. Controversial issues and actions resulting in significant social and economic consequences may lead managers to select less precautionary actions to minimize impacts. By doing so, risk is increased that sufficient rebuilding progress may not be made.

Scientific Uncertainty and Unpredictable Natural Events

Another challenge for managers is adapting to uncertain, unpredictable events. Scientific uncertainty often impedes policy actions (Watson-Wright 2005). Unpredictable factors, such as changes in economic conditions, weather events (e.g., hurricanes), and red tide, can also complicate rebuilding strategies and model projections and result in additional levels of uncertainty. Although scientists accept uncertainty when conducting research and stock assessments, uncertainty can reduce both political support and cooperation when implementing management policies (Watson-Wright 2005). Further, fishers often contend that management measures should not be implemented because of uncertainty.

A case in point is the 1999 red snapper stock assessment on which the current red snapper rebuilding plan is based. This assessment provided highly uncertain management benchmarks and was sensitive to model inputs. In contrast, estimates of stock productivity based on the most recent assessment were much less sensitive than in past assessments (SEDAR 2005). Despite the improvements in model sensitivity, both assessments concluded red snapper were severly overfished and undergoing overfishing. Although the latest stock assessment is much improved, considerable uncertainty still exists with regard to the stock–recruitment relationship, which indicates stock productivity is at its highest when the stock is most depleted (SEDAR 2005). The stock–recruitment relationship is considered the greatest source of uncertainty in future management projections and projections are only considered plausible over a short time frame (5–10 years) (SEDAR 2005).

Another source of uncertainty is the estimation of directed fishery discards. Estimates of discards and dead discards were reviewed during the SEDAR 7 data workshop (SEDAR 2005). Because only a short time series was available to estimate commercial discards and because the SEDAR data workshop panel believed recreational discards were much higher than estimated by the Marine Recreational Fisheries Statistics Survey (MRFSS), the red snapper stock assessment did not use commercial logbook or MRFSS discard estimates. Instead, discards were assumed to be due to the minimum size limit and were estimated from the predicted length composition of the catch (see Porch 2007, this volume). Comparison of MRFSS discard estimates with observer data since completion of the assessment has validated that discards appear to be underreported in MRFSS. However, because bycatch is an important component in the red snapper stock assessment, additional research is necessary to address scientific uncertainty and to improve the accuracy of discard estimates.

Recreational and commercial landings and effort data are another source of uncertainty that will need to be addressed in the future. The recent National Research Council (NRC) report on collection of recreational catch data has highlighted numerous deficiencies in recreational catch and effort estimates (NRC 2006). These deficiencies could lead to management measures that are either too restrictive or too lenient. NOAA Fisheries Service is working on implementing the recommendations of the NRC, which will help to improve recreational data collection. Additionally, observer coverage on vessels to monitor directed and incidental catches is not at levels sufficient to produce consistently reliable data for current and future monitoring needs. Also, scientists and managers often use self-reported commercial logbook and for-hire data; biases associated with these data collections programs are not well understood.

Regulatory Effectiveness

Once managers have approved and implemented regulatory changes, the next challenge for managers is ensuring regulatory measures achieve biological, social, and economic objectives. Often management measures fail to fully achieve biological goals, such as reductions in landings or bycatch, because it is difficult for scientists and managers to predict and accurately quantify changes in fisher behavior after implementation of regulations. A case in point is the implementation of BRD. Previous research indicated BRD would achieve reductions in red snapper bycatch in excess of 50% (Watson et al. 1999; GMFMC 2004). Bycatch reduction devices were selected by managers as a preferred alternative for managing GOM shrimp trawl bycatch because at the time they were believed to achieve the necessary reductions in red snapper bycatch, while also minimizing social and economic impacts. However, once BRD were fully implemented in the shrimp fishery, reductions in shrimp trawl bycatch were much less than previously estimated (Foster 2004).

Constraining red snapper landings to specified TAC levels also challenges managers, especially when TAC are reduced to relatively low levels when compared to historically allowable landings levels. Although managers have largely constrained fishery landings within specified TAC levels during the course of the last decade, improvements in red snapper population status have been small (i.e., SPR and stock biomass has not increased at projected rates). Since 1990, the Council has specified a TAC for the directed red snapper fishery that is allocated 51% to the commercial sector and 49% to the recreational sector. The commercial and recreational portions of the TAC are further specified as quotas. The commercial quota, as of 2007, is monitored in-season through submission of IFQ catch records. When commercial fishers exhaust their allocated IFQ shares, they must either stop fishing for the remainder of the year or buy/lease more shares from other IFQ participants. In contrast, the recreational fishery is not managed in-season because there is no timely method for monitoring landings, thereby preventing managers from shutting down the fishery if the quota is exceeded. Instead, managers must implement regulations that on average are expected to approximate the annual recreational quota. Although quota overruns have occurred in both the commercial and recreational fishery in recent years (SEDAR 2005), the likelihood of quota overruns is much greater for the recreational fishery because of the lack of in-season monitoring. Therefore, as the stock expands and availability of red snapper increases, managers will be challenged to restrict catch levels within the specified TAC.

Further challenging managers is the limited number of tools they have to restrict harvest (e.g., bag limits, closed seasons, size limits) and private angler effort. Limited access programs have been established for commercial reef fish vessels, charter vessels, and headboats to cap participation and effort, but the private sector is open access and has seen large increases in fishing effort directed toward red snapper in the past decade (Figure 2). As coastal populations grow, fishing pressure will be further increased on GOM fish populations, including red snapper. By 2030, which corresponds to the end of the red snapper rebuilding plan, the U.S. Census Bureau estimates the population in the five states bordering the GOM will have increased by 26.2 million people from 52.3 million in 2005 to 74.8 million in 2030.

Compatible Regulations

Regulatory consistency is a cornerstone to effective management. Not only do compatible regulations improve enforcement, they also reduce angler confusion and increase the likelihood of compliance. Inconsistency can result in incentives to harvest fish illegally, thereby compromising the effectiveness of regulations. The end result is regulations not achieving stated objectives, requiring managers to further restrict harvest in order to successfully rebuild the red snapper fishery. In the case of red snapper, state and federal fishing regulations are largely consistent, with the notable exceptions of the states of Texas

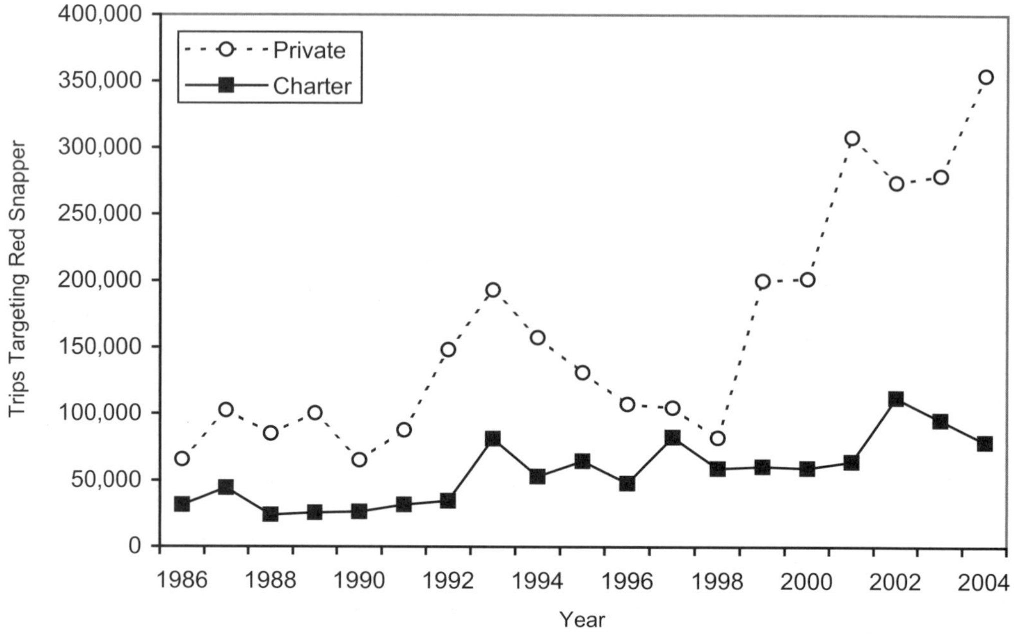

Figure 2. Estimated recreational trips targeting red snapper by mode (charter vessels and private/rental vessels), 1986–2004. Data from S. G. Holiman, personal communication. Methods for estimating target trips are described in Holiman 1996.

and Florida. Texas state waters currently remain open to recreational fishing year-round and the recreational fishing season in Florida state waters opens 6 d prior to the opening of the federal red snapper season. Although the latter inconsistency in regulations is due to differences in the timing of when federal and state policymakers took action, the former inconsistency is due to Texas' opposition to the existing closed season, which is during their winter tourist season. Although most red snapper landings occur in federal waters, inconsistent state regulations may require federal managers to implement more restrictive regulations to achieve necessary biological goals when rebuilding red snapper.

Technological Innovations

Global positioning systems (GPS), long-range navigation systems (LORAN), fish finders, fathometers, and other technological devices have revolutionized the fishing industry, which first relied on compasses and sextants to crudely navigate to fishing grounds. Improvements in position accuracy, reductions in costs, and increased availability of this technology have allowed anglers to find suitable fishing grounds more easily in recent decades. When coupled with the expansion of both artificial reef programs and oil and gas platforms throughout the GOM, accessibility to red snapper fishing locations has greatly increased. Existing technology and further technological improvements will greatly challenge management's ability to constrain fishing mortality in the future.

Conclusions

The sheer length of the red snapper rebuilding plan and the competing mandates fishery managers must consider complicate fishery management. Additionally, bycatch, political interests, scientific uncertainty, regulatory consistency, and technological changes require managers to continually adapt to changes in fishery conditions, unanticipated events (e.g., hurricanes, economic shifts), changes in legal requirements (e.g., amendments to the M-SFCMA), and new scientific advice. Recently stock assessment scientists have advised managers to focus on short-term directions (5–10 years) and how to achieve a more desired state, rather than focusing on longer-term management targets (SEDAR 2005). Because projections of future population status are often highly uncertain, and either may (or may not) represent future conditions, focusing on short-term directions allows for a more gradual and methodical approach to management. However, in doing so, managers must not lose sight of the end target and what it will take to rebuild the stock by 2032. Regular review of the rebuilding plan will allow managers to consider new information and to address unanticipated developments in the red snapper and shrimp fisheries. If necessary, these periodic reviews would also allow appropriate adjustments in red snapper regulations to be made should insufficient or unexpectedly rapid rebuilding progress occur. As periodic adjustments are made, managers will need to address the management challenges discussed herein and ensure sufficient progress is being made to end overfishing and rebuild the stock. Managers will also need to take advantage of the lessons learned from previous successes and failures to determine what management measures are most critical and necessary to address in order for red snapper populations to be rebuilt.

Acknowledgments

We would like to thank Heather Blough, David Nieland, and two anonymous reviewers for providing constructive input during the development of this manuscript. We would also like to thank Clay Porch and Stephen Holiman for supplying data for figures and discussions.

References

Collins, J. B. 1887. Notes on the red snapper fishery. Bulletin of the U.S. Fish Commission 6(1866): 299–300.

DOC (Department of Commerce). 1990. Magnuson Fishery Conservation and Management Act as amended through November 28, 1990. DOC, NOAA, National Marine Fisheries Service, Silver Spring, Maryland.

Foster, D.G. 2004. 1999–2003 Northcentral and Western Gulf of Mexico BRD performance report to SEDAR. SEDAR Data Workshop Report 38. http://www.sefsc.noaa.gov/sedar/download/SEDAR7_DW38.pdf?id=DOCUMENT. NMFS, Southeast Fisheries Science Center, Pascagoula, Mississippi.

Holiman, S. G. 1996. Estimating recreational effort using the Marine Recreational Fisheries Statistics Survey Data. National Marine Fisheries Service, NOAA Technical Memorandum. NMFS-SEFSC-389, St. Petersburg, Florida.

Hood, P. B., A. J. Strelcheck, and P. Steele. 2007. A history of red snapper management in the Gulf of Mexico. Pages 267–284 *in* W. F. Patterson, III, J. H. Cowan, Jr., G. R. Fitzhugh, and D. L. Nieland, editors. Red snapper ecology and fisheries in the U.S. Gulf of Mexico. American Fisheries Society, Symposium 60, Bethesda, Maryland.

GMFMC. 1981. Environmental impact statement and fishery management plan for the reef fish resources of the Gulf of Mexico. Gulf of Mexico Fishery Management Council, Tampa, Florida.

GMFMC. 1989. Amendment 1 to the reef fish fishery management plan. Gulf of Mexico Fishery Management Council, Tampa, Florida.

GMFMC. 2004. Amendment 22 to the reef fish fishery management plan to set red snapper sustainable fisheries act targets and thresholds, set a rebuilding plan, and establish bycatch reporting methodologies for the reef fish fishery. Gulf of Mexico Fishery Management Council, Tampa, Florida.

GMFMC. 2006. A framework measure to address bycatch reduction criterion for shrimp trawls in the Gulf of Mexico west of Cape San Blas, Florida under the fishery management plan for the shrimp fishery of the Gulf of Mexico, including environmental assessment, regulatory impact review, and regulatory flexibility analysis. Gulf of Mexico Fishery Management Council, Tampa, Florida.

GMFMC. 2007. Draft Amendment 27 to the reef fish fishery management plan and Amendment 14 to the shrimp fishery management plan. Gulf of Mexico Fishery Management Council, Tampa, Florida.

NOAA Fisheries. 2006. Environmental impact statement to evaluate alternatives to set Gulf of Mexico red snapper total allowable catch and reduce bycatch in the Gulf of Mexico directed and shrimp trawl fisheries. NOAA Fisheries Service, St. Petersburg, Florida.

Porch, C. E. 2007. An assessment of the red snapper fishery in the U.S. Gulf of Mexico using a spatially-explicit age-structured model. Pages 355–384 *in* W. F. Patterson, III, J. H. Cowan, Jr., G. R. Fitzhugh, and D. L. Nieland, editors. Red snapper ecology and fisheries in the U.S. Gulf of Mexico. American Fisheries Society, Symposium 60, Bethesda, Maryland.

Rice, J. C. In press. Will an ecosystem approach mitigate the factors of unsustainability? *In* J. J. Maguire, D. Greboval, editors. Collected papers from the 3rd workshop on factors of unsustainability. Siem Riep, Cambodia

Schirripa, M. J. and C. M. Legault. 1999. Status of the red snapper in U.S. waters of the Gulf of Mexico: updated through 1998. National Marine Fisheries Service, Contribution SFD-99/00–75, Miami, Florida.

SEDAR. 2005. SEDAR 7 Gulf of Mexico red snapper complete stock assessment report. Southeast Data, Assessment, and Review, Charleston, South Carolina.

Watson, J., D. Foster, S. Nichols, A. Shah, E. Scott-Denton, and J. Nance. 1999. The development of bycatch reduction technology in the southeastern United States shrimp fishery. Marine Technology Society Journal 33(2):55–56.

Watson-Wright, W. M. 2005. Policy and science: different roles in the pursuit of solutions to common problems. Marine Ecology Progress Series 300:291–296.

Wilson, C. A., D. L. Nieland, A. J. Fischer, and M. S. Baker, Jr. 2004. Red snapper *Lutjanus campechanus* in the northern Gulf of Mexico: Age and size composition of the commercial harvest and mortality of regulatory discards. Final Report to U.S. Department of Commerce, Marine Fisheries Initiative (MARFIN) Program, Cooperative Agreement NA17FF2007.